Contents

Independent Events ... 116
Order of Events ... 117
Practice Questions ... 118

S1 Section 3 — Binomial Distribution
Discrete Random Variables ... 120
The Binomial Probability Function ... 121
The Binomial Distribution ... 122
Using Binomial Tables ... 123
Mean and Variance of $B(n, p)$ ... 124
Binomial Distribution Problems ... 125
Practice Questions ... 126

S1 Section 4 — Normal Distribution
Normal Distributions ... 128
The Standard Normal Distribution, Z ... 129
Normal Distributions and Z-Tables ... 130
Practice Questions ... 132

S1 Section 5 — Estimation
Sampling and Statistics ... 134
Using Statistics ... 135
Standard Error and the Central Limit Theorem ... 136
Confidence Intervals for Normal Distributions ... 137
Confidence Intervals for Large n ... 138
Practice Questions ... 139

S1 Section 6 — Correlation and Regression
Correlation ... 141
Linear Regression ... 143
Practice Questions ... 145

S1 Practice Exams
S1 — Practice Exam One ... 147
S1 — Practice Exam Two ... 150

Statistical Tables ... 153

Mechanics M1

M1 Section 1 — Kinematics
Constant Acceleration Equations ... 158
Motion Graphs ... 160
Vectors ... 162
$i + j$ Vectors ... 165
Practice Questions ... 166

M1 Section 2 — Statics and Forces
Forces and Modelling ... 168
Forces are Vectors ... 171
Friction ... 174
Practice Questions ... 175

M1 Section 3 — Dynamics
Momentum ... 178
Newton's Laws ... 180
Friction and Inclined Planes ... 181
Connected Particles ... 183
Practice Questions ... 186

M1 Section 4 — Projectiles
Projectiles ... 189
Practice Questions ... 192

M1 Practice Exams
M1 — Practice Exam One ... 193
M1 — Practice Exam Two ... 196

Decision Maths D1

D1 Section 1 — Algorithms
Algorithms ... 199
Pseudo-Code and Flow Charts ... 200
Sorting ... 201
Practice Questions ... 205

D1 Section 2 — Algorithms on Graphs
Graphs ... 207
Minimum Spanning Trees ... 211
Dijkstra's Algorithm ... 214
Practice Questions ... 216

D1 Section 3 — The Route Inspection Problem
Traversable Graphs ... 218
Route Inspection Problems ... 219
Practice Questions ... 222

D1 Section 4 — Travelling Salesperson Problem
Travelling Salesperson Problem ... 224
Practice Questions ... 228

D1 Section 5 — Linear Programming
Linear Programming ... 230
Feasible Regions ... 231
Optimal Solutions ... 232
Optimal Integer Solutions ... 234
Practice Questions ... 235

D1 Section 6 — Matchings
Matchings ... 237
Alternating Paths ... 238
Maximum Matchings ... 239
Practice Questions ... 240

D1 Practice Exams
D1 — Practice Exam One ... 241
D1 — Practice Exam Two ... 244

C1 — Answers ... 248
C2 — Answers ... 262
S1 — Answers ... 276
M1 — Answers ... 293
D1 — Answers ... 307

Index ... 320

Editors:
Josephine Gibbons, Paul Jordin, Sharon Keeley-Holden, Simon Little, Sam Norman, Ali Palin, Andy Park, David Ryan, Lyn Setchell, Caley Simpson, Jane Towle, Jonathan Wray, Dawn Wright

Contributors:
Andy Ballard, Charley Darbishire, Claire Jackson, Tim Major, Mark Moody, Garry Rowlands, Mike Smith, Claire Thompson, Julie Wakeling, Chris Worth

Proofreaders:
Mona Allen, Janet Dickinson, Glenn Rogers, James Yates

Published by CGP

ISBN: 978 1 84762 581 6

Groovy Website: www.cgpbooks.co.uk

Printed by Elanders Ltd, Newcastle upon Tyne.

Based on the classic CGP style created by Richard Parsons.

A Few Definitions and Things

Yep, this is a pretty dull way to start a book. A list of definitions. But at least it gets it out of the way right at the beginning — it would be a bit mean of me to try and sneak it in halfway through and hope you wouldn't notice.

Polynomials

POLYNOMIALS are expressions of the form $a + bx + cx^2 + dx^3 + ...$

$5y^3 + 2y + 23$ ← Polynomial in the variable y.

$1 + x^2$

$z^4 + 3z - z^2 - 1$ ← Polynomial in the variable z.

An expression is made up of terms.
E.g: z^4, $3z$, $-z^2$ and -1

x, y and z are always **VARIABLES**. They're usually what you solve equations to find. They often have more than one possible value.

Letters like a, b, c are always **CONSTANTS**. Constants never change. They're fixed numbers — but can be represented by letters. π is a good example. You use the symbol π, but it's just a number = 3.1415...

Functions

FUNCTIONS take a value, do something to it, and output another value.

$f(x) = x^2 + 1$ ← function f takes a value, squares it and adds 1.

$g(x) = 2 - \sin 2x$ ← function g takes a value (in degrees), doubles it, takes the sine of it, then takes the value away from 2.

You can plug values into a function — just replace the variable with a certain number.

$f(-2) = (-2)^2 + 1 = 5$
$f(0) = (0)^2 + 1 = 1$
$f(252) = (252)^2 + 1 = 63505$

$g(-90°) = 2 - \sin(-180°) = 2 - 0 = 2$
$g(0°) = 2 - \sin 0° = 2 - 0 = 2$
$g(45°) = 2 - \sin 90° = 2 - 1 = 1$

Exam questions use functions all the time. They generally don't have that much to do with the actual question. It's just a bit of terminology to get comfortable with.

Multiplication and Division

There's three different ways of showing **MULTIPLICATION**:

1) with good old-fashioned "times" signs (×):

$$f(x) = (2x \times 6y) + (2x \times \sin x) + (z \times y)$$

The multiplication signs and the variable x are easily confused.

2) or sometimes just use a little dot:

$$f(x) = 2x.6y + 2x.\sin x + z.y$$

Dots are better for long expressions — they're less confusing and easier to read.

3) but you often don't need anything at all:

$$f(x) = 12xy + 2x\sin x + zy$$

And there's three different ways of showing **DIVISION**:

1) $\frac{x+2}{3}$

2) $(x + 2) \div 3$

3) $(x + 2)/3$

Equations and Identities

This is an **IDENTITY**:

$$x^2 - y^2 \equiv (x + y)(x - y)$$

Make up any values you like for x and y, and it's always true. The left-hand side always equals the right-hand side.

But this is an **EQUATION**:

$$y = x^2 + x$$

This has at most two possible solutions for each value of y. e.g. if y = 0, x can only be 0 or -1.

The difference is that the identity's true for all values of x and y, but the equation's only true for certain values.

NB: If it's an identity, use the ≡ sign instead of =.

Surds

A surd is a number like $\sqrt{2}$, $\sqrt[3]{12}$ or $5\sqrt{3}$ — one that's written with the $\sqrt{\ }$ sign. They're important because you can give exact answers where you'd otherwise have to round to a certain number of decimal places.

Surds are sometimes the only way to give an *Exact Answer*

Put $\sqrt{2}$ into a calculator and you'll get something like 1.414213562...
But square 1.414213562 and you get 1.999999999.

And no matter how many decimal places you use, you'll never get exactly 2.
The only way to write the exact, spot-on value is to use surds.

So, as you're not allowed a calculator for your C1 exam, leave your answer as a surd.

There are basically *Three Rules* for using *Surds*

There are three rules you'll need to know to be able to use surds properly. Check out the 'Rules of Surds' box below.

EXAMPLES (i) Simplify $\sqrt{12}$ and $\sqrt{\frac{3}{16}}$. (ii) Show that $\frac{9}{\sqrt{3}} = 3\sqrt{3}$. (iii) Find $(2\sqrt{5} + 3\sqrt{6})^2$.

(i) Simplifying surds means making the number in the $\sqrt{\ }$ sign smaller, or getting rid of a fraction in the $\sqrt{\ }$ sign.

$$\sqrt{12} = \sqrt{4 \times 3} = \sqrt{4} \times \sqrt{3} = 2\sqrt{3}$$

Using $\sqrt{ab} = \sqrt{a}\sqrt{b}$.

$$\sqrt{\frac{3}{16}} = \frac{\sqrt{3}}{\sqrt{16}} = \frac{\sqrt{3}}{4}$$

Using $\sqrt{\frac{a}{b}} = \frac{\sqrt{a}}{\sqrt{b}}$.

(ii) For questions like these, you have to write a number (here, it's 3) as $3 = (\sqrt{3})^2 = \sqrt{3} \times \sqrt{3}$

$$\frac{9}{\sqrt{3}} = \frac{3 \times 3}{\sqrt{3}} = \frac{3 \times \sqrt{3} \times \sqrt{3}}{\sqrt{3}} = 3\sqrt{3}$$

Cancelling $\sqrt{3}$ from the top and bottom lines.

(iii) Multiply surds very carefully — it's easy to make a silly mistake.

$$\begin{aligned}(2\sqrt{5} + 3\sqrt{6})^2 &= (2\sqrt{5} + 3\sqrt{6})(2\sqrt{5} + 3\sqrt{6}) \\ &= (2\sqrt{5})^2 + 2 \times (2\sqrt{5}) \times (3\sqrt{6}) + (3\sqrt{6})^2 \\ &= (2^2 \times \sqrt{5}^2) + (2 \times 2 \times 3 \times \sqrt{5} \times \sqrt{6}) + (3^2 \times \sqrt{6}^2) \\ &= 20 + 12\sqrt{30} + 54 \\ &= 74 + 12\sqrt{30}\end{aligned}$$

$= 4 \times 5 = 20$

$= 12\sqrt{5}\sqrt{6} = 12\sqrt{30}$

$= 9 \times 6 = 54$

Rules of Surds

There's not really very much to remember.

$$\sqrt{ab} = \sqrt{a}\sqrt{b}$$

$$\sqrt{\frac{a}{b}} = \frac{\sqrt{a}}{\sqrt{b}}$$

$$a = (\sqrt{a})^2 = \sqrt{a}\sqrt{a}$$

Remove surds from fractions by *Rationalising the Denominator*

Surds are pretty darn complicated.

So they're the last thing you want at the bottom of a fraction.

But have no fear — Rationalise the Denominator...

Yup, you heard... (it means getting rid of the surds from the bottom of a fraction).

EXAMPLE Rationalise the denominator of $\frac{1}{1 + \sqrt{2}}$

Multiply the top and bottom by the denominator (but change the sign in front of the surd).

$$\frac{1}{1 + \sqrt{2}} \times \frac{1 - \sqrt{2}}{1 - \sqrt{2}}$$

$$\frac{1 - \sqrt{2}}{(1 + \sqrt{2})(1 - \sqrt{2})} = \frac{1 - \sqrt{2}}{1^2 + \sqrt{2} - \sqrt{2} - \sqrt{2}^2}$$

This works because: $(a + b)(a - b) = a^2 - b^2$

$$\frac{1 - \sqrt{2}}{1 - 2} = \frac{1 - \sqrt{2}}{-1} = -1 + \sqrt{2}$$

Surely the pun is mightier than the surd...

You'll need to work with surds in your non-calculator C1 exam, as roots are nigh on impossible (well, very tricky) to work out without a calculator. Learn the rules in the box so you can write them down without thinking — then get loads of practice.

AS Level
[ma]thematics

[…]/ — no question about that.
[…]) revise properly and practise hard.

[…]ning in modules C1, C2, S1, M1 and D1.
[…]e you step-by-step through loads of examples.

[…] every topic there are warm-up and exam-style
[…]ctice exams at the end of each module.

[…] the whole thing vaguely entertaining for you.

Complete Revision and Practice

Exam Board: AQA

Contents

Core Mathematics C1

C1 Section 1 — Algebra Fundamentals
A Few Definitions and Things 1
Surds 2
Multiplying Out Brackets 3
Taking Out Common Factors 4
Algebraic Fractions 5
Simplifying Expressions 6
Practice Questions 7

C1 Section 2 — Quadratics
Sketching Quadratic Graphs 8
Factorising a Quadratic 9
Completing the Square 11
The Quadratic Formula 13
The Discriminant 14
Cows 16
Practice Questions 17

C1 Section 3 — Simultaneous Equations & Inequalities
Simultaneous Equations 19
Simultaneous Equations with Quadratics 20
Geometric Interpretation 21
Linear Inequalities 22
Quadratic Inequalities 23
Practice Questions 24

C1 Section 4 — More Polynomials
Factorising Cubics 26
Algebraic Division 27
The Remainder and Factor Theorems 28
Practice Questions 29

C1 Section 5 — Coordinate Geometry
Coordinate Geometry 30
Parallel and Perpendicular Lines 32
Curve Sketching 33
Circles 34
Practice Questions 36

C1 Section 6 — Differentiation
Differentiation 38
Finding Tangents and Normals 40
Stationary Points 41
Increasing and Decreasing Functions 42
Real-life Problems 43
Practice Questions 44

C1 Section 7 — Integration
Integration 47
Areas Under Curves 49
Areas Between Curves 50
Practice Questions 51

C1 Practice Exams
C1 — Practice Exam One 53
C1 — Practice Exam Two 55

Core Mathematics C2

C2 Section 1 — Algebra and Functions
Laws of Indices 57
Graph Transformations 58
Practice Questions 59

C2 Section 2 — Sequences and Series
Sequences 60
Arithmetic Progressions 62
Arithmetic Series and Sigma Notation 63
Geometric Progressions 64
Sequence and Series Problems 67
Binomial Expansions 68
Practice Questions 71

C2 Section 3 — Trigonometry
The Trig Formulas You Need to Know 74
Using the Sine and Cosine Rules 75
Arc Length and Sector Area 76
Graphs of Trig Functions 78
Transformed Trig Functions 79
Solving Trig Equations in a Given Interval 80
Practice Questions 84

C2 Section 4 — Exponentials and Logarithms
Logs 87
Exponentials and Logs 88
Practice Questions 90

C2 Section 5 — Differentiation
Differentiation 92
Practice Questions 94

C2 Section 6 — Integration
Integration 95
The Trapezium Rule 96
Practice Questions 98

C2 Practice Exams
C2 — Practice Exam One 100
C2 — Practice Exam Two 102

Statistics S1

S1 Section 1 — Numerical Measures
Location: Mean, Median and Mode 104
Dispersion: Range and Interquartile Range 106
Dispersion: Standard Deviation 107
Dispersion and Outliers 108
Linear Scaling 109
Practice Questions 110

S1 Section 2 — Probability
Random Events and Probabilities 112
Tree Diagrams 114
Conditional Probability 115

Multiplying Out Brackets

In this horrific nightmare that is AS-level maths, you need to manipulate and simplify expressions all the time.

Remove brackets by **Multiplying** them out

Here are the basic types you have to deal with. You'll have seen them before. But there's no harm in reminding you, eh?

Multiply Your Brackets Here — we do all shapes and sizes

Single Brackets

$a(b + c + d) = ab + ac + ad$

Squared Brackets

$(a + b)^2 = (a + b)(a + b) = a^2 + 2ab + b^2$

Use the middle stage until you're comfortable with it.
Just never make this mistake: $(a + b)^2 = a^2 + b^2$

Double Brackets

$(a + b)(c + d) = ac + ad + bc + bd$

Long Brackets

Write it out again with each term from one bracket separately multiplied by the other bracket.

$(x + y + z)(a + b + c + d)$

$= x(a + b + c + d) + y(a + b + c + d) + z(a + b + c + d)$

Then multiply out each of these brackets, one at a time.

Single Brackets

$3xy(x^2 + 2x - 8)$ — Multiply all the terms inside the brackets by the bit outside — separately.

$(3xy \times x^2) + (3xy \times 2x) + (3xy \times (-8))$ — I've put brackets round each bit to make it easier to read.

All the stuff in the brackets now needs sorting out. Work on each bracket separately.

$(3x^3y) + (6x^2y) + (-24xy)$ — Multiply the numbers first, then put the x's and other letters together.

$3x^3y + 6x^2y - 24xy$

Squared Brackets

Either write it as two brackets and multiply it out...

$(2y^2 + 3x)^2$

$(2y^2 + 3x)(2y^2 + 3x)$

$2y^2.2y^2 + 2y^2.3x + 3x.2y^2 + 3x.3x$ — The dot just means 'multiplied by' — the same as the × sign.

$4y^4 + 6xy^2 + 6xy^2 + 9x^2$ — From here on it's simplification — nothing more, nothing less.

$4y^4 + 12xy^2 + 9x^2$

...or do it in one go.

$(2y^2)^2 + 2(2y^2)(3x) + (3x)^2$

$a^2 \quad 2ab \quad b^2$

$4y^4 + 12xy^2 + 9x^2$

Long Brackets

$(2x^2 + 3x + 6)(4x^3 + 6x^2 + 3)$

$2x^2(4x^3 + 6x^2 + 3) + 3x(4x^3 + 6x^2 + 3) + 6(4x^3 + 6x^2 + 3)$ — Each term in the first bracket has been multiplied by the second bracket.

$(8x^5 + 12x^4 + 6x^2) + (12x^4 + 18x^3 + 9x) + (24x^3 + 36x^2 + 18)$ — Now multiply out each of these brackets.

$8x^5 + 24x^4 + 42x^3 + 42x^2 + 9x + 18$ — Then simplify it all...

Go forth and multiply out brackets...

OK, so this is obvious, but I'll say it anyway — if you've got 3 or more brackets together, multiply them out 2 at a time. Then you'll be turning a really hard problem into two easy ones. You can do that loads in maths. In fact, writing the same thing in different ways is what maths is about. That and sitting in classrooms with tacky 'maths can be fun' posters...

Taking Out Common Factors

Common factors need to be hunted down, and taken outside the brackets. They are a danger to your exam mark.

Spot those Common Factors

A bit which is in each term of an expression is a common factor.

Spot Those Common Factors $2x^3z + 4x^2yz + 14x^2y^2z$

Look for any bits that are in each term.

Numbers: there's a common factor of 2 here because 2 divides into 2, 4 and 14.

Variables: there's at least an x^2 in each term and there's a z in each term.

So there's a common factor of $2x^2z$ in this expression.

And Take Them Outside a Bracket

If you spot a common factor you can "take it out":

$2x^2z(x + 2y + 7y^2)$

Write the common factor outside a bracket.

and put what's left of each term inside the bracket.

Afterwards, always multiply back out to check you did it right:

Check by Multiplying Out Again

$2x^2z(x + 2y + 7y^2) = 2x^3z + 4x^2yz + 14x^2y^2z$

But it's not just numbers and variables you need to look for...

Brackets: $(y + a)^2(x - a)^3 + (x - a)^2$

$(x-a)^2$ is a common factor — it comes out to give:

$(x - a)^2((y + a)^2(x - a) + 1)$

Look for Common Factors when Simplifying Expressions

EXAMPLE Simplify... $(x + 1)(x - 2) + (x + 1)^2 - x(x + 1)$

There's an $(x + 1)$ factor in each term, so we can take this out as a common factor (hurrah).

$(x + 1)\{(x - 2) + (x + 1) - x\}$

The terms inside the big bracket are the old terms with an $(x + 1)$ removed.

At this point you should check that this multiplies out to give the original expression. (You can just do this in your head, if you trust it.)

Then simplify the big bracket's innards:

$(x + 1)(\cancel{x} - 2 + x + 1 - \cancel{x})$

$= (x + 1)(x - 1)$

$= x^2 - 1$

Get this answer by multiplying out the two brackets (or by using the "difference of two squares").

Bored of spotting trains or birds? Try common factors...

You'll be doing this business of taking out common factors a lot — so get your head round this. It's just a case of looking for things that are in all the different terms of an expression, i.e. bits they have in common. And if something's in all the different terms, save yourself some time and ink, and write it once — instead of two, three or more times.

Algebraic Fractions

No one likes fractions. But just like Mondays, you can't put them off forever. Face those fears. Here goes...

The first thing you've got to know about fractions:

$$\frac{a}{x} + \frac{b}{x} + \frac{c}{x} \equiv \frac{a+b+c}{x}$$

You can just add the stuff on the top lines because the bottom lines are all the same.

x is called a common denominator — a fancy way of saying 'the bottom line of all the fractions is x'.

Add fractions by putting them over a **Common Denominator...**

Finding a common denominator just means 'rewriting some fractions so all their bottom lines are the same'.

EXAMPLE Simplify $\frac{1}{2x} - \frac{1}{3x} + \frac{1}{5x}$

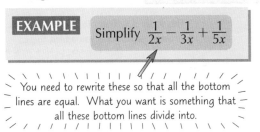

You need to rewrite these so that all the bottom lines are equal. What you want is something that all these bottom lines divide into.

Put It over a Common Denominator

30 is the lowest number that 2, 3 and 5 go into.
So the common denominator is $30x$.

$$\frac{15}{30x} - \frac{10}{30x} + \frac{6}{30x}$$

Always check that these divide out to give what you started with.

$$\frac{15-10+6}{30x} = \frac{11}{30x}$$

...even **horrible** looking ones

Yep, finding a common denominator even works for those fraction nasties — like these:

EXAMPLE Simplify $\frac{3}{x+2} + \frac{5}{x-3}$

Find the Common Denominator

Take all the individual 'bits' from the bottom lines and multiply them together.
Only use each bit once unless something on the bottom line is squared.

The individual 'bits' here are $(x + 2)$ and $(x - 3)$.

$$(x+2)(x-3)$$

Put Each Fraction over the Common Denominator

Make the denominator of each fraction into the common denominator.

$$\frac{3(x-3)}{(x+2)(x-3)} + \frac{5(x+2)}{(x+2)(x-3)}$$

Multiply the top and bottom lines of each fraction by whatever makes the bottom line the same as the common denominator.

Combine into One Fraction

Once everything's over the common denominator you can just add the top lines together.

$$= \frac{3(x-3)+5(x+2)}{(x+2)(x-3)}$$

All the bottom lines are the same — so you can just add the top lines.

All you need to do now is tidy up the top.

$$= \frac{3x-9+5x+10}{(x+2)(x-3)} = \frac{8x+1}{(x+2)(x-3)}$$

So much prettier now all the terms are together. Simple.

Well put me over a common denominator and pickle my walrus...

Adding fractions — turning lots of fractions into one fraction. Sounds pretty good to me, since it means you don't have to write as much. Better do it carefully, though — otherwise you can watch those marks shoot straight down the toilet.

Simplifying Expressions

I know this is basic stuff but if you don't get really comfortable with it you <u>will</u> make silly mistakes. You will.

Cancelling stuff on the top and bottom lines

Cancelling stuff is good — because it means you've got rid of something, and you don't have to write as much.

EXAMPLE Simplify $\frac{ax+ay}{az}$

You can do this in two ways. Use whichever you prefer — but make sure you understand the ideas behind both.

Factorise — then Cancel

$$\frac{ax+ay}{az} = \frac{a(x+y)}{az}$$ ← Factorise the top line.

Cancel the 'a'. → $$\frac{\cancel{a}(x+y)}{\cancel{a}z} = \frac{x+y}{z}$$

Split into Two Fractions — then Cancel

$$\frac{ax+ay}{az} = \frac{ax}{az} + \frac{ay}{az}$$ ← This is an okay thing to do — just think what you'd get if you added these.

$$= \frac{\cancel{a}x}{\cancel{a}z} + \frac{\cancel{a}y}{\cancel{a}z} = \frac{x}{z} + \frac{y}{z}$$ ← This answer's the same as the one from the first box — honest. Check it yourself by adding the fractions.

Simplifying complicated-looking Brackets

EXAMPLE Simplify the expression $(x-y)(x^2+xy+y^2)$

There's only one thing to do here.... Multiply out those brackets!

$$(x-y)(x^2+xy+y^2) = x(x^2+xy+y^2) - y(x^2+xy+y^2)$$ ← Multiplying each term in the first bracket by the second bracket.

$$= (x^3+x^2y+xy^2) - (x^2y+xy^2+y^3)$$ ← Multiplying out each of these two brackets.

$$= x^3+x^2y+xy^2-x^2y-xy^2-y^3$$ ← Don't forget these become minus signs because of the minus sign in front of the bracket.

And then the x^2y and the xy^2 terms disappear...

$$= x^3 - y^3$$

Sometimes you just have to do Anything you can think of and Hope...

Sometimes it's not easy to see what you're supposed to do to simplify something.
When this happens — just do anything you can think of and see what 'comes out in the wash'.

EXAMPLE Simplify $4x + \frac{4x}{x+1} - 4(x+1)$

There's nothing obvious to do — so do what you can. Try adding them as fractions...

$$4x + \frac{4x}{x+1} - 4(x+1) = \frac{(x+1)\times 4x}{x+1} + \frac{4x}{x+1} - \frac{(x+1)\times 4x(x+1)}{x+1}$$ ← The common denominator is $(x+1)$.

$$= \frac{4x^2+4x+4x-4(x+1)^2}{x+1}$$ ← Still looks horrible. So work out the brackets — but don't forget the minus signs.

$$= \frac{4x^2+4x+4x-4x^2-8x-4}{x+1}$$

$$= -\frac{4}{x+1}$$ ← Aha — everything disappears to leave you with this. And this is definitely simpler than it looked at the start.

Don't look at me like that...

Choose a word, any word at all. Like "Simple". Now stare at it. Keep staring at it. Does it look weird? No? Stare a bit longer. Now does it look weird? Yes? Why is that? I don't understand.

C1 Section 1 — Practice Questions

So that was the first section. Now, before you get stuck into Section Two, test yourself with these questions.

Warm-up Questions

1) Pick out the constants and the variables from the following equations:
   a) $(ax+6)^2 = 2b+3$ b) $f(x) = 12a + 3b^3 - 2$
   c) $y = \frac{-b \pm \sqrt{b^2 - 4ac}}{2a}$ d) $\frac{dy}{dx} = x^2 + ax + 2$
2) Which of these are identities (i.e. true for all variable values)?
   A $(x+b)(y-b) = xy + b(y-x) - b^2$ B $(2y + x^2) = 10$
   C $a^2 - b^2 = (a-b)(a+b)$ D $a^3 + b^3 = (a+b)(a^2 - ab + b^2)$
3) Find exact answers to these equations:
   a) $x^2 - 5 = 0$ b) $(x+2)^2 - 3 = 0$
4) Simplify: a) $\sqrt{28}$ b) $\sqrt{\frac{5}{36}}$ c) $\sqrt{18}$ d) $\sqrt{\frac{9}{16}}$
5) Show that a) $\frac{8}{\sqrt{2}} = 4\sqrt{2}$, and b) $\frac{\sqrt{2}}{2} = \frac{1}{\sqrt{2}}$
6) Find $(6\sqrt{3} + 2\sqrt{7})^2$
7) Rationalise the denominator of: $\frac{2}{3+\sqrt{7}}$
8) Remove the brackets and simplify the following expressions:
   a) $(a+b)(a-b)$ b) $(a+b)(a+b)$
   c) $35xy + 25y(5y+7x) - 100y^2$ d) $(x+3y+2)(3x+y+7)$
9) Take out the common factors from the following expressions:
   a) $2x^2y + axy + 2xy^2$ b) $a^2x + a^2b^2x^2$
   c) $16y + 8yx + 56x$ d) $x(x-2) + 3(2-x)$
10) Put the following expressions over a common denominator:
   a) $\frac{2x}{3} + \frac{y}{12} + \frac{x}{5}$ b) $\frac{5}{xy^2} - \frac{2}{x^2y}$ c) $\frac{1}{x} + \frac{x}{x+y} + \frac{y}{x-y}$
11) Simplify these expressions:
   a) $\frac{2a}{b} - \frac{a}{2b}$ b) $\frac{2p}{p+q} + \frac{2q}{p-q}$ c) "A bird in the hand is worth two in the bush"

Exam Questions

1 Simplify

a) $(5\sqrt{3})^2$ *(1 mark)*

b) $(5+\sqrt{6})(2-\sqrt{6})$ *(2 marks)*

2 Express $\frac{5+\sqrt{5}}{3-\sqrt{5}}$ in the form $a + b\sqrt{5}$, where a and b are integers. *(4 marks)*

3 Write $\sqrt{18} - 2\sqrt{8} + \frac{4}{\sqrt{2}}$ in the form $n\sqrt{2}$, where n is an integer. *(4 marks)*

4 Show that $\frac{4\sqrt{32}}{\sqrt{2}} - 2\sqrt{3} + \frac{4\sqrt{5}}{\sqrt{20}} + \sqrt{12}$ is an integer. *(5 marks)*

Sketching Quadratic Graphs

A quadratic equation is a polynomial of the form $y = ax^2 + bx + c$, where $a \neq 0$. Quadratics pop up absolutely everywhere in AS maths, so you need to be totally down with them. If a question doesn't seem to make sense, or you can't see how to go about solving a problem, try drawing a graph — it helps if you can actually see what the problem is, rather than just reading about it.

Sketch the graphs of the following quadratic equations:

① $y = 2x^2 - 4x + 3$

② $y = 8 - 2x - x^2$

Quadratic graphs are *Always* u-shaped or n-shaped

A The first thing you need to know is whether the graph's going to be u-shaped or n-shaped (upside down). To decide, look at the coefficient of x^2.

$y = 2x^2 - 4x + 3$

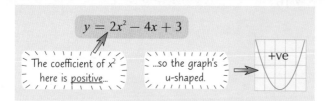

The coefficient of x^2 here is positive... ...so the graph's u-shaped.

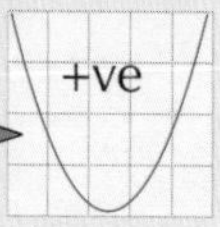

$y = 8 - 2x - x^2$

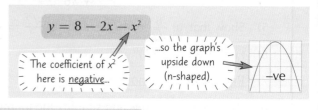

The coefficient of x^2 here is negative... ...so the graph's upside down (n-shaped).

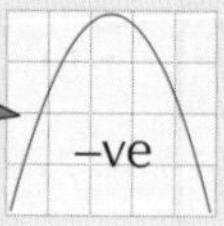

B Now find the places where the graph crosses the axes (both the y-axis and the x-axis).

(i) Put $x = 0$ to find where it meets the y-axis.

$y = 2x^2 - 4x + 3$

$y = (2 \times 0^2) - (4 \times 0) + 3$ so $y = 3$ ← That's where it crosses the y-axis.

(ii) Solve $y = 0$ to find where it meets the x-axis.

$2x^2 - 4x + 3 = 0$

$b^2 - 4ac = -8 < 0$ ← You could use the formula. But first check $b^2 - 4ac$ to see if $y = 0$ has any roots.

So it has no solutions, and doesn't cross the x-axis.

For more info, see p. 14.

(i) Put $x = 0$.

$y = 8 - 2x - x^2$

$y = 8 - (2 \times 0) - 0^2$ so $y = 8$

(ii) Solve $y = 0$.

$8 - 2x - x^2 = 0$ ← This equation factorises easily...

$\Rightarrow (2 - x)(x + 4) = 0$

$\Rightarrow x = 2$ or $x = -4$

C Finally, find the minimum or maximum (i.e. the vertex).

Since $y = 2(x - 1)^2 + 1$ ← By completing the square (see p. 11-12).

the minimum value is $y = 1$, which occurs at $x = 1$.

The minimum or maximum of the graph is always at $x = \frac{-b}{2a}$

The maximum value is halfway between the roots — the graph's symmetrical.

The maximum value is at $x = -1$

So the maximum is $y = 8 - (2 \times -1) - (-1)^2$

i.e. the graph has a maximum at the point (–1, 9).

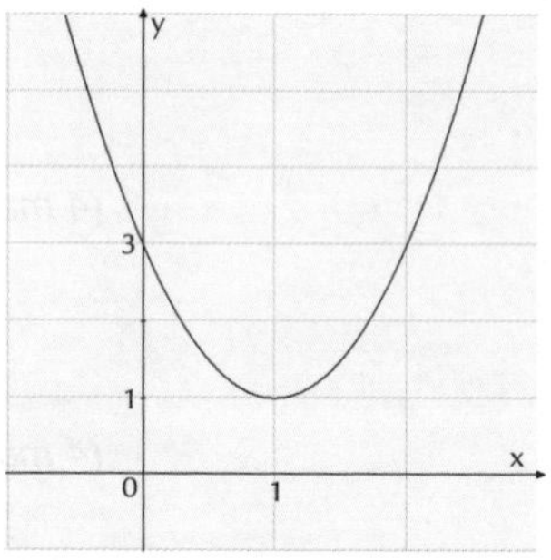

Sketching Quadratic Graphs

A) up or down — decide which direction the curve points in.

B) axes — find where the curve crosses them.

C) max / min — find the vertex.

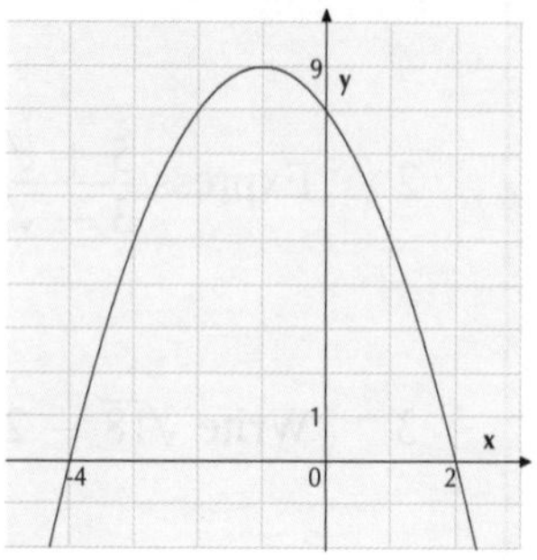

Van Gogh, Monet — all the greats started out sketching graphs...

So there are three steps here to learn. Simple enough. You can do the third step (finding the max/min point) by either a) completing the square, which is covered a bit later, or b) using the fact that the graph's symmetrical — so once you've found the points where it crosses the x-axis, the point halfway between them will be the max/min. It's all laughs here...

Factorising a Quadratic

Factorising a quadratic means putting it into two brackets — and is useful if you're trying to draw a graph of a quadratic or solve a quadratic equation. It's pretty easy if $a = 1$ (in ax^2+bx+c form), but can be a real pain otherwise.

$$x^2 - x - 12 = (x - 4)(x + 3)$$

Factorising's not so bad when a = 1

EXAMPLE Solve $x^2 - 8 = 2x$ by factorising.

A) Put into $ax^2+bx+c=0$ Form

$x^2 - 2x - 8 = 0$ ⟸ So $a = 1$, $b = -2$, $c = -8$.

Write down the two brackets with x's in: $x^2 - 2x - 8 = (x \quad)(x \quad)$

B) Find the Two Numbers

Find two numbers that multiply together to make c but which also add or subtract to give b (you can ignore any minus signs for now).

1 and 8 multiply to give 8 — and add / subtract to give 9 and 7.
2 and 4 multiply to give 8 — and add / subtract to give 6 and 2.

This is the value for b you're after — so this is the right combination: 2 and 4.

C) Find the Signs

Now all you have to do is put in the plus or minus signs.

If c is negative, then the signs must be different.

$x^2 - 2x - 8 = (x \quad 4)(x \quad 2)$

$x^2 - 2x - 8 = (x + 2)(x - 4)$

It must be +2 and –4 because $2 \times (-4) = -8$ and $2 + (-4) = 2 - 4 = -2$

D) Solve the Equation

All you've done so far is to factorise the equation — you've still got to solve it.

$(x + 2)(x - 4) = 0$

$\Rightarrow x + 2 = 0$ or $x - 4 = 0$

$\Rightarrow x = -2$ or $x = 4$

Don't forget this last step. The factors aren't the answer.

Factorising Quadratics

A) Rearrange the equation into the standard $ax^2 + bx + c$ form.

B) Write down the two brackets: $(x \quad)(x \quad)$

C) Find two numbers that multiply to give 'c' and add / subtract to give 'b' (ignoring signs).

D) Put the numbers in the brackets and choose their signs.

Another Example...

EXAMPLE Solve $x^2 + 4x - 21 = 0$ by factorising.

This equation is already in the standard format — you can write down the brackets straight away.

$x^2 + 4x - 21 = (x \quad)(x \quad)$

1 and 21 multiply to give 21 — and add / subtract to give 22 and 20.
3 and 7 multiply to give 21 — and add / subtract to give 10 and 4.

This is the value of 'b' you're after — 3 and 7 are the right numbers.

$x^2 + 4x - 21 = (x + 7)(x - 3)$

And solving the equation to find x gives... $\Rightarrow x = -7$ or $x = 3$

Scitardauq Gnisirotcaf — you should know it backwards...

Factorising quadratics — this is very basic stuff. You've really got to be comfortable with it. If you're even slightly rusty, you need to practise it until it's second nature. Remember why you're doing it — you don't factorise simply for the pleasure it gives you — it's so you can solve quadratic equations. Well, that's the theory anyway...

Factorising a Quadratic

It's not over yet...

Factorising a quadratic when $a \neq 1$

These can be a real pain. The basic method's the same as on the previous page — but it can be a bit more awkward.

EXAMPLE Factorise $3x^2 + 4x - 15$

A) Write Down Two Brackets

As before, write down two brackets — but instead of just having x in each, you need two things that will multiply to give $3x^2$.

$3x^2 + 4x - 15 = (3x \quad)(x \quad)$

It's got to be 3x and x here.

B) The Fiddly Bit

You need to find two numbers that multiply together to make 15 — but which will give you $4x$ when you multiply them by x and $3x$, and then add / subtract them.

$(3x \quad 1)(x \quad 15) \Rightarrow x$ and $45x$ which then add or subtract to give $46x$ and $44x$.

$(3x \quad 15)(x \quad 1) \Rightarrow 15x$ and $3x$ which then add or subtract to give $18x$ and $12x$.

$(3x \quad 3)(x \quad 5) \Rightarrow 3x$ and $15x$ which then add or subtract to give $18x$ and $12x$.

$(3x \quad 5)(x \quad 3) \Rightarrow 5x$ and $9x$ which then add or subtract to give $14x$ and $4x$.

This is the value you're after — so this is the right combination.

C) Add the Signs

You know the brackets must be like these... $(3x \quad 5)(x \quad 3) = 3x^2 + 4x - 15$

So all you have to do is put in the plus or minus signs.

'c' is negative — that means the signs in the brackets are different.

You've only got two choices — if you're unsure, just multiply them out to see which one's right.

$(3x + 5)(x - 3) = 3x^2 - 4x - 15$

or...

$(3x - 5)(x + 3) = 3x^2 + 4x - 15$ ⟸ So it's this one.

*Sometimes it's best just to **Cheat** and use the **Formula***

Check out p. 13 for all you need to know about the Quadratic Formula.

Here are two final points to bear in mind:

1) It won't always factorise.
2) Sometimes factorising is so messy that it's easier to just use the quadratic formula...

So if the question doesn't tell you to factorise, don't assume it will factorise.
And if it's something like this thing below, don't bother trying to factorise it...

EXAMPLE Solve $6x^2 + 87x - 144 = 0$

This will actually factorise, but there are 2 possible bracket forms to try.

$(6x \quad)(x \quad)$ or $(3x \quad)(2x \quad)$ And for each of these, there are 8 possible ways of making 144 to try.

And you can quote me on that...

"He who can properly do quadratic equations is considered a god."
Plato

"Quadratic equations are the music of reason."
James J Sylvester

Completing the Square

Completing the Square is a handy little trick that you should definitely know how to use.
It can be a bit fiddly — but it gives you loads of information about a quadratic really quickly.

Take any old quadratic and put it in a **Special Form**

Completing the square can be really confusing. For starters, what does "Completing the Square" mean? What is the square? Why does it need completing? Well, there is some logic to it:

1) The square is something like this: $(x + \text{something})^2$ It's basically the factorised equation (with the factors both the same), but there's something missing...

2) ...So you need to 'complete' it by adding a number to the square to make it equal to the original equation. $(x + \text{something})^2 + d$

You'll start with something like this... $2x^2 + 8x - 5$ → ...sort the x-coefficients... $2(x+2)^2 + ?$ → ...and you'll end up with something like this. $2(x+2)^2 - 13$

Lovely!

Make completing the square a bit **Easier**

There are only a few stages to completing the square — if you can't be bothered trying to understand it, just learn how to do it. But I reckon it's worth spending a bit more time to get your head round it properly.

A) Take Out a Factor of 'a' — take a factor of a out of the x^2 and x terms.

$f(x) = 2x^2 + 3x - 5$ ← This is in the form $ax^2 + bx + c$

This '2' is an 'a'.

$f(x) = 2\left(x^2 + \frac{3}{2}x\right) - 5$ ← Check that the bracket multiplies out to what you had before.

This is $\frac{b}{a}$

B) Rewrite the Bracket — rewrite the bracket as one bracket squared.

The number in the brackets is always half the old number in front of the x. $\frac{b}{2a}$

$f(x) = 2\left(x + \frac{3}{4}\right)^2 + d$ ← d is a number you have to find to make the new form equal to the old one.

Don't forget the 'squared' sign.

C) Complete the Square — find d.

To do this, make the old and new equations equal each other...

$$2\left(x + \frac{3}{4}\right)^2 + d = 2x^2 + 3x - 5$$

...and you can find d.

$$2x^2 + 3x + \frac{9}{8} + d = 2x^2 + 3x - 5$$

The x^2 and x bits are the same on both sides, so they can disappear.

$$\frac{9}{8} + d = -5$$
$$\Rightarrow d = -\frac{49}{8}$$

D) So the Answer is: $f(x) = 2x^2 + 3x - 5 = 2\left(x + \frac{3}{4}\right)^2 - \frac{49}{8}$

Completing the Square

A) THE BIT IN THE BRACKETS IS ALWAYS — $a\left(x + \frac{b}{2a}\right)^2$

B) CALL THE NUMBER AT THE END d — $a\left(x + \frac{b}{2a}\right)^2 + d$

C) MAKE THE TWO FORMS EQUAL — $ax^2 + bx + c = a\left(x + \frac{b}{2a}\right)^2 + d$

Complete your square — it'd be root not to...

Remember — you're basically trying to write the expression as one bracket squared, but it doesn't quite work. So you have to add a number (d) to make it work. It's a bit confusing at first, but once you've learnt it you won't forget it in a hurry.

Completing the Square

Once you've completed the square, you can very quickly say loads about a quadratic function. And it all relies on the fact that a squared number can never be less than zero... ever.

Completing the square can sometimes be ***Useful***

This is a quadratic written as a completed square. As it's a quadratic function and the coefficient of x^2 is positive, it's a u-shaped graph.

$$f(x) = 3x^2 - 6x - 7 = 3(x-1)^2 - 10$$

This is a square — it can never be negative. The smallest it can be is 0.

A Find the Minimum — make the bit in the brackets equal to zero.

When the squared bit is zero, $f(x)$ reaches its minimum value. This means the graph reaches its lowest point.

$f(x) = 3(x-1)^2 - 10$ — This number here is the minimum.

$f(1) = 3(1-1)^2 - 10$

$f(1) = 3(0)^2 - 10 = -10$

$f(1)$ means using $x = 1$ in the function.

So the minimum is -10, when $x = 1$

B Where Does f(x) Cross the x-axis? — i.e. find x.

Make the completed square function equal zero. → $3(x-1)^2 - 10 = 0$

Solve it to find where $f(x)$ crosses the x-axis.

$\Rightarrow (x-1)^2 = \frac{10}{3}$

da-de-dah ... rearranging again.

$\Rightarrow x - 1 = \pm\sqrt{\frac{10}{3}}$

$\Rightarrow x = 1 \pm \sqrt{\frac{10}{3}}$

So $f(x)$ crosses the x-axis when...

$x = 1 + \sqrt{\frac{10}{3}}$ or $x = 1 - \sqrt{\frac{10}{3}}$

These notes are all about graphs with positive coefficients in front of the x^2. But if the coefficient is negative, then the graph is flipped upside down (n-shaped, not u-shaped).

With this information, you can easily sketch the graph...

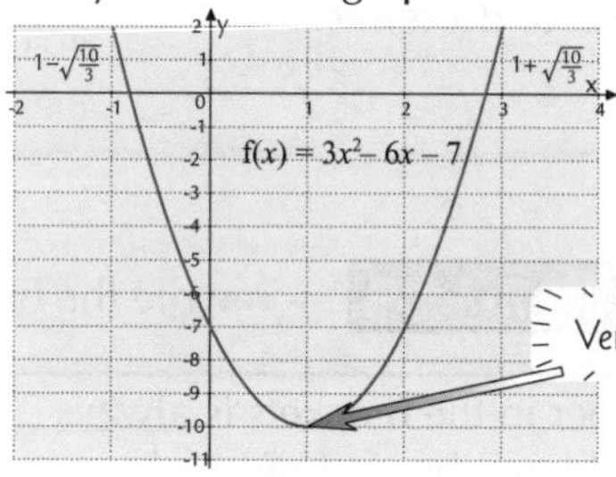

Vertex is at (1, –10).

In general, the graph of the completed square $y = (x-a)^2 + b$ is a translation of the graph of $y = x^2$. It's shifted 'a' spaces horizontally and 'b' spaces vertically.

Some functions don't have ***Real Roots***

By completing the square, you can also quickly tell if the graph of a quadratic function ever crosses the x-axis. It'll only cross the x-axis if the function changes sign (i.e. goes from positive to negative or vice versa). Take this function...

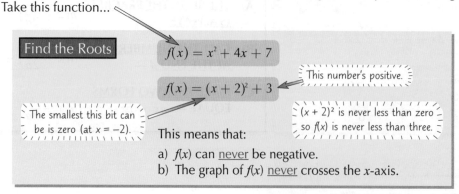

Find the Roots

$f(x) = x^2 + 4x + 7$

$f(x) = (x+2)^2 + 3$ — This number's positive.

The smallest this bit can be is zero (at $x = -2$).

$(x+2)^2$ is never less than zero so $f(x)$ is never less than three.

This means that:

a) $f(x)$ can never be negative.

b) The graph of $f(x)$ never crosses the x-axis.

If the coefficient of x^2 is negative, you can do the same sort of thing to check whether $f(x)$ ever becomes positive.

Don't forget — two wrongs don't make a root...

You'll be pleased to know that that's the end of me trying to tell you how to do something you probably really don't want to do. Now you can push it to one side and run off to roll around in a bed of nettles... much more fun.

The Quadratic Formula

Unlike factorising, the quadratic formula always works... no ifs, no buts, no butts, no nothing...

The **Quadratic Formula** — a reason to be cheerful, but careful...

If you want to solve a quadratic equation $ax^2 + bx + c = 0$, then the answers are given by this formula:

$$x = \frac{-b \pm \sqrt{b^2 - 4ac}}{2a}$$

The formula's a godsend — but use the power wisely...

If any of the coefficients (i.e. if a, b or c) in your quadratic equation are negative — be especially careful.

Always take things nice and slowly — don't try to rush it.

It's a good idea to write down what a, b and c are before you start plugging them into the formula.

There are a couple of minus signs in the formula — which can catch you out if you're not paying attention.

I shall teach you the ways of the **Formula**

EXAMPLE: Solve the quadratic equation $3x^2 - 4x = 8$, leaving your answer in surd form.

The mention of surds is a big clue that you should use the formula.

A Rearrange the Equation

Get the equation in the standard $ax^2 + bx + c = 0$ form.

$3x^2 - 4x = 8$

$3x^2 - 4x - 8 = 0$

B Find a, b and c

Write down the coefficients a, b and c — making sure you don't forget minus signs.

$3x^2 - 4x - 8 = 0$

$a = 3 \quad b = -4 \quad c = -8$

C Stick Them in the Formula

Very carefully, plug these numbers into the formula. It's best to write down each stage as you do it.

$$x = \frac{-b \pm \sqrt{b^2 - 4ac}}{2a}$$

$$x = \frac{-(-4) \pm \sqrt{(-4)^2 - 4 \times 3 \times (-8)}}{2 \times 3}$$

$$x = \frac{4 \pm \sqrt{16 + 96}}{6}$$

$$x = \frac{4 \pm \sqrt{112}}{6}$$

Simplify your answer as much as possible, using the rules of surds (see page 2).

$$x = \frac{2 \pm 2\sqrt{7}}{3}$$

The $\pm$ sign means that we have two different expressions for x — which you get by replacing the $\pm$ with + and −.

$$x = \frac{2 + 2\sqrt{7}}{3} \quad \text{or} \quad x = \frac{2 - 2\sqrt{7}}{3}$$

Using this magic formula, I shall take over the world... ha ha ha...

Okay, maybe it's not quite that good... but it's really important. So learn it properly — which means spending enough time until you can just say it out loud the whole way through, with no hesitations. Or perhaps you could try singing it as loud as you can to the tune of your favourite cheesy song. Sha-la-la-la-la-la-la-ha... La-di-da... Sha-la-la-la-la-la-la-ha...

The Discriminant

By using part of the quadratic formula, you can quickly tell if a quadratic equation has two solutions, one solution, or no solutions at all. Tell me more, I hear you cry...

How Many Roots? Check the $b^2 - 4ac$ bit...

$$x = \frac{-b \pm \sqrt{b^2 - 4ac}}{2a}$$

When you try to find the roots of a quadratic function, this bit in the square-root sign ($b^2 - 4ac$) can be positive, zero, or negative. It's this that tells you if a quadratic function has two distinct roots, two equal roots, or no roots.

The $b^2 - 4ac$ bit is called the discriminant.

Because — if the discriminant is positive, the formula will give you two different values — when you add or subtract the $\sqrt{b^2 - 4ac}$ bit.

But if it's zero, you'll only get one value, since adding or subtracting zero doesn't make any difference.

And if it's negative, you don't get any (real) values because you can't take the square root of a negative number.

Well, not in C1. In some areas of maths, you can actually take the square root of negative numbers and get 'imaginary' numbers. That's why we say no 'real' roots — because there are 'imaginary' roots!

It's good to be able to picture what this means:

A root is just the value of x when $y = 0$, so it's where the graph touches or crosses the x-axis.

$b^2 - 4ac > 0$
Two distinct real roots

So the graph crosses the x-axis twice and these are the roots:

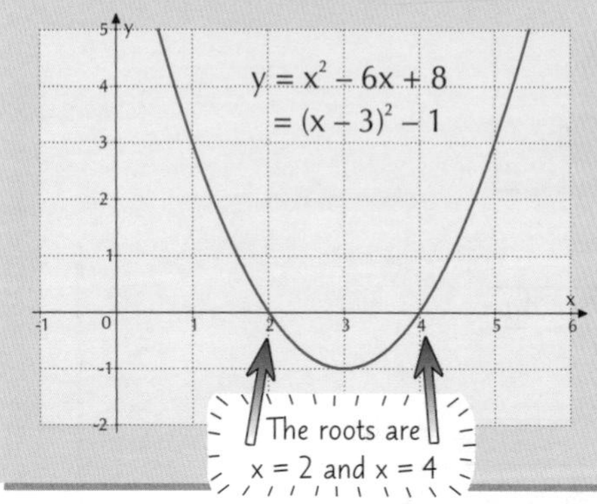

$b^2 - 4ac = 0$
Two equal roots
(Sometimes just called one root)

The graph just touches the x-axis from above (or from below if the x^2 coefficient is negative).

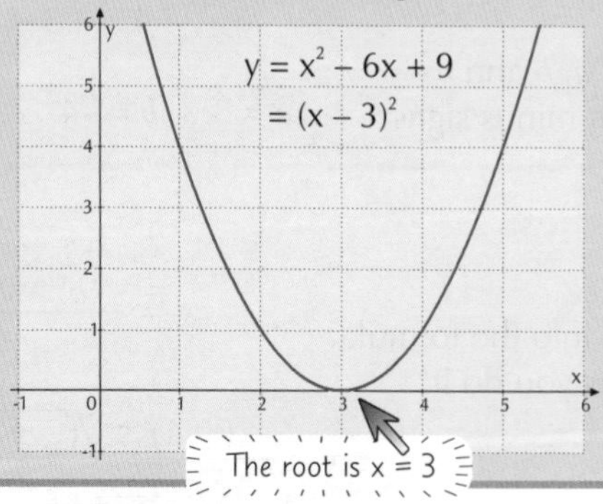

$b^2 - 4ac < 0$
No roots

The graph doesn't touch the x-axis at all.

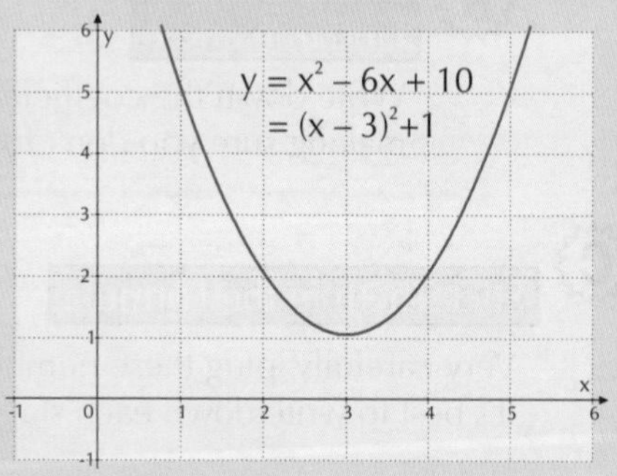

Identify a, b and c to find the Discriminant

The first thing you have to do when you're given a quadratic is to work out what a, b and c are. Make sure you get them the right way round — it's easy to get mixed up if the quadratic's in a different order.

EXAMPLE Find the discriminant of $15 - x - 2x^2$. How many real roots does $15 - x - 2x^2$ have?

First, identify a, b and c: $a = -2$, $b = -1$ and $c = 15$ (NOT $a = 15$, $b = -1$ and $c = -2$).

Then put these values into the formula for the discriminant:

$$b^2 - 4ac = (-1)^2 - (4 \times -2 \times 15) = 1 + 120 = 121.$$

The discriminant is > 0, so $15 - x - 2x^2$ has two distinct real roots.

ha ha ha ha haaaaaa... ha ha ha... ha ha ha... ha ha ha........

The Discriminant

The discriminant often comes up in exam questions — but sometimes they'll be sneaky and not actually tell you that's what you have to find. Any question that mentions roots of a quadratic will probably mean that you need to find the discriminant.

a, b *and* ***c*** *might be* ***Unknown***

In exam questions, you're often given a quadratic where one or more of a, b and c are given in terms of an unknown (usually k, but sometimes p or q). This means that you'll end up with an equation or inequality for the discriminant in terms of the unknown — you might have to solve it to find the value or range of values of the unknown.

EXAMPLE Find the range of values of k for which: a) f(x) has 2 distinct roots, b) f(x) has 1 root, c) f(x) has no real roots, where $f(x) = 3x^2 + 2x + k$.

First, decide what a, b and c are: $a = 3, b = 2, c = k$

Then work out what the discriminant is: $b^2 - 4ac = 2^2 - 4 \times 3 \times k = 4 - 12k$

These calculations are exactly the same. You don't need to do them if you've done a) because the only difference is the (in)equality symbol.

a) Two distinct roots means:

$$b^2 - 4ac > 0 \Rightarrow 4 - 12k > 0 \Rightarrow 4 > 12k \Rightarrow k < \frac{1}{3}$$

b) One root means:

$$b^2 - 4ac = 0 \Rightarrow 4 - 12k = 0 \Rightarrow 4 = 12k \Rightarrow k = \frac{1}{3}$$

c) No roots means:

$$b^2 - 4ac < 0 \Rightarrow 4 - 12k < 0 \Rightarrow 4 < 12k \Rightarrow k > \frac{1}{3}$$

You might have to ***Solve*** *a* ***Quadratic Inequality*** *to find k*

When you put your values of a, b and c into the formula for the discriminant, you might end up with a quadratic inequality in terms of k. You'll have to solve this to find the range of values of k — there's more on this on p. 23.

EXAMPLE The equation $kx^2 + (k + 3)x + 4 = 0$ has two distinct real solutions. Show that $k^2 - 10k + 9 > 0$, and find the set of values of k which satisfy this inequality.

Identify a, b and c: $a = k$, $b = (k + 3)$ and $c = 4$

Then put these values into the formula for the discriminant:

$$b^2 - 4ac = (k + 3)^2 - (4 \times k \times 4) = k^2 + 6k + 9 - 16k = k^2 - 10k + 9.$$

The original equation has two distinct real solutions, so the discriminant must be > 0.

So $k^2 - 10k + 9 > 0$.

Now, to find the set of values for k, you have to factorise the quadratic:

$$k^2 - 10k + 9 = (k - 1)(k - 9).$$

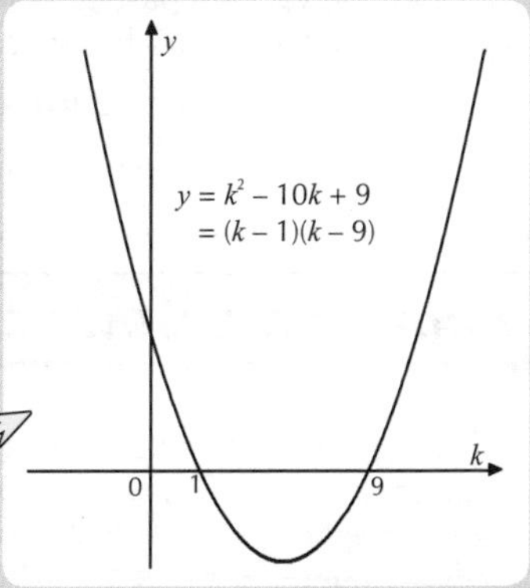

The solutions of this equation are $k = 1$ and $k = 9$. From the graph, you can see that this is a u-shaped quadratic which is > 0 when

$k < 1$ or when $k > 9$.

I'll try not to discriminate...

Don't panic if you're not sure how to solve quadratic inequalities — they're covered in more detail on p. 23. Chances are you'll get a discriminant question in the exam, so you need to know what to do. Although it might be tempting to hide under your exam desk and hope it doesn't find you, there's no escaping these questions — so get practising.

Cows

The stuff on this page isn't strictly on the syllabus. But I've included it anyway because I reckon it's really important stuff that you ought to know.

There are loads of **Different Types** of Cows

Dairy Cattle

Every day a dairy cow can produce up to 128 pints of milk — which can be used to make 14 lbs of cheese, 5 gallons of ice cream, or 6 lbs of butter.

The Jersey
The Jersey is a small breed best suited to pastures in high rainfall areas. It is kept for its creamy milk.

Advantages
1) *Can produce creamy milk until old age.*
2) *Milk is the highest in fat of any dairy breed (5.2%).*
3) *Fairly docile, although bulls can't be trusted.*

Disadvantages
1) *Produces less milk than most other breeds.*

The Holstein-Friesian
This breed can be found in many areas.
It is kept mainly for milk.

Advantages
1) *Produce more milk than any breed.*
2) *The breed is large, so bulls can be sold for beef.*

Disadvantages
1) *Milk is low in fat (3.5%).*

Beef Cattle

Cows are sedentary animals who spend up to 8 hours a day chewing the cud while standing still or lying down to rest after grazing. Getting fat for people to eat.

The Angus
The Angus is best suited to areas where there is moderately high rainfall.

Advantages
1) *Early maturing.*
2) *High ratio of meat to body weight.*
3) *Forages well.*
4) *Adaptable.*

The Hereford
The Hereford matures fairly early, but later than most shorthorn breeds. All Herefords have white faces, and if a Hereford is crossbred with any other breed of cow, all the offspring will have white or partially white faces.

Advantages
1) *Hardy.*
2) *Adaptable to different feeds.*

Disadvantages
1) *Susceptible to eye diseases.*

You will often see cows with pieces of grass sticking out of their mouths.

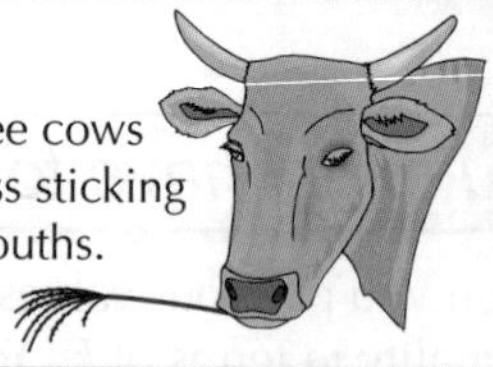

SOME IMPORTANT FACTS TO REMEMBER:
- A newborn calf can walk on its own an hour after birth.
- A cow's teeth are only on the bottom of her mouth.
- While some cows can live up to 40 years, they generally don't live beyond 20.

Milk comes from **Cows**

This is really important — try not to forget it.

Milk is an emulsion of butterfat suspended in a solution of water (roughly 80%), lactose, proteins and salts. Cow's milk has a specific gravity of around 1.03.
It's pasteurised by heating it to 63° C for 30 minutes. It's then rapidly cooled and stored below 10° C.

Louis Pasteur began his experiments into 'pasteurisation' in 1856.
By 1946, the vacuum pasteurisation method had been perfected, and in 1948, UHT (ultra heat-treated) pasteurisation was introduced.

$$\text{cow} + \text{grass} = \text{fat cow}$$
$$\text{fat cow} + \text{milking machine} \Rightarrow \text{milk}$$

Pantomime Cows aren't **Real**

If you go to see a pantomime around Christmas time, you may see a cow on stage. Don't get concerned about animal rights and exploitation of animals — it's not a real cow. Pantomime cows are just two people wearing a cow costume. Sometimes it'll be a pantomime horse instead.

The Cow
The cow is of the bovine ilk;
One end is moo, the other, milk.
— Ogden Nash

Famous Cows and Cow Songs

Famous Cows

1) Ermintrude from the Magic Roundabout.
2) The Laughing Cow.
3) Other TV commercial cows — Anchor, Dairylea
4) The cow that jumped over the moon.
5) Greek mythology was full of gods turning themselves and their girlfriends into cattle.

Cows in Pop Music
1) Boom Boom Cow — Black Eyed Peas
2) Saturday Night at the Moo-vies — The Drifters
3) I Kissed a Cow — Katy Perry
4) Take a Cow — Rihanna
5) Cows Don't Lie — Shakira

Where's me Jersey — I'm Friesian...

Cow-milking — an underrated skill, in my opinion. As Shakespeare once wrote, 'Those who can milk cows are likely to get pretty good grades in maths exams, no word of a lie'. Well, he probably would've written something like that if he was into cows. And he would've written it because cows are helpful when you're trying to work out what a question's all about — and once you know that, you can decide the best way forward. And if you don't believe me, remember the saying of the ancient Roman Emperor Julius Caesar — 'If in doubt, draw a cow'.

C1 Section 2 — Practice Questions

Mmmm, well, quadratic equations — not exactly designed to make you fall out of your chair through laughing so hard, are they? But (and that's a huge 'but') they'll get you plenty of marks come that fine morning when you march confidently into the exam hall — if you know what you're doing. Time for some practice questions methinks...

Warm-up Questions

1) Factorise the following expressions. While you're doing this, sing a jolly song to show how much you enjoy it.
   a) $x^2 + 2x + 1$, b) $x^2 - 13x + 30$, c) $x^2 - 4$, d) $3 + 2x - x^2$,
   e) $2x^2 - 7x - 4$, f) $5x^2 + 7x - 6$.

2) Solve the following equations. And sing verse two of your jolly song.
   a) $x^2 - 3x + 2 = 0$, b) $x^2 + x - 12 = 0$, c) $2 + x - x^2 = 0$, d) $x^2 + x - 16 = x$,
   e) $3x^2 - 15x - 14 = 4x$, f) $4x^2 - 1 = 0$, g) $6x^2 - 11x + 9 = 2x^2 - x + 3$.

3) Rewrite these quadratics by completing the square. Then state their maximum or minimum value and the value of x where this occurs. Also, say if and where they cross the x-axis — just for a laugh, like.
   a) $x^2 - 4x - 3$, b) $3 - 3x - x^2$, c) $2x^2 - 4x + 11$, d) $4x^2 - 28x + 48$.

4) How many roots do these quadratics have? Sketch their graphs.
   a) $x^2 - 2x - 3 = 0$, b) $x^2 - 6x + 9 = 0$, c) $2x^2 + 4x + 3 = 0$.

5) Solve these quadratic equations, leaving your answers in surd form where necessary.
   a) $3x^2 - 7x + 3 = 0$, b) $2x^2 - 6x - 2 = 0$, c) $x^2 + 4x + 6 = 12$.

Have a peek at p.23 for help on solving a quadratic inequality.

6) If the quadratic equation $x^2 + kx + 4 = 0$ has two roots, what are the possible values of k?

The warm-up questions will only get you as far as the third floor. And you'll have to use the stairs.
To get all the way to the top floor, you need to take the express lift that is this lovely set of exam questions...

Exam Questions

1 The equation $x^2 + 2kx + 4k = 0$, where k is a non-zero integer, has equal roots.

Find the value of k. *(4 marks)*

2 The equation $px^2 + (p + 3)x + 4 = 0$ has 2 distinct real solutions for x (p is a constant).

a) Show that $p^2 - 10p + 9 > 0$. *(3 marks)*

b) Hence find the range of possible values for p. *(4 marks)*

C1 Section 2 — Practice Questions

Two exam questions are <u>never enough</u> — so here are a few more...

3 Given that

$$5x^2 + nx + 14 \equiv m(x + 2)^2 + p,$$

find the values of the integers m, n and p. *(3 marks)*

4 a) Rewrite $x^2 - 12x + 15$ in the form $(x - a)^2 + b$, for integers a and b. *(2 marks)*

b) (i) Find the minimum value of $x^2 - 12x + 15$. *(1 mark)*

(ii) State the value of x at which this minimum occurs. *(1 mark)*

5 a) Use the quadratic formula to solve the equation $x^2 - 14x + 25 = 0$.
Leave your answer in simplified surd form. *(3 marks)*

b) Sketch the curve of $y = x^2 - 14x + 25$, giving the coordinates of the point where the curve crosses the x- and y-axis. *(3 marks)*

c) Hence solve the inequality $x^2 - 14x + 25 \leq 0$. *(1 mark)*

6 a) (i) Express $10x - x^2 - 27$ in the form $-(m - x)^2 + n$, where m and n are integers. *(2 marks)*

(ii) Hence show that $10x - x^2 - 27$ is always negative. *(1 mark)*

b) (i) State the coordinates of the maximum point of the curve $y = 10x - x^2 - 27$. *(2 marks)*

(ii) Sketch the curve, showing where the curve crosses the y-axis. *(2 marks)*

(iii) Write down the equation of the line of symmetry of the curve. *(1 mark)*

Simultaneous Equations

Solving simultaneous equations means finding the answers to two equations at the same time — i.e. finding values for x and y for which both equations are true. And it's one of those things that you'll have to do again and again — so it's definitely worth practising them until you feel really confident.

① $3x + 5y = -4$
② $-2x + 3y = 9$

This is how simultaneous equations are usually shown. It's a good idea to label them as equation ① and equation ② — so you know which one you're working with.

But they'll look different sometimes, maybe like this. Make sure you rearrange them as '$ax + by = c$'.

$4 + 5y = -3x$
$-2x = 9 - 3y$

rearrange as $ax + by = c$

$3x + 5y = -4$
$-2x + 3y = 9$

Solving them by Elimination

Elimination is a lovely method. It's really quick when you get the hang of it — you'll be doing virtually all of it in your head.

EXAMPLE:

① $3x + 5y = -4$
② $-2x + 3y = 9$

To get the x's to match, you need to multiply the first equation by 2 and the second by 3:

① ×2 $6x + 10y = -8$
② ×3 $-6x + 9y = 27$

Add the equations together to eliminate the x's.

① + ② $19y = 19$
$y = 1$

So y is 1. Now stick that value for y into one of the equations to find x:

$y = 1$ in ① $\Rightarrow 3x + 5 = -4$
$3x = -9$
$x = -3$

So the solution is $x = -3$, $y = 1$.

A Match the Coefficients

Multiply the equations by numbers that will make either the x's or the y's match in the two equations. (Ignoring minus signs.)

Go for the lowest common multiple (LCM). e.g. LCM of 2 and 3 is 6.

B Eliminate to Find One Variable

If the coefficients are the same sign, you'll need to subtract one equation from the other.

If the coefficients are different signs, you need to add the equations.

C Find the Variable You Eliminated

When you've found one variable, put its value into one of the original equations so you can find the other variable.

But you should always...

D Check Your Answer

...by putting these values into the other equation.

② $-2x + 3y = 9$
$x = -3$
$y = 1$

$-2 \times (-3) + 3 \times 1 = 6 + 3 = 9$

If these two numbers are the same, then the values you've got for the variables are right.

Elimination Method

1) Match the coefficients
2) Eliminate and then solve for one variable
3) Find the other variable (that you eliminated)
4) Check your answer

Eliminate your social life — do AS-level maths

This is a fairly basic method that won't be new to you. So make sure you know it. The only possibly tricky bit is matching the coefficients — work out the lowest common multiple of the coefficients of x, say, then multiply the equations to get this number in front of each x.

Simultaneous Equations with Quadratics

Elimination is great for simple equations. But it won't always work. Sometimes one of the equations has not just x's and y's in it — but bits with x^2 and y^2 as well. When this happens, you can only use the substitution method.

Use Substitution if one equation is Quadratic

EXAMPLE: $-x + 2y = 5$ Ⓛ ← The linear equation — with only x's and y's in.

$x^2 + y^2 = 25$ Ⓠ ← The quadratic equation — with some x^2 and y^2 bits in.

Rearrange the linear equation so that either x or y is on its own on one side of the equals sign.

Ⓛ $-x + 2y = 5$

$\Rightarrow x = 2y - 5$

Substitute this expression into the quadratic equation...

Sub into Ⓠ: $x^2 + y^2 = 25$

$\Rightarrow (2y - 5)^2 + y^2 = 25$

...and then rearrange this into the form $ax^2 + bx + c = 0$, so you can solve it — either by factorising or using the quadratic formula.

$\Rightarrow (4y^2 - 20y + 25) + y^2 = 25$

$\Rightarrow 5y^2 - 20y = 0$

$\Rightarrow 5y(y - 4) = 0$

$\Rightarrow y = 0$ or $y = 4$

Finally put both these values back into the linear equation to find corresponding values for x:

When $y = 0$: $-x + 2y = 5$ Ⓛ

$\Rightarrow x = -5$

When $y = 4$: $-x + 2y = 5$ Ⓛ

$\Rightarrow -x + 8 = 5$

$\Rightarrow x = 3$

So the solutions to the simultaneous equations are: $x = -5$, $y = 0$ and $x = 3$, $y = 4$.

As usual, check your answers by putting these values back into the original equations.

One Quadratic and One Linear Eqn

1) **Isolate variable in linear equation**
Rearrange the linear equation to get either x or y on its own.

2) **Substitute into quadratic equation**
— to get a quadratic equation in just one variable.

3) **Solve to get values for one variable**
— either by factorising or using the quadratic formula.

4) **Stick these values in the linear equation**
— to find corresponding values for the other variable.

Check Your Answer

$x = -5, y = 0$: $-(-5) + 2 \times 0 = 5$ ✓

$(-5)^2 + 0^2 = 25$ ✓

$x = 3, y = 4$: $-(3) + 2 \times 4 = 5$ ✓

$3^2 + 4^2 = 25$ ✓

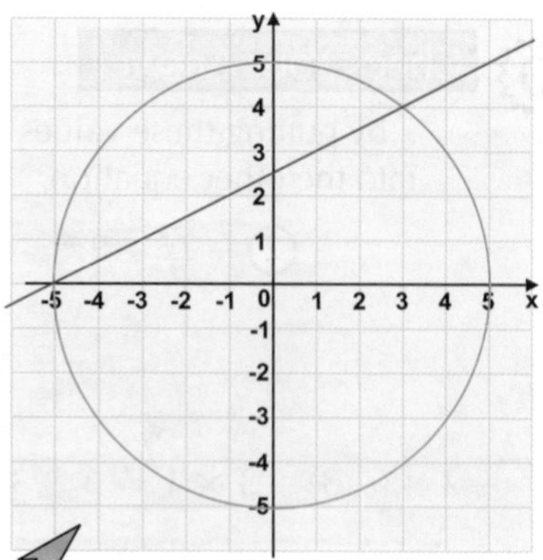

$y = x^2$ — a match-winning substitution...

The quadratic equation above is actually a circle about the origin with radius 5. The linear equation is just a standard straight line. So what you're actually finding here are the two points where the line passes through the circle. And these turn out to be (–5, 0) and (3, 4). Good, eh? There's more on the equation of a circle on page 34. Check it out.

Geometric Interpretation

When you have to interpret something geometrically, you have to draw a picture to show what's going on.

Two Solutions — Two points of Intersection

EXAMPLE

$y = x^2 - 4x + 5$ — ①
$y = 2x - 3$ — ②

SOLUTION

Substitute expression for y from ② into ①: $2x - 3 = x^2 - 4x + 5$

Rearrange and solve:
$$x^2 - 6x + 8 = 0$$
$$(x-2)(x-4) = 0$$
$$x = 2 \text{ or } x = 4$$

In ② gives:
$$x = 2 \Rightarrow y = 2 \times 2 - 3 = 1$$
$$x = 4 \Rightarrow y = 2 \times 4 - 3 = 5$$

There are 2 pairs of solutions: $x = 2, y = 1$ and $x = 4, y = 5$

Geometric Interpretation

So from solving the simultaneous equations, you know that the graphs meet in two places — the points (2, 1) and (4, 5).

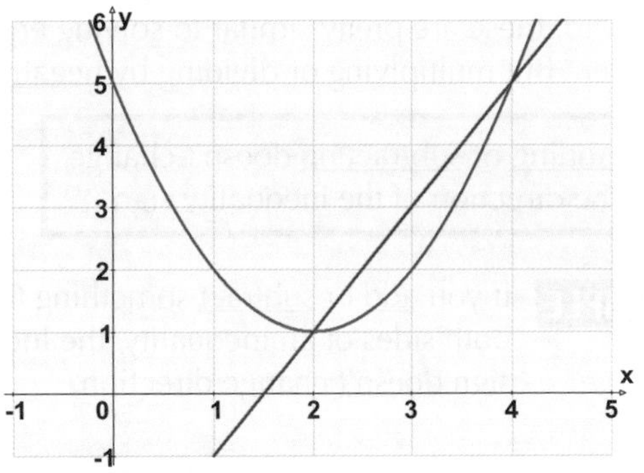

One Solution — One point of Intersection

EXAMPLE

$y = x^2 - 4x + 5$ — ①
$y = 2x - 4$ — ②

SOLUTION

Substitute ② in ①: $2x - 4 = x^2 - 4x + 5$

Rearrange and solve:
$$x^2 - 6x + 9 = 0$$
$$(x-3)^2 = 0$$
$$x = 3$$

Double root — i.e. you only get 1 solution from the quadratic.

In Equation ② gives:
$$y = 2 \times 3 - 4$$
$$y = 2$$

There's 1 solution: $x = 3, y = 2$

Geometric Interpretation

Since the equations have only one solution, the two graphs only meet at one point — (3, 2). The straight line is a tangent to the curve.

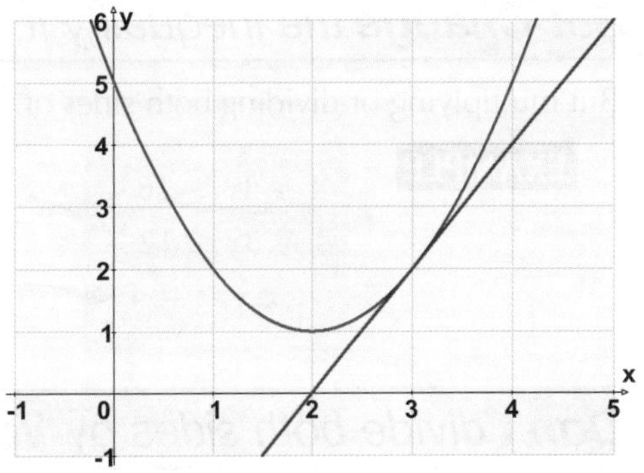

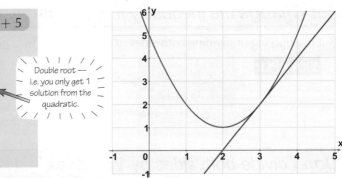

No Solutions means the Graphs Never Meet

EXAMPLE

$y = x^2 - 4x + 5$ — ①
$y = 2x - 5$ — ②

SOLUTION

Substitute ② in ①: $2x - 5 = x^2 - 4x + 5$

Rearrange and try to solve with the quadratic formula: $x^2 - 6x + 10 = 0$
$$b^2 - 4ac = (-6)^2 - 4 \cdot 10$$
$$= 36 - 40 = -4$$

$b^2 - 4ac < 0$, so the quadratic has no roots.
So the simultaneous equations have no solutions.

Geometric Interpretation

The equations have no solutions — the graphs never meet.

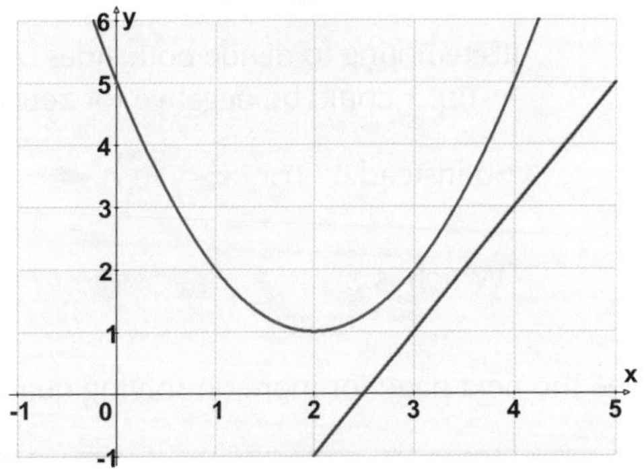

Geometric Interpretation? Frankly my dear, I don't give a damn...

This stuff also works the other way round. If you are given the graphs of two equations, then the points where they intersect are the solutions to the simultaneous equations — which is pretty handy cos then you can avoid doing lots of algebra. Nice.

Linear Inequalities

Solving inequalities is very similar to solving equations. You've just got to be really careful that you keep the inequality sign pointing the right way.

> Find the ranges of x that satisfy these inequalities:
> (i) $x - 3 < -1 + 2x$ (ii) $8x + 2 \geq 2x + 17$ (iii) $4 - 3x \leq 16$ (iv) $36x < 6x^2$

Sometimes the inequality sign **Changes Direction**

Like I said, these are pretty similar to solving equations — because whatever you do to one side, you have to do to the other. But multiplying or dividing by negative numbers changes the direction of the inequality sign.

> Adding or subtracting doesn't change the direction of the inequality sign

> Multiplying or dividing by something positive doesn't affect the inequality sign

EXAMPLE If you add or subtract something from both sides of an inequality, the inequality sign doesn't change direction.

$$x - 3 < -1 + 2x$$
$$\Rightarrow x - 2 < 2x$$
$$\Rightarrow -2 < x$$

And this is the same as...

$$x > -2$$

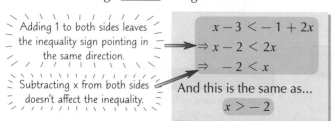

Adding 1 to both sides leaves the inequality sign pointing in the same direction.

Subtracting x from both sides doesn't affect the inequality.

EXAMPLE Multiplying or dividing both sides of an inequality by a positive number doesn't affect the direction of the inequality sign.

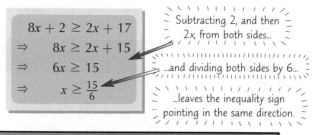

$$8x + 2 \geq 2x + 17$$
$$\Rightarrow 8x \geq 2x + 15$$
$$\Rightarrow 6x \geq 15$$
$$\Rightarrow x \geq \frac{15}{6}$$

Subtracting 2, and then 2x, from both sides...

...and dividing both sides by 6...

...leaves the inequality sign pointing in the same direction.

But **Change** the inequality if you **Multiply** or **Divide** by something **Negative**

But multiplying or dividing both sides of an inequality by a negative number changes the direction of the inequality.

EXAMPLE

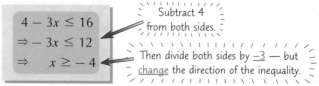

$$4 - 3x \leq 16$$
$$\Rightarrow -3x \leq 12$$
$$\Rightarrow x \geq -4$$

Subtract 4 from both sides.

Then divide both sides by -3 — but change the direction of the inequality.

> The reason for the sign changing direction is because it's just the same as swapping everything from one side to the other:
> $-3x \leq 12 \Rightarrow -12 \leq 3x \Rightarrow x \geq -4$

Don't divide both sides by **Variables** — like x and y

You've got to be really careful when you divide by things that might be negative — well basically, don't do it.

EXAMPLE $36x < 6x^2$

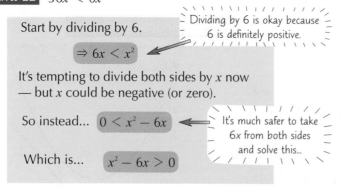

Start by dividing by 6.

$$\Rightarrow 6x < x^2$$

Dividing by 6 is okay because 6 is definitely positive.

It's tempting to divide both sides by x now — but x could be negative (or zero).

So instead... $0 < x^2 - 6x$

It's much safer to take 6x from both sides and solve this...

Which is... $x^2 - 6x > 0$

See the next page for more on solving quadratic inequalities.

Two Types of Inequality Sign

There are two kinds of inequality sign:

Type 1: $<$ — **less than**

$>$ — **greater than**

Type 2: $\leq$ — **less than or equal to**

$\geq$ — **greater than or equal to**

Whatever type the question uses — use the same kind all the way through your answer.

So no one knows we've arrived safely — splendid...

So just remember — inequalities are just like normal equations except that you have to reverse the sign when multiplying or dividing by a negative number. And don't divide both sides by variables. (You should know not to do this with normal equations anyway because the variable could be zero.) OK — lecture's over.

Quadratic Inequalities

With quadratic inequalities, you're best off drawing the graph and taking it from there.

Draw a **Graph** to solve a **Quadratic** inequality

EXAMPLE Find the ranges of x which satisfy these inequalities:

① $-x^2 + 2x + 4 \geq 1$ ② $2x^2 - x - 3 > 0$

First rewrite the inequality with zero on one side.

$$-x^2 + 2x + 3 \geq 0$$

Then draw the graph of: $y = -x^2 + 2x + 3$

So find where it crosses the x-axis (i.e. where $y = 0$):

$$-x^2 + 2x + 3 \Rightarrow x^2 - 2x - 3 = 0$$
$$\Rightarrow (x + 1)(x - 3) = 0$$
$$\Rightarrow x = -1 \text{ or } x = 3$$

And the coefficient of x^2 is negative, so the graph is n-shaped. So it looks like this:

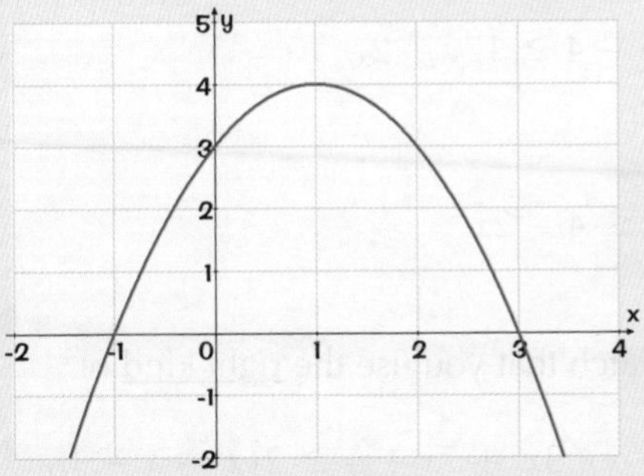

You're interested in when this is positive or zero, i.e. when it's above the x-axis.

From the graph, this is when x is between −1 and 3 (including those points). So your answer is...

$$-x^2 + 2x + 4 \geq 1 \text{ when } -1 \leq x \leq 3.$$

This one already has zero on one side, so draw the graph of $y = 2x^2 - x - 3$.

Find where it crosses the x-axis:

$$2x^2 - x - 3 = 0$$
$$\Rightarrow (2x - 3)(x + 1) = 0$$
$$\Rightarrow x = \tfrac{3}{2} \text{ or } x = -1$$

Factorise it to find the roots.

And the coefficient of x^2 is positive, so the graph is u-shaped. And looks like this:

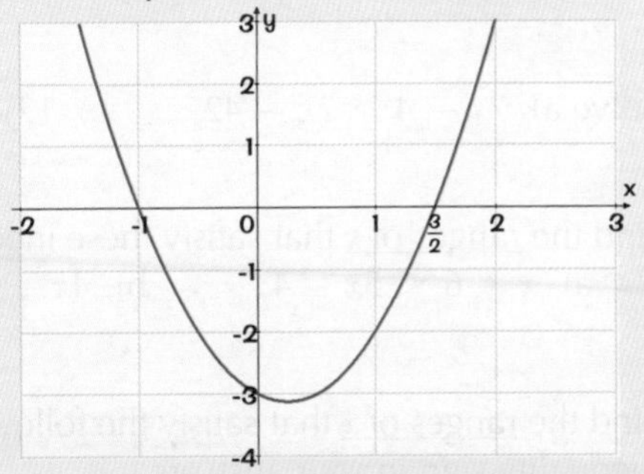

You need to say when this is positive. Looking at the graph, there are two parts of the x-axis where this is true — when x is less than −1 and when x is greater than 3/2. So your answer is:

$$2x^2 - x - 3 > 0 \text{ when } x < -1 \text{ or } x > \tfrac{3}{2}.$$

EXAMPLE (REVISITED) On the last page you had to solve $36x < 6x^2$.

$$36x < 6x^2$$
equation 1 ⟹ $\Rightarrow 6x < x^2$
$$\Rightarrow 0 < x^2 - 6x$$

So draw the graph of

$$y = x^2 - 6x = x(x - 6)$$

And this is positive when $x < 0$ or $x > 6$.

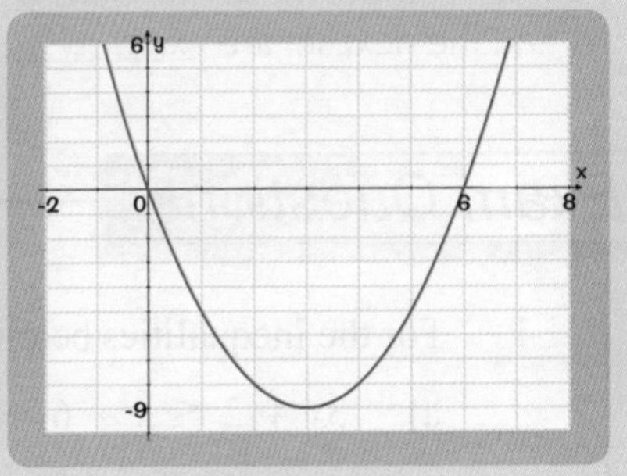

If you divide by x in equation 1, you'd only get half the solution — you'd miss the $x < 0$ part.

That's nonsense — I can see perfectly...

Call me sad, but I reckon these questions are pretty cool. They look a lot more difficult than they actually are and you get to draw a picture. Wow! When you do the graph, the important thing is to find where it crosses the x-axis (you don't need to know where it crosses the y-axis) and make sure you draw it the right way up. Then you just need to decide which bit of the graph you want. It'll either be the range(s) of x where the graph is below the x-axis or the range(s) where it's above. And this depends on the inequality sign.

C1 Section 3 — Practice Questions

What's that I hear you cry? You want practice questions — and lots of them. Well, it just so happens I've got a few here. For quadratic inequalities, my advice is, 'if you're not sure, draw a picture — even if it's not accurate'. And as for simultaneous equations — well, just don't rush them.

Warm-up Questions

1) Solve these sets of simultaneous equations:
   a) $3x - 4y = 7$ and $-2x + 7y = -22$, b) $2x - 3y = \frac{11}{12}$ and $x + y = -\frac{7}{12}$

2) Find where possible (and that's a bit of a clue) the solutions to these sets of simultaneous equations. Interpret your answers geometrically.
   a) $y = x^2 - 7x + 4$, $2x - y - 10 = 0$ b) $y = 30 - 6x + 2x^2$, $y = 2(x + 11)$ c) $x^2 + 2y^2 - 3 = 0$, $y = 2x + 4$

3) A bit trickier — find where the following lines meet:
   a) $y = 3x - 4$ and $y = 7x - 5$
   b) $y = 13 - 2x$ and $7x - y - 23 = 0$
   c) $2x - 3y + 4 = 0$ and $x - 2y + 1 = 0$

4) Solve a) $7x - 4 > 2x - 42$, b) $12y - 3 \leq 4y + 4$, c) $9y - 4 \geq 17y + 2$.

5) Find the ranges of x that satisfy these inequalities:
   a) $x + 6 < 5x - 4$, b) $4x - 2 > x - 14$, c) $7 - x \leq 4 - 2x$

6) Find the ranges of x that satisfy the following inequalities. (And watch that you use the right kind of inequality sign in your answers.)
   a) $3x^2 - 5x - 2 \leq 0$, b) $x^2 + 2x + 7 > 4x + 9$, c) $3x^2 + 7x + 4 \geq 2(x^2 + x - 1)$.

7) Find the ranges of x that satisfy these jokers:
   a) $x^2 + 3x - 1 \geq x + 2$, b) $2x^2 > x + 1$, c) $3x^2 - 12 < x^2 - 2x$

I know, I know. Those questions weren't enough for you. Not to worry, there are plenty more — and the next set are exam-style questions. Try to contain your excitement.

Exam Questions

1 For the inequalities below, find the set of values for x:

a) $3x + 2 \leq x + 6$, *(2 marks)*

b) $20 - x - x^2 > 0$, *(4 marks)*

c) $3x + 2 \leq x + 6$ and $20 - x - x^2 > 0$. *(1 mark)*

C1 Section 3 — Practice Questions

The world is full of inequality and injustice. We need to put a stop to it now. The first thing we need to do is... Oh, sorry, that's not what they meant by "solve the inequality". And I'd just come up with a solution for world peace.

2 Solve the inequalities:

a) $3 \le 2p + 5 \le 15$, *(3 marks)*

b) $q^2 - 9 > 0$. *(4 marks)*

3 a) Factorise $3x^2 - 13x - 10$. *(1 mark)*

b) Hence, or otherwise, solve $3x^2 - 13x - 10 \le 0$. *(3 marks)*

4 Find the coordinates of intersection for the following curve and line:

$$x^2 + 2y^2 = 36, \quad x + y = 6$$

(6 marks)

5 The curve C has equation $y = -x^2 + 3$ and the line l has equation $y = -2x + 4$.

a) Find the coordinates of the point (or points) of intersection of C and l. *(4 marks)*

b) Sketch the graphs of C and l on the same axes, clearly showing where the graphs intersect the x- and y- axes. *(5 marks)*

6 The line l has equation $y = 2x - 3$ and the curve C has equation $y = (x + 2)(x - 4)$.

a) Sketch the line l and the curve C on the same axes, showing the coordinates of the x- and y- intercepts. *(5 marks)*

b) Show that the x-coordinates of the points of intersection of l and C satisfy the equation $x^2 - 4x - 5 = 0$. *(2 marks)*

c) Hence, or otherwise, find the points of intersection of l and C. *(4 marks)*

Factorising Cubics

Factorising a quadratic function is okay — but you might also be asked to factorise a cubic (something with x^3 in it). And that takes a bit more time — there are more steps, so there are more chances to make mistakes.

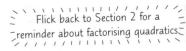

Flick back to Section 2 for a reminder about factorising quadratics.

Factorising a cubic given *One Factor*

$$f(x) = 2x^3 + x^2 - 8x - 4$$

Factorising a cubic means exactly what it meant with a quadratic — putting brackets in.
When they ask you to factorise a cubic equation, they'll usually tell you one of the factors.

EXAMPLE Given that $(x+2)$ is a factor of $f(x) = 2x^3 + x^2 - 8x - 4$, express $f(x)$ as the product of three linear factors.

① The first step is to find a quadratic factor. So write down the factor you know, along with another set of brackets.

$$(x+2)(\quad\quad) = 2x^3 + x^2 - 8x - 4$$

Put the x^2 bit in this new set of brackets.
These have to multiply together to give you this.

$$(x+2)(2x^2 \quad\quad -2) = 2x^3 + x^2 - 8x - 4$$

② Find the number for the second set of brackets.
These have to multiply together to give you this.

$$(x+2)(2x^2 \quad\quad -2) = 2x^3 + x^2 - 8x - 4$$

③ These multiplied give you $-2x$, but there's $-8x$ in $f(x)$ — so you need an 'extra' $-6x$.
And that's what this $-3x$ is for.

$$(x+2)(2x^2 - 3x - 2) = 2x^3 + x^2 - 8x - 4$$

You only need $-3x$ because it's going to be multiplied by 2 — which makes $-6x$.

Factorising Cubics

1) **Write down the factor $(x-k)$.**
2) **Put in the x^2 term.**
3) **Put in the constant.**
4) **Put in the x term by comparing the number of x's on both sides.**
5) **Check there are the same number of x^2's on both sides.**
6) **Factorise the quadratic you've found — if that's possible.**

If every term in the cubic contains an 'x' (i.e. $ax^3 + bx^2 + cx$) then just take out x as your first factor before factorising the remaining quadratic as usual.

④ Before you go any further, check that there are the same number of x^2's on both sides.

$4x^2$ from here...

$$(x+2)(2x^2 - 3x - 2) = 2x^3 + x^2 - 8x - 4$$

...and $-3x^2$ from here... ...add together to give this x^2.

If this is okay, factorise the quadratic into two linear factors.

$$(2x^2 - 3x - 2) = (2x + 1)(x - 2)$$

And so... $2x^3 + x^2 - 8x - 4 = (x+2)(2x+1)(x-2)$

If you wanted to solve a cubic, you'd do it exactly the same way — put it in the form $ax^3 + bx^2 + cx + d = 0$ and factorise.

I love the smell of fresh factorised cubics in the morning...

Factorising cubics is exactly the same as learning to unicycle... It's impossible at first. But when you finally manage it, it's really easy from then onwards and you'll never forget it. Probably. To tell the truth, I can't unicycle at all. So don't believe a word I say.

Algebraic Division

Algebraic division is one of those things that you have to learn when you're studying maths.
You'll probably never use it again once you've gone out into the 'real world', but hey ho... such is life.

Do *Algebraic Division* by means of *Subtraction*

$$(2x^3 - 3x^2 - 3x + 7) \div (x - 2) = ?$$

The trick with this is to see how many times you can subtract $(x - 2)$ from $(2x^3 - 3x^2 - 3x + 7)$.
The idea is to keep subtracting lumps of $(x - 2)$ until you've got rid of all the powers of x.

Do the subtracting in *Stages*

At each stage, always try to get rid of the highest power of x.
Then start again with whatever you've got left.

1) Start with $2x^3 - 3x^2 - 3x + 7$, and subtract $2x^2$ lots of $(x - 2)$ to get rid of the x^3 term.

$$(2x^3 - 3x^2 - 3x + 7) - 2x^2(x - 2)$$
$$(2x^3 - 3x^2 - 3x + 7) - 2x^3 + 4x^2$$
$$= x^2 - 3x + 7$$

$2x^3 \div x = 2x^2$

This is what's left — so now you have to get rid of the x^2 term.

2) Now start again with $x^2 - 3x + 7$. The highest power of x is the x^2 term. So subtract x lots of $(x - 2)$ to get rid of that.

$$(x^2 - 3x + 7) - x(x - 2)$$
$$(x^2 - 3x + 7) - x^2 + 2x$$
$$= -x + 7$$

Now start again with this — and get rid of the x term.

3) All that's left now is $-x + 7$. Get rid of the $-x$ by subtracting -1 times $(x - 2)$.

$$(-x + 7) - (-1(x - 2))$$
$$(-x + 7) + x - 2$$
$$= 5$$

There are no more powers of x to get rid of — so stop here.

The remainder's 5.

Interpreting the results...

Time to work out exactly what all that meant...

Started with: $2x^3 - 3x^2 - 3x + 7$

Subtracted: $2x^2(x - 2) + x(x - 2) - 1(x - 2)$

$= (x - 2)(2x^2 + x - 1)$

Remainder: $= 5$

So... $2x^3 - 3x^2 - 3x + 7 = (x - 2)(2x^2 + x - 1) + 5$

...or to put that another way...

$$\frac{2x^3 - 3x^2 - 3x + 7}{x - 2} = 2x^2 + x - 1 \text{ with remainder } 5.$$

$2x^2 + x - 1$ is called the quotient.

Algebraic Division

$$(ax^3 + bx^2 + cx + d) \div (x - k) = ?$$

1) SUBTRACT a multiple of $(x - k)$ to get rid of the highest power of x.
2) REPEAT step 1 until you've got rid of all the powers of x.
3) WORK OUT how many lumps of $(x - k)$, you've subtracted, and the REMAINDER.

Algebraic division is a beautiful thing that we should all cherish...

Revising algebraic division isn't the most enjoyable way to spend an afternoon, it's true, but it's in the specification, and so you need to be comfortable with it. It involves the same process you use when you're doing long division with numbers — so if you're having trouble following the above, do $4863 \div 7$ really slowly. What you're doing at each stage is subtracting multiples of 7, and you do this until you can't take any more 7s away, which is when you get your remainder.

The Remainder and Factor Theorems

The Remainder Theorem and the Factor Theorem are easy, and possibly quite useful.

The **Remainder Theorem** is an easy way to work out **Remainders**

When you divide $f(x)$ by $(x - a)$, the remainder is $f(a)$.

So in the example on the previous page, you could have worked out the remainder dead easily.

1) $f(x) = 2x^3 - 3x^2 - 3x + 7$.
2) You're dividing by $(x - 2)$, so $a = 2$.
3) So the remainder must be $f(2) = (2 \times 8) - (3 \times 4) - (3 \times 2) + 7 = 5$.

Careful now... when you're dividing by something like $(x + 7)$, a is negative — so here, $a = -7$.

If you want the remainder after dividing by something like $(ax - b)$, there's an extension to the remainder theorem...

When you divide $f(x)$ by $(ax - b)$, the remainder is $f\left(\frac{b}{a}\right)$.

EXAMPLE Find the remainder when you divide $2x^3 - 3x^2 - 3x + 7$ by $2x - 1$.

$f(x) = 2x^3 - 3x^2 - 3x + 7$. You're dividing by $2x - 1$, so $a = 2$ and $b = 1$.

So the remainder must be: $f\left(\frac{1}{2}\right) = 2\left(\frac{1}{8}\right) - 3\left(\frac{1}{4}\right) - 3\left(\frac{1}{2}\right) + 7 = 5$

The **Factor Theorem** is just the Remainder Theorem with a **Zero Remainder**

If you get a remainder of zero when you divide $f(x)$ by $(x - a)$, then $(x - a)$ must be a factor. That's the Factor Theorem.

If $f(x)$ is a polynomial, and $f(a) = 0$, then $(x - a)$ is a factor of $f(x)$.

In other words: If you know the roots, you also know the factors — and vice versa.

EXAMPLE Show that $(2x + 1)$ is a factor of $f(x) = 2x^3 - 3x^2 + 4x + 3$.

The question's giving you a big hint here. Notice that $2x + 1 = 0$ when $x = -\frac{1}{2}$. So plug this value of x into $f(x)$. If you show that $f(-\frac{1}{2}) = 0$, then the factor theorem says that $(x + \frac{1}{2})$ is a factor — which means that $2 \times (x + \frac{1}{2}) = (2x + 1)$ is also a factor.

$f(x) = 2x^3 - 3x^2 + 4x + 3$ and so $f\left(-\frac{1}{2}\right) = 2 \times \left(-\frac{1}{8}\right) - 3 \times \frac{1}{4} + 4 \times \left(-\frac{1}{2}\right) + 3 = 0$

So, by the factor theorem, $(x + \frac{1}{2})$ is a factor of $f(x)$, and so $(2x + 1)$ is also a factor.

(x – 1) is a Factor if the coefficients **Add Up To 0**

This works for all polynomials — no exceptions. It could save a fair whack of time in the exam.

EXAMPLE Factorise the polynomial $f(x) = 6x^2 - 7x + 1$

The coefficients (6, –7 and 1) add up to 0. That means $f(1) = 0$, and so $(x - 1)$ is a factor. Easy.

Then just factorise it like any quadratic to get this: $f(x) = 6x^2 - 7x + 1 = (6x - 1)(x - 1)$

Factorising a **Cubic** given **No Factors**

If the question doesn't give you any factors, the best way to find a factor of a cubic is to guess — use trial and error.

First, add up the coefficients to check if $(x - 1)$ is a factor.

If that doesn't work, keep trying small numbers (find $f(-1)$, $f(2)$, $f(-2)$, $f(3)$, $f(-3)$ and so on) until you find a number that gives you zero when you put it in the cubic. Call that number k. $(x - k)$ is a factor of the cubic.

Then finish factorising the cubic using the method on page 26.

C1 Section 4 — Practice Questions

It's time for some more practice questions. Before you jump in at the deep end though, make sure you stretch properly, and then do a couple of laps around the desk to make sure you're all warmed up. We don't want you getting cramp now, do we?

Warm-up Questions

1) Write the following functions f(x) in the form $f(x) = (x + 2)g(x) +$ remainder (where g(x) is a quadratic):
a) $f(x) = 3x^3 - 4x^2 - 5x - 6$, b) $f(x) = x^3 + 2x^2 - 3x + 4$, c) $f(x) = 2x^3 + 6x - 3$

2) Find the remainder when the following are divided by: (i) $(x + 1)$, (ii) $(x - 1)$
a) $f(x) = 6x^3 - x^2 - 3x - 12$, b) $f(x) = x^4 + 2x^3 - x^2 + 3x + 4$, c) $f(x) = x^5 + 2x^2 - 3$

3) Find the remainder when $f(x) = 3x^3 + 7x^2 - 12x + 14$ is divided by:
a) $x + 2$ b) $2x + 4$ c) $x - 3$ d) $2x - 6$

4) Which of the following are factors of $f(x) = 2x^3 - 6x + 4$?
a) $x - 1$ b) $x + 1$ c) $x - 2$ d) $2x - 2$

5) Find the values of c and d so that $2x^3 + 5x^2 + cx + d$ is exactly divisible by $(x - 2)(x + 3)$.

Now you're all hot and sweaty, you're ready for the main event — exam-style questions.

Exam Questions

1 $f(x) = 2x^3 - 5x^2 - 4x + 3$

a) Find the remainder when f(x) is divided by

(i) $(x - 1)$ *(2 marks)*

(ii) $(2x + 1)$ *(2 marks)*

b) Show using the factor theorem that $(x + 1)$ is a factor of f(x). *(2 marks)*

c) Factorise f(x) completely. *(4 marks)*

2 $f(x) = (4x^2 + 3x + 1)(x - p) + 5$, where p is a constant.

a) State the value of f(p). *(1 mark)*

b) Find the value of p, given that when f(x) is divided by $(x + 1)$, the remainder is -1. *(2 marks)*

c) Find the remainder when f(x) is divided by $(x - 1)$. *(1 mark)*

Coordinate Geometry

Welcome to geometry club... nice — today I shall be mostly talking about straight lines...

Finding the equation of a line **Through Two Points**

If you get through your exam without having to find the equation of a line through two points, I'm a Dutchman.

EXAMPLE Find the equation of the line that passes through the points (–3, 10) and (1, 4), and write it in the forms:

$y - y_1 = m(x - x_1)$ $y = mx + c$ $ax + by + c = 0$

— where a, b and c are integers.

You might be asked to write the equation of a line in any of these forms — but they're all similar.
Basically, if you find an equation in one form — you can easily convert it into either of the others.

The **Easiest** to find is **$y - y_1 = m(x - x_1)$**...

Point 1 is (–3, 10) and Point 2 is (1, 4).

Label the Points Label Point 1 as (x_1, y_1) and Point 2 as (x_2, y_2).

Point 1 — $(x_1, y_1) = (-3, 10)$

Point 2 — $(x_2, y_2) = (1, 4)$

It doesn't matter which way round you label them.

Find the Gradient Find the gradient of the line m — this is $m = \frac{y_2 - y_1}{x_2 - x_1}$.

$$m = \frac{4 - 10}{1 - (-3)} = \frac{-6}{4} = -\frac{3}{2}$$

Be careful here, y goes on the top, x on the bottom.

Write Down the Equation Write down the equation of the line, using the coordinates x_1 and y_1 — this is just $y - y_1 = m(x - x_1)$.

$x_1 = -3$ and $y_1 = 10$ ⟹

$$y - 10 = -\tfrac{3}{2}(x - (-3))$$
$$y - 10 = -\tfrac{3}{2}(x + 3)$$

...and **Rearrange** this to get the other two forms:

For the form $y = mx + c$, take everything except the y over to the right.

$$y - 10 = -\tfrac{3}{2}(x + 3)$$
$$\Rightarrow y = -\tfrac{3}{2}x - \tfrac{9}{2} + 10$$
$$\Rightarrow y = -\tfrac{3}{2}x + \tfrac{11}{2}$$

To find the form $ax + by + c = 0$, take everything over to one side — and then get rid of any fractions.

Multiply the whole equation by 2 to get rid of the 2's on the bottom line.

$$y = -\tfrac{3}{2}x + \tfrac{11}{2}$$
$$\Rightarrow \tfrac{3}{2}x + y - \tfrac{11}{2} = 0$$
$$\Rightarrow 3x + 2y - 11 = 0$$

Equations of Lines

1) **LABEL the points (x_1, y_1) and (x_2, y_2).**
2) **GRADIENT — find it and call it m.**
3) **WRITE DOWN THE EQUATION using $y - y_1 = m(x - x_1)$**
4) **CONVERT to one of the other forms, if necessary.**

If you end up with an equation like $\frac{3}{2}x - \frac{4}{3}y + 6 = 0$, where you've got a 2 and a 3 on the bottom of the fractions — multiply everything by the lowest common multiple of 2 and 3, i.e. 6.

There ain't nuffink to this geometry lark, Mister...

This is the sort of stuff that looks hard but is actually pretty easy. Finding the equation of a line in that first form really is a piece of cake — the only thing you have to be careful of is when a point has a negative coordinate (or two). In that case, you've just got to make sure you do the subtractions properly when you work out the gradient. See, this stuff ain't so bad...

Coordinate Geometry

Now you're all clued up on the equations of straight lines, it's time to move onto line segments. Instead of going on forever, a line segment is the part of a line between two end points, and there's all sorts of cool stuff you can find out about them.

Find the **Mid-Point** by finding the **Average** of the **End Points**

In the exam you could be asked to find the mid-point of a line segment.
To do this, just add the coordinates of the end-points of the line segment together, then divide by two:

EXAMPLE Points A and B are given by the coordinates (7, 4) and (–1, –2) respectively. M is the mid-point of the line segment AB. Find the coordinates of M.

Take the coordinates of A and B and add them together:

$$(7, 4) + (-1, -2) = (7 - 1, 4 - 2) = (6, 2)$$

Now divide by two: $\left(\frac{6}{2}, \frac{2}{2}\right) = (3, 1)$

So the mid-point of AB has coordinates (3, 1)

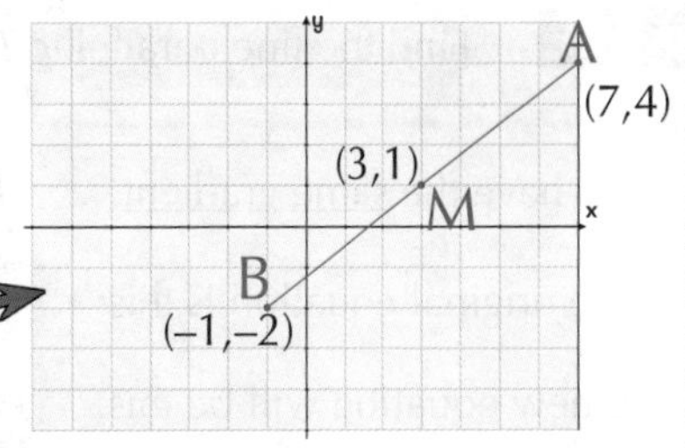

Use **Pythagoras** to find the **Distance** between two points

You may also be asked to find the distance between two points, (i.e. the length of a line segment). Luckily, there's a formula you can use:

$$d = \sqrt{(x_2 - x_1)^2 + (y_2 - y_1)^2}$$

The formula comes from using Pythagoras' theorem.

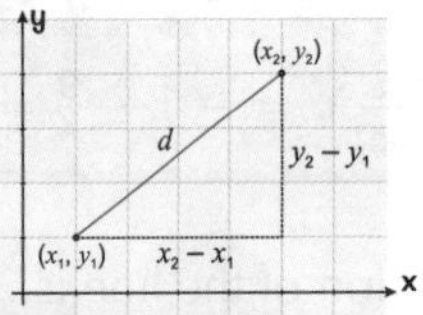

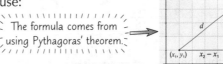

EXAMPLE A line segment has endpoints P and Q, which have coordinates (6, 2) and (–1, 0) respectively. Find the length of PQ.

You'd get exactly the same answer if you took Q as (x_1, y_1) and P as (x_2, y_2).

Take point P as (x_1, y_1) and Q as (x_2, y_2).

So $x_1 = 6$, $x_2 = -1$, $y_1 = 2$ and $y_2 = 0$. Plugging these into the formula gives:

$$d = \sqrt{(-1-6)^2 + (0-2)^2} = \sqrt{(-7)^2 + (-2)^2} = \sqrt{49+4} = \sqrt{53}$$

So the length of PQ is $\sqrt{53}$

EXAMPLE The point U has coordinates $(3, k)$, and the point V has coordinates (15, 6). UV has length 13. Find all possible values of k.

Substituting into the equation gives: $13 = \sqrt{(15-3)^2 + (6-k)^2} = \sqrt{12^2 + (6-k)^2}$

Squaring both sides:

$13^2 = 12^2 + (6-k)^2$

$169 = 144 + (6-k)^2$

$25 = (6-k)^2$

$25 = 36 - 12k + k^2$

$k^2 - 12k + 11 = 0$

$(k-1)(k-11) = 0$

So $k = 1$ or $k = 11$

CGP — Coordinate Geometry Practitioners...

As long as you know how to add, subtract, square and square root then you should be fine with the stuff on this page. If not, then you could always try drawing the line segment and measuring it with a ruler — although it probably won't get you the right answer. Or any marks. Actually, you're best off steering clear of that method altogether.

Parallel and Perpendicular Lines

This page is based around two really important facts that you've got to know — one about parallel lines, one about perpendicular lines. It's really a page of unparalleled excitement...

Two more lines...

Line l_1
$3x - 4y - 7 = 0$
$y = \frac{3}{4}x - \frac{7}{4}$

Line l_2
$x - 3y - 3 = 0$
$y = \frac{1}{3}x - 1$

...and two points...

Point A (3, –1)

Point B (–2, 4)

Parallel lines have equal *Gradient*

That's what makes them parallel — the fact that the gradients are the same.

EXAMPLE Find the line parallel to l_1 that passes through the point A (3, –1).

Parallel lines have the same gradient.

The original equation is this: $y = \frac{3}{4}x - \frac{7}{4}$

So the new equation will be this: $y = \frac{3}{4}x + c$

We know that the line passes through A, so at this point x will be 3, and y will be –1.

We just need to find c.

Stick these values into the equation to find c.

$$-1 = \frac{3}{4} \times 3 + c$$

$$\Rightarrow c = -1 - \frac{9}{4} = -\frac{13}{4}$$

So the equation of the line is... $y = \frac{3}{4}x - \frac{13}{4}$

And if you're only given the $ax + by + c = 0$ form it's even easier:

The original line is: $3x - 4y - 7 = 0$

So the new line is: $3x - 4y - k = 0$

Then just use the values of x and y at the point A to find k...

$$3 \times 3 - 4 \times (-1) - k = 0$$
$$\Rightarrow 13 - k = 0$$
$$\Rightarrow k = 13$$

So the equation is: $3x - 4y - 13 = 0$

The gradient of a *Perpendicular* line is: *–1 ÷ the Other Gradient*

Finding perpendicular lines (or 'normals') is just as easy as finding parallel lines — as long as you remember the gradient of the perpendicular line is –1 ÷ the gradient of the other one.

EXAMPLE Find the line perpendicular to l_2 that passes through the point B (–2, 4).

l_2 has equation: $y = \frac{1}{3}x - 1$

So if the equation of the new line is $y = mx + c$, then

$$m = -1 \div \frac{1}{3}$$
$$\Rightarrow m = -3$$

Since the gradient of a perpendicular line is: –1 ÷ the other one.

Also... $4 = (-3) \times (-2) + c$

$$\Rightarrow c = 4 - 6 = -2$$

Putting the coordinates of B(–2, 4) into $y = mx + c$.

So the equation of the line is...

$y = -3x - 2$

Or if you start with: l_2 $x - 3y - 3 = 0$

To find a perpendicular line, swap these two numbers around, and change the sign of one of them. (So here, 1 and –3 become 3 and 1.)

So the new line has equation...

$3x + y + d = 0$

Or you could have used $-3x - y + d = 0$.

But... $3 \times (-2) + 4 + d = 0$

$$\Rightarrow d = 2$$

Using the coordinates of point B.

And so the equation of the perpendicular line is...

$3x + y + 2 = 0$

Wowzers — parallel lines on the same graph dimension...

This looks more complicated than it actually is. All you're doing is finding the equation of a straight line through a certain point — the only added complication is that you have to find the gradient first. And there's another way to remember how to find the gradient of a normal — just remember that the gradients of perpendicular lines multiply together to make –1.

Curve Sketching

A picture speaks a thousand words... and graphs are what pass for pictures in maths. They're dead useful in getting your head round tricky questions, and time spent learning how to sketch graphs is time well spent.

The graph of $y = kx^n$ is a different shape for different k and n

Usually, you only need a rough sketch of a graph — so just knowing the basic shapes of these graphs will do.

n positive and even

You get a u-shape or an n-shape.

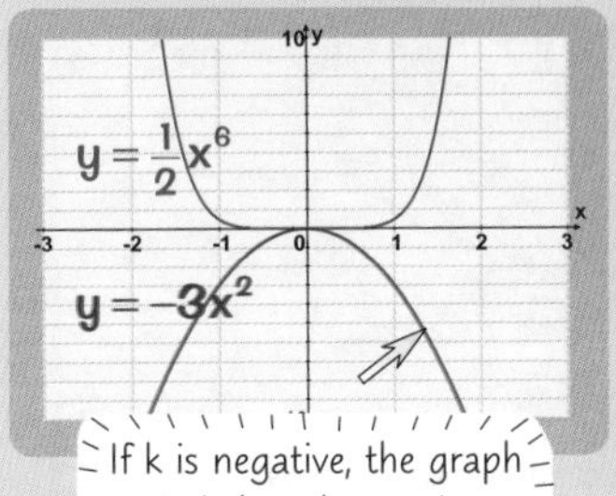

n positive and odd

You get a 'corner-to-corner' shape.

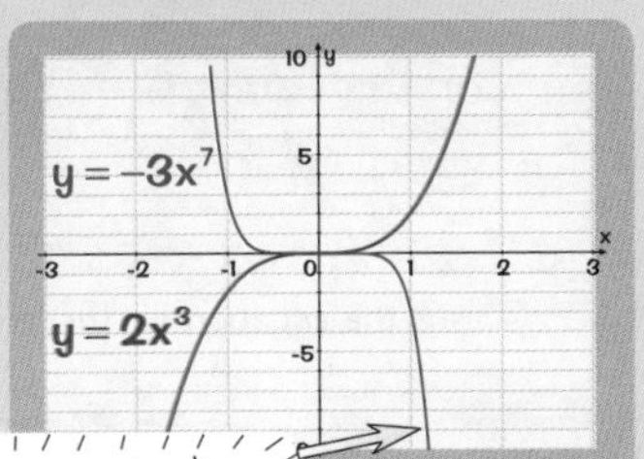

n negative and even

You get a graph with two bits next to each other.

Both graphs have asymptotes of x = 0 and y = 0

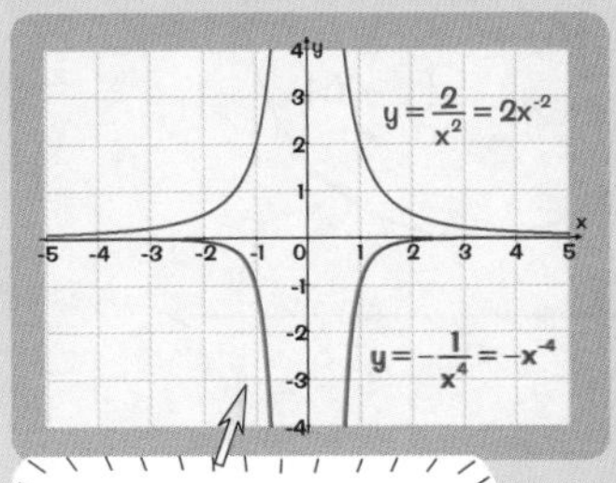

n negative and odd

You get a graph with two bits opposite each other.

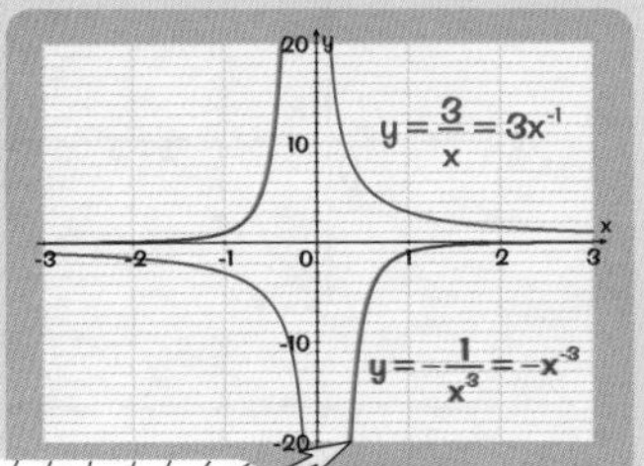

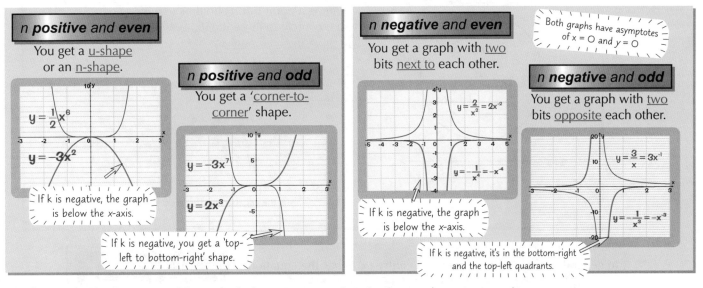

An asymptote of a curve is a line which the curve gets infinitely close to, but never touches.

If you know the Factors of a cubic — the graph's easy to Sketch

A cubic function has an x^3 term in it, and all cubics have 'bottom-left to top-right' shape — or a 'top-left to bottom-right' shape if the coefficient of x^3 is negative.

If you know the factors of a cubic, the graph is easy to sketch — just find where the function is zero.

EXAMPLE Sketch the graphs of the following cubic functions.

(i) $f(x) = x(x - 1)(2x + 1)$ (ii) $g(x) = (1 - x)(x^2 - 2x + 2)$ (iii) $h(x) = (x - 3)^2(x + 1)$ (iv) $m(x) = (2 - x)^3$

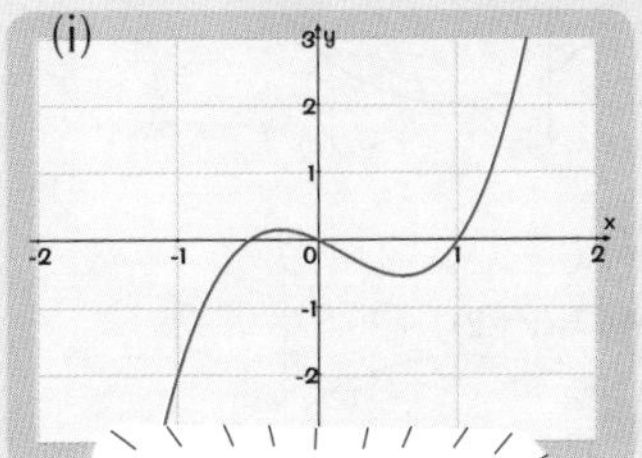

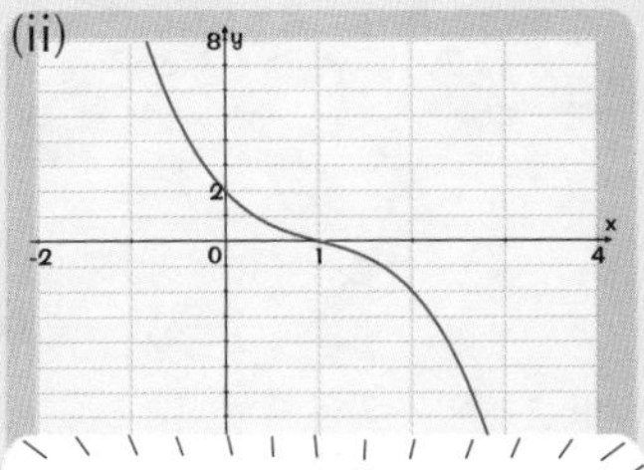

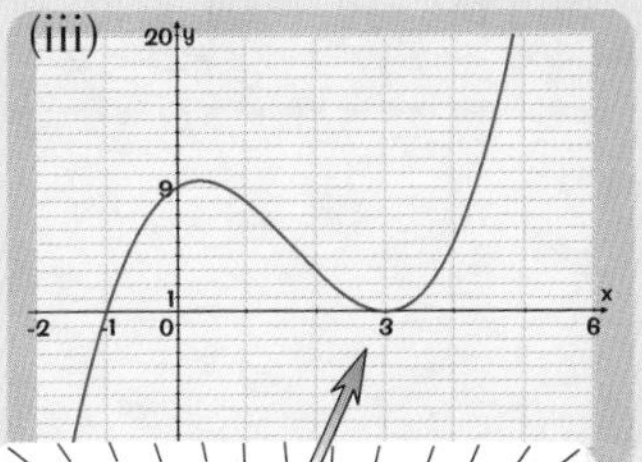

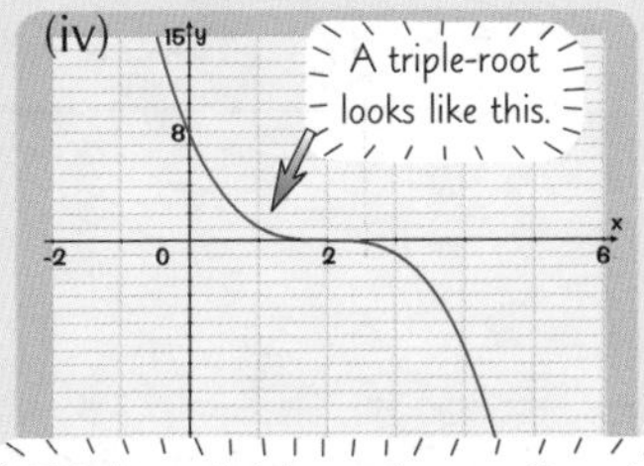

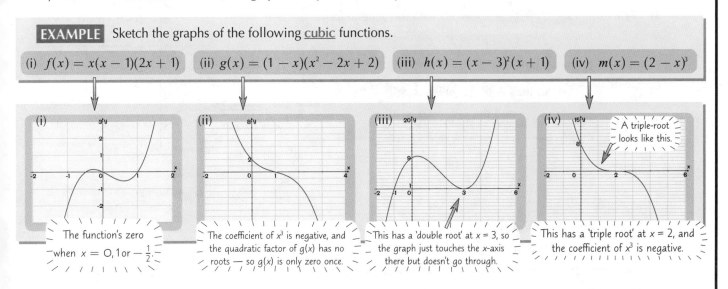

Graphs, graphs, graphs — you can never have too many graphs...

It may seem like a lot to remember, but graphs can really help you get your head round a question — a quick sketch can throw a helluva lot of light on a problem that's got you completely stumped. So being able to draw these graphs won't just help with an actual graph-sketching question — it could help with loads of others too. Got to be worth learning.

Circles

I always say a <u>beautiful shape</u> deserves a <u>beautiful formula</u>, and here you've got one of my favourite double-acts...

Equation of a circle: $(x - a)^2 + (y - b)^2 = r^2$

The equation of a circle looks complicated, but it's all based on Pythagoras' theorem. Take a look at the circle below, with centre (6, 4) and radius 3.

The circle $(x - a)^2 + (y - b)^2 = r^2$ is a translation of the circle of radius r centred on the origin. It moves 'a' to the right and 'b' up.

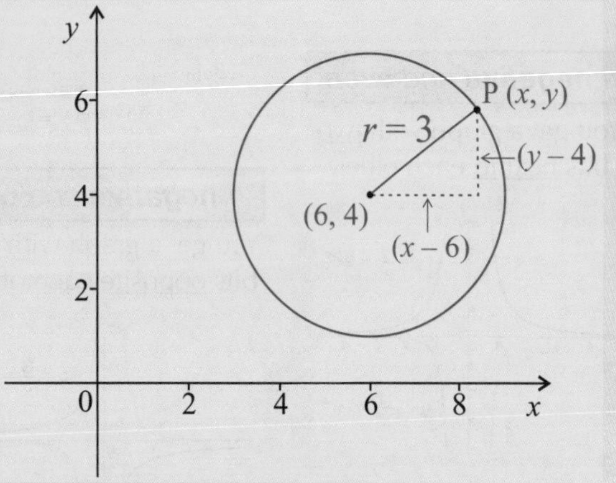

Joining a point P (x, y) on the circumference of the circle to its centre (6, 4), we can create a <u>right-angled triangle</u>.

Now let's see what happens if we use <u>Pythagoras' theorem</u>:

$$(x - 6)^2 + (y - 4)^2 = 3^2$$

or: $$\mathbf{(x - 6)^2 + (y - 4)^2 = 9}$$

This is the equation for the circle. It's as easy as that.

In general, a circle with radius r and centre (a, b) has the equation: $(x - a)^2 + (y - b)^2 = r^2$

EXAMPLE:

i) What is the centre and radius of the circle with equation $(x - 2)^2 + (y + 3)^2 = 16$

ii) Write down the equation of the circle with centre (–4, 2) and radius 6.

SOLUTION:

i) Comparing $(x - 2)^2 + (y + 3)^2 = 16$ with the general form:

$$(x - a)^2 + (y - b)^2 = r^2$$

then $a = 2$, $b = -3$ and $r = 4$.

So the centre (a, b) is: (2, –3)
and the radius (r) is: 4.

And as if by magic, here it is.

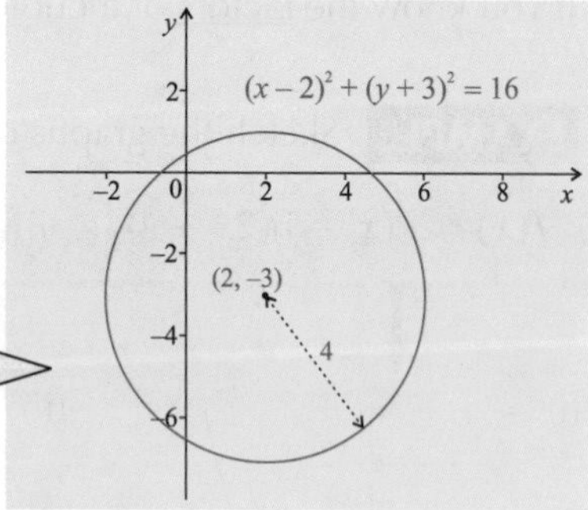

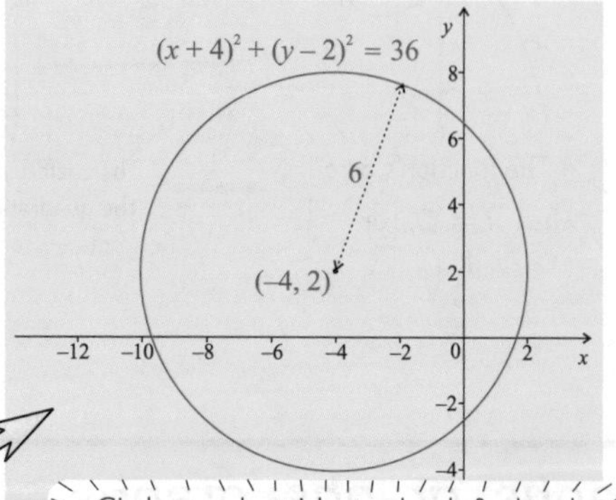

ii) The question says, 'Write down...', so you know you don't need to do any working.
The centre of the circle is (–4, 2), so $a = -4$ and $b = 2$.
The radius is 6, so $r = 6$.
Using the general equation for a circle $(x - a)^2 + (y - b)^2 = r^2$ you can write: $(x + 4)^2 + (y - 2)^2 = 36$

Circles can be tricky to sketch freehand. A compass is definitely the way forward.

This is pretty much all you need to learn. Everything on the next page uses stuff you should know already.

Circles

Rearrange the equation into the familiar form

Sometimes you'll be given an equation for a circle that doesn't look much like $(x - a)^2 + (y - b)^2 = r^2$.
This is a bit of a pain, because it means you can't immediately tell what the **radius** is or where the **centre** is.
But all it takes is a bit of **rearranging**.

Let's take the equation: $x^2 + y^2 - 6x + 4y + 4 = 0$

You need to get it into the form $(x - a)^2 + (y - b)^2 = r^2$.

This is just like completing the square.

$x^2 + y^2 - 6x + 4y + 4 = 0$

$x^2 - 6x + y^2 + 4y + 4 = 0$

$(x - 3)^2 - 9 + (y + 2)^2 - 4 + 4 = 0$

$(x - 3)^2 + (y + 2)^2 = 9$ ⟹ This is the recognisable form, so the centre is **(3, –2)** and the radius is $\sqrt{9}$ **= 3**.

Have a look at C1 Section 2 for more on completing the square.

Don't forget the Properties of Circles

You will have seen the circle rules at GCSE. You'll sometimes need to dredge them up in your memory for these circle questions. Here's a reminder of a few useful ones.

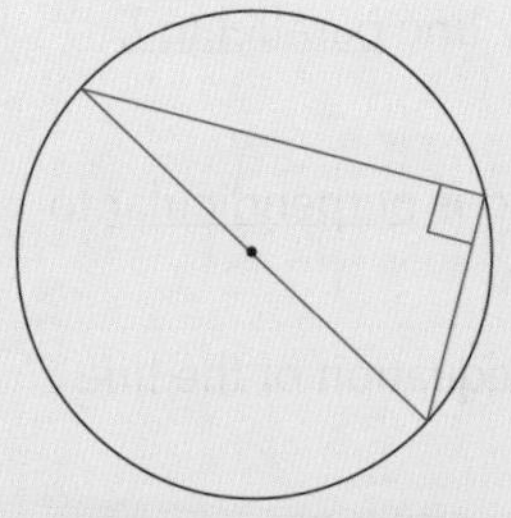

The angle in a semicircle is a right angle.

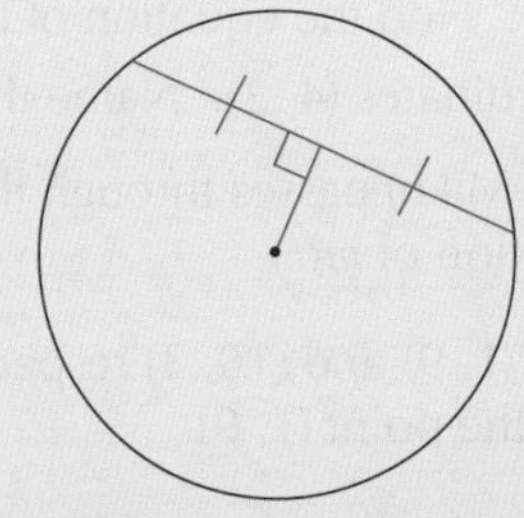

The perpendicular from the centre to a chord bisects the chord.

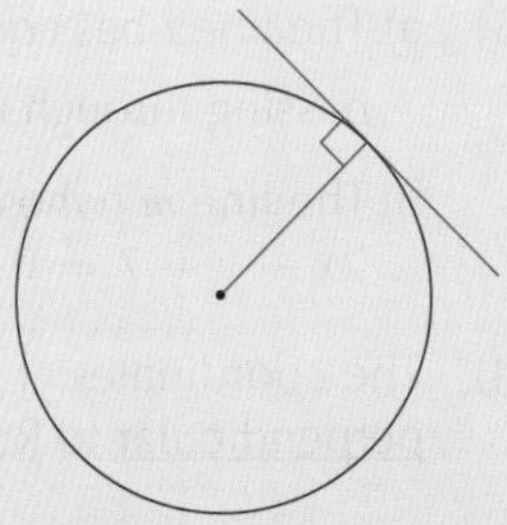

A radius and tangent to the same point will meet at right angles.

Use the Gradient Rule for Perpendicular Lines

Remember that the tangent at a given point will be perpendicular to the radius at that same point.

EXAMPLE:

Point A (6, 4) lies on a circle with the equation $x^2 + y^2 - 4x - 2y - 20 = 0$.

i) Find the centre and radius of the circle.

ii) Find the equation of the tangent to the circle at A.

The radius is the normal to the circle. If you're asked to find the equation of the normal to the circle at a given point, use the gradient of the radius rather than the tangent in $y - y_1 = m(x - x_1)$.

SOLUTION:

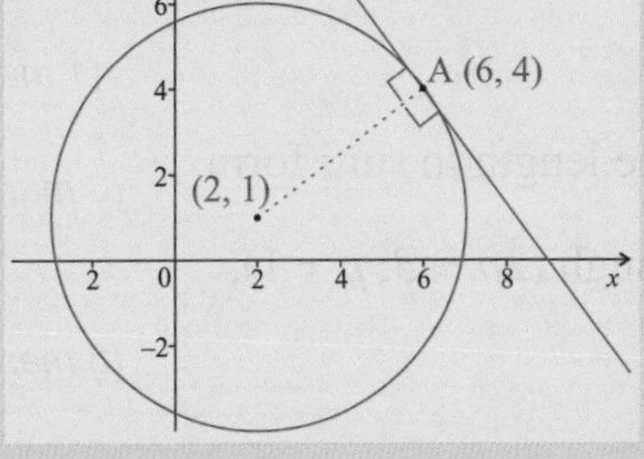

i) Rearrange the equation to show it as the sum of 2 squares:

$x^2 + y^2 - 4x - 2y - 20 = 0$

$x^2 - 4x + y^2 - 2y - 20 = 0$

$(x - 2)^2 - 4 + (y - 1)^2 - 1 - 20 = 0$

$(x - 2)^2 + (y - 1)^2 = 25$

This shows the centre is (2, 1) and the radius is 5.

ii) The tangent is at right angles to the radius at (6, 4).

Gradient of radius at (6, 4) $= \dfrac{4 - 1}{6 - 2} = \dfrac{3}{4}$

Gradient of tangent $= \dfrac{-1}{\frac{3}{4}} = -\dfrac{4}{3}$

Using $y - y_1 = m(x - x_1)$

$y - 4 = -\frac{4}{3}(x - 6)$

$3y - 12 = -4x + 24$

$3y + 4x - 36 = 0$

So the chicken comes from the egg, and the egg comes from the chicken...

Well folks, at least it makes a change from all those straight lines and quadratics.
I reckon if you know the **formula** and **what it means**, you should be absolutely **fine** with questions on circles.

C1 Section 5 — Practice Questions

There you go then. Another section down. And in a way it was quite exciting, I'm sure you'll agree. Though as you're probably aware, we mathematicians take our excitement from wherever we can get it. Anyway, I'll leave you alone now to savour these practice questions. Don't skip the warm-up — you don't want to hurt yourself...

Warm-up Questions

1) Find the equations of the straight lines that pass through the points
   a) (2, –1) and (–4, –19), b) $\left(0, -\frac{1}{3}\right)$ and $\left(5, \frac{2}{3}\right)$.

   Write each of them in the forms
   i) $y - y_1 = m(x - x_1)$,
   ii) $y = mx + c$,
   iii) $ax + by + c = 0$, where a, b and c are integers.

2) Point A has coordinates (2, 5) and point B has coordinates (12, –1). M is the mid-point of the line segment AB. Find:
   a) The coordinates of M,
   b) The length of AM, leaving your answer in surd form.

3) a) The line l has equation $y = \frac{3}{2}x - \frac{2}{3}$. Find the equation of the lovely, cuddly line parallel to l, passing through the point with coordinates (4, 2). Name this line Lilly.

   b) The line m (whose name is actually Mike) passes through the point (6, 1) and is perpendicular to $2x - y - 7 = 0$. What is the equation of m?

4) The coordinates of points R and S are (1, 9) and (10, 3) respectively. Find the equation of the line perpendicular to RS, passing through the point (1, 9).

5) Admit it — you love curve-sketching. We all do — and like me, you probably can't get enough of it. So more power to your elbow, and sketch these cubic graphs:
   a) $y = (x - 4)^3$, b) $y = (3 - x)(x + 2)^2$,
   c) $y = (1 - x)(x^2 - 6x + 8)$, d) $y = (x - 1)(x - 2)(x - 3)$.

6) Give the radius and the coordinates of the centre of the circles with the following equations:
   a) $x^2 + y^2 = 9$ b) $(x - 2)^2 + (y + 4)^2 = 4$ c) $x(x + 6) = y(8 - y)$

And the excitement continues with another thrilling selection of exam questions.
I hope that you get as much pleasure out of them as I have.

Exam Questions

1 Point A has coordinates $(1, k)$ and point B has coordinates (6, 3). The line AB has equation $5y - x = 9$.

   a) Find the value of k.
   (1 mark)

   b) Find the length and the mid-point of AB. Leave your answer for the length in surd form.
   (3 marks)

   c) Point C has coordinates (–1, 0). Given that the line BC passes through $(2p + 3, p + 1)$, where p is a constant, find p.
   (5 marks)

C1 Section 5 — Practice Questions

2 The circle Q has the equation $x^2 - 6x + y^2 - 8y = k$.

a) Given that Q has radius 2 and centre $C = (3, 4)$, find k. *(3 marks)*

b) Find the coordinates of the points, A and B, where the line $x + y = 9$ intersects the circle. *(4 marks)*

c) Find the coordinates of the mid point of the line AB. *(1 marks)*

d) Find the perpendicular distance from AB to C. *(2 marks)*

3 The line l passes through the point $S\ (7, -3)$ and has gradient -2.

a) Find an equation of l, giving your answer in the form $y = mx + c$. *(3 marks)*

b) The point T has coordinates (5, 1). Show that T lies on l. *(1 mark)*

4 C is a circle with the equation: $x^2 + y^2 - 2x - 10y + 21 = 0$.

a) Find the centre and radius of C. Where appropriate, give your answers in surd form. *(5 marks)*

The line joining $P(3, 6)$ and $Q(q, 4)$ is a diameter of C.

b) Show that $q = -1$. *(3 marks)*

c) Find the equation of the tangent to C at Q, giving your answer in the form $ax + by + c = 0$, where a, b and c are integers. *(5 marks)*

5 The line l has equation $4x + 3y = 15$.

a) Find the gradient of l. *(2 marks)*

b) The point R lies on l and has coordinates (3, 1). Find the equation of the line which passes through the point R and is perpendicular to l, giving your answer in the form $y = mx + c$. *(3 marks)*

6 The circle C has centre (4, 10) and touches the y-axis, as shown.

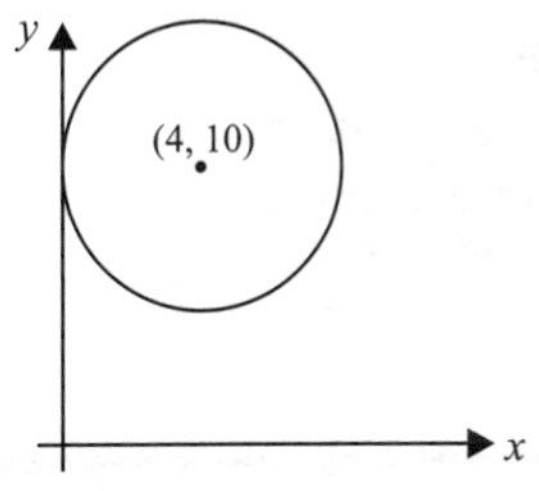

a) Write down the equation of the circle in the form $(x - a)^2 + (y - b)^2 = r^2$. *(2 marks)*

b) Show algebraically that the point (8, 10) lies on the circle. *(1 mark)*

c) Find the equation of the tangent to the circle at the point $(1, 10 + \sqrt{7})$. Give your answer in the form $Ax + By = C$, where A, B and C are constants to be determined. *(4 marks)*

Differentiation

Brrrrrr... differentiation is a bad one — it really is. Not because it's that hard, but because it comes up all over the place in exams. So if you don't know it perfectly, you're asking for trouble. Differentiation is a great way to work out gradients of graphs. You take a function, differentiate it, and you can quickly tell how steep a graph is. It's magic.

Use this formula to differentiate *Powers of x*

$$\frac{d}{dx}(x^n) = nx^{n-1}$$

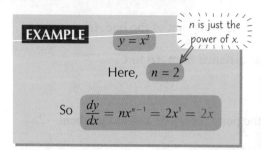

EXAMPLE $y = x^2$ (*n* is just the power of *x*.)

Here, $n = 2$

So $\frac{dy}{dx} = nx^{n-1} = 2x^1 = 2x$

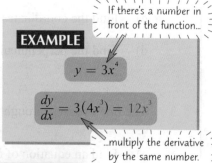

EXAMPLE (If there's a number in front of the function...)

$y = 3x^4$

$\frac{dy}{dx} = 3(4x^3) = 12x^3$

(...multiply the derivative by the same number.)

Derivatives of Functions

1) $\frac{dy}{dx}$ **means 'the derivative of** y **with respect to** x**'.**
2) **Derivative just means 'the thing you get when you differentiate something'.**
3) **For a function** $f(x)$**, the derivative is written** $f'(x)$ **(pronounced "f-dash of x").**
4) **The derivative of** $f(x)$ **at a particular point is the gradient of the tangent to the graph of** $y = f(x)$ **at that point. It also shows the rate of change of** $f(x)$ **at that point — see the next page.**
5) **A constant always differentiates to zero — see below.**

Differentiate each term in an equation *Separately*

This formula is better than cake — even better than that really nice sticky black chocolate one from that place in town. Even if there are loads of terms in the equation, it doesn't matter. Differentiate each bit separately and you'll be fine. Here is an example...

EXAMPLE $y = 6x^2 + 4x + 1$ $= 6x^2 + 4x + x^0$ ($x^0 = 1$)

$\frac{dy}{dx} = 6(2x) + 4(x^0) + 0x^{-1}$ (Differentiate each bit separately...) ($= 0$)

(...and add or subtract the results.) $\frac{dy}{dx} = 12x + 4$

Dario Gradient — differentiating Crewe from the rest...

If you're going to bother doing maths, you've got to be able to differentiate things. Simple as that. But luckily, once you can do the simple stuff, you should be all right. Big long equations are just made up of loads of simple little terms, so they're not really that much harder. Learn the formula, and make sure you can use it by practising all day and all night forever.

Differentiation

Differentiation is what you do if you need to find a gradient. Excited yet?

Differentiate to find **Gradients...**

EXAMPLE Find the gradient of the graph $y = x^2$ at $x = 1$ and $x = -2$...

You need the gradient of the graph of...

$$y = x^2$$

So differentiate this function to get...

$$\frac{dy}{dx} = 2x$$

Now when $x = 1, \frac{dy}{dx} = 2,$

And so the gradient of the graph at $x = 1$ is 2.

And when $x = -2, \frac{dy}{dx} = -4,$

So the gradient of the graph at $x = -2$ is -4.

Use differentiation to find the gradient of a curve — which is the same as the gradient of the tangent at any given point.

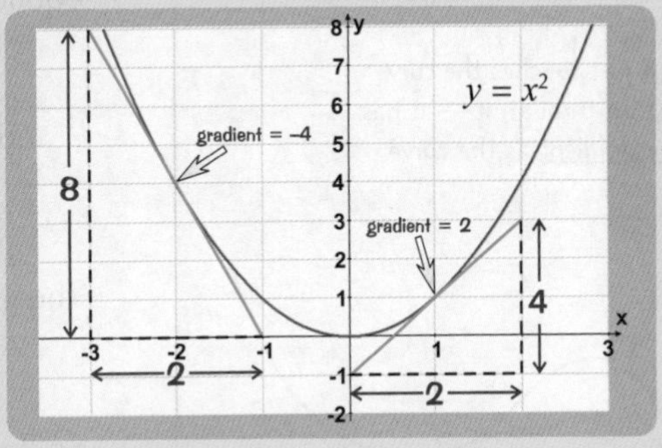

A tangent is a line that just touches a curve without actually going through it.

...which tell you **Rates of Change...**

So, you've differentiated an equation and found the gradient at a point — which is really useful because this tells you the rate of change of the curve at that point (e.g. from distance vs. time graphs you can work out speed).

EXAMPLE A sports car pulls off from a junction and drives away, travelling s metres in t seconds. For the first 10 seconds, its path can be described by the equation $s = 2t^2$.

Find: a) the speed of the car after 8 seconds and b) the car's acceleration during this period.

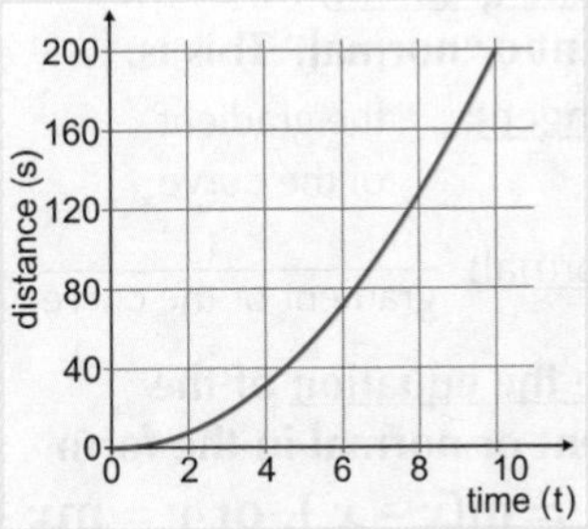

a) You can work out the speed by calculating the gradient of the curve $s = 2t^2$.

Differentiate to give: $\frac{ds}{dt} = 4t$ When $t = 8$, $\frac{ds}{dt} = 32$

So, the car is travelling at 32 ms^{-1} after 8 seconds.

b) Acceleration is the rate that speed (v) changes (i.e. it is the gradient of $v = 4t$).

So differentiate again to find the acceleration: $\frac{d^2s}{dt^2} = 4$

This means that the car's acceleration during this period is 4 ms^{-2}

This is called a second-order derivative — because you've differentiated twice

Help me Differentiation — You're my only hope...

There's not much hard maths on this page — but there are a couple of very important ideas that you need to get your head round pretty darn soon. Understanding that differentiating gives the gradient of the graph is more important than washing regularly — AND THAT'S IMPORTANT. The other thing on the page you need to know is that the gradient tells you the rate of change of a function — which is also vital when working out what a question is after.

Finding Tangents and Normals

What's a tangent? Beats me. Oh no, I remember, it's one of those thingies on a curve. Ah, yes... I remember now...

Tangents **Just** touch a curve

To find the equation of a tangent or a normal to a curve, you first need to know its gradient — so differentiate. Then complete the line's equation using the coordinates of one point on the line.

EXAMPLE Find the tangent to the curve $y = (4 - x)(x + 2)$ at the point (2, 8).

Tangents and Normals...

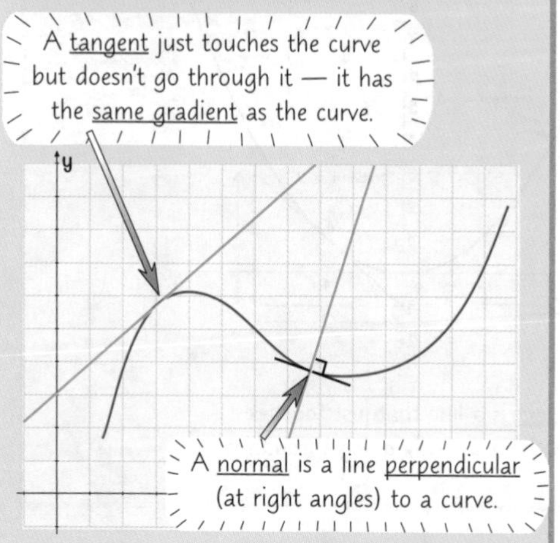

A tangent just touches the curve but doesn't go through it — it has the same gradient as the curve.
y
A normal is a line perpendicular (at right angles) to a curve.

To find the curve's (and the tangent's) gradient, first write the equation in a form you can differentiate...

$$y = 8 + 2x - x^2$$

...and then differentiate it.

$$\frac{dy}{dx} = 2 - 2x$$

The gradient of the tangent will be the gradient of the curve at $x = 2$.

At $x = 2$, $\frac{dy}{dx} = -2$,

So the tangent has equation,

$$y - y_1 = -2(x - x_1)$$

in $y - y_1 = m(x - x_1)$ form. See page 30.

And since it passes through the point (2, 8), this becomes

$y - 8 = -2(x - 2)$, or $y = -2x + 12$.

You can also write it in $y = mx + c$ form.

Normals are at **Right Angles** to a curve

EXAMPLE Find the normal to the curve $y = \frac{(x + 2)(x + 4)}{6}$ at the point (4, 8).

There's more info on parallel and perpendicular lines on p. 32.

Write the equation of the curve in a form you can differentiate.

$$y = \frac{x^2 + 6x + 8}{6} = \frac{x^2}{6} + x + \frac{4}{3}$$

Differentiate it...

$$\frac{dy}{dx} = \frac{1}{6}(2x) + 1 + 0 = \frac{x}{3} + 1$$

Find the gradient at the point you're interested in. At $x = 4$,

$$\frac{dy}{dx} = \frac{4}{3} + 1 = \frac{7}{3}$$

So the gradient of the normal is $-\frac{3}{7}$.

Because the gradient of the normal multiplied by the gradient of the curve must be -1.

And the equation of the normal is $y - y_1 = -\frac{3}{7}(x - x_1)$.

Finally, since the normal goes through the point (4, 8), the equation of the normal must be $y - 8 = -\frac{3}{7}(x - 4)$, or after rearranging, $y = -\frac{3}{7}x + \frac{68}{7}$.

Finding Tangents and Normals

1) **Differentiate the function.**
2) **Find the gradient, m, of the tangent or normal. This is,**
   for a tangent: the gradient of the curve
   for a normal: $\frac{-1}{\text{gradient of the curve}}$
3) **Write the equation of the tangent or normal in the form $y - y_1 = m(x - x_1)$, or $y = mx + c$.**
4) **Complete the equation of the line using the coordinates of a point on the line.**

Repeat after me... "I adore tangents and normals..."

Examiners can't stop themselves saying the words 'Find the tangent...' and 'Find the normal...'. They love the words. These phrases are music to their ears. They can't get enough of them. I just thought it was my duty to tell you that. And so now you know, you'll definitely be wanting to learn how to do the stuff on this page. Of course you will.

Stationary Points

Differentiation is how you find gradients of curves. So you can use differentiation to find a stationary point (where a graph 'levels off') — that means finding where the gradient becomes zero.

*Stationary Points are when the gradient is **Zero***

EXAMPLE Find the stationary points on the curve $y = 2x^3 - 3x^2 - 12x + 5$, and work out the nature of each one.

A stationary point can be...

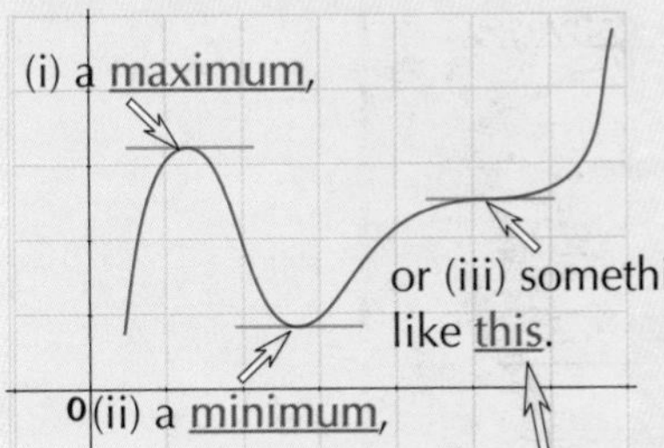

At stationary points, the gradient = 0, which means $\frac{dy}{dx} = 0$.

This kind of stationary point is called a 'point of inflection'.

You need to find where $\frac{dy}{dx} = 0$. So first, differentiate the function.

$$y = 2x^3 - 3x^2 - 12x + 5 \Rightarrow \frac{dy}{dx} = 6x^2 - 6x - 12$$

This is the expression for the gradient.

And then set this derivative equal to zero.

$$6x^2 - 6x - 12 = 0$$
$$\Rightarrow x^2 - x - 2 = 0 \Rightarrow (x - 2)(x + 1) = 0$$
$$\Rightarrow x = 2 \text{ or } x = -1$$

So the graph has two stationary points, at $x = 2$ and $x = -1$.

The stationary points are actually at (2, –15) and (–1, 12).

Substitute the x values into the function to find the y-coordinates.

*Decide if it's a **Maximum** or a **Minimum** by differentiating **Again***

Once you've found where the stationary points are, you have to decide whether each of them is a maximum or minimum — this is all a question means when it says, '...determine the nature of the turning points'.

A turning point is another name for a maximum or a minimum.

To decide whether a stationary point is a maximum or a minimum — just differentiate again to find $\frac{d^2y}{dx^2}$.

For a function f(x), $\frac{d^2y}{dx^2}$ can be written f″(x)

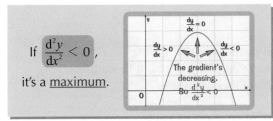

If $\frac{d^2y}{dx^2} < 0$, it's a maximum.

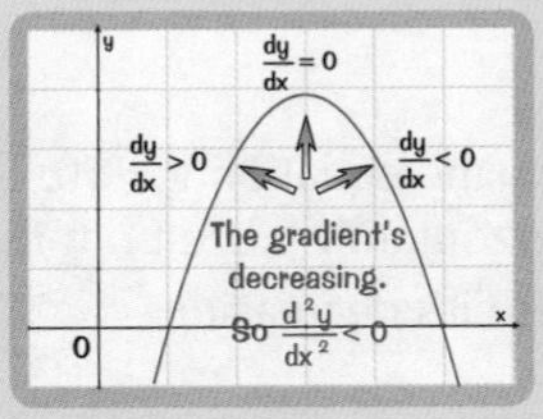

If $\frac{d^2y}{dx^2} > 0$, it's a minimum.

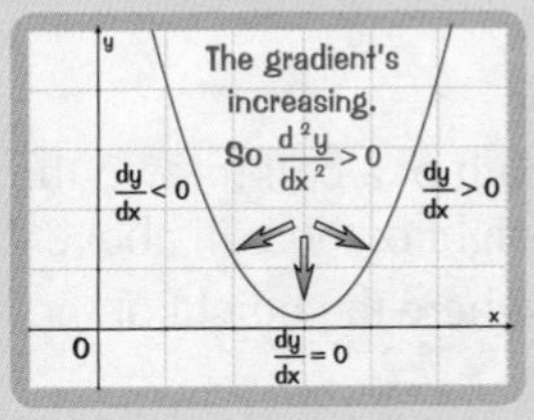

But if $\frac{d^2y}{dx^2} = 0$, you can't tell what type of stationary point it is.

You've just found that $\frac{dy}{dx} = 6x^2 - 6x - 12$.

So differentiating again gives $\frac{d^2y}{dx^2} = 12x - 6$.

Stick in the x-coordinates of the stationary points.

At $x = -1$, $\frac{d^2y}{dx^2} = -18$, which is negative — so $x = -1$ is a maximum.

And at $x = 2$, $\frac{d^2y}{dx^2} = 18$, which is positive — so $x = 2$ is a minimum.

And since a cubic graph (where the coefficient of x^3 is positive) goes from bottom-left to top-right...

...you can draw a rough sketch of the graph, even though the roots would be hard to find.

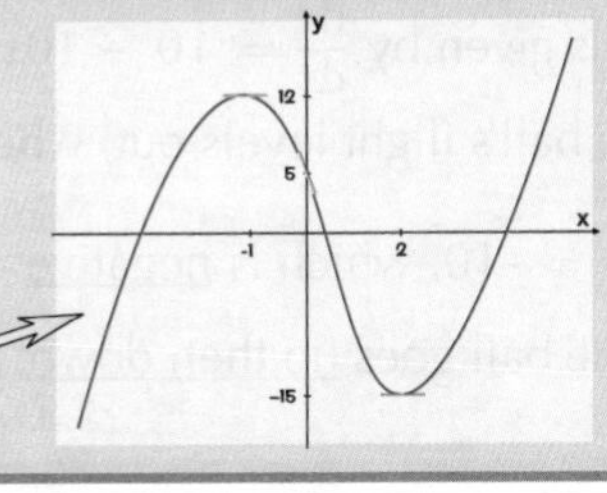

Stationary Points

1) **Find stationary points by solving $\frac{dy}{dx} = 0$.**

2) **Differentiate again to decide whether a point is a maximum or a minimum.**

3) **If $\frac{d^2y}{dx^2} < 0$ — it's a maximum.**

If $\frac{d^2y}{dx^2} > 0$ — it's a minimum.

An anagram of differentiation is "Perfect Insomnia Cure"...

No joke, is it — this differentiation business — but it's a dead important topic in maths. It's so important to know how to find whether a stationary point is a max or a min — but it can get a bit confusing. Try remembering MINMAX — which is short for 'MINUS means a MAXIMUM'. Or make up some other clever way to remember what means what.

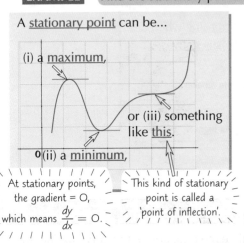

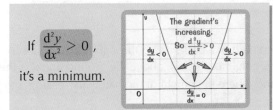

Increasing and Decreasing Functions

Differentiation is all about finding gradients. Which means that you can find out where a graph is going up... ...and where it's going down. Lovely.

*Find out if a function is **Increasing** or **Decreasing***

You can use differentiation to work out exactly where a function is increasing or decreasing — and how quickly.

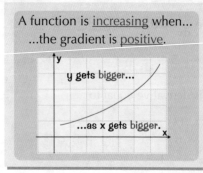

A function is increasing when... ...the gradient is positive.

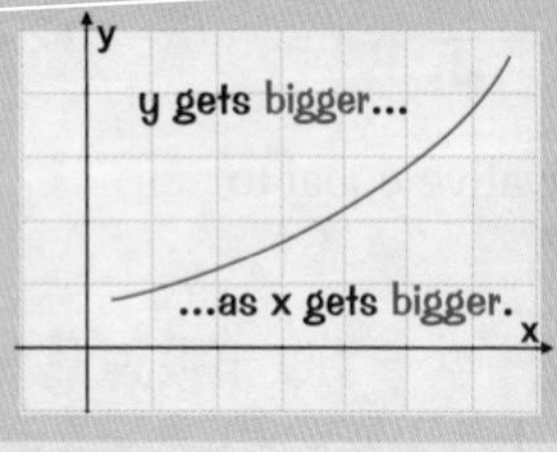

y
y gets bigger...
...as x gets bigger.
x

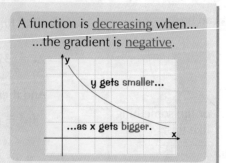

A function is decreasing when... ...the gradient is negative.

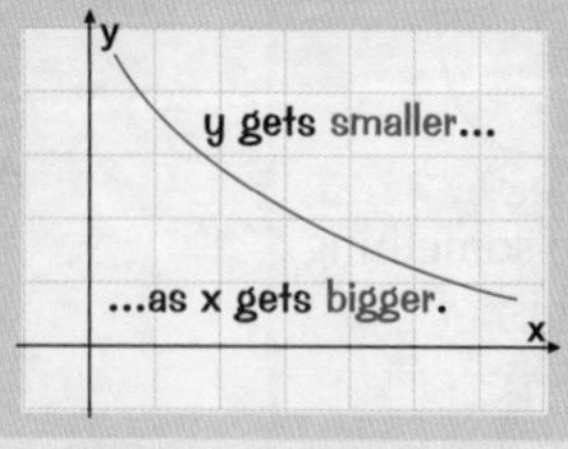

y
y gets smaller...
...as x gets bigger.
x

And there's more...

The bigger the gradient, the faster y changes with x.

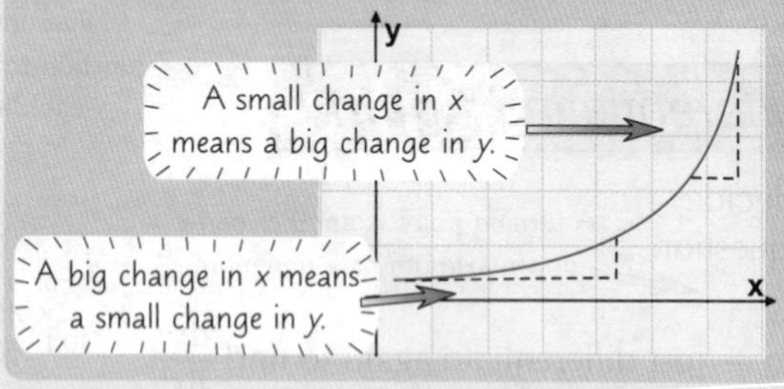

y
A small change in x means a big change in y.
A big change in x means a small change in y.
x

Differentiation and Gradients

Differentiate the equation of the curve to find an expression for its gradient.

1) An increasing function has a positive gradient.
2) A decreasing function has a negative gradient.

EXAMPLE The path of a ball thrown through the air is described by the equation $y = 10x - 5x^2$, where y is the height of the ball above the ground and x is the horizontal distance from its starting point. Find where the height of the ball is increasing and where it's decreasing.

You have the equation for the path of the ball, and you need to know where y is increasing and where it's decreasing. That makes this a question about gradients — so differentiate.

$$y = 10x - 5x^2 \text{ so } \frac{dy}{dx} = 10 - 10x$$

This is an increasing function when: $10 - 10x > 0$, i.e. when $x < 1$, so the ball's height is increasing for $0 \leq x < 1$.

And it's a decreasing function when: $10 - 10x < 0$, i.e. when $x > 1$, so its height decreases for $x > 1$ (until it lands).

Just to check: The gradient of the ball's path is given by $\frac{dy}{dx} = 10 - 10x$.

There's a turning point (i.e. the ball's flight levels out) when $x = 1$.

Differentiating again gives $\frac{d^2y}{dx^2} = -10$, which is negative — and so the turning point is a maximum.

This is what you'd expect — the ball goes up then down, not the other way round.

Decreasing function — also known as ironing...

Basically, you can tell whether a function is getting bigger or smaller by looking at the derivative. To make it more interesting, I wrote it as a nursery rhyme: the f(*duke of york*) = 10 000*x*, and when they were up the derivative was positive, and when they were down the derivative was negative, and when they were only halfway up at a stationary point the derivative was neither negative nor positive, it was zero. Catchy eh?

Real-life Problems

Differentiation isn't just mathematical daydreaming. It can be applied to real-life problems. For instance, you can use differentiation to find out the maximum possible volume of a box, given a limited amount of cardboard. Thrilling.

Differentiation works for **Volume** and **Area** too

To find the maximum for a shape's volume, all you need is an equation for the volume in terms of only one variable — then just differentiate as normal. But examiners don't hand it to you on a plate — there's usually one too many variables chucked in. So you've got to know how to manipulate the information to get rid of that unwanted variable.

EXAMPLE

A jewellery box with a lid and dimensions $2x$ cm by x cm by y cm is made using a total of 108 cm^2 of wood.

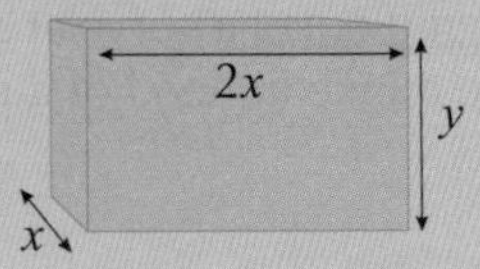

a) Show that the volume of the box can be expressed as: $V = 36x - \frac{4x^3}{3}$.

b) Use calculus to find the maximum volume.

a) You know the basic equation for volume: $V = \text{width} \times \text{height} \times \text{depth}$
$= x \times 2x \times y$

But the question asks for volume in terms of x only — you don't want that pesky y in there. So you need to find y in terms of x and substitute that in. Use the given dimensions and surface area value to find y in terms of x:

① First, write an expression for the surface area:

$$2 \times [(2x \times x) + (2x \times y) + (x \times y)] = 108$$

$$\Rightarrow \; 2x^2 + 3xy = 54$$

Be careful when adding up the sides — here there's a lid so there are two of each side, but sometimes you'll get an open-topped shape.

② Then, rearrange to find an expression for y:

$$y = \frac{54 - 2x^2}{3x}$$

$$y = \frac{18}{x} - \frac{2x}{3}$$

③ Finally, substitute the new expression for y into the equation for V, so that it's all in terms of x...

$$V = x \times 2x \times y = 2x^2 \times \left(\frac{18}{x} - \frac{2x}{3}\right)$$

... and voila, the form the question asks for appears, as if by mathgic...

$$V = 36x - \frac{4x^3}{3}$$

b) Now it's just differentiating as normal, hurrah...

① Differentiate and find x when $\frac{dV}{dx} = 0$:

$$\frac{dV}{dx} = 36 - 4x^2$$

$$36 - 4x^2 = 0 \Rightarrow 36 = 4x^2 \Rightarrow 9 = x^2 \Rightarrow x = 3$$

The other solution, $x = -3$ isn't relevant in this context.

② Check $x = 3$ is a maximum:

$$\frac{d^2V}{dx^2} = -8x = -8 \times 3 = -24$$

$\frac{d^2V}{dx^2} < 0$ so yes, it's a maximum

③ Calculate V for $x = 3$:

$$V = (36 \times 3) - \frac{4 \times 3^3}{3}$$

$$= 108 - 36 = 72 \text{ cm}^3$$

C1 Section 6 — Practice Questions

That's what differentiation is all about. Yes, there are fiddly things to remember — but overall, it's not as bad as all that. And just think of all the lovely marks you'll get if you can answer questions like these in the exam...

Warm-up Questions

1) An easy one to start with. Write down the formula for differentiating any power of x.

2) Differentiate these functions with respect to x:
   a) $y = x^2 + 2$, b) $y = x^4 + x$,

3) What's the connection between the gradient of a curve at a point and the gradient of the tangent to the curve at the same point? (That sounds like a joke in need of a punchline — but sadly, this is no joke.)

4) Find the gradients of these graphs at $x = 2$:

a) 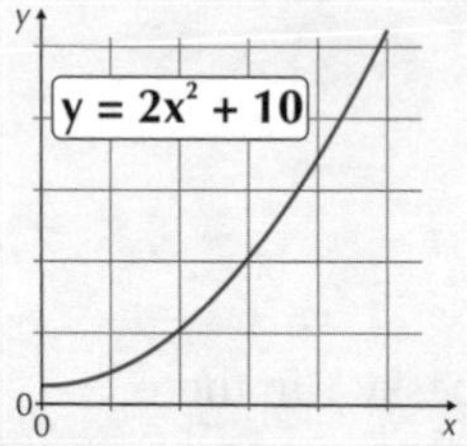

b) 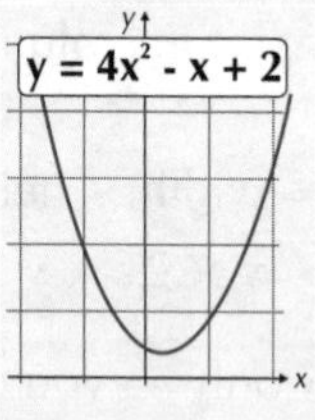

c) 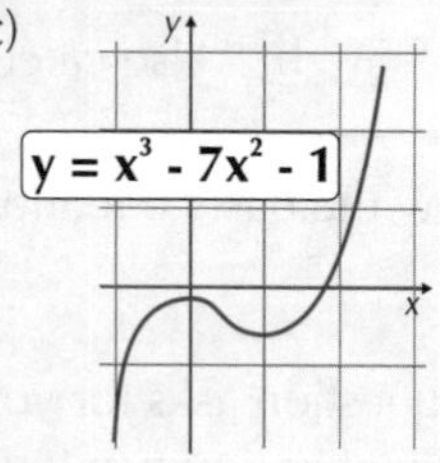

5) 1 litre of water is poured into a bowl.
   The volume (v) of water in the bowl (in ml) is defined by the function: $v = 17t^2 - 10t$
   Find the rate at which water is poured into the bowl when $t = 4$ seconds.

6) Yawn, yawn. Find the equations of the tangent and the normal to the curve $y = x^2 - 3x - 10$ at $x = 5$.

7) Show that the lines $y = \frac{x^3}{3} - 2x^2 - 4x + \frac{86}{3}$ and $y = \frac{x}{4} + 1$ both go through the point (4, 2), and are perpendicular at that point. Good question, that — nice and exciting, just the way you like 'em.

8) a) Write down what a stationary point is.

   b) Find the stationary points of the graph of $y = x^3 - 6x^2 + 9x + 1$.

9) Find when $y = 6(x + 2)(x - 3)$ is increasing and decreasing.

10) The height (h m) a firework can reach is related to the mass (m g) of fuel it carries as shown below:

$$h = \frac{m^2}{10} - \frac{m^3}{300}$$

Find the mass of fuel required to achieve the maximum height.

C1 Section 6 — Practice Questions

I know, I know, one page of differentiation questions just isn't enough to quench your maths thirst. Have a few more glugs of the sweet, sweet, algebraic nectar. Mmm, numbers...

Exam Questions

1 Given that $y = x^7 + x + \sqrt{2}$, find:

a) $\frac{dy}{dx}$ when $x = 1$. *(2 marks)*

b) $\frac{d^2y}{dx^2}$ when $x = 1$. *(3 marks)*

2 The curve C is given by the equation $y = 2x^3 - 4x^2 - 4x + 12$.

a) Find $\frac{dy}{dx}$. *(2 marks)*

b) Write down the gradient of the tangent to the curve at the point where $x = 2$. *(1 mark)*

c) Hence or otherwise find an equation for the normal to the curve at this point. *(3 marks)*

3 The curve C is given by the equation $y = mx^3 - x^2 + 8x + 2$, for a constant m.

a) Find $\frac{dy}{dx}$. *(2 marks)*

The point P lies on C, and has the x-value 5. The normal to C at P is parallel to the line given by the equation $y + 4x - 3 = 0$.

b) Find the gradient of curve C at P. *(3 marks)*

Hence or otherwise, find:

c) (i) the value of m. *(3 marks)*

(ii) the y-value at P. *(2 marks)*

4 For $x \geq 0$ and $y \geq 0$, x and y satisfy the equation $2x - y = 6$.

a) If $W = x^2y^2$, show that $W = 4x^4 - 24x^3 + 36x^2$. *(2 marks)*

b) (i) Show that $\frac{dW}{dx} = k(2x^3 - 9x^2 + 9x)$, and find the value of the integer k. *(4 marks)*

(ii) Find the value of $\frac{dW}{dx}$ when $x = 1$. *(1 mark)*

c) Find $\frac{d^2W}{dx^2}$ and give its value when $x = 1$. *(2 marks)*

C1 Section 6 — Practice Questions

5 a) Find $\frac{dy}{dx}$ for the curve $y = 6 + \frac{2x^3 - 6x^2 + 6x}{3}$.

(3 marks)

b) Hence find the coordinates of the stationary point on the curve.

(3 marks)

6 a) Determine the coordinates of the stationary points for the curve $y = x(x - 1)^2$.

(5 marks)

b) Find whether each of these points is a maximum or minimum.

(3 marks)

7 Ayesha is building a closed-back bookcase. She uses a total of 6 m^2 of wood (not including shelving) to make a bookcase that is x metres high, $\frac{x}{2}$ metres wide and d metres deep, as shown.

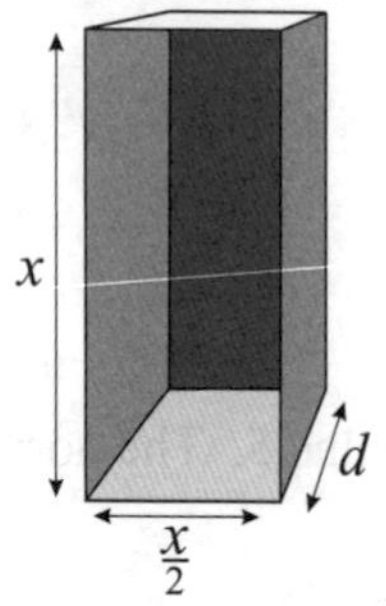

a) Show that the full capacity of the bookcase is given by: $V = x - \frac{x^3}{12}$.

(4 marks)

b) Find the value of x for which V is stationary.

(4 marks)

c) Show that this is a maximum point and hence calculate the maximum V.

(4 marks)

8 You are given the curve $f(x) = x^3 - 3x^2 - 1$.

a) Find the coordinates of any stationary points on the curve.

(5 marks)

b) Determine the nature of these stationary points.

(3 marks)

c) Show that $f(x)$ is increasing when $x = 4$.

(1 mark)

Integration

Integration is the 'opposite' of differentiation — and so if you can differentiate, you can be pretty confident you'll be able to integrate too. There's just one extra thing you have to remember — the constant of integration...

You need the constant because there's **More Than One** right answer

When you integrate something, you're trying to find a function that returns to what you started with when you differentiate it. And when you add the constant of integration, you're just allowing for the fact that there's more than one possible function that does this...

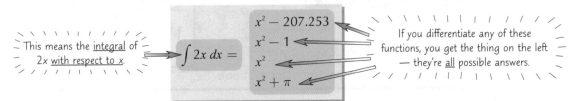

This means the integral of $2x$ with respect to x.

$$\int 2x\,dx = \quad x^2 - 207.253,\quad x^2 - 1,\quad x^2,\quad x^2 + \pi$$

If you differentiate any of these functions, you get the thing on the left — they're all possible answers.

So the answer to this integral is actually...

$$\int 2x\,dx = x^2 + C$$

The 'C' just means 'any number'. This is the constant of integration.

You only need to add a constant of integration to indefinite integrals like these ones. Definite integrals are integrals with limits (or little numbers) next to the integral sign, but we'll come to that later. For a sneak preview, flick ahead to p. 49.

Up the power by **One** — then **Divide** by it

The formula below tells you how to integrate any power of x (except x^{-1}).

This is an indefinite integral — it doesn't have any limits (numbers) next to the integral sign.

$$\int x^n\,dx = \frac{x^{n+1}}{n+1} + C$$

In a nutshell, this says:

To integrate a power of x: (i) Increase the power by one — then divide by it.
and (ii) Stick a constant on the end.

EXAMPLES Use the integration formula...

① For regular powers of x,

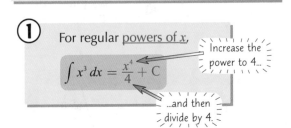

$$\int x^3\,dx = \frac{x^4}{4} + C$$

Increase the power to 4...

...and then divide by 4.

② For constants,

Remember, $x^0 = 1$.

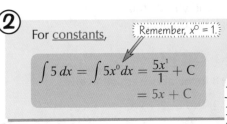

$$\int 5\,dx = \int 5x^0\,dx = \frac{5x^1}{1} + C = 5x + C$$

In general, $\int k\,dx = kx + C$ — which is what you would expect, as when you differentiate $kx + C$, you get k. Good eh?

③ And for more complicated looking stuff:

$$\int (4x^3 - x^2 + 6x - 2)\,dx = \frac{4x^4}{4} - \frac{x^3}{3} + \frac{6x^2}{2} - 2x + C$$
$$= x^4 - \frac{x^3}{3} + 3x^2 - 2x + C$$

Do each of these bits separately.

CHECK YOUR ANSWERS: You can check you've integrated properly by differentiating the answer — you should end up with the thing you started with.

Indefinite integrals — joy without limits...

This integration lark isn't so bad then — there's only a couple of things to remember and then you can do it no problem. But that constant of integration catches loads of people out — it's so easy to forget — and you'll definitely lose marks if you do forget it. You have been warned. Other than that, there's not much to it. Hurray.

Integration

By now, you're probably aware that maths isn't something you do unless you're a bit of a thrill-seeker. You know, sometimes they even ask you to find a curve with a certain derivative that goes through a certain point.

*You sometimes need to find the **Value** of the **Constant of Integration***

When they tell you something else about the curve in addition to its derivative, you can work out the value of that constant of integration. Usually the something is the coordinates of one of the points the curve goes through.

Really Important Bit...

When you differentiate y, you get $\frac{dy}{dx}$.

And when you integrate $\frac{dy}{dx}$, you get y

(if you ignore the constant of integration).

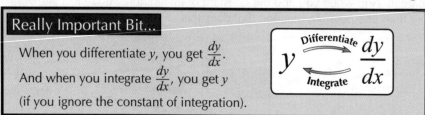

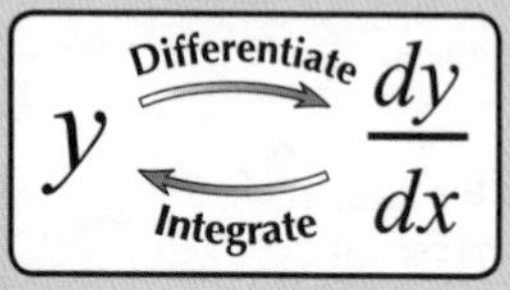

EXAMPLE The curve $f(x)$ goes through the point (2, 8) and $f'(x) = 6x(x - 1)$.

Find $f(x)$.

$f'(x)$ is just another way of saying dy/dx. When you integrate $f'(x)$ you get $f(x)$ and when you differentiate $f(x)$ you get $f'(x)$.

You know the derivative $f'(x)$ and need to find the function $f(x)$ — so integrate.

Remember:
Even if you don't have any extra information about the curve — you still have to add a constant when you work out an integral without limits.

$$f'(x) = 6x(x-1) = 6x^2 - 6x$$

So integrating both sides gives...

$$f(x) = \int (6x^2 - 6x)dx$$
$$\Rightarrow f(x) = \frac{6x^3}{3} - \frac{6x^2}{2} + C$$
$$\Rightarrow f(x) = 2x^3 - 3x^2 + C$$

Don't forget the constant of integration.

Check this is correct by differentiating it and making sure you get what you started with.

$$f(x) = 2x^3 - 3x^2 + C = 2x^3 - 3x^2 + Cx^0$$
$$f'(x) = 2(3x^2) - 3(2x^1) + C(0x^{-1})$$
$$f'(x) = 6x^2 - 6x$$

A constant always differentiates to zero.

So this function's got the correct derivative — but you haven't finished yet.

You now need to find C — and you do this by using the fact that it goes through the point (2, 8).

$$f(x) = 2x^3 - 3x^2 + C$$

Putting $x=2$ and $f(x)=8$ in the above equation gives...

$$8 = (2 \times 2^3) - (3 \times 2^2) + C$$
$$\Rightarrow 8 = 16 - 12 + C$$
$$\Rightarrow C = 4$$

So the answer you need is this one:

$$f(x) = 2x^3 - 3x^2 + 4$$

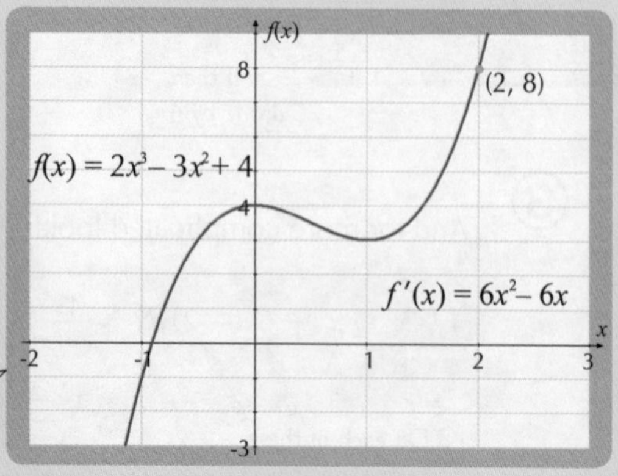

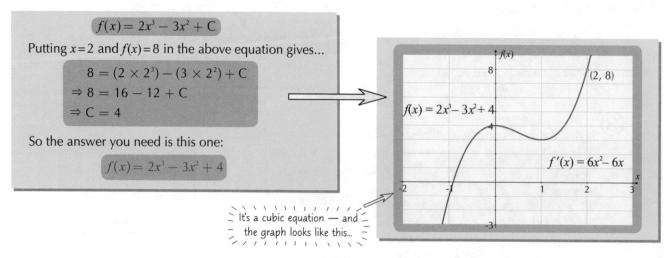

It's a cubic equation — and the graph looks like this...

Maths and alcohol don't mix — so never drink and derive...

That's another page under your belt and — go on, admit it — there was nothing too horrendous on it. If you can do the stuff from the previous page and then substitute some numbers into an equation, you can do everything from this page too. So if you think this is boring, you'd be right. But if you think it's much harder than the stuff before, you'd be wrong.

Areas Under Curves

Some integrals have limits (i.e. little numbers) next to the integral sign. You integrate them in exactly the same way — but you don't need a constant of integration. Much easier. And scrummier and yummier too.

A Definite Integral finds the Area Under a Curve

This definite integral tells you the area between the graph of $y = x^3$ and the x-axis between $x = -2$ and $x = 2$:

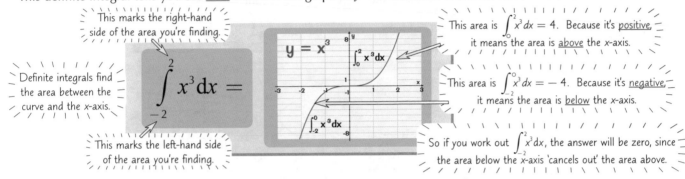

This marks the right-hand side of the area you're finding.

Definite integrals find the area between the curve and the x-axis.

$$\int_{-2}^{2} x^3 dx =$$

This marks the left-hand side of the area you're finding.

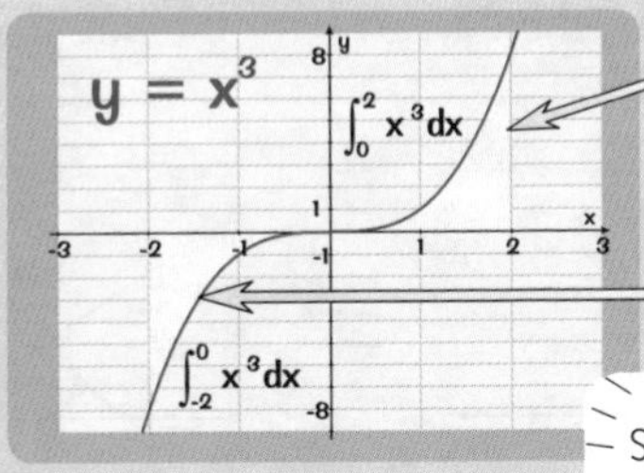

This area is $\int_0^2 x^3 dx = 4$. Because it's positive, it means the area is above the x-axis.

This area is $\int_{-2}^0 x^3 dx = -4$. Because it's negative, it means the area is below the x-axis.

So if you work out $\int_{-2}^2 x^3 dx$, the answer will be zero, since the area below the x-axis 'cancels out' the area above.

Do the integration in the same way — then use the Limits

Finding a definite integral isn't really any harder than an indefinite one — there's just an extra stage you have to do. After you've integrated the function you have to work out the value of this new function by sticking in the limits.

EXAMPLE Evaluate $\int_1^3 (x^2 + 2)dx$.

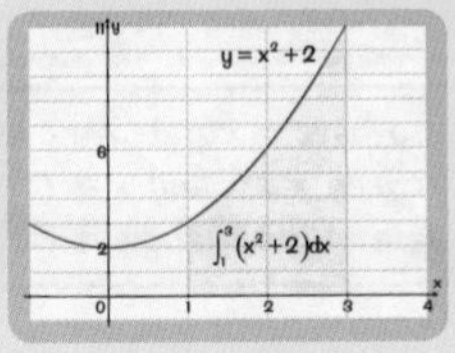

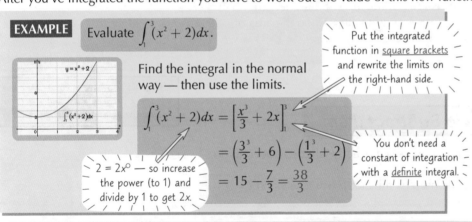

Find the integral in the normal way — then use the limits.

$$\int_1^3 (x^2+2)dx = \left[\frac{x^3}{3} + 2x\right]_1^3$$
$$= \left(\frac{3^3}{3} + 6\right) - \left(\frac{1^3}{3} + 2\right)$$
$$= 15 - \frac{7}{3} = \frac{38}{3}$$

Put the integrated function in square brackets and rewrite the limits on the right-hand side.

You don't need a constant of integration with a definite integral.

$2 = 2x^0$ — so increase the power (to 1) and divide by 1 to get $2x$.

Definite Integrals
After you've integrated the function — put both the limits in and find the values. Then subtract what the bottom limit gave you from what the top limit gave you.

EXAMPLE The graph shows the function $y = -x^2 + 5x - 4$. Find the area of the shaded region bounded by $y = -x^2 + 5x - 4$, the x-axis and the points A and B, as shown on the graph.

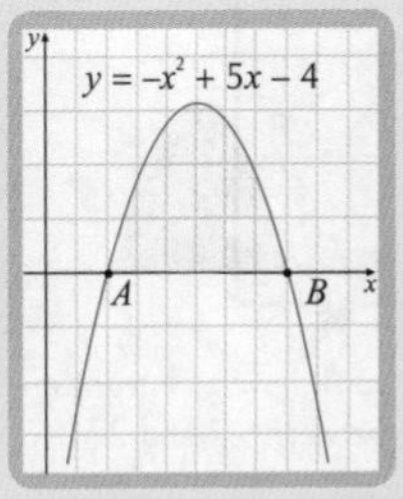

Before you find the shaded area, you need to calculate the x-values of A and B. These will be the limits of your integral.

A and B are the two points where the curve crosses the x-axis. They occur when $y = 0$, i.e. when $-x^2 + 5x - 4 = 0$.

$-x^2 + 5x - 4 = 0 \Rightarrow x^2 - 5x + 4 = 0$

Multiplying throughout by –1 so that the coefficient of x^2 is 1 makes factorising a bit easier.

$\Rightarrow (x - 1)(x - 4) = 0 \Rightarrow x = 1, x = 4.$

So the graph crosses the y-axis when $x = 1$ and $x = 4$. The required area is then given by:

$$\int_1^4 (-x^2 + 5x - 4)dx = \left[-\frac{x^3}{3} + \frac{5x^2}{2} - 4x\right]_1^4 = \left(-\frac{4^3}{3} + \frac{5(4)^2}{2} - 4(4)\right) - \left(-\frac{1^3}{3} + \frac{5(1)^2}{2} - 4(1)\right)$$
$$= -\frac{64}{3} + \frac{80}{2} - 16 + \frac{1}{3} - \frac{5}{2} + 4 = 7 - \frac{5}{2} = \frac{9}{2}$$

A few nasty fractions, I know — but just think of the sense of satisfaction you'll get at the end.

My hobbies? Well I'm really inte grating. Especially carrots.

It's still integration — but this time you're putting two numbers into an expression afterwards. So although this may not be the wild and crazy fun-packed time your teachers promised you when they were trying to persuade you to take AS maths, you've got to admit that a lot of this stuff is pretty similar — and if you can do one bit, you can use that to do quite a few other bits too. Maths is like that. But I admit it's probably not as much fun as a big banana-and-toffee cake.

Areas Between Curves

With a bit of thought, you can use integration to find all kinds of areas — even ones that look quite tricky at first. The best way to work out what to do is draw a picture. Then it'll seem easier. I promise you it will.

Sometimes you have to **Add** integrals...

This looks pretty hard — until you draw a picture and see what it's all about.

EXAMPLE Find the area enclosed by the curve $y = x^2$, the line $y = 2 - x$ and the x-axis.

Find out where the graphs meet by solving $x^2 = 2 - x$ — they meet at $x = 1$ (they also meet at $x = -2$, but this isn't in A).

You have to find area A — but you'll need to split it into two smaller pieces.

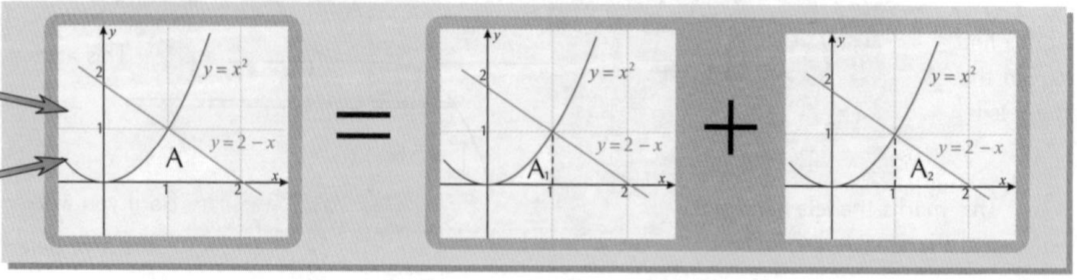

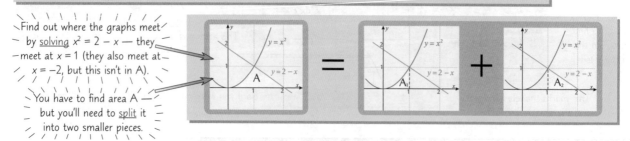

And it's pretty clear from the picture that you'll have to find the area in two lumps, A_1 and A_2.

The first area you need to find is A_1:

$$A_1 = \int_0^1 x^2 dx = \left[\frac{x^3}{3}\right]_0^1 = \left(\frac{1}{3} - 0\right) = \frac{1}{3}$$

The other area you need is A_2:

A_2 is just a triangle, with base length $2 - 1 = 1$ and height $= 1$. So the area of the triangle is $\frac{1}{2} \times b \times h = \frac{1}{2} \times 1 \times 1 = \frac{1}{2}$.

And the area the question actually asks for is $A_1 + A_2$. This is

$$A = A_1 + A_2 = \frac{1}{3} + \frac{1}{2} = \frac{5}{6}$$

You could also have integrated the line $y = 2 - x$ between $x = 1$ and $x = 2$, but finding the area of the triangle is easier.

...sometimes you have to **Subtract** them

Again, it's best to look at the pictures to work out exactly what you need to do.

EXAMPLE Find the area enclosed by the curves $y = x^2 + 1$ and $y = 9 - x^2$.

Solve $x^2 + 1 = 9 - x^2$ to find where the curves meet.

$$x^2 + 1 = 9 - x^2 \Rightarrow 2x^2 = 8 \Rightarrow x^2 = 4 \Rightarrow x = \pm 2$$

So you'll have to integrate between −2 and 2.

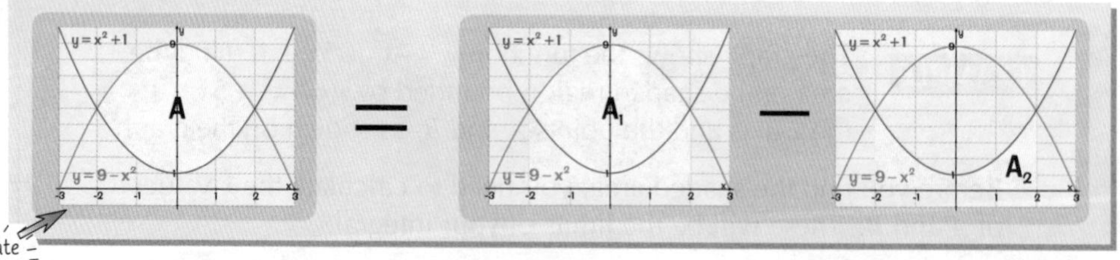

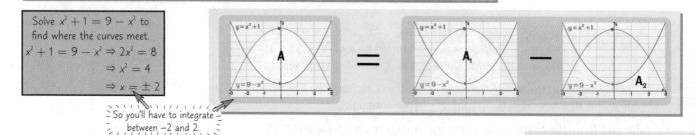

The area under the green curve A_1 is:

$$A_1 = \int_{-2}^{2} (9 - x^2)dx = \left[9x - \frac{x^3}{3}\right]_{-2}^{2} = \left(18 - \frac{2^3}{3}\right) - \left(-18 - \frac{(-2)^3}{3}\right) = \left(18 - \frac{8}{3}\right) - \left(-18 - \left(\frac{-8}{3}\right)\right) = \frac{46}{3} - \left(-\frac{46}{3}\right) = \frac{92}{3}$$

The area under the red curve is:

$$A_2 = \int_{-2}^{2} (x^2 + 1)dx = \left[\frac{x^3}{3} + x\right]_{-2}^{2} = \left(\frac{2^3}{3} + 2\right) - \left(\frac{(-2)^3}{3} + (-2)\right) = \left(\frac{8}{3} + 2\right) - \left(-\frac{8}{3} - 2\right) = \frac{28}{3}$$

And the area you need is the difference between these:

$$A = A_1 - A_2 = \frac{92}{3} - \frac{28}{3} = \frac{64}{3}$$

Instead of integrating before subtracting — you could try 'subtracting the curves', and then integrating. This last area A is also:

$$A = \int_{-2}^{2} \{(9 - x^2) - (x^2 + 1)\} dx$$

And so, our hero integrates the area between two curves, and saves the day...

That's the basic idea of finding the area enclosed by two curves and lines — draw a picture and then break the area down into smaller, easier chunks. Questions like this aren't hard — but they can sometimes take quite a long time. Great.

C1 Section 7 — Practice Questions

Integration is pretty much the opposite of differentiation. So if you can differentiate, then chances are you can integrate as well. Which brings us (kind of) neatly on to these questions. Come and have a go, if you think you're hard enough...

Warm-up Questions

1) Write down the steps involved in integrating a power of x.

2) What's an indefinite integral? Why do you have to add a constant of integration when you find an indefinite integral?

3) How can you check whether you've integrated something properly? (Without asking someone else.)

4) How can you tell whether an integral is a definite one or an indefinite one? (It's easy really — it just sounds difficult.)

5) Integrate these: a) $\int 10x^4 dx$, b) $\int (3x + 5x^2) dx$, c) $\int (x^2(3x + 2)) dx$.

6) Work out the equation of the curve that has derivative $\frac{dy}{dx} = 6x - 7$ and goes through the point (1, 0).

7) a) Find the equation of the curve that has derivative $\frac{dy}{dx} = 3x^3 + 2$ and goes through the point (1, 0).

b) How would you change the equation if the curve had to go through the point (1, 2) instead?

(Don't start the whole question again.)

8) What does a definite integral represent on a graph?

9) Evaluate the following definite integrals:

a) $\int_0^1 (4x^3 + 3x^2 + 2x + 1) dx$, b) $\int_{-3}^{3} (9 - x^2) dx$.

10) Use integration to find the yellow area in each of these graphs:

a)

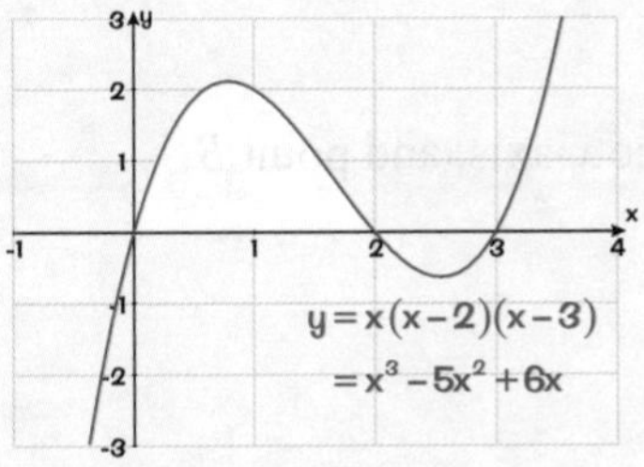

b)

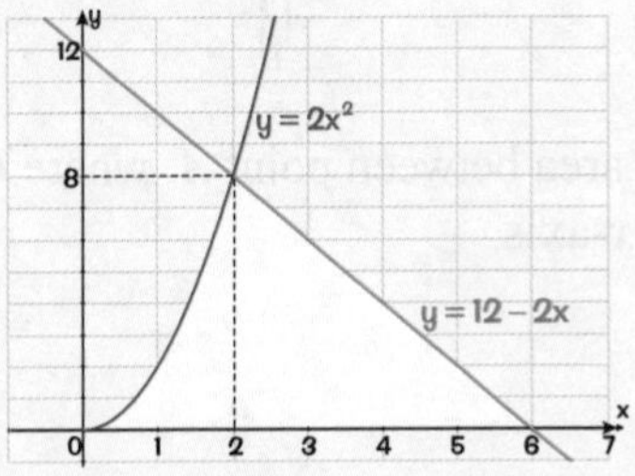

c)

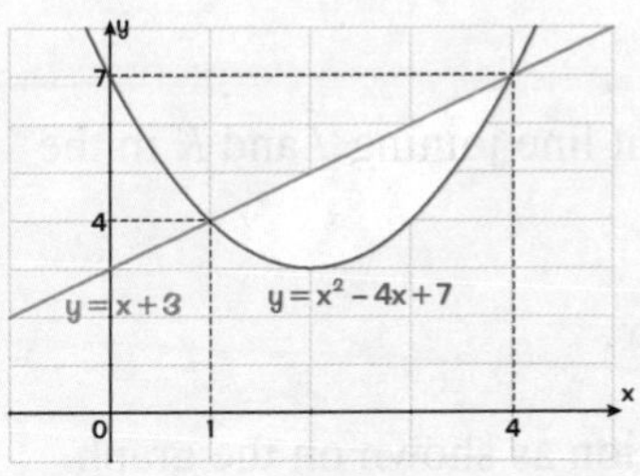

C1 Section 7 — Practice Questions

My dad used to call exams 'umbrellas' because he thought it sounded less scary.
So here are some umbrella-style questions. Also useful if it's raining...

Exam Questions

1 a) Show that $(5 + 2\sqrt{x})(5 - 2\sqrt{x})$ can be written in the form $a + bx$, stating the values of the constants a and b.

(2 marks)

b) Find $\int [(5 + 2\sqrt{x})(5 - 2\sqrt{x})]dx$.

(2 marks)

2 $f'(x) = (x - 1)(3x - 1)$ where $x > 0$

a) The curve C is given by $f(x)$ and goes through point P (3, 10). Find $f(x)$.

(6 marks)

b) The equation for the normal to C at the point P can be written in the form $y = \frac{a - x}{b}$ where a and b are integers. Find the values of a and b.

(4 marks)

3 Find the value of $\int_0^2 (2x - 6x^2 + 1)dx$.

(5 marks)

4 The diagram shows the graph of the curve C, defined by $y = x(x - 1)^2$.

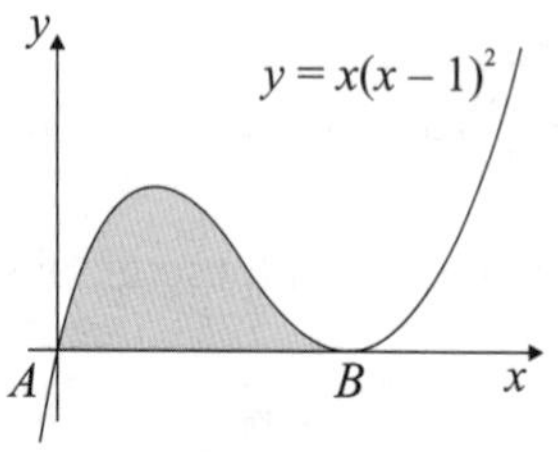

Calculate the shaded area between point A, where C intersects the x-axis, and point B, where C touches the x-axis.

(5 marks)

5 The diagram shows the graph of the curve $y = (x + 2)(x - 2)$.
Points J (–2, 0) and K (1, –3) lie on the curve.

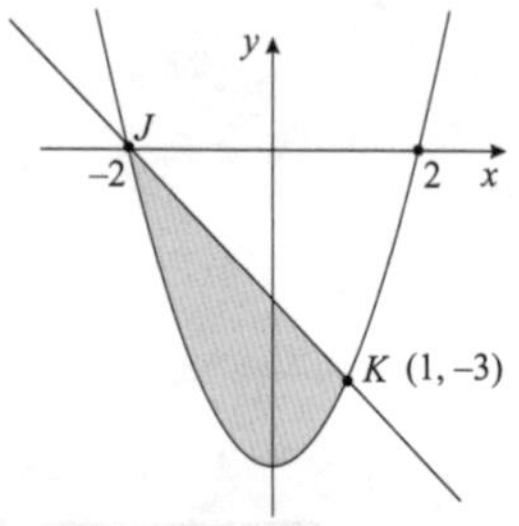

a) Find the equation of the straight line joining J and K in the form $y = mx + c$.

(2 marks)

b) Calculate $\int_{-2}^{1} (x + 2)(x - 2)dx$.

(4 marks)

c) Find the area of the shaded region as shown on the graph.

(3 marks)

General Certificate of Education
Advanced Subsidiary (AS) and Advanced Level

Core Mathematics C1 — Practice Exam One

Time Allowed: 1 hour 30 min

Calculators may **not** be used for this exam

There are 75 marks available for this paper.

1 The equation $x^2 + (k + 1)x + 4k = 0$, where k is a constant, has equal roots.

a) Show that k satisfies the equation $k^2 - 14k + 1 = 0$. *(3 marks)*

b) (i) Express $k^2 - 14k + 1$ in the form $(x - p)^2 + q$, where p and q are integers. *(2 marks)*

(ii) Hence state the minimum value of $k^2 - 14k + 1$. *(1 mark)*

(iii) Write down the value of k for which this minimum value occurs. *(1 mark)*

2 The points A and B have coordinates (1, 6) and (–2, 9) respectively. Find:

a) the equation of L, the line which passes through A and B, *(3 marks)*

b) the coordinates of M, the mid-point of the line segment AB, *(2 marks)*

c) the equation of the line perpendicular to L which passes through M, *(3 marks)*

d) the coordinates of the points where L intersects the curve C, given by $y = x^2 + 3x - 5$. *(5 marks)*

3 A circle, C, has equation $x^2 - 6x + y^2 = 3$.

a) Find the exact coordinates of the points where C intersects the x-axis. *(4 marks)*

b) Rearrange the equation of C into the form $(x - a)^2 + (y - b)^2 = c$. *(2 marks)*

c) Write down the radius, r, and the coordinates of Q, the centre of the circle.
Where appropriate, give your answers in simplified surd form. *(3 marks)*

d) (i) Find the distance from Q to the point P with coordinates (5, 4).
Give your answer in simplified surd form. *(3 marks)*

(ii) Does P lie inside C? Give a reason for your answer. *(2 marks)*

4 a) Express $(5\sqrt{5} + 2\sqrt{3})^2$ in the form $a + b\sqrt{c}$, where a, b and c are integers to be found. *(4 marks)*

b) Rationalise the denominator of $\dfrac{10}{\sqrt{5}+1}$. *(3 marks)*

5 a) Rewrite the following equation in the form $f(x) = 0$, where $f(x)$ is of the form $f(x) = ax^3 + bx^2 + cx + d$:

$$(x - 1)(x^2 + x + 1) = 2x^2 - 17$$

(2 marks)

b) Use the factor theorem to show that $(x + 2)$ is a factor of $f(x)$. *(2 marks)*

c) Express $f(x)$ in the form $f(x) = (x + 2)(x^2 + px + q)$, where p and q are constants. *(3 marks)*

d) Show that $f(x) = 0$ has only one root. *(3 marks)*

e) Consider the equation $y = f(x)$.

(i) Find $\dfrac{dy}{dx}$. *(2 marks)*

(ii) Find the tangent to the curve $y = f(x)$ at $x = 1$. *(4 marks)*

(iii) Show that $f(x)$ is increasing when $3x^2 - 4x > 0$. *(1 mark)*

(iv) Solve the quadratic inequality $3x^2 - 4x > 0$. *(2 marks)*

6 a) A curve, C, has equation $y = -x^2 + x + 6$. Show that C has a stationary point at $(0.5, 6.25)$. *(4 marks)*

b) The diagram shows the graphs of C and the line L, given by the equation $y = x + 2$.

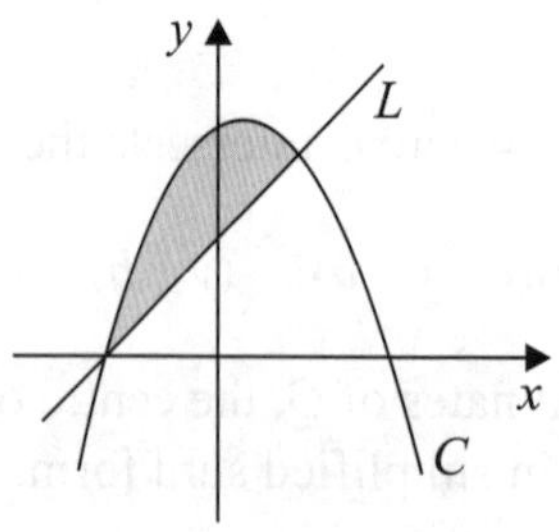

(i) Determine the coordinates of the points of intersection of C and L. *(5 marks)*

(ii) Hence find the area of the shaded region bound by C and L. *(6 marks)*

General Certificate of Education
Advanced Subsidiary (AS) and Advanced Level

Core Mathematics C1 — Practice Exam Two

Time Allowed: 1 hour 30 min

Calculators may **not** be used for this exam

There are 75 marks available for this paper.

1 a) Simplify $(\sqrt{3}+1)(\sqrt{3}-1)$. *(2 marks)*

b) Rationalise the denominator of the expression $\frac{\sqrt{3}-1}{\sqrt{3}+1}$. *(3 marks)*

c) Solve the equation $x\sqrt{2} = \sqrt{72} - 4\sqrt{8}$. *(3 marks)*

2 The sides of a triangle ABC lie on lines given as follows:

side AB lies on $y = 3$, side BC lies on $2x - 3y - 21 = 0$, side AC lies on $3x + 2y - 12 = 0$

a) Find the coordinates of the vertices of the triangle. *(5 marks)*

b) Show that the triangle is right-angled. *(2 marks)*

c) Find the length of the side AC, leaving your answer in simplified surd form. *(2 marks)*

3 a) Express $x^2 - 7x + 17$ in the form $(x - m)^2 + n$, where m and n are constants. *(3 marks)*

b) Hence state the maximum value of $f(x) = \frac{1}{x^2 - 7x + 17}$. *(2 marks)*

c) Describe the geometrical transformation which maps the graph of $y = x^2$ onto the graph of $y = x^2 - 7x + 17$. *(3 marks)*

4 The function $f(x)$ is given by $f(x) = 3jx - jx^2 + 1$, where j is a constant. Given that the equation $f(x) = 0$ has no real roots,

a) Show that $9j^2 + 4j < 0$. *(3 marks)*

b) Hence find the range of possible values of j. *(3 marks)*

5 A curve has the equation $y = f(x)$, where $f(x) = x^3 - 3x + 2$.

a) (i) Find $f'(x)$. *(2 marks)*

(ii) Hence find the coordinates of any stationary points on the curve. *(3 marks)*

(iii) For each stationary point you have found, determine whether it is a maximum or minimum. *(3 marks)*

b) (i) Find the remainder when $f(x)$ is divided by $(x - 3)$. *(2 marks)*

(ii) Show that $(x - 1)^2(x + 2)$ is a factorisation of $x^3 - 3x + 2$. *(2 marks)*

c) Sketch the graph of $f(x)$, indicating any intercepts with the x- and y-axes and any stationary points. *(4 marks)*

6 The diagram shows the graphs of the curve $y = x^2 - 4x + 6$ and the line $y = -x + 4$.

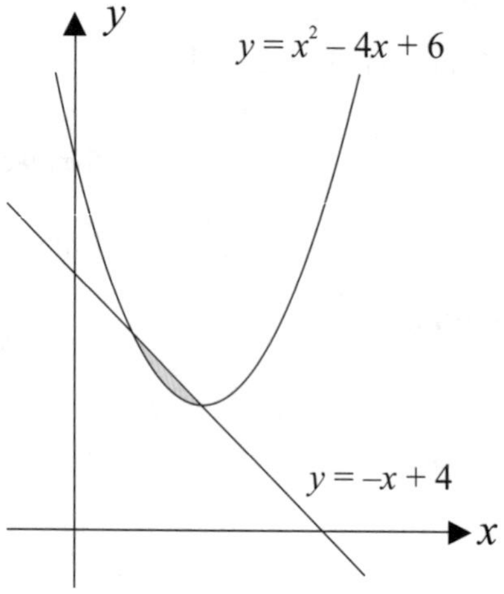

a) Find $\frac{dy}{dx}$ for the curve $y = x^2 - 4x + 6$. *(2 marks)*

b) Find the coordinates of the points where the line and curve intersect. *(5 marks)*

c) Evaluate the integral $\int_1^2 (x^2 - 4x + 6)\,dx$. *(3 marks)*

d) Hence, or otherwise, show that the total area enclosed by the curve $y = x^2 - 4x + 6$ and the line $y = -x + 4$ is equal to $\frac{1}{6}$. *(4 marks)*

7 The points A, B and C have coordinates (2, 1), (0, –5) and (4, –1) respectively. A, B and C lie on a circle of diameter AB, as shown.

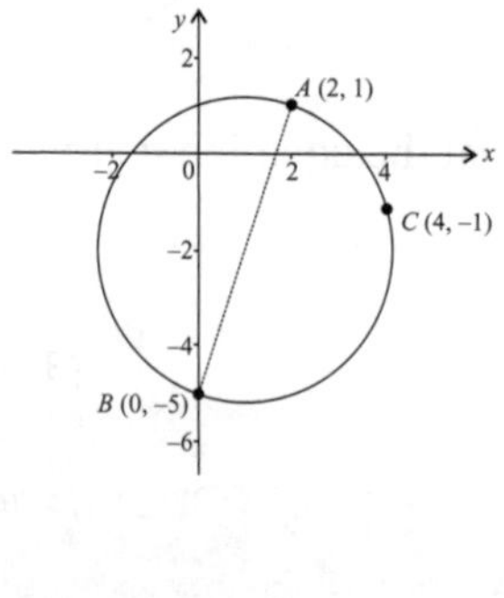

a) Find the centre and radius of the circle. *(4 marks)*

b) Show that the equation of the circle can be written in the form:
$x^2 + y^2 - 2x + 4y - 5 = 0$ *(3 marks)*

c) The tangent at A and the normal at C cross at D.
Find the coordinates of D. *(7 marks)*

Laws of Indices

You use the laws of indices a helluva lot in C2 — when you're integrating, differentiating and ...er... well loads of other places. So take the time to get them sorted now.

Three mega-important Laws of Indices

You must know these three rules. I can't make it any clearer than that.

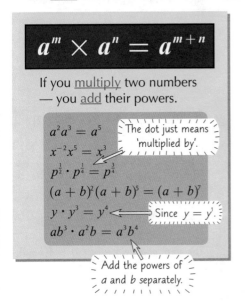

$$a^m \times a^n = a^{m+n}$$

If you multiply two numbers — you add their powers.

$a^2a^3 = a^5$

$x^{-2}x^5 = x^3$

$p^{\frac{1}{2}} \cdot p^{\frac{1}{4}} = p^{\frac{3}{4}}$

$(a+b)^2(a+b)^5 = (a+b)^7$

$y \cdot y^3 = y^4$

$ab^3 \cdot a^2b = a^3b^4$

The dot just means 'multiplied by'.

Since $y = y^1$.

Add the powers of a and b separately.

$$\frac{a^m}{a^n} = a^{m-n}$$

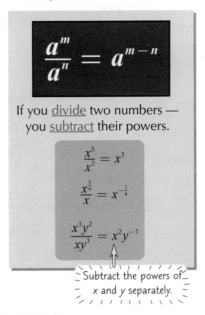

If you divide two numbers — you subtract their powers.

$\frac{x^5}{x^2} = x^3$

$\frac{x^{\frac{3}{4}}}{x} = x^{-\frac{1}{4}}$

$\frac{x^3y^2}{xy^3} = x^2y^{-1}$

Subtract the powers of x and y separately.

$$(a^m)^n = a^{mn}$$

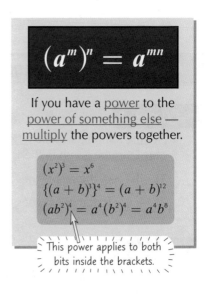

If you have a power to the power of something else — multiply the powers together.

$(x^2)^3 = x^6$

$\{(a+b)^3\}^4 = (a+b)^{12}$

$(ab^2)^4 = a^4(b^2)^4 = a^4b^8$

This power applies to both bits inside the brackets.

Other important stuff about Indices

You can't get very far without knowing this sort of stuff. Learn it — you'll definitely be able to use it.

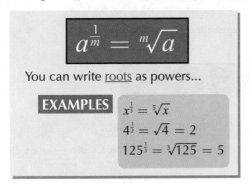

$$a^{\frac{1}{m}} = \sqrt[m]{a}$$

You can write roots as powers...

EXAMPLES

$x^{\frac{1}{5}} = \sqrt[5]{x}$

$4^{\frac{1}{2}} = \sqrt{4} = 2$

$125^{\frac{1}{3}} = \sqrt[3]{125} = 5$

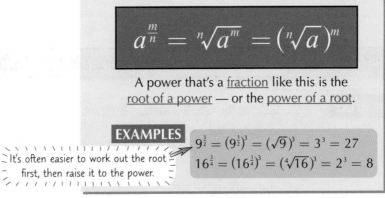

$$a^{\frac{m}{n}} = \sqrt[n]{a^m} = (\sqrt[n]{a})^m$$

A power that's a fraction like this is the root of a power — or the power of a root.

EXAMPLES

$9^{\frac{3}{2}} = (9^{\frac{1}{2}})^3 = (\sqrt{9})^3 = 3^3 = 27$

$16^{\frac{3}{4}} = (16^{\frac{1}{4}})^3 = (\sqrt[4]{16})^3 = 2^3 = 8$

It's often easier to work out the root first, then raise it to the power.

$$a^{-m} = \frac{1}{a^m}$$

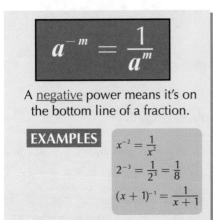

A negative power means it's on the bottom line of a fraction.

EXAMPLES

$x^{-2} = \frac{1}{x^2}$

$2^{-3} = \frac{1}{2^3} = \frac{1}{8}$

$(x+1)^{-1} = \frac{1}{x+1}$

$$a^0 = 1$$

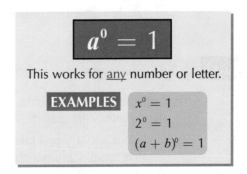

This works for any number or letter.

EXAMPLES

$x^0 = 1$

$2^0 = 1$

$(a+b)^0 = 1$

Indices, indices — de fish all live indices...

What can I say that I haven't said already? Blah, blah, important. Blah, blah, learn these. Blah, blah, use them all the time. Mmm, that's about all that needs to be said really. So I'll be quiet and let you get on with what you need to do.

Graph Transformations

Suppose you start with any old function f(x). Then you can <u>transform</u> (change) it in three ways — by <u>translating</u> it, <u>stretching</u> or <u>reflecting</u> it.

$y = f(x)$

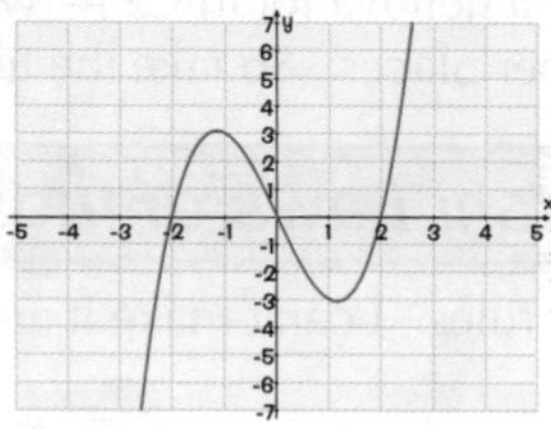

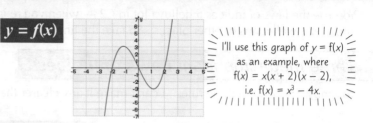

I'll use this graph of $y = f(x)$ as an example, where $f(x) = x(x + 2)(x - 2)$, i.e. $f(x) = x^3 - 4x$.

Translations are caused by *Adding* things

$y = f(x) + a$

<u>Adding</u> a number to the <u>whole function</u> <u>translates</u> the graph in the <u>y-direction</u>.

1) If $a > 0$, the graph goes <u>upwards</u>.
2) If $a < 0$, the graph goes <u>downwards</u>.

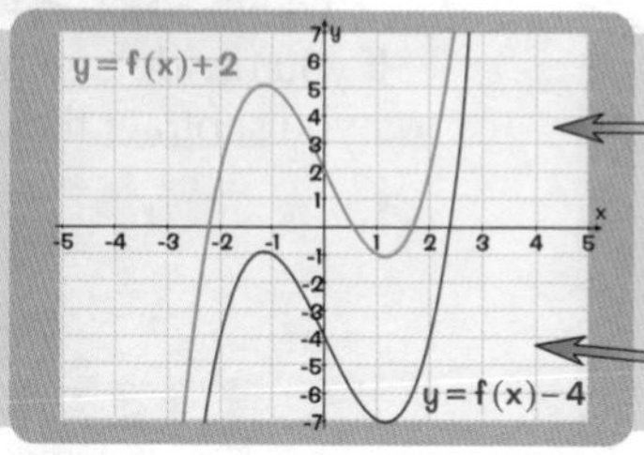

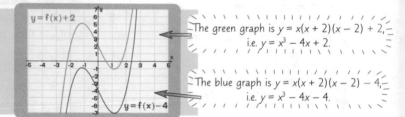

The green graph is $y = x(x + 2)(x - 2) + 2$, i.e. $y = x^3 - 4x + 2$.

The blue graph is $y = x(x + 2)(x - 2) - 4$, i.e. $y = x^3 - 4x - 4$.

$y = f(x + a)$

Writing '$x + a$' instead of 'x' means the graph moves <u>sideways</u> ("translated in the <u>x-direction</u>").

1) If $a > 0$, the graph goes to the <u>left</u>.
2) If $a < 0$, the graph goes to the <u>right</u>.

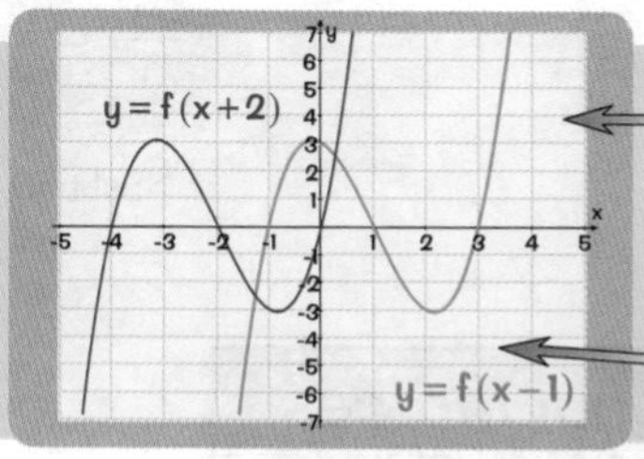

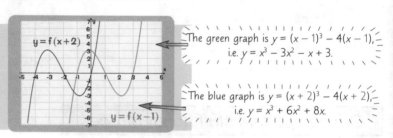

The green graph is $y = (x - 1)^3 - 4(x - 1)$, i.e. $y = x^3 - 3x^2 - x + 3$.

The blue graph is $y = (x + 2)^3 - 4(x + 2)$, i.e. $y = x^3 + 6x^2 + 8x$.

Translations are often written in the form $\begin{bmatrix} p \\ q \end{bmatrix}$, where the graph is shifted 'p' spaces horizontally and 'q' spaces vertically. If p is positive then the graph moves to the right, if it is negative then the graph moves to the left. If q is positive then the graph moves upwards, if it is negative then it moves downwards.

Stretches and *Reflections* are caused by *Multiplying* things

$y = af(x)$

<u>Multiplying</u> the <u>whole function</u> <u>stretches</u>, <u>squeezes</u> or <u>reflects</u> the graph <u>vertically</u>.

1) <u>Negative</u> values of 'a' <u>reflect</u> the basic shape in the <u>x-axis</u>.
2) If $a > 1$ or $a < -1$ (i.e. $|a| > 1$) the graph is <u>stretched vertically</u>.
3) If $-1 < a < 1$ (i.e. $|a| < 1$) the graph is <u>squashed vertically</u>.

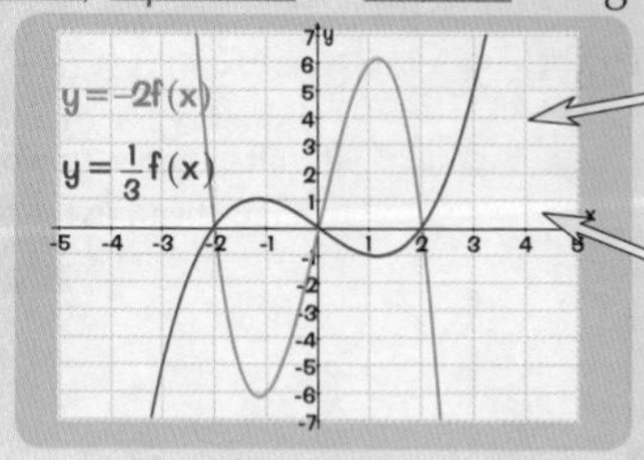

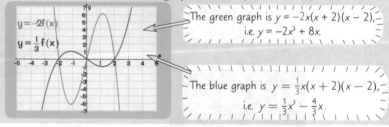

The green graph is $y = -2x(x + 2)(x - 2)$, i.e. $y = -2x^3 + 8x$.

The blue graph is $y = \frac{1}{3}x(x + 2)(x - 2)$, i.e. $y = \frac{1}{3}x^3 - \frac{4}{3}x$.

$y = f(ax)$

Writing 'ax' instead of 'x' <u>stretches</u>, <u>squeezes</u> or <u>reflects</u> the graph <u>horizontally</u>.

1) <u>Negative</u> values of 'a' <u>reflect</u> the basic shape in the <u>y-axis</u>.
2) If $a > 1$ or $a < -1$ (i.e. if $|a| > 1$) the graph is <u>squashed horizontally</u>.
3) If $-1 < a < 1$ (i.e. if $|a| < 1$) the graph is <u>stretched horizontally</u>.

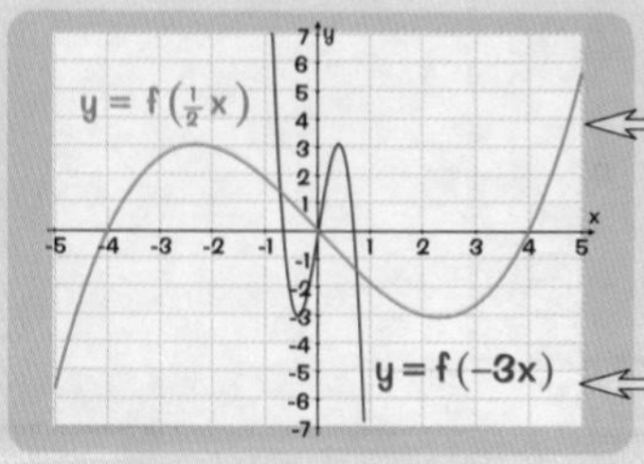

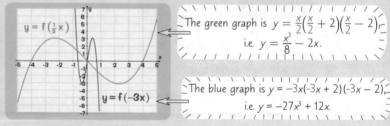

The green graph is $y = \frac{x}{2}\left(\frac{x}{2} + 2\right)\left(\frac{x}{2} - 2\right)$, i.e. $y = \frac{x^3}{8} - 2x$.

The blue graph is $y = -3x(-3x + 2)(-3x - 2)$, i.e. $y = -27x^3 + 12x$.

C2 Section 1 — Practice Questions

Well, what a lovely little section that was. And, to finish it off, here's a lovely little page of questions.

Warm-up Questions

1) Simplify these:

a) $x^3.x^5$ b) $a^7.a^8$ c) $\frac{x^8}{x^2}$ d) $(a^2)^4$ e) $(xy^2).(x^3yz)$ f) $\frac{a^2b^4c^6}{a^3b^2c}$

2) Work out the following:

a) $16^{1/2}$ b) $8^{1/3}$ c) $16^{3/4}$ d) x^0 e) $49^{-1/2}$

Two questions are never enough for anybody, so it's time to shimmy on over to the exam style ones.

Exam Questions

1 a) Write down the value of $27^{\frac{1}{3}}$. *(1 mark)*

b) Find the value of $27^{\frac{4}{3}}$. *(2 marks)*

2 a) Describe the geometrical transformation that maps the graph of $y = f(x)$ onto the graph of:

(i) $y = f\left(\frac{1}{2}x\right)$, *(1 mark)*

(ii) $y = f(x - 4)$, *(1 mark)*

(iii) $y = -f(x)$. *(1 mark)*

b) The curve $y = f(x)$ is translated vertically two units upwards. State the equation of the curve after it has been transformed, in terms of $f(x)$. *(1 mark)*

3 Given that $10000\sqrt{10} = 10^k$, find the value of k. *(3 marks)*

4 Factorise completely

$$2x^4 - 32x^2.$$

(3 marks)

5 Write

$$\frac{x + 5x^3}{\sqrt{x}}$$

in the form $x^m + 5x^n$, where m and n are constants. *(2 marks)*

Sequences

A sequence is a list of numbers that follow a certain pattern. Sequences can be finite or infinite (infinity — oooh), and they're usually generated in one of two ways. And guess what? You have to know everything about them.

A Sequence can be defined by its nth Term

You almost definitely covered this stuff at GCSE, so no excuses for mucking it up.

The point of all this is to show how you can work out any value (the nth term) from its position in the sequence (n).

EXAMPLE Find the nth term of the sequence 5, 8, 11, 14, 17, ...

1st	2nd	3rd	4th	5th
5	8	11	14	17

+3 +3 +3 +3

Each term is 3 more than the one before it. That means that you need to start by multiplying n by 3.

Take the first term (where $n = 1$). If you multiply n by 3, you still have to add 2 to get 5.

The same goes for $n = 2$. To get 8 you need to multiply n by 3, then add 2.
Every term in the sequence is worked out exactly the same way.

So n^{th} term is $3n + 2$.

You can define a sequence by a Recurrence Relation too

Don't be put off by the fancy name — recurrence relations are pretty easy really.

The main thing to remember is:
a_k just means the kth term of the sequence

The next term in the sequence is a_{k+1}. You need to describe how to work out a_{k+1} if you're given a_k.

EXAMPLE Find the recurrence relation of the sequence 5, 8, 11, 14, 17, ...

From the example above, you know that each term equals the one before it, plus 3.

This is written like this: $\mathbf{a_{k+1} = a_k + 3}$

So, if k = 5, $a_k = a_5$ which stands for the 5th term, and $a_{k+1} = a_6$ which stands for the 6th term.

In everyday language, $a_{k+1} = a_k + 3$ means that the sixth term equals the fifth term plus 3.

BUT $a_{k+1} = a_k + 3$ on its own isn't enough to describe 5, 8, 11, 14, 17, ...
For example, the sequence 87, 90, 93, 96, 99, ... also has each term being 3 more than the one before.

The description needs to be more specific, so you've got to give one term in the sequence, as well as the recurrence relation. You almost always give the first value, a_1.

Putting all of this together gives 5, 8, 11, 14, 17, ... as $a_{k+1} = a_k + 3$, $a_1 = 5$.

Sequences

Some sequences involve **Multiplying**

You've done the easy 'adding' business. Now it gets really tough — multiplying. Are you sure you're ready for this...

EXAMPLE A sequence is defined by $a_{k+1} = 2a_k - 1$, $a_2 = 5$. List the first five terms.

OK, you're told the second term, $a_2 = 5$. Just plug that value into the equation, and carry on from there.

$a_3 = 2 \times 5 - 1 = 9$ ← From the equation $a_k = a_2$ so $a_{k+1} = a_3$

$a_4 = 2 \times 9 - 1 = 17$ ← Now use a_3 to find $a_{k+1} = a_4$ and so on...

$a_5 = 2 \times 17 - 1 = 33$

Now to find the first term, a_1:

$a_2 = 2a_1 - 1$ ← Just make $a_k = a_1$

$5 = 2a_1 - 1$

$2a_1 = 6$

$a_1 = 3$

So the first five terms of the sequence are 3, 5, 9, 17, 33 .

Some Sequences have a **Certain Number** of terms — others go on **Forever**

Some sequences are only defined for a certain number of terms.

It's the $1 \leq k \leq 20$ bit that tells you it's finite.

For example, $a_{k+1} = a_k + 3$, $a_1 = 1$, $1 \leq k \leq 20$ will be 1, 4, 7, 10, ..., 58 and will contain 20 terms.
This is a finite sequence.

Other sequences don't have a specified number of terms and could go on forever.

For example, $u_{k+1} = u_k + 2$, $u_1 = 5$, will be 5, 7, 9, 11, 13, ... and won't have a final term.
This is an infinite sequence.

While others are periodic, and just revisit the same values over and over again.

For example, $u_k = u_{k-3}$, $u_1 = 1$, $u_2 = 4$, $u_3 = 2$, will be 1, 4, 2, 1, 4, 2, 1, 4, 2,...
This is a periodic sequence with period 3.

Convergent Sequences have a **Limit**

The terms of some sequences get closer and closer to a limit without ever reaching it — these sequences are called convergent sequences. If a sequence does not approach a limit, it is called a divergent sequence.
For a convergent sequence defined by a relation $a_{k+1} = f(a_k)$, the limit can be found by solving $L = f(L)$.

EXAMPLE A sequence is defined by the relation $a_{k+1} = -\frac{1}{2}a_k + 2$.
Find L, the limit of a_k as k tends to infinity.

'As k tends to infinity' just means 'as k gets really ruddy big'. You might also see it written as $k \to \infty$.

Here, $f(a_k) = -\frac{1}{2}a_k + 2$

Now take $L = f(L)$, which gives $L = -\frac{1}{2}L + 2$

Rearranging: $\frac{3}{2}L = 2 \Rightarrow L = \frac{4}{3}$ So, the limit of the sequence as k tends to infinity is $\frac{4}{3}$.

Like maths teachers, sequences can go on and on and on and on...

If you know the formula for the nth term, you can work out any term using a single formula, so it's kind of easy. If you only know a recurrence relation, then you can only work out the next term. So if you want the 20th term, and you only know the first one, then you have to use the recurrence relation 19 times. (So it'd be quicker to work out a formula really.)

Arithmetic Progressions

Right, you've got basic sequences tucked under your belt now — time to step it up a notch (sounds painful). When the terms of a sequence progress by adding a fixed amount each time, this is called an arithmetic progression.

It's all about Finding the nth Term

The first term of a sequence is given the symbol a. The amount you add each time is called the common difference, or d. The position of any term in the sequence is called n.

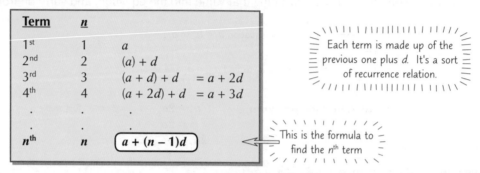

Term	n		
1st	1	a	
2nd	2	$(a) + d$	
3rd	3	$(a + d) + d$	$= a + 2d$
4th	4	$(a + 2d) + d$	$= a + 3d$
.	.	.	
.	.	.	
n^{th}	n	$a + (n - 1)d$	

Each term is made up of the previous one plus d. It's a sort of recurrence relation.

This is the formula to find the nth term

EXAMPLE Find the 20th term of the arithmetic progression 2, 5, 8, 11, ... and find the formula for the nth term.

Here $a = 2$ and $d = 3$.

To get d, just find the difference between two terms next to each other — e.g. $11 - 8 = 3$

So 20th term $= a + (20 - 1)d$
$= 2 + 19 \times 3$
$= 59$

The general term is the nth term, i.e. $a + (n - 1)d$
$= 2 + (n - 1)3$
$= 3n - 1$

A Series is when you Add the Terms to Find the Total

S_n is the total of the first n terms of the arithmetic progression:

$$S_n = a + (a + d) + (a + 2d) + (a + 3d) + \dots + (a + (n - 1)d)$$

There's a really neat version of the same formula too:

$$S_n = n \times \frac{(a + l)}{2}$$

The l stands for the last value in the progression. You work it out as $l = a + (n - 1)d$

Nobody likes formulas, so think of it as the average of the first and last terms multiplied by the number of terms.

EXAMPLE Find the sum of the series with first term 3, last term 87 and common difference 4.

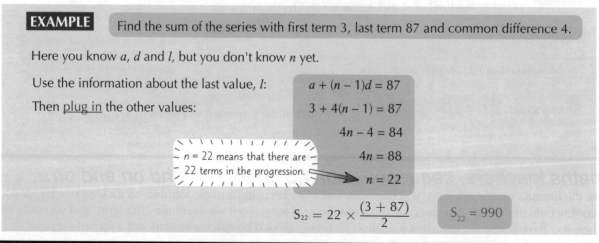

Here you know a, d and l, but you don't know n yet.

Use the information about the last value, l: $a + (n - 1)d = 87$

Then plug in the other values: $3 + 4(n - 1) = 87$

$4n - 4 = 84$

$4n = 88$

$n = 22$

$n = 22$ means that there are 22 terms in the progression.

$$S_{22} = 22 \times \frac{(3 + 87)}{2}$$

$S_{22} = 990$

Arithmetic Series and Sigma Notation

They Won't always give you the Last Term...

...but don't panic — there's a formula to use when the last term is unknown. But you knew I'd say that, didn't you?

You know $l = a + (n-1)d$ and $S_n = n \times \frac{(a+l)}{2}$.

Plug l into S_n and rearrange to get the formula in the box:

$$S_n = \frac{n}{2}[2a + (n-1)d]$$

EXAMPLE For the sequence –5, –2, 1, 4, 7, ... find the sum of the first 20 terms.

So $a = -5$ and $d = 3$.
The question says $n = 20$ too.

$$S_{20} = \frac{20}{2}[2 \times -5 + (20-1) \times 3]$$
$$= 10[-10 + 19 \times 3]$$
$$S_{20} = 470$$

There's Another way of Writing Series, too

So far, the letter S has been used for the sum. The Greeks did a lot of work on this — their capital letter for S is Σ or sigma. This is used today, together with the general term, to mean the sum of the series.

EXAMPLE Find $\sum_{n=1}^{15}(2n+3)$

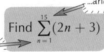

...and ending with $n = 15$

Starting with $n = 1$...

This means you have to find the sum of the first 15 terms of the series with n^{th} term $2n + 3$.

The first term ($n = 1$) is 5, the second term ($n = 2$) is 7, the third is 9, ... and the last term ($n = 15$) is 33.

In other words, you need to find 5 + 7 + 9 + ... + 33. This gives $a = 5$, $d = 2$, $n = 15$ and $l = 33$.

You know all of a, d, n and l, so you can use either formula:

$$S_n = n\frac{(a+l)}{2}$$
$$S_{15} = 15\frac{(5+33)}{2}$$
$$S_{15} = 15 \times 19$$
$$S_{15} = 285$$

It makes no difference which method you use.

$$S_n = \frac{n}{2}[2a + (n-1)d]$$
$$S_{15} = \frac{15}{2}[2 \times 5 + 14 \times 2]$$
$$S_{15} = \frac{15}{2}[10 + 28]$$
$$S_{15} = 285$$

Use Arithmetic Progressions to add up the First n Whole Numbers

The sum of the first n natural numbers looks like this:

$$S_n = 1 + 2 + 3 + ... + (n-2) + (n-1) + n$$

So $a = 1$, $l = n$ and also $n = n$.
Now just plug those values into the formula:

Natural numbers are just positive whole numbers.

$$S_n = n \times \frac{(a+l)}{2} \Longrightarrow S_n = \frac{1}{2}n(n+1)$$

EXAMPLE Add up all the whole numbers from 1 to 100.

Sounds pretty hard, but all you have to do is stick it into the formula:

$S_{100} = ½ \times 100 \times 101$. So $S_{100} = 5050$

It's pretty easy to prove this:

1) Say, $S_n = 1 + 2 + 3 + ... + (n-2) + (n-1) + n$ ①
2) ① is just addition, so it's also true that:
   $S_n = n + (n-1) + (n-2) + ... + 3 + 2 + 1$ ②
3) Add ① and ② together to get:
   $2S_n = (n+1) + (n+1) + (n+1) + ... + (n+1) + (n+1) + (n+1)$
   $\Rightarrow 2S_n = n(n+1) \Rightarrow S_n = \frac{1}{2}n(n+1)$. Voilà.

This sigma notation is all Greek to me... *(Ho ho ho)*

A sequence is just a list of numbers (with commas between them) — a series on the other hand is when you add all the terms together. It doesn't sound like a big difference, but mathematicians get all hot under the collar when you get the two mixed up. Remember that BlackADDer was a great TV series — not a TV sequence. (Sounds daft, but I bet you remember it now.)

Geometric Progressions

You have a geometric progression when each term in a sequence is found by multiplying the previous term by a (constant) number. Let me explain...

Geometric Progressions Multiply by a Constant each time

Geometric progressions work like this: the next term in the sequence is obtained by multiplying the previous one by a constant value. Couldn't be easier.

$u_1 = a \quad = a$

$u_2 = a \times r \quad = ar$

$u_3 = a \times r \times r \quad = ar^2$

$u_4 = a \times r \times r \times r \quad = ar^3$

The first term (u_1) is called 'a'.

The number you multiply by each time is called 'the common ratio', symbolised by 'r'.

Here's the formula describing any term in the geometric progression: $u_n = ar^{n-1}$

EXAMPLE

There is a chessboard with a 1p piece on the first square, 2p on the second square, 4p on the third, 8p on the fourth and so on until the board is full. Calculate how much money is on the board.

This is a geometric progression, where you get the next term in the sequence by multiplying the previous one by 2.

So $a = 1$ (because you start with 1p on the first square) and $r = 2$.

So $u_1 = 1,\ u_2 = 2,\ u_3 = 4,\ u_4 = 8...$

To be continued... (once we've gone over how to sum the terms of a geometric progression).

A Sequence becomes a Series when you Add the Terms to find the Total

S_n stands for the sum of the first n terms of the geometric progression.
In the example above, you're told to work out S_{64} (because there are 64 squares on a chessboard).

To work out the formula for the sum of a G.P. you use two series and subtract.

For a G.P.: $S_n = a + ar + ar^2 + ar^3 + ... + ar^{n-1}$

Multiplying by r gives: $rS_n = ar + ar^2 + ar^3 + ... + ar^{n-2} + ar^{n-1} + ar^n$

Subtracting gives: $S_n - rS_n = a - ar^n$

Factorising: $(1-r)S_n = a(1-r^n) \Longrightarrow S_n = \dfrac{a(1-r^n)}{1-r}$

If the series were subtracted the other way around you'd get $S_n = \dfrac{a(r^n-1)}{r-1}$. Both versions are correct.

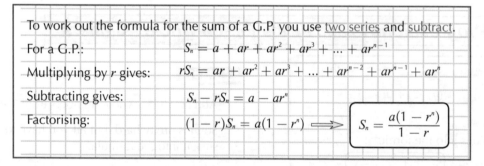

So, back to the chessboard example:

$a = 1,\ r = 2,\ n = 64 \quad S_{64} = \dfrac{1(1-2^{64})}{1-2}$

$S_{64} = 1.84 \times 10^{19}$ pence or £1.84×10^{17}

The whole is more than the sum of the parts — hmm, not in maths, it ain't...

You really need to understand the difference between arithmetic and geometric progressions — it's not hard, but it needs to be fixed firmly in your head. There are only a few formulas for sequences and series (the nth term of a sequence, the sum of the first n terms of a series), and these are in the formula book they give you — but make sure you know how to use them.

Geometric Progressions

Geometric progressions can either **Grow** or **Shrink**

In the chessboard example, each term was bigger than the previous one: 1, 2, 4, 8, 16, …
You can create a series where each term is smaller than the previous one by using a small value of r.

EXAMPLE If $a = 20$ and $r = \frac{1}{5}$, write down the first five terms of the sequence and the 20th term.

Each term is the previous one multiplied by r.

$u_1 = 20$
$u_2 = 20 \times \frac{1}{5} = 4$
$u_3 = 4 \times \frac{1}{5} = 0.8$
$u_4 = 0.8 \times \frac{1}{5} = 0.16$
$u_5 = 0.16 \times \frac{1}{5} = 0.032$

$u_{20} = ar^{19}$
$= 20 \times \left(\frac{1}{5}\right)^{19}$
$= 1.048576 \times 10^{-12}$

The sequence is tending towards zero, but won't ever get there.

In general, for each term to be smaller than the one before, you need $|r| < 1$.
A sequence with $|r| < 1$ is called convergent, since the terms converge to a limit.
Any other sequence (like the chessboard example on page 64) is called divergent.

$|r|$ means the modulus (or size) of r, ignoring the sign of the number. So $|r| < 1$ means that $-1 < r < 1$.

A **Convergent Series** has a **Sum to Infinity**

In other words, if you just kept adding terms to a convergent series, you'd get closer and closer to a certain number, but you'd never actually reach it.

If $|r| < 1$ and n is very, very big, then r^n will be very, very small — or to put it technically, $r^n \to 0$. (Try working out $(½)^{100}$ on your calculator if you don't believe me.)

This means $(1 - r^n)$ is really, really close to 1.

So, as $n \to \infty$, $S_n \to \frac{a}{1-r}$.

It's easier to remember as $S_\infty = \frac{a}{1-r}$

S_∞ just means 'sum to infinity'.

EXAMPLE If $a = 2$ and $r = ½$, find the sum to infinity of the geometric series.

$u_1 = 2$	⟹	$S_1 = 2$
$u_2 = 2 \times \frac{1}{2} = 1$	⟹	$S_2 = 2 + 1 = 3$
$u_3 = 1 \times \frac{1}{2} = \frac{1}{2}$	⟹	$S_3 = 2 + 1 + \frac{1}{2} = 3\frac{1}{2}$
$u_4 = \frac{1}{2} \times \frac{1}{2} = \frac{1}{4}$	⟹	$S_4 = 2 + 1 + \frac{1}{2} + \frac{1}{4} = 3\frac{3}{4}$
$u_5 = \frac{1}{4} \times \frac{1}{2} = \frac{1}{8}$	⟹	$S_5 = 2 + 1 + \frac{1}{2} + \frac{1}{4} + \frac{1}{8} = 3\frac{7}{8}$
$u_6 = \frac{1}{8} \times \frac{1}{2} = \frac{1}{16}$	⟹	$S_6 = 2 + 1 + \frac{1}{2} + \frac{1}{4} + \frac{1}{8} + \frac{1}{16} = 3\frac{15}{16}$

These values are getting smaller each time.

These values are getting closer (converging) to 4.

So, the sum to infinity is 4.

You can show this graphically:

The line on the graph is getting closer and closer to 4, but it'll never actually get there.

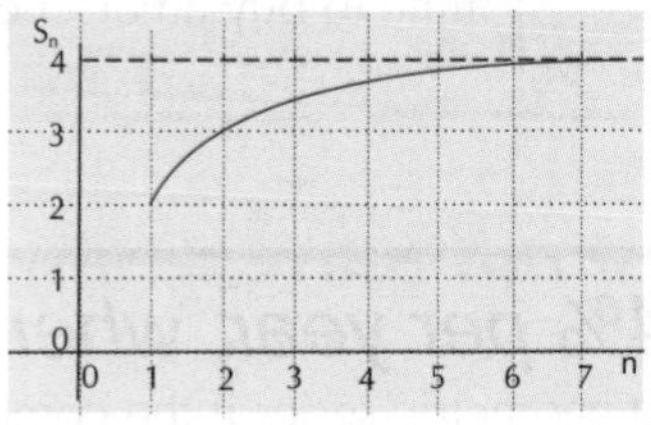

Of course, you could have saved yourself a lot of bother by using the sum to infinity formula:

$$S_\infty = \frac{a}{1-r} = \frac{2}{1-\frac{1}{2}} = 4$$

Geometric Progressions

A **Divergent** series **Doesn't** have a sum to infinity

EXAMPLE If $a = 2$ and $r = 2$, find the sum to infinity of the series.

$u_1 = 2 \Longrightarrow S_1 = 2$

$u_2 = 2 \times 2 = 4 \Longrightarrow S_2 = 2 + 4 = 6$

$u_3 = 4 \times 2 = 8 \Longrightarrow S_3 = 2 + 4 + 8 = 14$

$u_4 = 8 \times 2 = 16 \Longrightarrow S_4 = 2 + 4 + 8 + 16 = 30$

$u_5 = 16 \times 2 = 32 \Longrightarrow S_5 = 2 + 4 + 8 + 16 + 32 = 62$

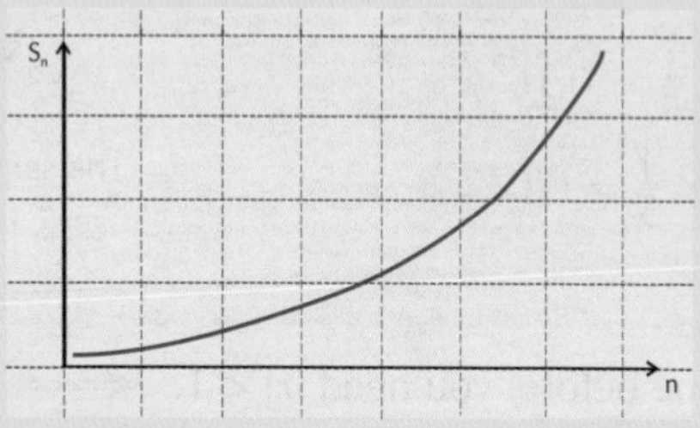

This is an exponential graph — see C2 Section 4.

As $n \to \infty$, $S_n \to \infty$ in a big way. So big, in fact, that eventually you can't work it out — so don't bother.

There is no sum to infinity for a divergent series.

EXAMPLE When a baby is born, £3000 is invested in an account with a fixed interest rate of 4% per year.

a) What will the account be worth at the start of the seventh year?

b) What age will the child be, to the nearest year, when the account has doubled in value?

a) $u_1 = a = 3000$

$u_2 = 3000 + (4\% \text{ of } 3000)$ ← This is the interest.

$= 3000 + (0.04 \times 3000)$

$= 3000(1 + 0.04)$

$= 3000 \times 1.04$ ← So, r = 1.04

$u_3 = u_2 \times 1.04$

$= (3000 \times 1.04) \times 1.04$

$= 3000 \times (1.04)^2$

$u_4 = 3000 \times (1.04)^3$

. . . ← I've missed out some steps here — check that you understand what's happened.

. . .

$u_7 = 3000 \times (1.04)^6$

$= £3795.96$ (to the nearest penny)

b) You need to know when $u_n > 6000$ ← double the original value.

From part a) you can tell that $u_n = 3000 \times (1.04)^{n-1}$

So $3000 \times (1.04)^{n-1} > 6000$

$(1.04)^{n-1} > 2$

To complete this you need to use logs (see C2 Section 4):

$\log(1.04)^{n-1} > \log 2$

$(n - 1)\log(1.04) > \log 2$

$n - 1 > \dfrac{\log 2}{\log 1.04}$

$n - 1 > 17.67$

$n > 18.67$ (to 2 d.p.)

So u_{19} (the amount at the start of the 19th year) will be more than double the original amount — plenty of time to buy a Porsche for the 21st birthday.

So tell me — if my savings earn 4% per year, when will I be rich...

Now here's a funny thing — you can have a convergent geometric series if the common ratio is small enough. I find this odd — that I can keep adding things to a sum forever, but the sum never gets really really big.

Sequence and Series Problems

Now that you've got your toolbox of formulas:

Arithmetic Progressions

$u_n = a + (n-1)d$

$S_n = \frac{n}{2}[2a + (n-1)d]$

Geometric Progressions

$u_n = ar^{n-1}$

$S_n = \frac{a(1-r^n)}{1-r}$

$S_\infty = \frac{a}{1-r}$

It's time to mix things up a bit and take on the big boys — some serious series problems...

*You might need to use **Simultaneous Equations** to solve some problems...*

EXAMPLE The sum of the first five terms of an arithmetic progression is 5, and the sixth term is 13. Find the eighth term in the progression.

The question gives you a particular term and a sum of terms, so use the formulas $u_n = a + (n-1)d$ and $S_n = \frac{n}{2}[2a + (n-1)d]$.

Plugging the values given in the question gives: $u_6 = a + (6-1)d = 13$ and $S_5 = 5 = \frac{5}{2}[2a + (5-1)d]$

Which simplify to: $a + 5d = 13$, call this ①
$a + 2d = 1$, call this ②

Solve the simultaneous equations: ① – ② $\Rightarrow$ $3d = 12$ $\Rightarrow$ $d = 4$

$d = 4$ in ② $\Rightarrow$ $a = 1 - 8 = -7$

Finally, use $a + (n-1)d$ to find the 8th term: $u_8 = a + (n-1)d = -7 + (7 \times 4) = 21$

*... and **Quadratic Equations** to solve others*

EXAMPLE The sum to infinity of a geometric series is –32. The second term of the progression is –8. Find the common ratio of the progression.

Use the formula for the *n*th term of a geometric progression: $u_n = ar^{n-1}$

and the sum to infinity of a geometric series: $S_\infty = \frac{a}{1-r}$

Plugging in the information from the question gives: $u_2 = ar = -8$, call this ①

$S_\infty = -32 = \frac{a}{1-r}$, call this ②

Rearrange ① into the form $a = -\frac{8}{r}$ in order to eliminate a from ②:

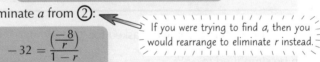

If you were trying to find a, then you would rearrange to eliminate r instead.

$$-32 = \frac{\left(\frac{-8}{r}\right)}{1-r}$$

$$-32(1-r) = \frac{(-8)}{r}$$

$$-32r + 32r^2 = -8$$

$$4r^2 - 4r + 1 = 0$$

$$(2r-1)^2 = 0$$

$$\Rightarrow 2r = 1, \quad r = \frac{1}{2}$$

This quadratic only has one root. If you get a quadratic with two roots, pick the one for which $|r| < 1$, as a series with a sum to infinity must be convergent.

At last – the series finalé...

Make sure you know which formulas apply to arithmetic and which to geometric progressions. You might have to use different combinations of them in the exam to get your simultaneous equations. Just look at what you are given in the question and what you are being asked to do, and use the corresponding formulas. That's all there is to it really.

Binomial Expansions

If you're feeling a bit stressed, just take a couple of minutes to relax before trying to get your head round this page — it's a bit of a stinker in places. Have a cup of tea and think about something else for a couple of minutes. Ready...

Writing **Binomial Expansions** *is all about* **Spotting Patterns**

Doing binomial expansions just involves multiplying out brackets. It would get nasty when you raise the brackets to higher powers — but once again I've got a cunning plan...

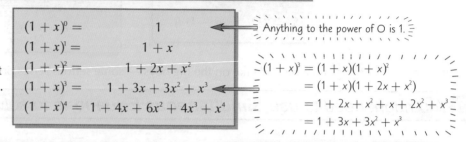

$$(1+x)^0 = 1$$
$$(1+x)^1 = 1 + x$$
$$(1+x)^2 = 1 + 2x + x^2$$
$$(1+x)^3 = 1 + 3x + 3x^2 + x^3$$
$$(1+x)^4 = 1 + 4x + 6x^2 + 4x^3 + x^4$$

Anything to the power of 0 is 1.

$$\begin{aligned}(1+x)^3 &= (1+x)(1+x)^2\\ &= (1+x)(1+2x+x^2)\\ &= 1+2x+x^2+x+2x^2+x^3\\ &= 1+3x+3x^2+x^3\end{aligned}$$

A Frenchman named Pascal spotted the pattern in the coefficients and wrote them down in a triangle.
So it was called 'Pascal's Triangle' (imaginative, eh?).
The pattern's easy — each number is the sum of the two above it.

So, the next line will be: **1 5 10 10 5 1**
giving $\mathbf{(1+x)^5 = 1 + 5x + 10x^2 + 10x^3 + 5x^4 + x^5}$.

1
1 1
1 2 1
1 3 3 + 1
1 4 6 = 4 1

You **Don't** *need to write out Pascal's Triangle for* **Higher Powers**

There's a formula for the numbers in the triangle. The formula looks horrible (one of the worst in AS maths) so don't try to learn it letter by letter — look for the patterns in it instead. Here's an example:

EXAMPLE Expand $(1+x)^{20}$, giving the first four terms only.

So you can use this formula for any power, the power is called n. In this example $n = 20$.

$$(1+x)^n = 1 + \frac{n}{1}x + \frac{n(n-1)}{1\times 2}x^2 + \boxed{\frac{n(n-1)(n-2)}{1\times 2\times 3}x^3} + \ldots\ldots\ldots\ldots + x^n$$

Here's a closer look at the term in the black box:

$$\frac{n(n-1)(n-2)}{1\times 2\times 3}x^3$$

There are three things multiplied together on the top row. If n=20, this would be 20×19×18.

Start here. The power of x is 3 and everything else here is based on 3.

There are three integers here multiplied together. 1×2×3 is written as 3! and called 3 factorial.

This means, if $n = 20$ and you were asked for 'the term in x^7' you should write $\frac{20\times 19\times 18\times 17\times 16\times 15\times 14}{1\times 2\times 3\times 4\times 5\times 6\times 7}x^7$.

This can be simplified to $\frac{20!}{7!13!}x^7$

$20\times 19\times 18\times 17\times 16\times 15\times 14 = \frac{20!}{13!}$ because it's the numbers from 20 to 1 multiplied together, divided by the numbers from 13 to 1 multiplied together.

Believe it or not, there's an even shorter form: $\frac{20!}{7!13!}$ is written as ${}^{20}C_7$ or $\binom{20}{7}$

$${}^nC_r = \binom{n}{r} = \frac{n!}{r!(n-r)!}$$

Your calculator will probably have an nCr button for working this out. To calculate $\binom{20}{7}$ you'd put in 20 [nCr] 7 =.

So, to finish the example, $(1+x)^{20} = 1 + \frac{20}{1}x + \frac{20\times 19}{1\times 2}x^2 + \frac{20\times 19\times 18}{1\times 2\times 3}x^3 + \ldots = 1 + 20x + 190x^2 + 1140x^3 + \ldots$

Binomial Expansions

It's slightly more complicated when the **Coefficient** of x isn't 1

EXAMPLE What is the term in x^5 in the expansion of $(1 - 3x)^{12}$?

The term in x^5 will be as follows:

$$\frac{12 \times 11 \times 10 \times 9 \times 8}{1 \times 2 \times 3 \times 4 \times 5}(-3x)^5$$

$$= \frac{12!}{5!7!}(-3)^5x^5 = -\frac{12!}{5!7!} \times 3^5x^5 = -192456x^5$$

Watch out — the –3 is included here with the x.

Tip — the digits on the bottom of the fraction should always add up to the number on the top.

Note that $(-3)^{\text{even}}$ will always be positive and $(-3)^{\text{odd}}$ will always be negative.

Some **Binomials** contain **More Complicated Expressions**

The binomials so far have all had a 1 in the brackets — things get tricky when there's a number other than 1. Don't panic, though. The method is the same as before once you've done a bit of factorising.

EXAMPLE What is the coefficient of x^4 in the expansion of $(2 + 5x)^7$?

Factorising $(2 + 5x)$ gives $2\left(1 + \frac{5}{2}x\right)$

So, $(2 + 5x)^7$ gives $2^7\left(1 + \frac{5}{2}x\right)^7$

It's really easy to forget the first bit (here it's 2^7) — you've been warned...

$$(2 + 5x)^7 = 2^7\left(1 + \frac{5}{2}x\right)^7$$

$$= 2^7\left[1 + 7\left(\frac{5}{2}x\right) + \frac{7 \times 6}{1 \times 2}\left(\frac{5}{2}x\right)^2 + \frac{7 \times 6 \times 5}{1 \times 2 \times 3}\left(\frac{5}{2}x\right)^3 + \frac{7 \times 6 \times 5 \times 4}{1 \times 2 \times 3 \times 4}\left(\frac{5}{2}x\right)^4 + \ldots\right]$$

Here's the one you want.

The coefficient of x^4 will be $2^7 \times \frac{7!}{4!3!}\left(\frac{5}{2}\right)^4 = 175000$

Don't forget the 2^7.

So, there's no need to work out all of the terms.
In fact, you could have gone directly to the term in x^4 by using the method on page 68.

Note: the question asked for the coefficient of x^4 in the expansion, so don't include any x's in your answer. If you'd been asked for the term in x^4 in the expansion, then you should have included the x^4 in your answer.

Always read the question very carefully.

Binomial Expansions

Fed up of binomials yet? Good. Me neither.

Some **Binomials** are **More Complicated** still

Sometimes you'll see a binomial with a <u>power</u> of x inside the bracket, or even with <u>two brackets</u> together. Cripes.

EXAMPLE Expand $\left(1+\frac{2}{x^2}\right)^4$.

This is the same method as the first example on the previous page, just, you know, worse.

The trick here is to break the problem down into stages.
First, expand $(1+x)^4$, using the coefficients from Pascal's Triangle, but write $\left(\frac{2}{x^2}\right)$ in place of every x...

$$\left(1+\frac{2}{x^2}\right)^4 = 1+4\left(\frac{2}{x^2}\right)+6\left(\frac{2}{x^2}\right)^2+4\left(\frac{2}{x^2}\right)^3+\left(\frac{2}{x^2}\right)^4$$

...then deal with the powers of x and tidy it up:

$$= 1+4\left(\frac{2}{x^2}\right)+6\left(\frac{4}{x^4}\right)+4\left(\frac{8}{x^6}\right)+\left(\frac{16}{x^8}\right) = 1+\frac{8}{x^2}+\frac{24}{x^4}+\frac{32}{x^6}+\frac{16}{x^8}$$

EXAMPLE Find the coefficient of x^3 in the expansion of $(1+x)(1+2x)^4$.

First, find the expansion of $(1+2x)^4$ using the method above:

$$(1+2x)^4 = 1+4(2x)+6(2x)^2+4(2x)^3+(2x)^4$$
$$= 1+8x+24x^2+32x^3+16x^4$$

Then instead of multiplying the whole expansion by $(1+x)$, just look at the bits which will multiply together to give a term with x^3 in, i.e.

$$(1\times 32x^3)+(x\times 24x^2) = 32x^3+24x^3 = 56x^3$$

So the coefficient of x^3 in this expansion is 56.

Sometimes you'll just have to use the **Formula**

All of the binomials so far have been of the form $(c+dx)^n$, where c and d are constants. Things won't always be so nice though — you might be asked to expand something of the form $(a+b)^n$, where a and b could be <u>anything at all</u> (even functions of x). Fortunately, there's a <u>formula</u> to help you do it:

$$(a+b)^n = a^n+\binom{n}{1}a^{n-1}b+\binom{n}{2}a^{n-2}b^2+\dots+\binom{n}{r}a^{n-r}b^r+\dots+b^n$$

Don't worry, you don't need to memorise this formula — you'll be given it in the exam.

I know, it's an absolute shocker, but it will make questions a lot easier if you can get your head around it. An example will help:

EXAMPLE Expand and simplify $\left(2x+\frac{1}{x}\right)^5$.

Comparing to the formula, $a=2x$, $b=\frac{1}{x}$ and $n=5$.
So use the formula with these values in place of a and b:

Remember $^nC_r = \binom{n}{r} = \frac{n!}{r!(n-r)!}$

$$\left(2x+\frac{1}{x}\right)^5 = (2x)^5+\binom{5}{1}(2x)^4\left(\frac{1}{x}\right)+\binom{5}{2}(2x)^3\left(\frac{1}{x}\right)^2+\binom{5}{3}(2x)^2\left(\frac{1}{x}\right)^3+\binom{5}{4}(2x)\left(\frac{1}{x}\right)^4+\left(\frac{1}{x}\right)^5$$

$$= 32x^5+5(16x^4)\left(\frac{1}{x}\right)+10(8x^3)\left(\frac{1}{x^2}\right)+10(4x^2)\left(\frac{1}{x^3}\right)+5(2x)\left(\frac{1}{x^4}\right)+\frac{1}{x^5}$$

$$= 32x^5+80x^3+80x+\frac{40}{x}+\frac{10}{x^3}+\frac{1}{x^5}$$

I never pay for food — I use the buy-no-meal formula...

That ^nC_r button is pretty bloomin' useful (but remember that it could be called something else on your calculator) — it saves a lot of button pressing and errors. Just make sure you put your numbers in the right way round. But that's all you need to know on binomial expansion, and on sequences and series in general — so it was reasonably short, if not exactly sweet.

C2 Section 2 — Practice Questions

Well there's not a whole heap to learn in this section, but it's still important you work through these revision questions just to make sure you're as happy as a ferret in a trouser shop with Sequences and Series. Dead easy marks to pick up in the exam — so you'd feel pretty daft if you didn't learn this stuff. Come on then, get practising...

Warm-up Questions

1) Find the nth term for the following sequences:

a) 2, 6, 10, 14, ... b) 0.2, 0.7, 1.2, 1.7, ... c) 21, 18, 15, 12, ... d) 76, 70, 64, 58, ...

2) Describe the sequence 32, 37, 42, 47, ... using a recurrence relation.

3) Find the last term of the sequence that starts with 3, has a common difference of 0.5 and has 25 terms.

4) Find the common difference in a sequence that starts with –2, ends with 19 and has 29 terms.

5) Find the sum of the series that starts with 7, ends with 35 and has 8 terms.

6) Find the sum of the series that begins with 5, 8, ... and ends with 65.

7) A series has first term 7 and 5th term 23. Find:

a) the common difference, b) the 15th term, c) the sum of the first 10 terms.

8) A series has seventh term 36 and tenth term 30. Find the sum of the first five terms and the nth term.

9) Find: a) $\sum_{n=1}^{20}(3n - 1)$, b) $\sum_{n=1}^{10}(48 - 5n)$.

10) For the geometric progression 2, –6, 18, ..., find:

a) the 10th term, b) the sum of the first 10 terms.

11) Find the sum of the first 12 terms of the following geometric series:

a) 2 + 8 + 32 + ... b) 30 + 15 + 7.5 + ...

12) Find the common ratio for the following geometric series.
State which ones are convergent and which are divergent.

a) 1 + 2 + 4 +... b) 81 + 27 + 9 + ... c) $1 + \frac{1}{3} + \frac{1}{9} + ...$ d) $4 + 1 + \frac{1}{4} + ...$

13) For the geometric progression 24, 12, 6, ..., find:

a) the common ratio, b) the seventh term,

c) the sum of the first 10 terms, d) the sum to infinity.

14) A geometric progression begins 2, 6, ... Which term of the geometric progression equals 1458?

15) Write down the sixth row of Pascal's triangle. (Hint: it starts with a '1'.)

16) Give the first four terms in the expansion of $(1 + x)^{12}$.

17) What is the term in x^4 in the expansion of $(1 - 2x)^{16}$?

18) Find the coefficient of x^2 in the expansion of $(2 + 3x)^5$.

C2 Section 2 — Practice Questions

I like sequences, they're shiny and colourful and sparkly and... wait, what? ... Oh. Fiddlesticks.

Exam Questions

1 A sequence is defined by the recurrence relation: $h_{n+1} = 2h_n + 2$ when $n \geq 1$.

a) Given that $h_1 = 5$, find the values of h_2, h_3, and h_4. *(3 marks)*

b) Calculate the value of $\sum_{r=3}^{6} h_r$. *(3 marks)*

2 A sequence $a_1, a_2, a_3, \ldots$ is defined by $a_1 = k$, $a_{n+1} = 3a_n + 11$, $n \geq 1$, where k is a constant.

a) Show that $a_4 = 27k + 143$. *(3 marks)*

b) Find the value of k, given that $\sum_{r=1}^{4} a_r = 278$. *(3 marks)*

3 a) Find the binomial expansion of $\left(\frac{1}{2x} + \frac{x}{2}\right)^3$. Simplify your answer. *(3 marks)*

b) Hence find the coefficient of x in the expansion of $(2 + x^2)\left(\frac{1}{2x} + \frac{x}{2}\right)^3$. *(3 marks)*

4 A geometric series has the first term 12 and is defined by: $u_{n+1} = 12 \times 1.3^n$.

a) Is the series convergent or divergent? *(1 mark)*

b) Find the values of the 3rd and 10th terms. *(2 marks)*

5 Find the coefficients of x, x^2, x^3 and x^4 in the binomial expansion of $(4 + 3x)^{10}$. *(4 marks)*

6 A geometric series has second term 250 and fourth term 10.

a) Find the two possible values of r. *(4 marks)*

b) You are given that $r > 0$. Find:

(i) the first term of the series, *(1 mark)*

(ii) the eighth term of the series, *(2 marks)*

(iii) the sum of the first eight terms of the series.
Give your answer correct to one decimal place. *(2 marks)*

C2 Section 2 — Practice Questions

7 In a geometric series, $a = 20$ and $r = \frac{3}{4}$.

Find values for the following, giving your answers to 3 significant figures where necessary:

a) S_∞ *(2 marks)*

b) u_{15} *(2 marks)*

8 u_n is the nth term of the sequence defined by $u_{n+1} = au_n + b$, where a and b are constants.

The first three terms of the sequence are $u_1 = 10$, $u_2 = -1$ and $u_3 = \frac{6}{5}$. Find:

a) a and b. *(5 marks)*

b) the value of u_5. *(2 marks)*

c) L, the limit of u_n as n tends to infinity. *(2 marks)*

9 The first term of an arithmetic sequence is 22 and the common difference is –1.1.

a) Find the value of the 31st term. *(2 marks)*

b) If the kth term of the sequence is 0, find k. *(2 marks)*

c) The sum to n terms of the sequence is S_n.
Find the value of n at which S_n first becomes negative. *(4 marks)*

10 Two different geometric series have the same second term and sum to infinity:

$$u_2 = 5 \text{ and } S_\infty = 36.$$

a) Show that $36r^2 - 36r + 5 = 0$, where r represents the two possible ratios. *(4 marks)*

b) Hence find the values of r, and the corresponding first terms, for both geometric series. *(4 marks)*

11 a) Use binomial expansion to express $(1 + 3x)^5$ in the form $1 + ax + bx^2 + cx^3 + dx^4 + ex^5$, where a, b, c, d and e are constants. *(6 marks)*

b) Hence show that the coefficient of the x^2 term in the expansion of $(1 + 3x)^5(1 + x)$ is 105. *(3 marks)*

The Trig Formulas You Need to Know

There are some trig formulas you need to know for the exam.
So here they are — learn them or you're seriously stuffed. Worse than an aubergine.

The **Sine Rule** and **Cosine Rule** work for **Any** triangle

Remember, these three formulas work for ANY triangle, not just right-angled ones.

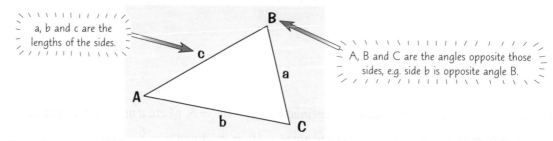

a, b and c are the lengths of the sides.

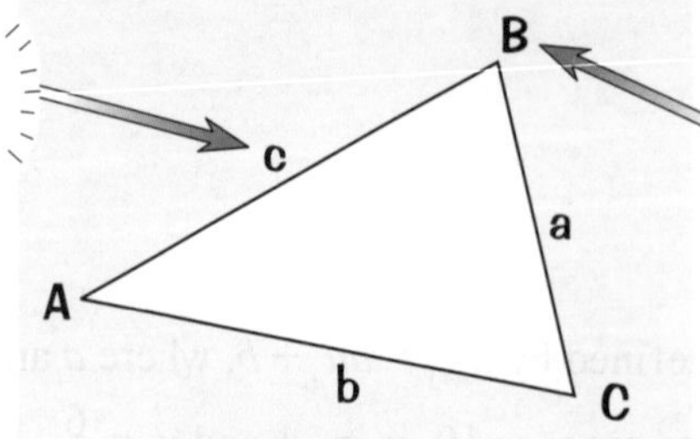

A, B and C are the angles opposite those sides, e.g. side b is opposite angle B.

THE SINE RULE

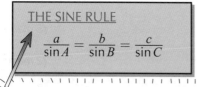

$$\frac{a}{\sin A} = \frac{b}{\sin B} = \frac{c}{\sin C}$$

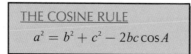

THE COSINE RULE

$$a^2 = b^2 + c^2 - 2bc\cos A$$

AREA OF ANY TRIANGLE

$$Area = \tfrac{1}{2}ab\sin C$$

Don't forget, this also works the other way up: $\frac{\sin A}{a} = \frac{\sin B}{b} = \frac{\sin C}{c}$

Sine Rule or **Cosine Rule** — which one is it...

To decide which of these two rules you need to use, look at what you already know about the triangle.

Sine Rule

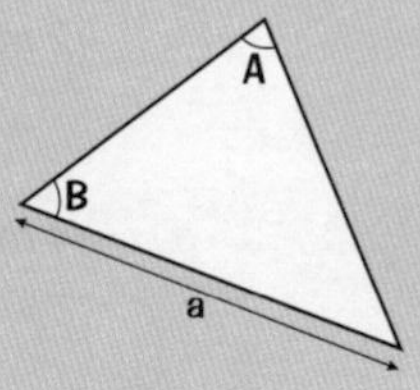

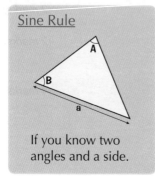

If you know two angles and a side.

Cosine Rule

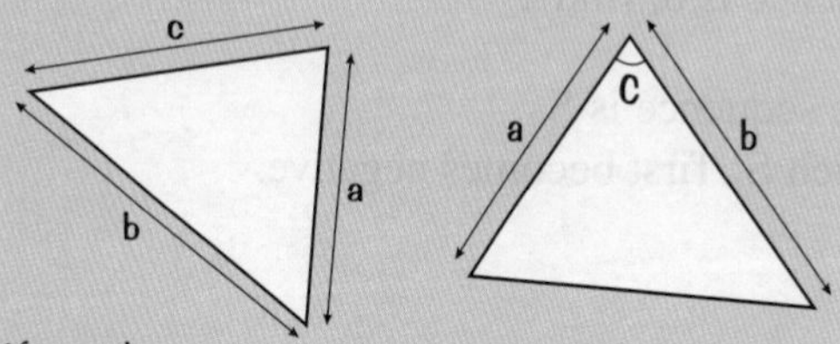

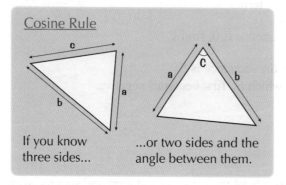

If you know three sides... ...or two sides and the angle between them.

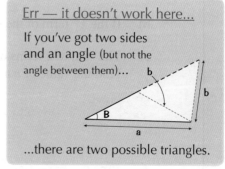

Err — it doesn't work here...

If you've got two sides and an angle (but not the angle between them)...

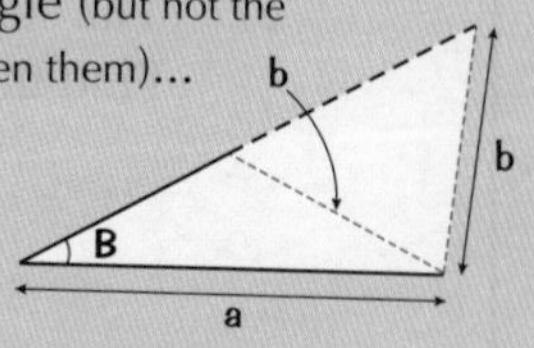

...there are two possible triangles.

The **Best** has been saved till last...

These two identities are really important. You'll need them loads.

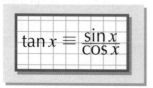

$$\tan x \equiv \frac{\sin x}{\cos x}$$

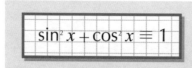

$$\sin^2 x + \cos^2 x \equiv 1$$

$$\Rightarrow \sin^2 x \equiv 1 - \cos^2 x$$
$$\cos^2 x \equiv 1 - \sin^2 x$$

Work out these two using $\sin^2 x + \cos^2 x \equiv 1$.

These two come up in exam questions all the time. Learn them.
Learnthemlearnthemlearnthemlearnthemlearnthemlear... okay, I'll stop now.

Tri angles — go on... you might like them.

Formulas and trigonometry go together even better than Richard and Judy. I can count 7 formulas on this page. That's not many, so please, just make sure you know them. If you haven't learnt them I will cry for you. I will sob.

Using the Sine and Cosine Rules

This page is about "solving" triangles, which just means finding all their sides and angles when you already know a few.

EXAMPLE Solve ΔABC, in which A = 40°, a = 27 m, B = 73°. Then find the area.

Draw a quick sketch first — don't worry if it's not deadly accurate, though.

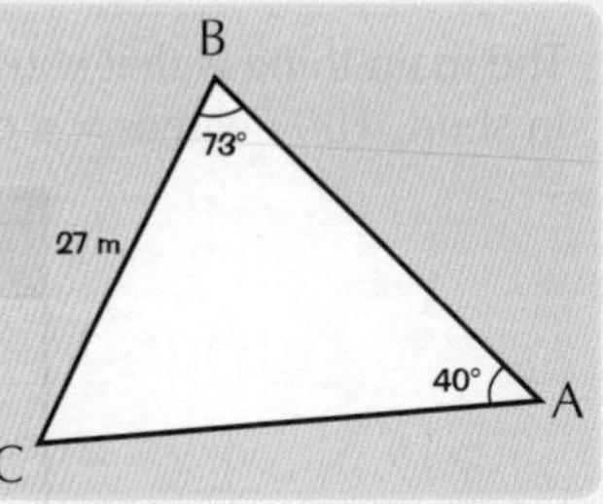

You're given 2 angles and a side, so you need the Sine Rule.

Make sure you put side a opposite angle A.

First of all, get the other angle: $\angle C = (180 - 40 - 73)° = 67°$

Then find the other sides, one at a time:

$$\frac{a}{\sin A} = \frac{b}{\sin B} \Rightarrow \frac{27}{\sin 40°} = \frac{b}{\sin 73°}$$
$$\Rightarrow b = \frac{\sin 73°}{\sin 40°} \times 27 = \underline{40.2\text{ m}}$$

$$\frac{c}{\sin C} = \frac{a}{\sin A} \Rightarrow \frac{c}{\sin 67°} = \frac{27}{\sin 40°}$$
$$\Rightarrow c = \frac{\sin 67°}{\sin 40°} \times 27 = \underline{38.7\text{ m}}$$

Now just use the formula to find its area.

$$\text{Area of } \Delta\text{ABC} = \tfrac{1}{2}ab\sin C$$
$$= \tfrac{1}{2} \times 27 \times 40.169 \times \sin 67°$$
$$= \underline{499.2\text{ m}^2}$$

Use a more accurate value for b here, rather than the rounded value 40.2.

EXAMPLE Find X, Y and z.

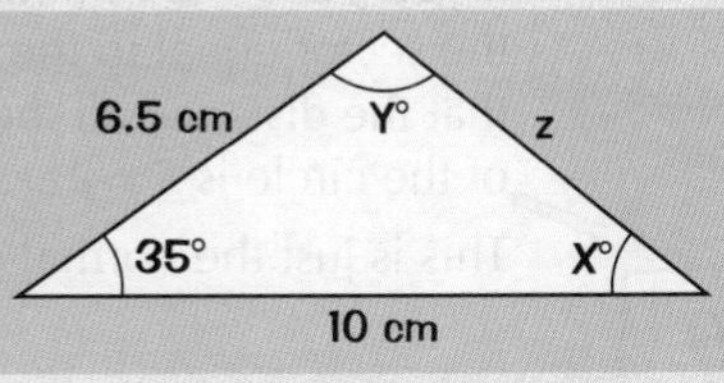

You've been given 2 sides and the angle between them, so you're first going to need the Cosine Rule to find side z.

$$a^2 = b^2 + c^2 - 2bc\cos A$$

$$z^2 = (6.5)^2 + 10^2 - 2(6.5)(10)\cos 35°$$
$$\Rightarrow z^2 = 142.25 - 130\cos 35°$$
$$\Rightarrow z^2 = 35.7602$$
$$\Rightarrow z = \underline{5.98\text{ cm}}$$

In this case, angle A is 35°, and side a is actually z.

Now that you've got all the sides and one angle, you can use the Sine Rule to find the other two angles.

$$\frac{a}{\sin A} = \frac{b}{\sin B} = \frac{c}{\sin C}$$

Remember — if $\frac{a}{\sin A} = \frac{b}{\sin B}$ then $\frac{\sin A}{a} = \frac{\sin B}{b}$

$$\frac{\sin X}{6.5} = \frac{\sin 35°}{5.9800}$$
$$\Rightarrow \sin X = 0.6235$$
$$\Rightarrow X = \sin^{-1} 0.6235$$
$$\Rightarrow X = \underline{38.6°}$$

$$\frac{\sin Y}{10} = \frac{\sin 35°}{5.9800}$$
$$\Rightarrow \sin Y = 0.9592$$
$$\Rightarrow Y = \sin^{-1} 0.9592$$
$$\Rightarrow Y = 73.6° \text{ or } \underline{106.4°}$$

This is the answer you need. Be careful: your calculator only gives you values for $\sin^{-1}$ between −90° and 90°. See page 80.

Check your answers by adding up all the angles in the triangle. If they don't add up to 180°, you've gone wrong somewhere.

Arc Length and Sector Area

Arc lengths and sector areas are easier than you'd think — once you've learnt two simple(ish) formulas.

Always work in **Radians** for **Arc Length** and **Sector Area Questions**

Remember — for arc length and sector area questions you've got to measure all the angles in radians.

The main thing is that you know how radians relate to degrees.
In short, 180 degrees = π radians. The table below shows you how to convert between the two units:

Converting angles	
Radians to degrees:	Degrees to radians:
Divide by π, multiply by 180.	Divide by 180, multiply by π.

Here's a table of some of the common angles you're going to need — in degrees and radians:

Degrees	0	30	45	60	90	120	180	270	360
Radians	0	$\frac{\pi}{6}$	$\frac{\pi}{4}$	$\frac{\pi}{3}$	$\frac{\pi}{2}$	$\frac{2\pi}{3}$	π	$\frac{3\pi}{2}$	2π

If you have part of a circle (like a section of pie chart), you can work out the length of the curved side, or the area of the 'slice of pie' — as long as you know the angle at the centre (θ) and the length of the radius (r). Read on...

You can find the **Length** of an **Arc** using a nice easy formula...

For a circle with a radius of r, where the angle θ is measured in radians, the arc length of the sector S is given by:

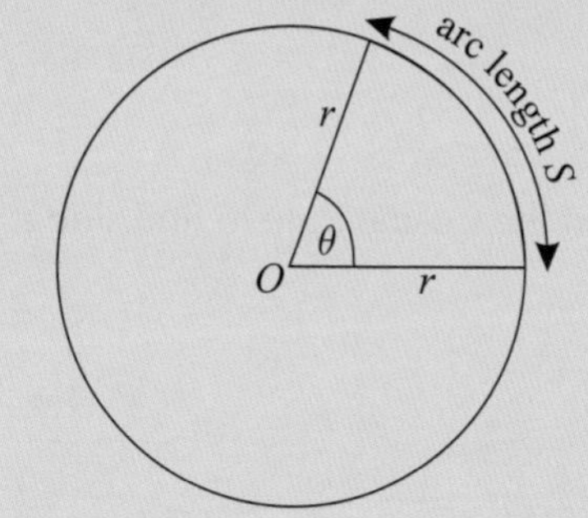

$$S = r\theta$$

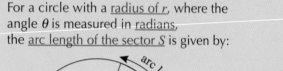

If you put $\theta = 2\pi$ in this formula (and so make the sector equal to the whole circle), you get that the distance all the way round the outside of the circle is $S = 2\pi r$.

This is just the normal circumference formula.

...and the area of a **Sector** using a similar formula

For a circle with a radius of r, where the angle θ is measured in radians, you can work out A, the area of the sector, using:

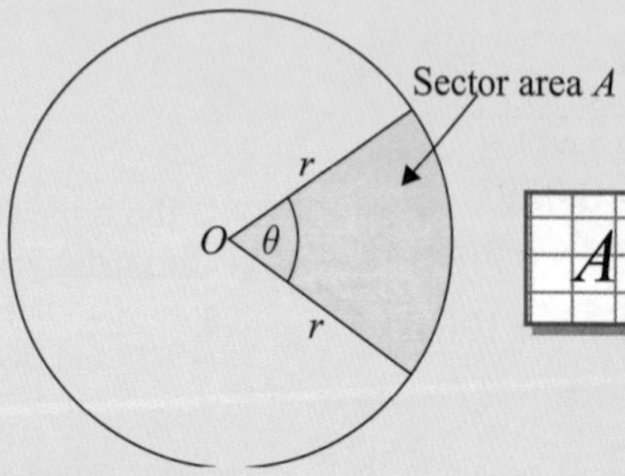

$$A = \frac{1}{2}r^2\theta$$

Again, if you put $\theta = 2\pi$ in the formula, you find that the area of the whole circle is $A = \frac{1}{2}r^2 \times 2\pi = \pi r^2$.

This is just the normal 'area of a circle' formula.

Arc Length and Sector Area

Questions on trigonometry quite often use the same angles — so it makes life easier if you know the sin, cos and tan of these commonly used angles. Or to put it another way, examiners expect you to know them — so learn them.

Draw Triangles to remember sin, cos and tan of the Important Angles

You should know the values of sin, cos and tan at 30°, 60° and 45°. But to help you remember, you can draw these two groovy triangles. It may seem a complicated way to learn a few numbers, but it does make it easier. Honest.

The idea is you draw the triangles below, putting in their angles and side lengths. Then you can use them to work out special trig values like sin 45° or cos 60° more accurately than any calculator (which only gives a few decimal places).

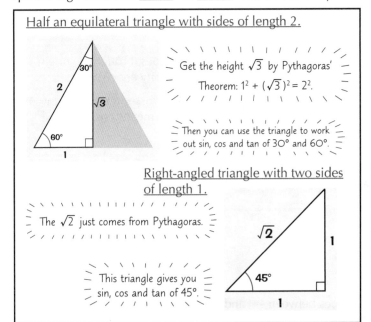

Half an equilateral triangle with sides of length 2.

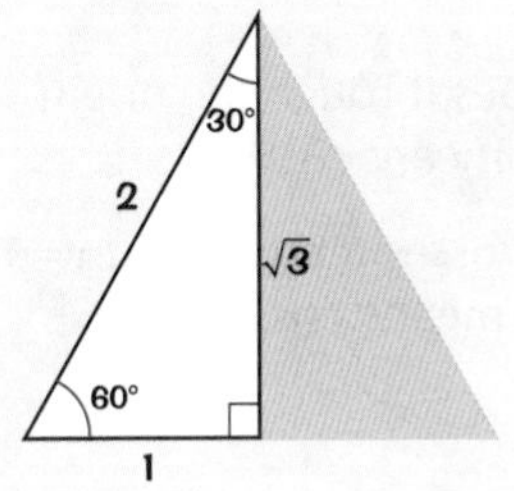

Get the height $\sqrt{3}$ by Pythagoras' Theorem: $1^2 + (\sqrt{3})^2 = 2^2$.

Then you can use the triangle to work out sin, cos and tan of 30° and 60°.

Right-angled triangle with two sides of length 1.

The $\sqrt{2}$ just comes from Pythagoras.

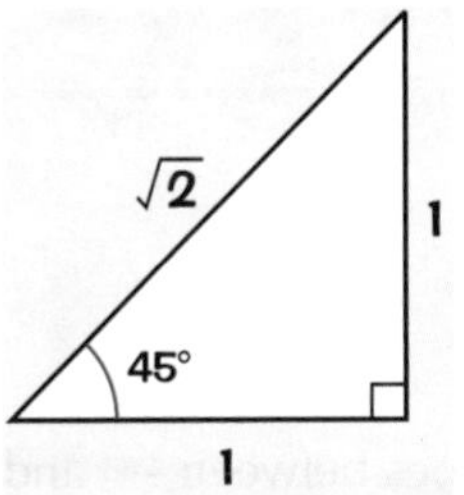

This triangle gives you sin, cos and tan of 45°.

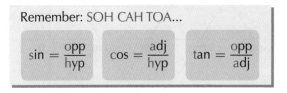

Remember: SOH CAH TOA...

$\sin = \frac{\text{opp}}{\text{hyp}}$ $\cos = \frac{\text{adj}}{\text{hyp}}$ $\tan = \frac{\text{opp}}{\text{adj}}$

Trig Values from Triangles

$\sin 30° = \frac{1}{2}$	$\sin 60° = \frac{\sqrt{3}}{2}$	$\sin 45° = \frac{1}{\sqrt{2}}$
$\cos 30° = \frac{\sqrt{3}}{2}$	$\cos 60° = \frac{1}{2}$	$\cos 45° = \frac{1}{\sqrt{2}}$
$\tan 30° = \frac{1}{\sqrt{3}}$	$\tan 60° = \sqrt{3}$	$\tan 45° = 1$

EXAMPLE Find the exact length L and area A in the diagram.

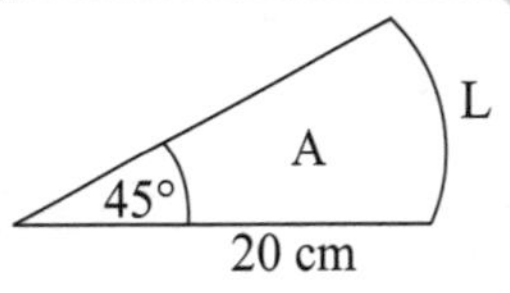

Right, first things first... it's an arc length and sector area, so you need the angle in radians.

$$45° = \frac{45 \times \pi}{180} = \frac{\pi}{4} \text{ radians}$$

Or you could just quote this if you've learnt the stuff on the previous page.

Now bung everything in your formulas: $L = r\theta = 20 \times \frac{\pi}{4} = 5\pi \text{ cm}$

$$A = \frac{1}{2}r^2\theta = \frac{1}{2} \times 20^2 \times \frac{\pi}{4} = 50\pi \text{ cm}^2$$

EXAMPLE Find the area of the shaded part of the symbol.

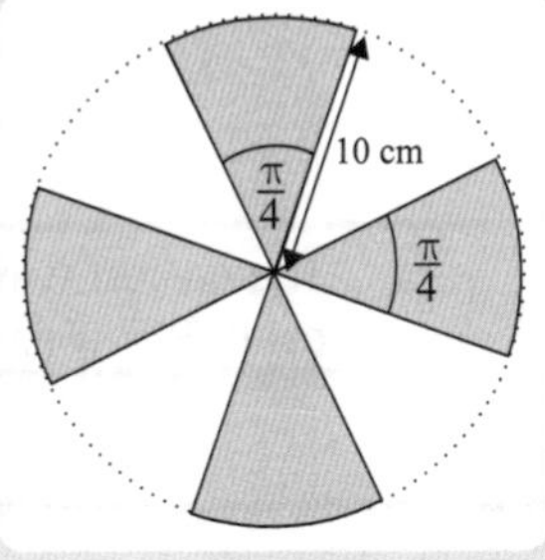

Instead, you could use the total angle of all the shaded sectors (π).

You need the area of the 'leaves' and so use the formula $\frac{1}{2}r^2\theta$.

Each leaf has area $\frac{1}{2} \times 10^2 \times \frac{\pi}{4} = \frac{25\pi}{2}$ cm²

So the area of the whole symbol = $4 \times \frac{25\pi}{2} = 50\pi$ cm²

π = 3.141592653589793238462643383279502884197169399... *(Make sure you know it)*

It's worth repeating, just to make sure — those formulas for arc length and sector area only work if the angle is in radians.

Graphs of Trig Functions

Before you leave this page, you should be able to close your eyes and picture these three graphs in your head, properly labelled and everything. If you can't, you need to learn them more. I'm not kidding.

sin x and cos x are always in the range –1 to 1

sin x and cos x are similar — they just bob up and down between –1 and 1.

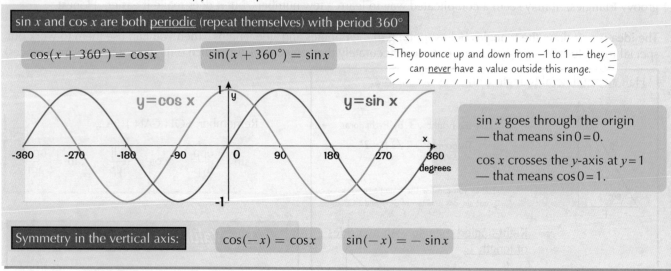

sin x and cos x are both periodic (repeat themselves) with period 360°

$\cos(x + 360°) = \cos x$ $\sin(x + 360°) = \sin x$

They bounce up and down from –1 to 1 — they can never have a value outside this range.

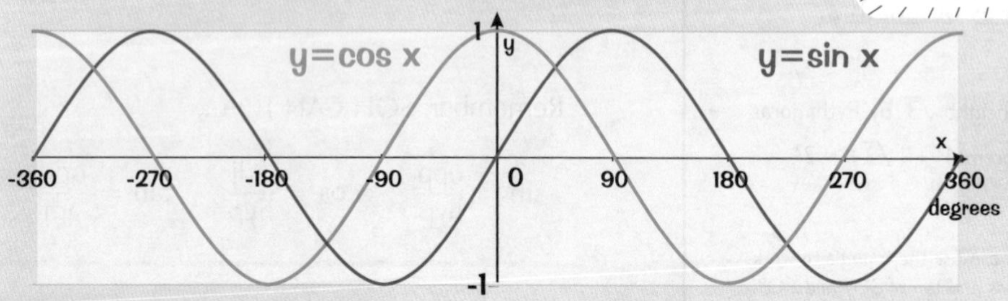

sin x goes through the origin — that means $\sin 0 = 0$.

cos x crosses the y-axis at $y = 1$ — that means $\cos 0 = 1$.

Symmetry in the vertical axis: $\cos(-x) = \cos x$ $\sin(-x) = -\sin x$

tan x can be Any Value at all

tan x is different from sin x or cos x.
It doesn't go gently up and down between –1 and 1 — it goes between –∞ and +∞.

TAN X IS ALSO PERIODIC — BUT WITH PERIOD 180°

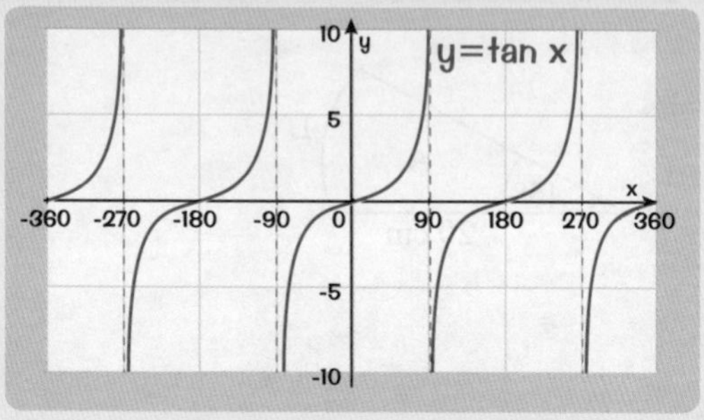

tan x is undefined at ±90°, ±270°, ±450°,...

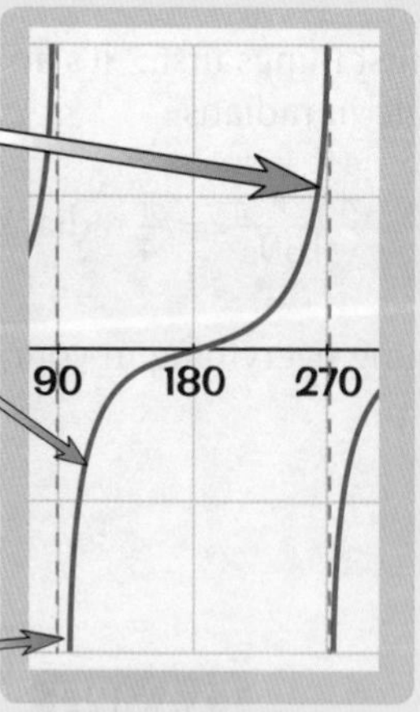

tan x goes from –∞ to +∞ every 180°

So it's got period 180°, and takes every possible value in each 180° interval.

As you approach one of these undefined points from the left, tan x just shoots up to infinity.
As you approach from the right, it drops to minus infinity.
The graph never ever touches these lines. But it does get infinitely close, if you see what I mean... These are called asymptotes.
90
180
270

$\tan(x + 180°) = \tan x$

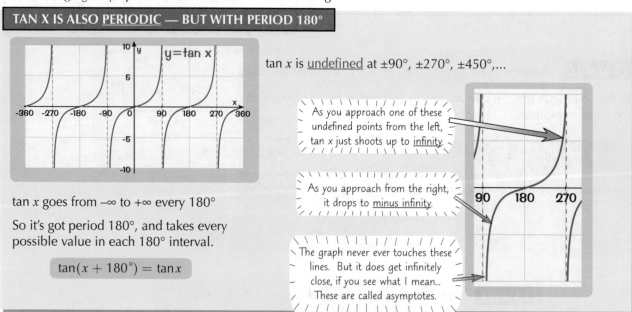

The easiest way to sketch any of these graphs is to plot the important points which happen every 90° (i.e. –180°, –90°, 0°, 90°, 180°, 270°, 360°...) and then just join the dots up.

Sin and cos can make your life worthwhile — give them a chance...

It's really really really really really important that you can draw the trig graphs on this page, and get all the labels right. Make sure you know what value sin, cos and tan have at the interesting points — i.e. 0°, 90°, 180°, 270°, 360°. It's easy to remember what the graphs look like, but you've got to know exactly where they're max, min, zero, etc.

Transformed Trig Functions

Transformed trigonometric graphs look much the same as the bog-standard ones, just a little <u>different</u>. There are three main types of transformation.

*There are 3 basic types of **Transformed Trig Graph**...*

$y = n \sin x$ — a ***Vertical Stretch*** or ***Squash***

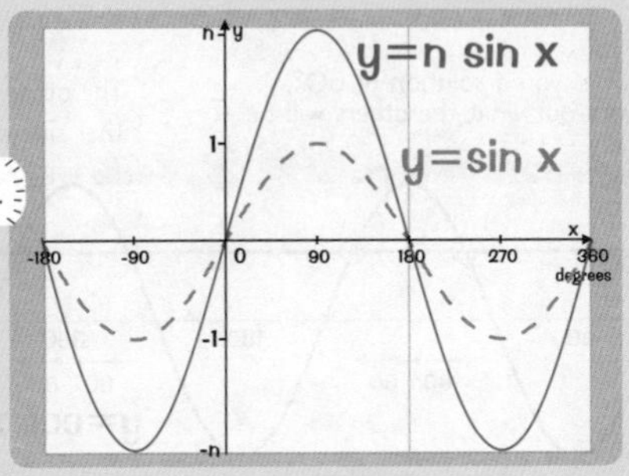

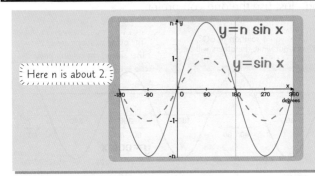

Here n is about 2.

If $n > 1$, the graph of $y = \sin x$ is <u>stretched vertically</u> by a factor of n.

If $0 < n < 1$, the graph is <u>squashed</u>.

And if $n < 0$, the graph is also <u>reflected</u> in the <u>x-axis</u>.

$y = \sin nx$ — a ***Horizontal Squash*** or ***Stretch***

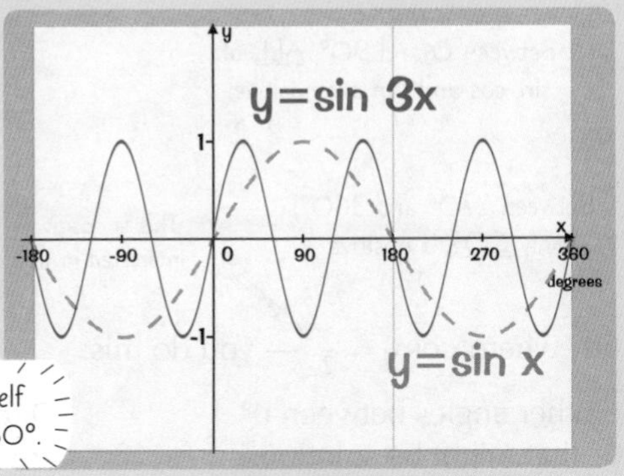

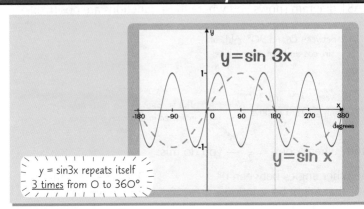

y = sin3x repeats itself <u>3 times</u> from 0 to 360°.

If $n > 1$, the graph of $y = \sin x$ is <u>squashed horizontally</u> by a factor of n.

If $0 < n < 1$, the graph is <u>stretched</u>.

And if $n < 0$, the graph is also <u>reflected</u> in the <u>y-axis</u>.

$y = \sin (x + c)$ — a ***Translation*** along the x-axis

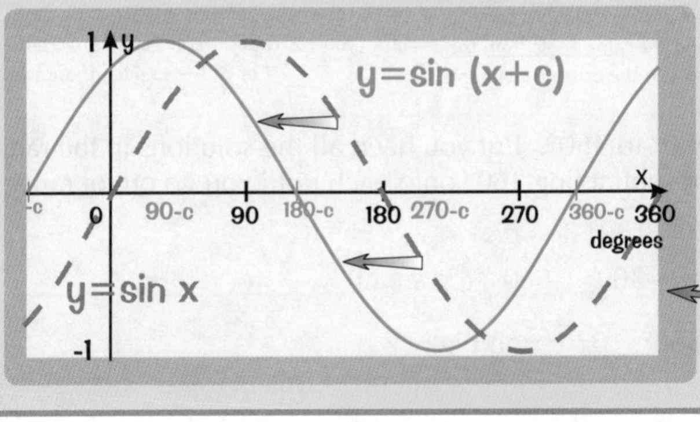

For $c > 0$, $\sin (x + c)$ is just $\sin x$ <u>shifted c to the left</u>.

Similarly, $\sin (x - c)$ is just $\sin x$ <u>shifted c to the right</u>.

For $y = \sin(x + c)$, the 'interesting' points are when $x + c = 0, 90°, 180°, 270°$, etc., i.e. when $x = -c, 90 - c, 180 - c, 270 - c, ...$

Curling up on the sofa with 2cos x — that's my idea of cosiness ☺

One thing you've really got to be careful about is making sure you move or stretch the graphs in the <u>right</u> direction. In that last example, the graph would have moved to the right if "c" was negative. And it gets confusing with the horizontal and vertical stretching and squashing — in the first two examples above, $n > 1$ means a <u>vertical stretch</u> but a <u>horizontal squash</u>.

Solving Trig Equations in a Given Interval

I used to really hate trig stuff like this. But once I'd got the hang of it, I just couldn't get enough. I stopped going out, lost interest in the opposite sex — the CAST method became my life. Learn it, but be careful. It's addictive.

There are **Two Ways** to find Solutions in an **Interval**...

EXAMPLE Solve $\cos x = \frac{1}{2}$ for $-360° \leq x \leq 720°$.

Like I said — there are two ways to solve this kind of question. Just use the one you prefer...

You can draw a **graph**...

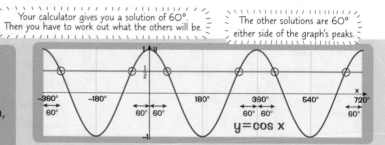

Your calculator gives you a solution of 60°. Then you have to work out what the others will be.

The other solutions are 60° either side of the graph's peaks.

1) Draw the graph of $y = \cos x$ for the range you're interested in...
2) Get the first solution from your calculator and mark this on the graph,
3) Use the symmetry of the graph to work out what the other solutions are:

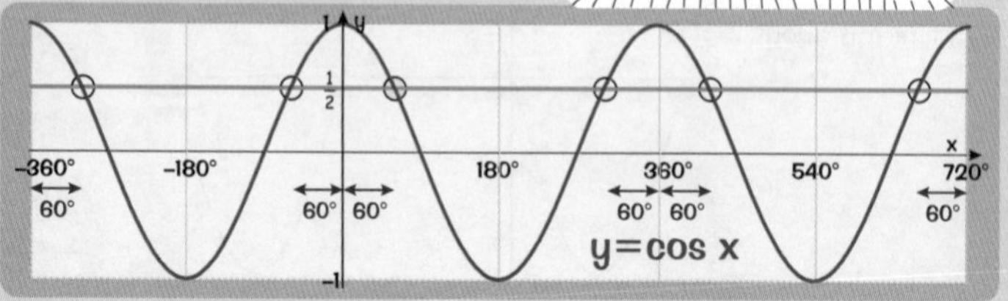

So the solutions are: –300°, –60°, 60°, 300°, 420° and 660°.

...or you can use the **CAST** diagram

CAST stands for COS, ALL, SIN, TAN — and the CAST diagram shows you where these functions are positive:

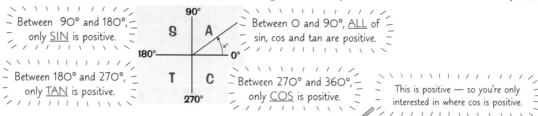

Between 90° and 180°, only SIN is positive.

Between 180° and 270°, only TAN is positive.

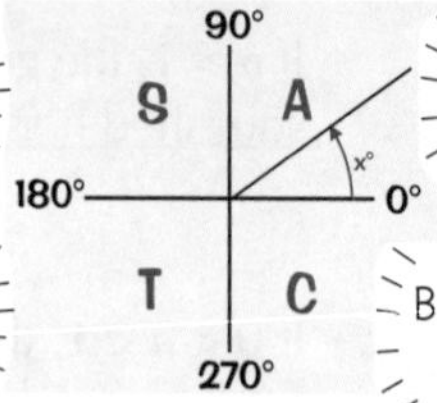

Between 0 and 90°, ALL of sin, cos and tan are positive.

Between 270° and 360°, only COS is positive.

This is positive — so you're only interested in where cos is positive.

First, to find all the values of x between 0° and 360° where $\cos x = \frac{1}{2}$ — you do this:

Put the first solution onto the CAST diagram.

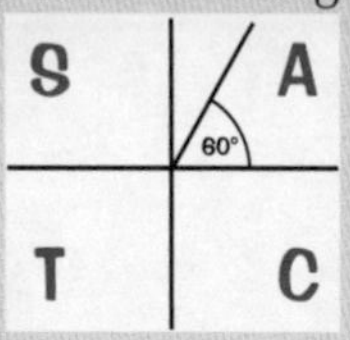

The angle from your calculator goes anticlockwise from the x-axis (unless it's negative — then it would go clockwise into the 4th quadrant).

Find the other angles between 0° and 360° that might be solutions.

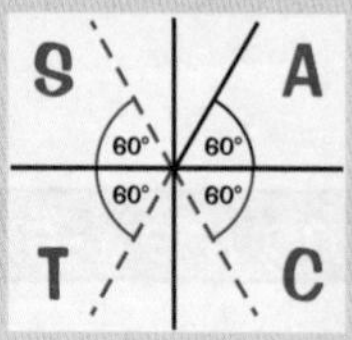

The other possible solutions come from making the same angle from the horizontal axis in the other 3 quadrants.

Ditch the ones that are the wrong sign.

$\cos x = ½$, which is positive. The CAST diagram tells you cos is positive in the 4th quadrant — but not the 2nd or 3rd — so ditch those two angles.

So you've got solutions 60° and 300° in the range 0° to 360°. But you need all the solutions in the range –360° to 720°. Get these by repeatedly adding or subtracting 360° onto each until you go out of range:

$x = 60° \Rightarrow$ (adding 360°) $x = 420°$, 780° (too big)
and (subtracting 360°) $x = -300°$, –660° (too small)

$x = 300° \Rightarrow$ (adding 360°) $x = 660°$, 1020° (too big)
and (subtracting 360°) $x = -60°$, –420° (too small)

So the solutions are: $x = -300°$, –60°, 60°, 300°, 420° and 660°.

And I feel that love is dead, I'm loving angles instead...

Suppose the first solution you get is negative, let's say $-d°$, then you'd measure it clockwise on the CAST diagram. So it'd be $d°$ in the 4th quadrant. Then you'd work out the other 3 possible solutions in exactly the same way, rejecting the ones which weren't the right sign. Got that? No? Got that? No? Got that? Yes? Good!

Solving Trig Equations in a Given Interval

Sometimes it's a bit more complicated. But only a bit.

*Sometimes you end up with **sin kx = number...***

For these, it's definitely easier to draw the graph rather than use the CAST method — that's one reason why being able to sketch trig graphs properly is so important.

EXAMPLE Solve: $\sin 3x = -\frac{1}{\sqrt{2}}$ for $0° \leq x \leq 360°$.

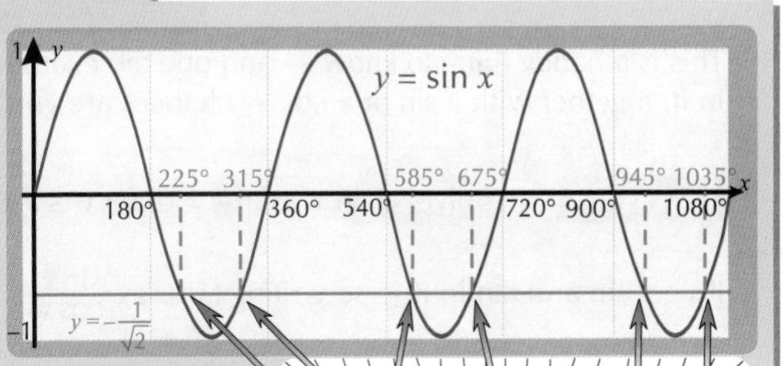

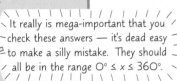

These are the solutions — but remember that this is for 3x. You'll need to divide by 3 to get your final answers.

1) You've got $3x$ instead of x, which means the range you need to find solutions in is $0° \leq 3x \leq 1080°$. So draw the graph of $y = \sin x$ between 0° and 1080°.
2) Use your calculator to find the first solution. You'll get $3x = -45°$ — but this is outside the range for $3x$, so use the pattern of the graph to find a solution in the range. As the sin curve repeats every 360°, there'll be a solution at 360 – 45 = 315°.
3) Now use your graph to find the other 5 solutions. You can see that there's another solution at 180 + 45 = 225°. Then add on 360° and 720° to both 225° and 315° to get:

   $3x = 225°, 315°, 585°, 675°, 945°$ and $1035°$.

   Divide by 3 to get the solutions for x:

   $x = 75°, 105°, 195°, 225°, 315°$ and $345°$.

   It really is mega-important that you check these answers — it's dead easy to make a silly mistake. They should all be in the range $0° \leq x \leq 360°$.
4) Check your answers by putting these values back into your calculator.

*...or **sin (x + k) = number***

All the steps in this example are just the same as in the one above.

EXAMPLE Solve $\sin(x + 60°) = \frac{3}{4}$ for $-360° \leq x \leq 360°$, giving your answers to 2 decimal places.

This is the graph of $y = \sin x$ between −300° and 420°.

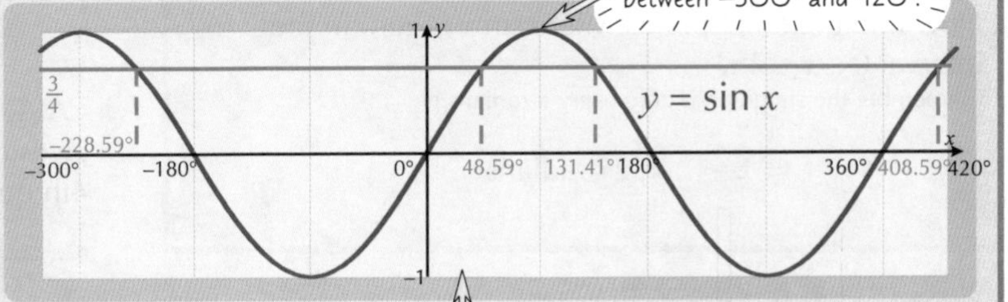

All the solutions on the graph are values of $x + 60°$. You need to subtract 60 to find the values of x — the solutions you want.

1) You've got $\sin(x + 60°)$ instead of $\sin x$ — so the range is $-300° \leq x + 60° \leq 420°$.
2) Use your calculator to get that first solution...

   $$\sin(x + 60°) = \frac{3}{4}$$
   $$\Rightarrow x + 60° = 48.59°$$
3) From the graph, you can see that the other solutions occur at 180 – 48.59, 360 + 48.59 and –180 – 48.59. The solutions are:

   $x + 60° = -228.58°, 48.59°, 131.41°$ and $408.59°$

   Subtract 60 to get the solutions for x:

   $x = -11.41°, -288.59°, 71.41°$ and $348.59°$
4) Check your answers by putting these values back into your calculator.

Live a life of sin (and cos and tan)...

Yep, the examples on this page are pretty fiddly. The most important bit is actually getting the sketch right for your new range. If you don't, you're in big trouble. Then you've just got to carefully use the sketch to work out the other solutions. It's tricky, but you'll feel better about yourself when you've got it mastered. Ah you will, you will, you will ...

Solving Trig Equations in a Given Interval

Now for something really exciting — trig identities. Mmm, well, maybe exciting was the wrong word. But they can be dead useful, so here goes...

*For equations with **tan x** in, it often helps to use this...*

$$\tan x \equiv \frac{\sin x}{\cos x}$$

This is a handy thing to know — and one the examiners love testing. Basically, if you've got a trig equation with a tan in it, together with a sin or a cos — chances are you'll be better off if you rewrite the tan using this formula.

EXAMPLE Solve: $3\sin x - \tan x = 0$, for $0 \le x \le 2\pi$.

It's got sin and tan in it — so writing tan x as $\frac{\sin x}{\cos x}$ is probably a good move:

$$3\sin x - \tan x = 0$$
$$\Rightarrow 3\sin x - \frac{\sin x}{\cos x} = 0$$

Get rid of the $\cos x$ on the bottom by multiplying the whole equation by $\cos x$.

$$\Rightarrow 3\sin x\cos x - \sin x = 0$$

Now — there's a common factor of $\sin x$. Take that outside a bracket.

$$\Rightarrow \sin x(3\cos x - 1) = 0$$

And now you're almost there. You've got two things multiplying together to make zero. That means either one or both of them is equal to zero themselves.

$$\Rightarrow \sin x = 0 \quad \text{or} \quad 3\cos x - 1 = 0$$

$\sin x = 0$

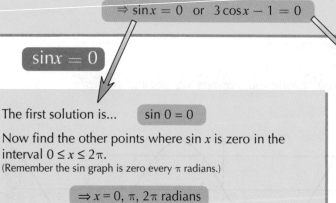

The first solution is... $\sin 0 = 0$

Now find the other points where $\sin x$ is zero in the interval $0 \le x \le 2\pi$.
(Remember the sin graph is zero every π radians.)

$$\Rightarrow x = 0,\ \pi,\ 2\pi \text{ radians}$$

$3\cos x - 1 = 0$

CAST gives any solutions in the interval $0 \le x \le 2\pi$.

Rearrange... $\cos x = \frac{1}{3}$

So the first solution is...

$$\cos^{-1}\frac{1}{3} = 1.231$$

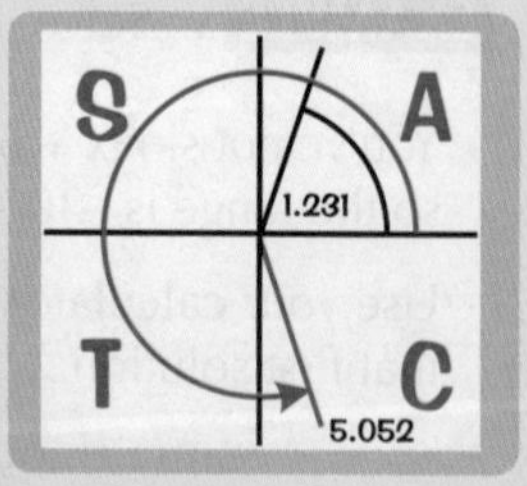

CAST (or the graph of cos x) gives another solution in the 4th quadrant...

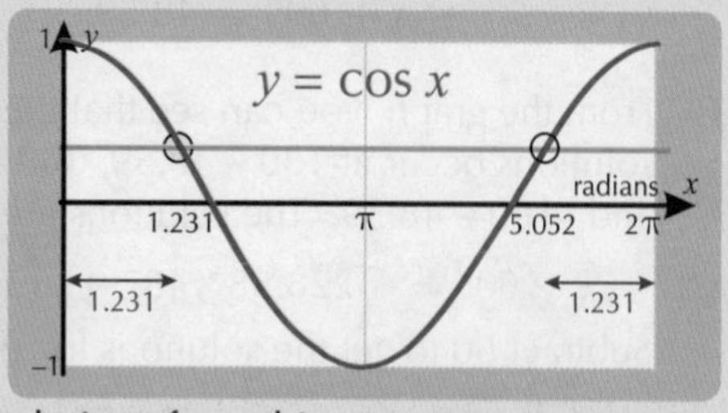

And the two solutions from this part are:

$$\Rightarrow x = 1.231,\ 5.052 \text{ radians}$$

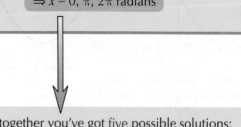

So altogether you've got <u>five</u> possible solutions:

$$\Rightarrow x = 0,\ 1.231,\ \pi,\ 5.052,\ 2\pi \text{ radians}$$

Trigonometry is the root of all evil...

What a page — you don't have fun like that every day, do you? No, trig equations are where it's at. This is a really useful trick, though — and can turn a nightmare of an equation into a bit of a pussycat. <u>Rewriting</u> stuff using <u>different</u> formulas is always worth trying if it feels like you're getting stuck — even if you're not sure why when you're doing it. You might have a flash of inspiration when you see the new version.

Solving Trig Equations in a Given Interval

Another trig identity — and it's a good 'un — examiners love it. And it's not difficult either.

And if you have a sin² x or a cos² x, think of this straight away...

$$\sin^2 x + \cos^2 x \equiv 1 \quad \Rightarrow \quad \begin{matrix} \sin^2 x \equiv 1 - \cos^2 x \\ \cos^2 x \equiv 1 - \sin^2 x \end{matrix}$$

Use this identity to get rid of a $\sin^2$ or a $\cos^2$ that's making things awkward...

EXAMPLE Solve: $2\sin^2 x + 5\cos x = 4$, for $0° \leq x \leq 360°$.

You can't do much while the equation's got both sin's and cos's in it. So replace the $\sin^2 x$ bit with $1 - \cos^2 x$.

$$2(1 - \cos^2 x) + 5\cos x = 4$$

Multiply out the bracket and rearrange it so that you've got zero on one side — and you get a quadratic in $\cos x$:

Now the only trig function is cos.

$$\Rightarrow 2 - 2\cos^2 x + 5\cos x = 4$$
$$\Rightarrow 2\cos^2 x - 5\cos x + 2 = 0$$

This is a quadratic in $\cos x$. It's easier to factorise this if you make the substitution $y = \cos x$.

$$2y^2 - 5y + 2 = 0$$
$$\Rightarrow (2y - 1)(y - 2) = 0$$
$$\Rightarrow (2\cos x - 1)(\cos x - 2) = 0$$

$2y^2 - 5y + 2 = (2y \;\; ?)(y \;\; ?)$
$= (2y - 1)(y - 2)$

Now one of the brackets must be 0. So you get 2 equations as usual:

You've already done this example on page 80.

$$2\cos x - 1 = 0 \quad \text{or} \quad \cos x - 2 = 0$$

$\cos x = \frac{1}{2} \Rightarrow x = 60°$ or $x = 300°$ and $\cos x = 2$

This is a bit weird. cos x is always between –1 and 1. So you don't get any solutions from this bracket.

So at the end of all that, the only solutions you get are $x = 60°$ and $x = 300°$. How boring.

Use the *Trig Identities* to prove something is the *Same* as something else

Another use for these trig identities is proving that two things are the same.

EXAMPLE Show that $\dfrac{\cos^2\theta}{1 + \sin\theta} \equiv 1 - \sin\theta$

The identity sign ≡ means that this is true for all θ, rather than just certain values.

Prove things like this by playing about with one side of the equation until you get the other side.

<u>Left-hand side</u>: $\dfrac{\cos^2\theta}{1 + \sin\theta}$

The only thing I can think of doing here is replacing $\cos^2\theta$ with $1 - \sin^2\theta$. (Which is good because it works.)

$$\equiv \frac{1 - \sin^2\theta}{1 + \sin\theta}$$

The next trick is the hardest to spot. Look at the top — does that remind you of anything?

The top line is a difference of two squares:

$$\equiv \frac{(1 + \sin\theta)(1 - \sin\theta)}{1 + \sin\theta}$$

$1 - a^2 = (1 + a)(1 - a)$
$\Rightarrow 1 - \sin^2\theta = (1 + \sin\theta)(1 - \sin\theta)$

$\equiv 1 - \sin\theta$, <u>the right-hand side</u>.

Trig identities — the path to a brighter future...

That was a pretty miserable section. But it's over. These trig identities aren't exactly a barrel of laughs, but they are a definite source of marks — you can bet your last penny they'll be in the exam. That substitution trick to get rid of a $\sin^2$ or a $\cos^2$ and end up with a quadratic in $\sin x$ or $\cos x$ is a real examiners' favourite. Those identities can be a bit daunting, but it's always worth having a few tricks in the back of your mind — always look for things that factorise, or fractions that can be cancelled down, or ways to use those trig identities. Ah, it's all good clean fun.

C2 Section 3 — Practice Questions

I know, I know — that was a long section. And, rather predictably, it's followed by lots of practice questions. Brace yourself for the warm-up — this is going to be a heavy session...

Warm-up Questions

1) What is tan x in terms of cos x and sin x? What is $\cos^2 x$ in terms of $\sin^2 x$?
2) Draw a triangle ΔXYZ with sides of length x, y and z. Write down the Sine and Cosine Rules for this triangle. Write down an expression for its area.
3) Solve : a) ΔABC in which A = 30°, C = 25°, b = 6 m, and find its area.
   b) ΔPQR in which p = 3 km, q = 23 km, R = 10°. (answers to 2 d.p.)
4) My pet triangle Freda has sides of length 10, 20 and 25. Find her angles (in degrees to 1 d.p.).
5) Find the 2 possible triangles ΔABC which satisfy b=5, a=3, A=35°. (This is tricky: Sketch it first, and see if you can work out how to make 2 different triangles satisfying the data given.)
6) Write down the exact values of cos 30°, sin 30°, tan 30°, cos 45°, sin 45°, tan 45°, cos 60°, sin 60° and tan 60°.
7) Sketch the graphs for sin$x$, cos$x$ and tan$x$. Make sure you label all the max/min/zero/undefined points.
8) Sketch the following graphs:
   a) $y = \frac{1}{2}\cos x$ (for $0° \leq x \leq 360°$) b) $y = \sin(x + 30°)$ (for $0° \leq x \leq 360°$) c) $y = \tan 3x$ (for $0° \leq x \leq 180°$)
9) a) Solve each of these equations for $0° \leq \theta \leq 360°$:
   (i) $\sin\theta = -\frac{\sqrt{3}}{2}$ (ii) $\tan\theta = -1$ (iii) $\cos\theta = -\frac{1}{\sqrt{2}}$
   b) Solve each of these equations for $-180° \leq \theta \leq 180°$ (giving your answer to 1 d.p.):
   (i) $\cos 4\theta = -\frac{2}{3}$ (ii) $\sin(\theta + 35°) = 0.3$ (iii) $\tan\left(\frac{1}{2}\theta\right) = 500$
10) Find all the solutions to $6\sin^2 x = \cos x + 5$ in the range $0° \leq x \leq 360°$ (answers to 1 d.p.).
11) Solve $3\tan x + 2\cos x = 0$ for $-90° \leq x \leq 90°$
12) Simplify: $(\sin y + \cos y)^2 + (\cos y - \sin y)^2$
13) Show that $\dfrac{\sin^4 x + \sin^2 x\cos^2 x}{\cos^2 x - 1} \equiv -1$

Well after that vigorous warm-up you should be more than ready to tackle this mental marathon of exam-style questions. Take your time to work out what each question wants — if it looks impossible, there's probably a way of simplifying it somehow. A picture paints a thousand words, so sketch it out if you're totally confused.

Exam Questions

1 The diagram shows the locations of two walkers, X and Y, after walking in different directions from the same start position.

X walked due south for 150 m. Y walked 250 m on a bearing of 100°.

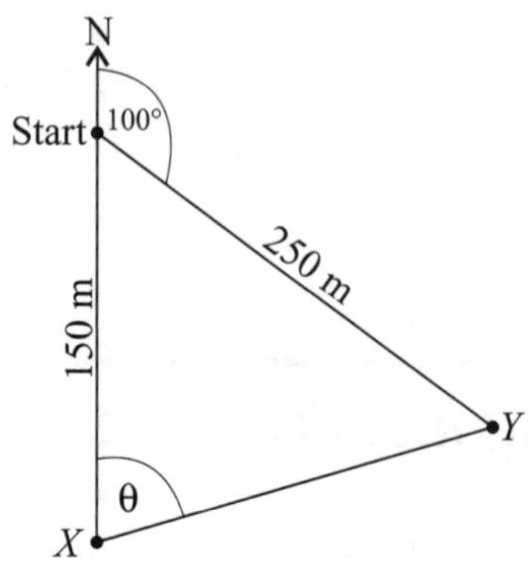

a) Calculate the distance between the two walkers, in m to the nearest m. *(2 marks)*

b) Show that $\dfrac{\sin\theta}{\sin 80°} = 0.93$ to 2 decimal places. *(3 marks)*

C2 Section 3 — Practice Questions

2 The shape ABC shown here is a concrete slab that forms part of a paved area around a circular pond. The curve AC is an arc of a circle with centre M and radius 2.20 m.

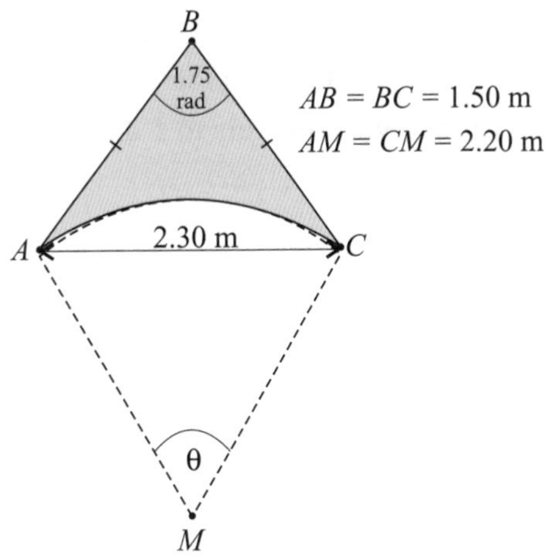

Find, to 3 significant figures:

a) The size of angle θ in radians, *(2 marks)*

b) The perimeter of the slab in m, *(3 marks)*

c) The area of the slab in m^2. *(5 marks)*

3 A piece of jewellery is made by joining a triangular piece of silver to a piece of glass to form a sector of a circle with radius 8 cm and angle $\frac{\pi}{4}$ radians, as shown below. The length of OX is 5 cm.

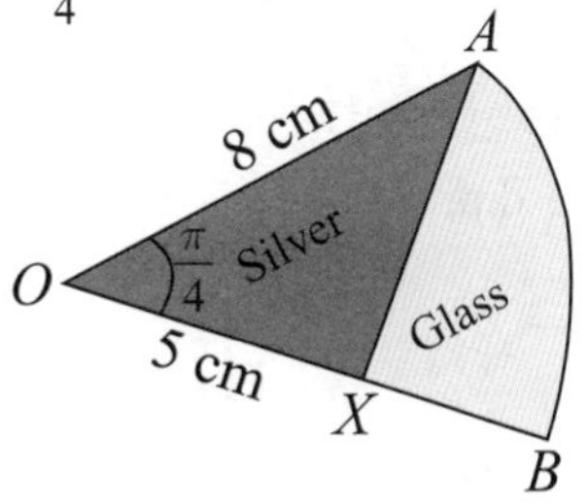

a) Find the area of the piece of glass used. *(3 marks)*

b) Find the perimeter of the piece of glass used. *(4 marks)*

4 a) (i) Sketch, for $0 \leq x \leq 360°$, the graph of $y = \cos(x + 60°)$. *(2 marks)*

(ii) Write down all the values of x, for $0 \leq x \leq 360°$, where $\cos(x + 60°) = 0$. *(2 marks)*

b) Sketch, for $0 \leq x \leq 180°$, the graph of $y = \sin 4x$. *(2 marks)*

c) Solve, for $0 \leq x \leq 180°$, the equation: $\sin 4x = 0.5$, giving your answers in degrees. *(4 marks)*

C2 Section 3 — Practice Questions

5 For an angle x, $3 \cos x = 2 \sin x$.

a) Find $\tan x$. *(2 marks)*

b) Hence, or otherwise, find all the values of x, in the interval $0 \leq x \leq 360°$, for which $3 \cos x = 2 \sin x$, giving your answers to 1 d.p. *(2 marks)*

6 Solve, for $0 \leq x \leq 2\pi$:

a) $\tan\left(x + \frac{\pi}{6}\right) = \sqrt{3}$. *(4 marks)*

b) $2 \cos\left(x - \frac{\pi}{4}\right) = \sqrt{3}$ *(4 marks)*

c) $\sin 2x = -\frac{1}{2}$ *(6 marks)*

7 a) Show that the equation:

$$2(1 - \cos x) = 3 \sin^2 x$$

can be written as

$$3 \cos^2 x - 2 \cos x - 1 = 0$$

(2 marks)

b) Use this to solve the equation

$$2(1 - \cos x) = 3 \sin^2 x$$

for $0 \leq x \leq 360°$, giving your answers to 1 d.p. *(6 marks)*

8 Find all the values of x, in the interval $0 \leq x \leq 2\pi$, for which:

$$2 - \sin x = 2 \cos^2 x$$

giving your answers in terms of π. *(6 marks)*

9 Solve the following equations, for $-\pi \leq x \leq \pi$:

a) $(1 + 2 \cos x)(3 \tan^2 x - 1) = 0$ *(6 marks)*

b) $3 \cos^2 x = \sin^2 x$ *(4 marks)*

Logs

Don't be put off by your parents or grandparents telling you that logs are hard. Logarithm is just a fancy word for power, and once you know how to use them you can solve all sorts of equations.

You need to be able to Switch between Different Notations

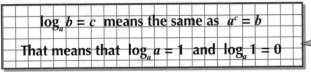

$\log_a b = c$ means the same as $a^c = b$

That means that $\log_a a = 1$ and $\log_a 1 = 0$

The little number 'a' after 'log' is called the base. Logs can be to any base, but base 10 is the most common. The button marked 'log' on your calculator uses base 10.

EXAMPLE Index notation: $10^2 = 100$ log notation: $\log_{10} 100 = 2$

The base goes here but it's usually left out if it's 10.

So the logarithm of 100 to the base 10 is 2, because 10 raised to the power of 2 is 100.

EXAMPLES

Write down the values of the following:

a) $\log_2 8$ b) $\log_9 3$ c) $\log_5 5$

a) 8 is 2 raised to the power of 3, so $2^3 = 8$ and $\log_2 8 = 3$

b) 3 is the square root of 9, or $9^{1/2} = 3$, so $\log_9 3 = 1/2$

c) Anything to the power of 1 is itself, so $\log_5 5 = 1$

Write the following using log notation:

a) $5^3 = 125$ b) $3^0 = 1$

You just need to make sure you get things in the right place.

a) 3 is the power or logarithm that 5 (the base) is raised to to get 125, so $\log_5 125 = 3$

b) You'll need to remember this one: $\log_3 1 = 0$

The Laws of Logarithms are Unbelievably Useful

Whenever you have to deal with logs, you'll end up using the laws below. That means it's no bad idea to learn them by heart right now.

Laws of Logarithms

$$\log_a x + \log_a y = \log_a (xy)$$

$$\log_a x - \log_a y = \log_a \left(\frac{x}{y}\right)$$

$$\log_a x^k = k \log_a x$$

So $\log_a \frac{1}{x} = -\log_a x$

You've got to be able to change the base of a log too.

Change of Base

$$\log_a x = \frac{\log_b x}{\log_b a}$$

So $\log_7 4 = \frac{\log_{10} 4}{\log_{10} 7} = 0.7124$

To check: $7^{0.7124} = 4$

Use the Laws to Manipulate Logs

EXAMPLE Write each expression in the form $\log_a n$, where n is a number.

a) $\log_a 5 + \log_a 4$ b) $2 \log_a 6 - \log_a 9$

a) $\log_a x + \log_a y = \log_a (xy)$

You just have to multiply the numbers together:

$\log_a 5 + \log_a 4 = \log_a (5 \times 4)$
$= \log_a 20$

b) $\log_a x^k = k \log_a x$

$2 \log_a 6 = \log_a 6^2 = \log_a 36$

$\log_a 36 - \log_a 9 = \log_a (36 \div 9)$
$= \log_a 4$

It's sometimes hard to see the wood for the trees — especially with logs...

Tricky, tricky, tricky... I think of $\log_a b$ as 'the power I have to raise a to if I want to end up with b' — that's all it is. And the log laws make a bit more sense if you think of 'log' as meaning 'power'. For example, you know that $2^a \times 2^b = 2^{a+b}$ — this just says that if you multiply the two numbers, you add the powers. Well, the first law of logs is saying the same thing. Any road, even if you don't really understand why they work, make sure you know the log laws like you know your own navel.

Exponentials and Logs

Okay, you've done the theory of logs. So now it's a bit of stuff about exponentials (the opposite of logs, kind of), and then it'll be time to get your calculator out for a bit of button pressing...

Graphs of a^x Never Reach Zero

All the graphs of $y = a^x$ (exponential graphs) where $a > 1$ have the same basic shape. The graphs for $a = 2$, $a = 3$ and $a = 4$ are shown on the right.

- All the a's are greater than 1 — so y increases as x increases.
- The bigger a is, the quicker the graphs increase.
- As x decreases, y decreases at a smaller and smaller rate — y will approach zero, but never actually get there.

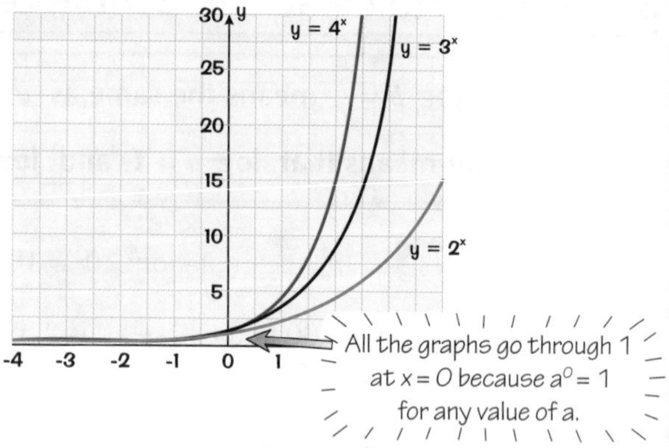

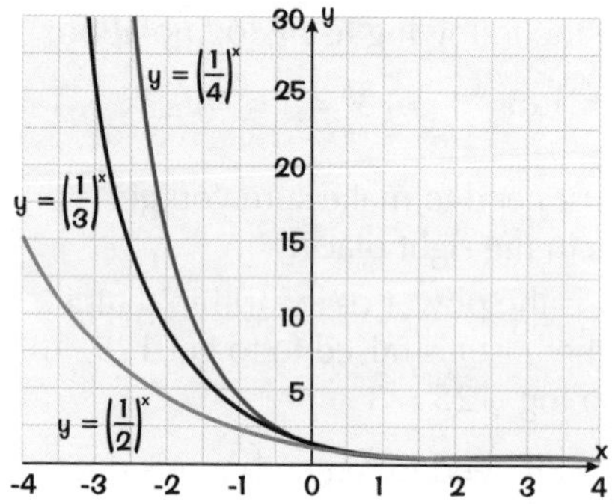

The graphs on the left are for $y = a^x$ where $a < 1$ (they're for $a = \frac{1}{2}, \frac{1}{3}$ and $\frac{1}{4}$).

- All the a's are less than 1 — meaning y decreases as x increases.
- As x increases, y decreases at a smaller and smaller rate — again, y will approach zero, but never actually get there.

You can use exponentials and logs to solve equations

EXAMPLE 1) Solve $2^{4x} = 3$ to 3 significant figures.

You want x on its own, so take logs of both sides (by writing 'log' in front of both sides): $\log 2^{4x} = \log 3$

Now use one of the laws of logs: $\log x^k = k \log x$: $4x \log 2 = \log 3$

You can now divide both sides by '4 log 2' to get x on its own: $x = \frac{\log 3}{4 \log 2}$

But $\frac{\log 3}{4 \log 2}$ is just a number you can find using a calculator: $x = 0.396$ (to 3 s.f.)

I don't know about you, but I enjoyed that more than the biggest, fastest rollercoaster. You want another? OK then...

EXAMPLE 2) Solve $7 \log_{10} x = 5$ to 3 significant figures.

You want x on its own, so begin by dividing both sides by 7: $\log_{10} x = \frac{5}{7}$

You now need to take exponentials of both sides by doing '10 to the power of both sides' (since the log is to base 10): $10^{\log_{10} x} = 10^{\frac{5}{7}}$

Logs and exponentials are inverse functions, so they cancel out: $x = 10^{\frac{5}{7}}$

Again, $10^{\frac{5}{7}}$ is just a number you can find using a calculator: $x = 5.18$ (to 3 s.f.)

Exponentials and Logs

Use the Calculator Log Button Whenever You Can

EXAMPLE Use logarithms to solve the following for x, giving the answers to 4 s.f.

a) $10^{3x} = 4000$ b) $7^x = 55$ c) $\log_2 x = 5$

You've got the magic buttons on your calculator, but you'd better follow the instructions and show that you know how to use the log rules covered earlier.

a) $10^{3x} = 4000$ — there's an 'unknown' in the power, so take logs of both sides.
(In theory, it doesn't matter what base you use, but your calculator has a '$\log_{10}$' button, so base 10 is usually a good idea. But whatever base you use, use the same one for both sides.)

So taking logs to base 10 of both sides of the above equation gives:

$\log 10^{3x} = \log 4000$

i.e. $3x \log 10 = \log 4000$

i.e. $3x = \log 4000$, so $x = 1.201$ (to 4 sig. fig.)

Since $\log_{10} 10 = 1$

b) $7^x = 55$. Again, take logs of both sides, and use the log rules: $x \log_{10} 7 = \log_{10} 55$, so $x = \frac{\log_{10} 55}{\log_{10} 7} = 2.059$

c) $\log_2 x = 5$ — to get rid of a log, you 'take exponentials', meaning you do '2 (the base) to the power of each side'.

Think of 'taking logs' and 'taking exponentials' as opposite processes — one cancels the other out: $2^{\log_2 x} = 2^5$

i.e. $x = 32$

You might have to Combine the Laws of Logs to Solve equations

If the examiners are feeling particularly mean, they might make you use more than one law to solve an equation.

EXAMPLE Solve the equation $\log_3(2 - 3x) - 2\log_3 x = 2$.

First, combine the log terms into one term (you can do this because they both have the same base): $\log_3 \frac{2-3x}{x^2} = 2$

Remember that $2\log x = \log x^2$.

Then take exponentials of both sides: $3^{\log_3 \frac{2-3x}{x^2}} = 3^2 \Rightarrow \frac{2-3x}{x^2} = 9$

Finally, rearrange the equation and solve for x: $2 - 3x = 9x^2 \Rightarrow 0 = 9x^2 + 3x - 2$

$\Rightarrow 0 = (3x - 1)(3x + 2)$

Ignore the negative solution because you can't take logs of a negative number.

So $x = \frac{1}{3}$.

Exponential Growth and Decay Applies to Real-life Problems

Logs can even be used to solve real-life problems.

EXAMPLE The radioactivity of a substance decays by 20 per cent over a year. The initial level of radioactivity is 400. Find the time taken for the radioactivity to fall to 200 (the half-life).

$R = 400 \times 0.8^T$ where R is the level of radioactivity at time T years.
We need $R = 200$, so solve $200 = 400 \times 0.8^T$

The 0.8 comes from 1 – 20% decay.

$0.8^T = \frac{200}{400} = 0.5 \Rightarrow T \log 0.8 = \log 0.5 \Rightarrow T = \frac{\log 0.5}{\log 0.8} = 3.106$ years

If in doubt, take the log of something — that usually works...

The thing about exponential growth is that it's really useful, as it happens in real life all over the place. Money in a bank account earns interest at a certain percentage per year, and so the balance rises exponentially (if you don't spend or save anything). Likewise, if you got 20% cleverer for every week you studied, that would also be exponential... and impressive.

C2 Section 4 — Practice Questions

Logs and exponentials are surprisingly useful things. As well as being in your exam they pop up all over the place in real life — savings, radioactive decay, growth of bacteria — all logarithmic.
And now for something (marginally) different:

Warm-up Questions

1) Write down the values of the following:
   a) $\log_3 27$ b) $\log_3 (1 \div 27)$ c) $\log_3 18 - \log_3 2$
2) Simplify the following:
   a) $\log 3 + 2 \log 5$ b) $\frac{1}{2} \log 36 - \log 3$ c) $\log 2 - \frac{1}{4} \log 16$
3) Simplify $\log_b (x^2 - 1) - \log_b (x - 1)$
4) a) Copy and complete the table for the function $y = 4^x$:

x	-3	-2	-1	0	1	2	3
y							

   b) Using suitable scales, plot a graph of $y = 4^x$ for $-3 < x < 3$.
   c) Use the graph to solve the equation $4^x = 20$.
5) Solve these little jokers:
   a) $10^x = 240$ b) $\log_{10} x = 5.3$
   c) $10^{2x+1} = 1500$ d) $4^{(x-1)} = 200$
6) Find the smallest integer P such that $1.5^P > 1\,000\,000$.

Time for some practice at the real thing. On your marks... Get set... Go.

Exam Questions

1 a) Write the following expressions in the form $\log_a n$, where n is an integer:

(i) $\log_a 20 - 2 \log_a 2$

(3 marks)

(ii) $\frac{1}{2} \log_a 16 + \frac{1}{3} \log_a 27$

(3 marks)

b) Find the value of:

(i) $\log_2 64$

(1 mark)

(ii) $2 \log_3 9$

(2 marks)

c) Calculate the value of the following, giving your answer to 4 d.p.

(i) $\log_6 25$

(1 mark)

(ii) $\log_3 10 + \log_3 2$

(2 marks)

C2 Section 4 — Practice Questions

2 a) Solve the equation

$$2^x = 9$$

giving your answer to 2 decimal places. *(3 marks)*

b) Hence, or otherwise, solve the equation

$$2^{2x} - 13(2^x) + 36 = 0$$

giving each solution to an appropriate degree of accuracy. *(5 marks)*

3 Solve the equation

$$\log_7 (y + 3) + \log_7 (2y + 1) = 1$$

where $y > 0$. *(5 marks)*

4 a) Solve the equation:

$$\log_3 x = -\frac{1}{2}$$

leaving your answer as an exact value. *(3 marks)*

b) Find x, where

$$2 \log_3 x = -4$$

leaving your answer as an exact value. *(2 marks)*

5 a) Find x, if

$$6^{(3x + 2)} = 9$$

giving your answer to 3 significant figures. *(3 marks)*

b) Find y, if

$$3^{(y^2 - 4)} = 7^{(y + 2)}$$

giving your answer to 3 significant figures. *(5 marks)*

6 For the positive integers p and q,

$$\log_4 p - \log_4 q = \frac{1}{2}$$

a) Show that $p = 2q$. *(3 marks)*

b) Solve the equation for p and q

$$\log_2 p + \log_2 q = 7$$

(5 marks)

Differentiation

Well, just when you think you've seen it all, along comes something fresh and exciting to knock you off your feet... Differentiation. Oh, hang on a tic...

Use the formula to differentiate Negative and Fractional Powers

1) In Core 2 you need to be able to do everything you learned about differentiation in Core 1 all over again — only this time with all sorts of crazy polynomials involving negative and fractional powers. You lucky jotters.
2) So before you go any further, it might be worth taking a goosey at C1 Section 6 (pages 38-43) and reminding yourself what this differentiating lark is all about.
3) In case you've forgotten it, or if you'd just really like to see it again, the formula for differentiating powers of x is shown below. This formula works just as well for negative and fractional powers as it does for regular ones.

$$\frac{d}{dx}(x^n) = nx^{n-1}$$

This is in no way a gratuitous way to fill some space on the page. No siree. Nope. Ok, it is.

4) Things are much easier to differentiate when they're written as powers of x — like writing $\sqrt{x}$ as $x^{\frac{1}{2}}$. So it's always best to do this to each term in an equation first, then use the formula to differentiate the equation.

Examples

See page 57 for more on negative powers.

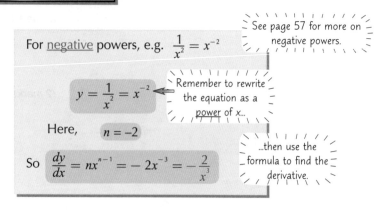

For negative powers, e.g. $\frac{1}{x^2} = x^{-2}$

$$y = \frac{1}{x^2} = x^{-2}$$

Remember to rewrite the equation as a power of x...

Here, $n = -2$

So $\frac{dy}{dx} = nx^{n-1} = -2x^{-3} = -\frac{2}{x^3}$

...then use the formula to find the derivative.

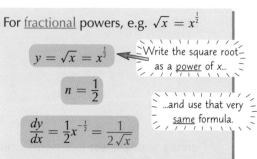

For fractional powers, e.g. $\sqrt{x} = x^{\frac{1}{2}}$

$$y = \sqrt{x} = x^{\frac{1}{2}}$$

Write the square root as a power of x...

$$n = \frac{1}{2}$$

$$\frac{dy}{dx} = \frac{1}{2}x^{-\frac{1}{2}} = \frac{1}{2\sqrt{x}}$$

...and use that very same formula.

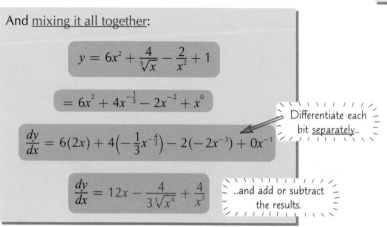

And mixing it all together:

$$y = 6x^2 + \frac{4}{\sqrt[3]{x}} - \frac{2}{x^2} + 1$$

$$= 6x^2 + 4x^{-\frac{1}{3}} - 2x^{-2} + x^0$$

$$\frac{dy}{dx} = 6(2x) + 4\left(-\frac{1}{3}x^{-\frac{4}{3}}\right) - 2(-2x^{-3}) + 0x^{-1}$$

Differentiate each bit separately...

$$\frac{dy}{dx} = 12x - \frac{4}{3\sqrt[3]{x^4}} + \frac{4}{x^3}$$

...and add or subtract the results.

Power Laws:
Differentiation's much easier if you know the Power Laws really well. Like knowing that

$x^1 = x$ and $\sqrt{x} = x^{\frac{1}{2}}$

See page 57 for more info.

Just like page 38 — but fractionally better, I reckon...

Hurrah, hurrah and thrice hurrah! A whole page of stuff that's no more difficult than stuff you've already done in C1. They don't come along that often, so enjoy it. Just make sure you're down with negative and fractional powers, and you'll be set.

Differentiation

I'll say it again for those of you at the back not paying attention.
You need to be able to do everything you did in Differentiation in C1, only with fractional and negative powers...

You still need to be able to find Tangents and Normals...

EXAMPLE Find the tangent to the curve $y = \frac{(x+2)(x+4)}{6\sqrt{x}}$ at the point (4, 4).

Write the equation of the curve in a form you can differentiate.

$$y = \frac{x^2+6x+8}{6x^{\frac{1}{2}}} = \tfrac{1}{6}x^{\frac{3}{2}} + x^{\frac{1}{2}} + \tfrac{4}{3}x^{-\frac{1}{2}}$$

Dividing each term on the top line by the bottom line.

Differentiate it...

$$\frac{dy}{dx} = \tfrac{1}{6}\left(\tfrac{3}{2}x^{\frac{1}{2}}\right) + \tfrac{1}{2}x^{-\frac{1}{2}} + \tfrac{4}{3}\left(-\tfrac{1}{2}x^{-\frac{3}{2}}\right)$$
$$= \tfrac{1}{4}\sqrt{x} + \frac{1}{2\sqrt{x}} - \frac{2}{3\sqrt{x}^3}$$

Find the gradient at the point you're interested in. At $x = 4$,

Remember — the gradient of a tangent is the gradient of the curve at that point.

$$\frac{dy}{dx} = \frac{1}{4}\times 2 + \frac{1}{2\times 2} - \frac{2}{3\times 8} = \frac{2}{3}$$

And so the equation of the tangent is $y - y_1 = \frac{2}{3}(x - x_1)$.

Since the tangent goes through the point (4, 4), the equation becomes $y - 4 = \frac{2}{3}(x-4)$, or after rearranging, $y = \frac{2}{3}x + \frac{4}{3}$.

...and Stationary Points on Curves

EXAMPLE Find any stationary points on the curve $y = 4x^2 + \frac{1}{x}$ and determine their nature.

Rewrite the equation as $y = 4x^2 + x^{-1}$, then differentiate to get: $\frac{dy}{dx} = 8x - x^{-2} = 8x - \frac{1}{x^2}$.

Make this equal to zero to find any stationary points:

$$8x - \frac{1}{x^2} = 0$$
$$\Rightarrow\ 8x = \frac{1}{x^2}\ \Rightarrow\ x^3 = \frac{1}{8}\ \Rightarrow\ x = \frac{1}{2}.$$

At stationary points, $\frac{dy}{dx} = 0$.

Taking the cube root only gives one solution, unlike taking the square root — so here there's only one stationary point.

Plug this back into the original equation to find the corresponding y-value: $y = 4\left(\frac{1}{2}\right)^2 + \frac{1}{\left(\frac{1}{2}\right)} = 3.$

So the coordinates of the stationary point are $\left(\frac{1}{2}, 3\right)$.

Differentiate again to determine the nature of the stationary point: $\frac{d^2y}{dx^2} = 8 + 2x^{-3} = 8 + \frac{2}{x^3}$.

When $x = \frac{1}{2}$, this becomes $\frac{d^2y}{dx^2} = 8 + 2x^{-3} = 8 + \frac{2}{\left(\frac{1}{2}\right)^3} = 8 + 16 = 24.$

24 > 0, so the stationary point is a minimum.

Remember, if $\frac{d^2y}{dx^2} > 0$, it's a minimum, and if $\frac{d^2y}{dx^2} < 0$, its a maximum.

Help me Differentiation — You're still my only hope...

Quite a pleasant little section this. No new stuff to learn really, just those pesky powers to deal with. Nice one.

C2 Section 5 — Practice Questions

Well, another section down. Have a go at these questions to make sure you understood it all, then it's on to the next one...

Warm-up Questions

1) Differentiate these functions with respect to x: a) $y = x^4 + \sqrt{x}$, b) $y = \frac{7}{x^2} - \frac{3}{\sqrt{x}} + 12x^3$.

2) Find the stationary points of the function $y = x^3 + \frac{3}{x}$.
Decide whether each stationary point is a minimum or a maximum.

3) Find the equations of the tangent and the normal to the curve $y = \sqrt{x^3} - 3x - 10$ at $x = 16$.

Exam Questions

1 Given that $y = x^7 + \frac{2}{x^3}$, find:

a) $\frac{dy}{dx}$ *(2 marks)*

b) $\frac{d^2y}{dx^2}$ *(2 marks)*

2 a) Show that the equation $\frac{x^2 + 3x^{\frac{3}{2}}}{\sqrt{x}}$ can be written in the form $x^p + 3x^q$, and state the values of p and q. *(3 marks)*

b) Now let $y = 3x^3 + 5 + \frac{x^2 + 3x^{\frac{3}{2}}}{\sqrt{x}}$. Find $\frac{dy}{dx}$, giving each coefficient in its simplest form. *(4 marks)*

3 Given the curve $y = \frac{1}{\sqrt{x}} + \frac{1}{x}$, find:

a) the gradient of the tangent to the curve at the point $\left(4, \frac{3}{4}\right)$. *(4 marks)*

b) the equation of the normal to the curve at this point. *(4 marks)*

4 A curve is given by the equation $y = x + \frac{1}{x^2}$.

a) Show that the curve has a stationary point when $x = \sqrt[3]{2}$. *(3 marks)*

b) Find the nature of this stationary point. *(3 marks)*

5 A steam train travels between Haverthwaite and Eskdale at a speed of x miles per hour and burns y units of coal, where y is given by: $2\sqrt{x} + \frac{27}{x}$, for $x > 2$.

a) Find the speed that gives the minimum coal consumption. *(5 marks)*

b) Find $\frac{d^2y}{dx^2}$, and hence show that this speed gives the minimum coal consumption. *(2 marks)*

c) Calculate the minimum coal consumption. *(1 mark)*

Integration

Well you can't have one without the other can you? So, here goes — integration of negative and fractional powers...

Up the power by **One** — then **Divide** by it

$$\int x^n \, dx = \frac{x^{n+1}}{n+1} + C$$

It's the Golden Rule of integrating polynomials, and, happily, it still applies to negative and fractional powers.

Examples

For negative powers,

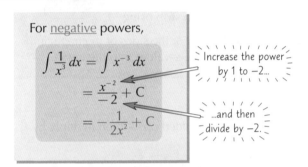

$$\int \frac{1}{x^3}\,dx = \int x^{-3}\,dx = \frac{x^{-2}}{-2} + C = -\frac{1}{2x^2} + C$$

Increase the power by 1 to −2...

...and then divide by −2.

For fractional powers,

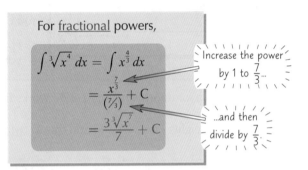

$$\int \sqrt[3]{x^4}\,dx = \int x^{\frac{4}{3}}\,dx = \frac{x^{\frac{7}{3}}}{(7/3)} + C = \frac{3\sqrt[3]{x^7}}{7} + C$$

Increase the power by 1 to $\frac{7}{3}$...

...and then divide by $\frac{7}{3}$.

And for more complicated looking stuff...

$$\int \left(3x^2 - \frac{2}{\sqrt{x}} + \frac{7}{x^2}\right) dx = \int \left(3x^2 - 2x^{-\frac{1}{2}} + 7x^{-2}\right) dx$$

$$= \frac{3x^3}{3} - \frac{2x^{\frac{1}{2}}}{(1/2)} + \frac{7x^{-1}}{-1} + C$$

$$= x^3 - 4\sqrt{x} - \frac{7}{x} + C$$

Do each of these bits separately.

The only time this method doesn't work is when you try to integrate $\frac{1}{x} = x^{-1}$. When you increase the power by 1 (to get zero) and then divide by zero, you get big problems. Luckily, you won't have to deal with this until C3. Phew.

As you may have expected, all of the things you learned about integration in C1 — finding areas under curves and whatnot — still apply. So, here's one final juicy example for you to get your teeth into:

EXAMPLE Find the area under the curve $y = \frac{15}{x^2} - \frac{30}{x^3}$ for $2 \le x \le 10$.

For this, you need to integrate from $x = 2$ up to $x = 10$.

$$A = \int_2^{10} \left(\frac{15}{x^2} - \frac{30}{x^3}\right) dx = 15\int_2^{10} (x^{-2} - 2x^{-3})\,dx$$

$$= 15\left[\frac{x^{-1}}{-1} - \frac{2x^{-2}}{(-2)}\right]_2^{10}$$

$$= 15\left[-\frac{1}{x} + \frac{1}{x^2}\right]_2^{10}$$

$$= 15\left\{\left(-\frac{1}{10} + \frac{1}{100}\right) - \left(-\frac{1}{2} + \frac{1}{4}\right)\right\} = 15 \times \frac{4}{25} = \frac{12}{5}$$

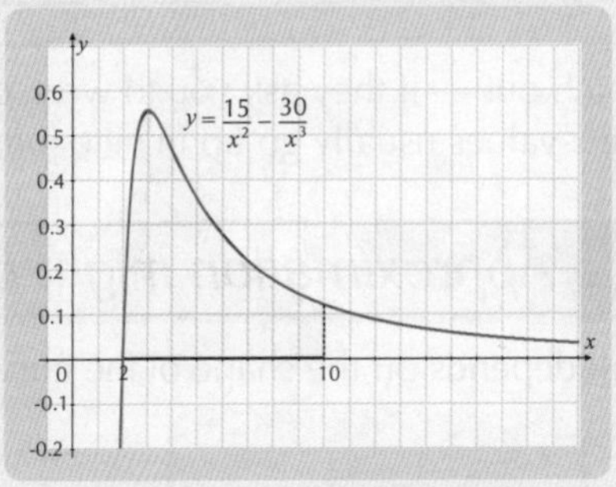

Integration — Int it great...

Well, what is there to say about integration that hasn't already been said? The fractions and negatives can make things a bit tricky — just be careful when dividing by fractions and keep track of where all the minus signs go and you should be fine.

The Trapezium Rule

Sometimes integrals can be just too hard to do using the normal methods — then you need to know other ways to solve them. That's where the Trapezium Rule comes in.

The **Trapezium Rule** is Used to Find the **Approximate Area** Under a Curve

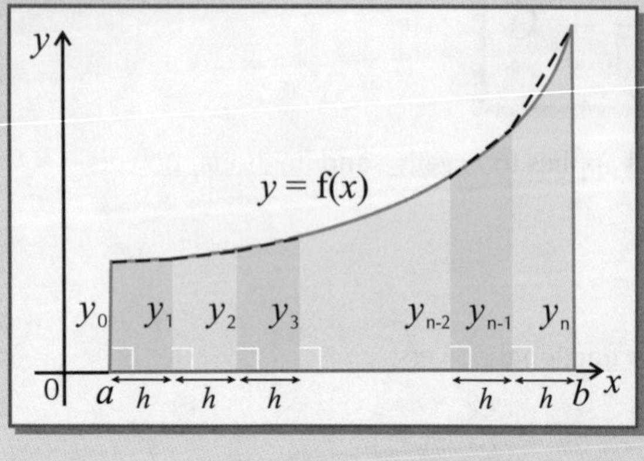

The area of each trapezium is $A = \frac{h}{2}(y_n + y_{n+1})$

The area represented by $\int_a^b y\,dx$ is approximately:

$$\int_a^b y\,dx \approx \frac{h}{2}[y_0 + 2(y_1 + y_2 + \ldots + y_{n-1}) + y_n]$$

where n is the number of strips or intervals and h is the width of each strip.

You can find the width of each strip using $h = \frac{(b-a)}{n}$

$y_0, y_1, y_2, \ldots, y_n$ are the heights of the sides of the trapeziums — you get these by putting the x-values into the curve.

So basically the formula for approximating $\int_a^b y\,dx$ works like this:

'Add the first and last heights $(y_0 + y_n)$ and add this to twice all the other heights added up — then multiply by $\frac{h}{2}$.'

EXAMPLE Find an approximate value for $\int_0^2 \sqrt{4 - x^2}\,dx$ using 4 strips. Give your answer to 4 s.f.

Start by working out the width of each strip: $h = \frac{(b-a)}{n} = \frac{(2-0)}{4} = 0.5$

This means the x-values are $x_0 = 0$, $x_1 = 0.5$, $x_2 = 1$, $x_3 = 1.5$ and $x_4 = 2$ (the question specifies 4 strips, so $n = 4$).

Set up a table and work out the y-values or heights using the equation in the integral.

x	$y = \sqrt{4 - x^2}$
$x_0 = 0$	$y_0 = \sqrt{4 - 0^2} = 2$
$x_1 = 0.5$	$y_1 = \sqrt{4 - 0.5^2} = \sqrt{3.75} = 1.936491673$
$x_2 = 1.0$	$y_2 = \sqrt{4 - 1.0^2} = \sqrt{3} = 1.732050808$
$x_3 = 1.5$	$y_3 = \sqrt{4 - 1.5^2} = \sqrt{1.75} = 1.322875656$
$x_4 = 2.0$	$y_4 = \sqrt{4 - 2.0^2} = 0$

Now put all the y-values into the formula with h and n:

$$\int_a^b y\,dx \approx \frac{0.5}{2}[2 + 2(1.9365 + 1.7321 + 1.3229) + 0]$$
$$\approx 0.25[2 + 2 \times 4.9915]$$
$$\approx 2.996 \text{ to 4 s.f.}$$

Watch out — if they ask you to work out a question with 5 y-values (or 'ordinates') then this is the same as 4 strips. The x-values usually go up in nice jumps — if they don't then check your calculations carefully.

The Approximation might be an **Overestimate** or an **Underestimate**

It all depends on the shape of the curve...

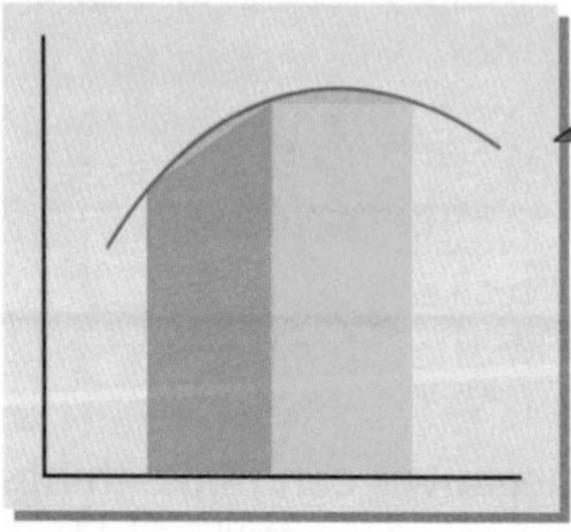

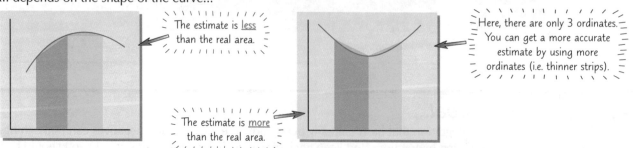

The estimate is less than the real area.

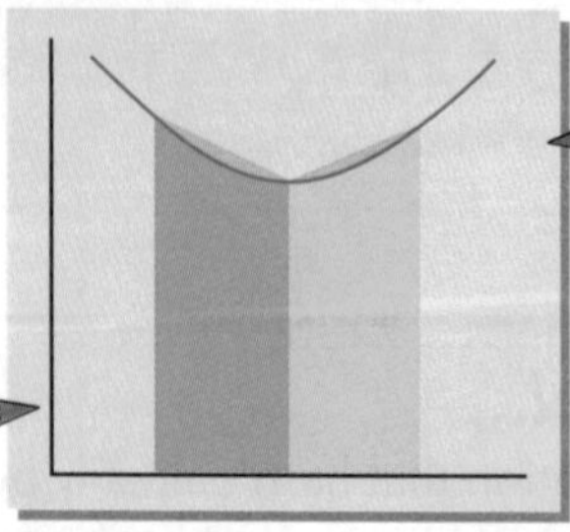

The estimate is more than the real area.

Here, there are only 3 ordinates. You can get a more accurate estimate by using more ordinates (i.e. thinner strips).

The Trapezium Rule

These are usually popular questions with examiners — as long as you're careful there are plenty of marks to be had.

The Trapezium Rule is in the Formula Booklet

...so don't try any heroics — always look it up and use it with these questions.

EXAMPLE Use the trapezium rule with 7 ordinates to find an approximation to $\int_1^{2.2} 2\log_{10}x\,dx$

Remember, 7 ordinates means 6 strips — so $n = 6$.

Calculate the width of the strips: $h = \frac{(b-a)}{n} = \frac{(2.2-1)}{6} = 0.2$

Set up a table and work out the y-values using $y = 2\log_{10}x$:

x	$y = 2\log_{10}x$
$x_0=1.0$	$y_0= 2\log_{10}1 = 0$
$x_1=1.2$	$y_1= 2\log_{10}1.2 = 0.15836$
$x_2=1.4$	$y_2=0.29226$
$x_3=1.6$	$y_3=0.40824$
$x_4=1.8$	$y_4=0.51055$
$x_5=2.0$	$y_5=0.60206$
$x_6=2.2$	$y_6=0.68485$

$y_6 = 2\log_{10}b = 0.68485$

Putting all these values in the formula gives:

$$\int_a^b y\,dx \approx \frac{0.2}{2}[0 + 2(0.15836 + 0.29226 + 0.40824 + 0.51055 + 0.60206) + 0.68485]$$

$$\approx 0.1 \times [0.68485 + 2 \times 1.97147]$$

$$\approx 0.462779$$

$$\approx 0.463 \text{ to 3 d.p.}$$

EXAMPLE Use the trapezium rule with 8 intervals to find an approximation to $\int_0^{\pi} \sin x\,dx$

Whenever you get a calculus question using trig functions, you have to use radians. You'll probably be given a limit with π in, which is a pretty good reminder.

There are 8 intervals, so $n = 8$.

Keep your x-values in terms of π.

Calculate the width of the strips: $h = \frac{(b-a)}{n} = \frac{(\pi-0)}{8} = \frac{\pi}{8}$

Set up a table and work out the y-values:

x	$y = \sin x$
$x_0= 0$	$y_0= \sin 0 = 0$
$x_1 = \frac{\pi}{8}$	$y_1=0.38268$
$x_2 = \frac{\pi}{4}$	$y_2=0.70711$
$x_3 = \frac{3\pi}{8}$	$y_3=0.92388$
$x_4 = \frac{\pi}{2}$	$y_4=1$
$x_5 = \frac{5\pi}{8}$	$y_5=0.92388$
$x_6 = \frac{3\pi}{4}$	$y_6=0.70711$
$x_7 = \frac{7\pi}{8}$	$y_7= 0.38268$
$x_8 = \pi$	$y_8= 0$

So, putting all this in the formula gives:

$$\int_a^b y\,dx \approx \frac{1}{2}\cdot\frac{\pi}{8}[0 + 2(0.383 + 0.707 + 0.924 + 1 + 0.924 + 0.707 + 0.383) + 0]$$

$$\approx \frac{\pi}{16} \times [2 \times 5.028]$$

$$\approx 1.97 \text{ to 3 s.f.}$$

These values are quicker to work-out if you know that the graph is symmetrical.

Maths rhyming slang #3: Dribble and drool — Trapezium rule...

Take your time with trapezium rule questions — it's so easy to make a mistake with all those numbers flying around. Make a nice table showing all your ordinates (careful — this is always one more than the number of strips). Then add up y_1 to y_{n-1} and multiply the answer by 2. Add on y_0 and y_n. Finally, multiply what you've got so far by the width of a strip and divide by 2. It's a good idea to write down what you get after each stage, by the way — then if you press the wrong button (easily done) you'll be able to pick up from where you went wrong. They're not hard — just fiddly.

C2 Section 6 — Practice Questions

Penguins evolved with a layer of blubber under their skin to keep them warm.
They should've saved themselves the effort and done these questions instead — mmm, toasty warm...

Warm-up Questions

1) Evaluate the following definite integrals:

a) $\int_1^2 \left(\frac{8}{x^5} + \frac{3}{\sqrt{x}}\right)dx$, b) $\int_1^6 \frac{3}{x^2}\,dx$.

2) Find area A in the diagram below:

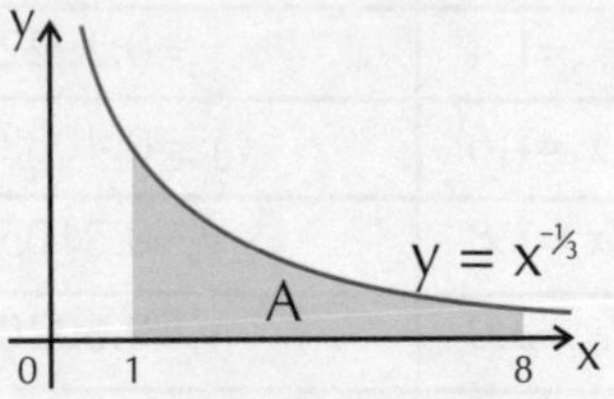

3) Use the trapezium rule with n intervals to estimate:

a) $\int_0^3 (9 - x^2)^{\frac{1}{2}} dx$ with $n = 3$, b) $\int_{0.2}^{1.2} x^{x^2} dx$ with $n = 5$.

They think it's all over...

Exam Questions

1 Find $f(x)$ in each case below. Give each term in its simplest form.

a) $f'(x) = x^{-\frac{1}{2}} + 4 - 5x^3$ *(3 marks)*

b) $f'(x) = 2x + \frac{3}{x^2}$ *(2 marks)*

c) $f'(x) = 6x^2 - \frac{1}{3\sqrt{x}}$ *(2 marks)*

2 a) Show that $(5 + 2\sqrt{x})^2$ can be written in the form $a + b\sqrt{x} + cx$, stating the values of the constants a, b and c. *(3 marks)*

b) Find $\int (5 + 2\sqrt{x})^2 dx$. *(3 marks)*

3 Find the value of $\int_2^7 (2x - 6x^2 + \sqrt{x})dx$. Give your answer to 4 d.p. *(5 marks)*

C2 Section 6 — Practice Questions

4 The curve C has the equation $y = f(x)$, $x > 0$. $f'(x)$ is given as $2x + 5\sqrt{x} + \frac{6}{x^2}$.

A point P on curve C has the coordinates (3, 7).

Find $f(x)$, giving your answer in its simplest form. *(6 marks)*

5 $f'(x) = \frac{1}{\sqrt{36x}} - 2\left(\sqrt{\frac{1}{x^3}}\right)$ where $x > 0$

a) Show that $f'(x) = Ax^{-\frac{1}{2}} - Bx^{-\frac{3}{2}}$ and give the values of A and B. *(3 marks)*

b) The curve C is given by $y = f(x)$ and goes through point P (1, 7).
Find the equation of the curve. *(4 marks)*

6 Curve C has equation $y = f(x)$, $x \neq 0$, where the derivative is given by $f'(x) = x^3 - \frac{2}{x^2}$.

The point P (1, 2) lies on C.

a) Find an equation for the tangent to C at the point P, giving your answer in the form $y = mx + c$, where m and c are integers. *(4 marks)*

b) Find $f(x)$. *(5 marks)*

7 a) Using the trapezium rule with n intervals, estimate the values of:

(i) $\int_2^8 \left(\sqrt{3x^3} + \frac{2}{\sqrt{x}}\right)dx$, $n = 3$ *(4 marks)*

(ii) $\int_1^5 \left(\frac{x^3 - 2}{4}\right)dx$, $n = 4$ *(4 marks)*

b) How could you change your application of the trapezium rule to get better approximations? *(1 mark)*

8 Complete the table and hence use the trapezium rule with 6 ordinates to estimate $\int_{1.5}^{4} y\,dx$.

x	$x_0 = 1.5$	$x_1 =$	$x_2 =$	$x_3 =$	$x_4 = 3.5$	$x_5 = 4.0$
$y = 3x - \sqrt{2^x}$	$y_0 =$	$y_1 = 4$	$y_2 = 5.12156$	$y_3 =$	$y_4 =$	$y_5 = 8.0$

(7 marks)

...it is now.

General Certificate of Education
Advanced Subsidiary (AS) and Advanced Level

Core Mathematics C2 — Practice Exam One

Time Allowed: 1 hour 30 min

Graphical calculators may be used for this exam.
Unless told otherwise, give any non-exact numerical answers to 3 significant figures.
There are 75 marks available for this paper.

1 a) Write down the value of $\log_a a$.

(1 mark)

b) Describe a single geometrical transformation that maps the graph of $y = 4^{2x}$ onto the graph of:

(i) 4^x,

(1 mark)

(ii) 4^{2x+1}.

(1 mark)

c) Solve the equation $4^{2x+1} = 9$.

(3 marks)

d) Given that:

$$\log_a x = \log_a 8 + 2(\log_a 4 - \log_a 2) - 1.$$

Express x in terms of a in a form not involving any logarithms.

(4 marks)

2 The diagram shows a sector of a circle of radius 4 cm and angle 0.2 radians.
Find:

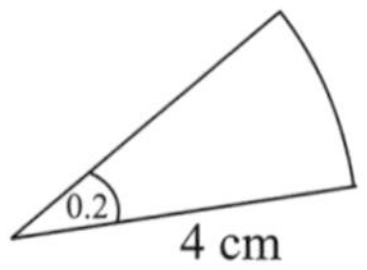

a) the area of the sector,

(2 marks)

b) the perimeter of the sector.

(3 marks)

3 a) A sequence is defined by the recurrence relation $x_{n+1} = 3x_n - 4$.
Given that the first term is 6, find x_4.

(2 marks)

b) The third term of an arithmetic progression is 9 and the seventh term is 33.

(i) Find the first term and the common difference.

(3 marks)

(ii) Find S_{12}, the sum of the first 12 terms in the series.

(3 marks)

(iii) Hence or otherwise find: $\sum_{1}^{12}(6n + 1)$.

(2 marks)

4 A geometric series has 3rd term 81. The sum to infinity of the series is 10 times greater than the first term of the series. Find:

a) the common ratio of the series,

(3 marks)

b) the first term of the series,

(2 marks)

c) the sum of the first seven terms of the series. Give your answer to 1 decimal place.

(2 marks)

5 a) Use the binomial expansion to expand $\left(1+\frac{3}{x^3}\right)^4$. *(5 marks)*

b) Hence find:

(i) $\frac{dy}{dx}$, given that $y=\left(1+\frac{3}{x^3}\right)^4$, *(4 marks)*

(ii) $\int_{-1}^{1}\left(1+\frac{3}{x^3}\right)^4 dx$. *(4 marks)*

6 a) Express $\frac{x^2+2x}{\sqrt{x}}$ in the form x^m+2x^n, where m and n are constants to be found. *(2 marks)*

b) Find $\frac{dy}{dx}$ for $y=\frac{x^2+2x}{\sqrt{x}}+3x^3-x$ (where $x>0$). *(4 marks)*

c) Hence show that the tangent to the curve $y=\frac{x^2+2x}{\sqrt{x}}+3x^3-x$ at the point where $x=1$ is $2y=21x-11$. *(4 marks)*

7 a) Use the trapezium rule with 4 ordinates (3 strips) to find an approximate value for

$$\int_0^3 \sqrt{3+x^2}\,dx.$$

Give your answer to 3 s.f. *(4 marks)*

b) How could you improve your estimate? *(1 mark)*

8 The diagram shows a triangle with sides of length 6 cm, 4 cm and x cm and interior angles of size $\alpha°$, $\beta°$ and 30°. Find:

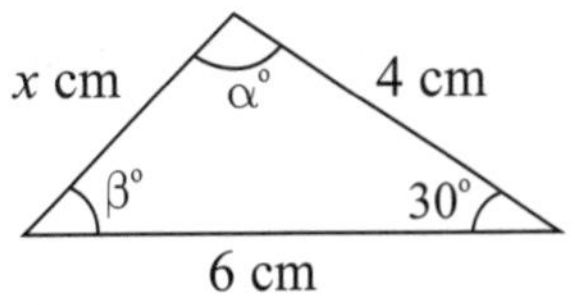

a) the length x, *(2 marks)*

b) α and β, *(3 marks)*

c) the area of the triangle. *(2 marks)*

9 a) Show that the equation

$$\tan^2\theta+\frac{\tan\theta}{\cos\theta}=1$$

can be written in the form $2\sin^2\theta+\sin\theta-1=0$. *(3 marks)*

b) Hence find all solutions to the equation $\tan^2\theta+\frac{\tan\theta}{\cos\theta}=1$ in the interval $0\le\theta\le360°$. *(5 marks)*

General Certificate of Education
Advanced Subsidiary (AS) and Advanced Level

Core Mathematics C2 — Practice Exam Two

Time Allowed: 1 hour 30 min

Graphical calculators may be used for this exam.
Unless told otherwise, give any non-exact numerical answers to 3 significant figures.
There are 75 marks available for this paper.

1 a) Write down the exact value of $36^{-\frac{1}{2}}$. *(2 marks)*

b) Simplify $\frac{x^6 \times x^3}{\sqrt{x^4}} \div x^{\frac{1}{2}}$. *(3 marks)*

c) Hence find $\int_0^1 \left(\frac{x^6 \times x^3}{\sqrt{x^4}} \div x^{\frac{1}{2}}\right) dx$. *(3 marks)*

2 a) Write down the value of $\log_3 3$. *(1 mark)*

b) Given that $\log_a \alpha = \log_a 4 + 3\log_a 2$, show that $\alpha = 32$. *(2 marks)*

c) Solve the equation $0.5^{2x} = 0.3$. *(3 marks)*

3 a) Write down the first four terms in the expansion of $(1 + ax)^{10}$, $a > 0$. *(2 marks)*

b) Find the coefficient of x^2 in the expansion of $(2 + 3x)^5$. *(2 marks)*

c) If the coefficients of x^2 in both expansions are equal, find the value of a. *(2 marks)*

4 a) Sketch the graph of $y = \cos x$ for $0° \le x \le 360°$. *(1 mark)*

b) Show that the equation $2\sin^2 x = 1 + \cos x$ may be written as a quadratic in $\cos x$. *(3 marks)*

c) Hence solve $2\sin^2 x = 1 + \cos x$, giving all values of x such that $0° \le x \le 360°$. *(4 marks)*

d) The curve $y = \cos x$ is translated through $\begin{bmatrix} 60° \\ 1 \end{bmatrix}$ to give the curve $y = f(x)$.

(i) Find an expression for $f(x)$. *(2 marks)*

(ii) Sketch the graph of $f(x)$ for $0° \le x \le 360°$. *(3 marks)*

(iii) Describe the geometrical transformation that maps the curve with equation $y = \cos x$ onto the curve with equation $y = \frac{1}{2}\cos x$. *(2 marks)*

5 A geometric series $u_1 + u_2 + ... + u_n$ has 3rd term $\frac{5}{2}$ and 6th term $\frac{5}{16}$.

a) Find the common ratio and the first term of the series.
Hence give the formula for the nth term of the series. *(4 marks)*

b) Find $\sum_{n=1}^{10} u_n$. Give your answer as a fraction in its simplest terms. *(3 marks)*

c) Show that the sum to infinity of the series is 20. *(2 marks)*

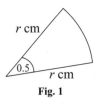

6

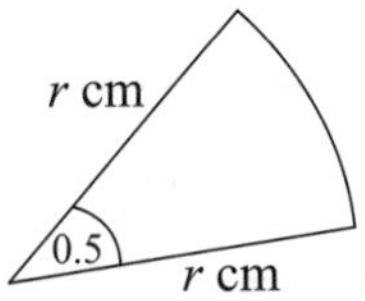

Fig. 1

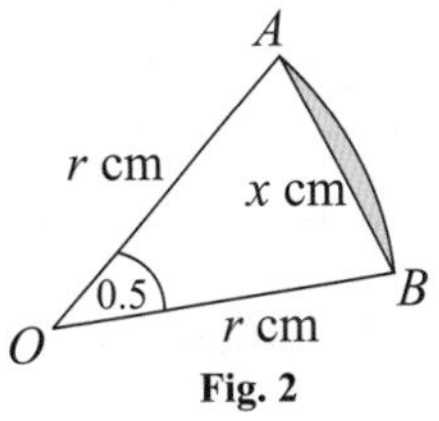

Fig. 2

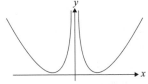

a) Fig. 1 shows a sector of a circle of radius r cm and angle 0.5 radians.
The perimeter of the sector is equal to the area of the sector. Find r. *(6 marks)*

b) Fig. 2 shows a triangle OAB inscribed on the sector. OAB has sides of length $OA = OB = r$ cm and $AB = x$ cm. Find:

(i) the length x, *(2 marks)*

(ii) The area of the shaded region shown in Fig. 2. *(5 marks)*

7 The diagram shows the graph of the curve given by the equation $y = \frac{1}{x^2} + x^2$.

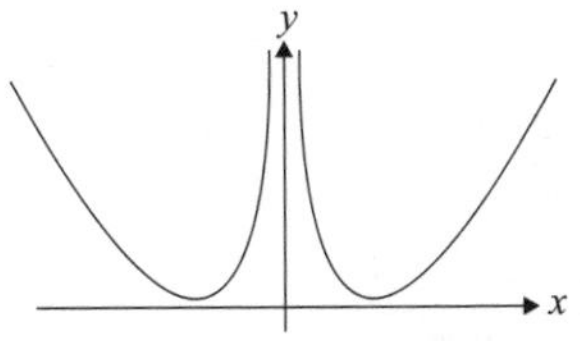

a) Find the coordinates of the stationary points on the curve. *(5 marks)*

b) Find the tangent to the curve at the point where $x = 2$. *(4 marks)*

c) Find the area bounded by the curve, the lines $x = 1$ and $x = 2$ and the x-axis. *(3 marks)*

8 The diagram shows the graph of $y = 2^{x^2}$.

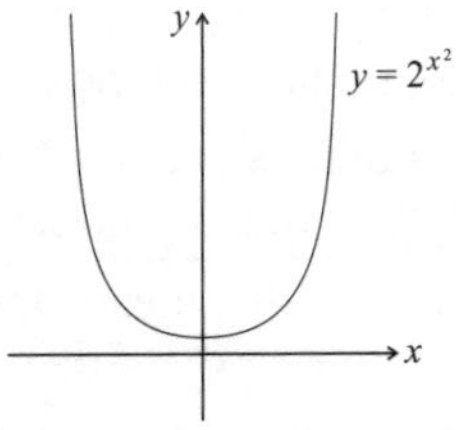

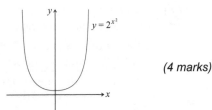

a) Use the trapezium rule with 5 ordinates (4 strips) to find an estimate for the area of the region bounded by the axes, the curve and the line $x = 2$. *(4 marks)*

b) State whether the estimate in a) is an overestimate or an underestimate, giving a reason for your answer. *(2 marks)*

Location: Mean, Median and Mode

The mean, median and mode are measures of location or central tendency (basically... where the centre of the data lies).

The Definitions are really GCSE stuff

You more than likely already know them. But if you don't, learn them now — you'll be needing them loads.

Mean = $\bar{x} = \frac{\Sigma x}{n}$ or $\frac{\Sigma fx}{\Sigma f}$

The Σ (sigma) things just mean you add stuff up — so Σx means you add up all the values of x.

where each x is a data value, f is the frequency of each x value (the number of times it occurs), and n is the total number of data values.

Median = middle data value when all the data values are placed in order of size.

So half the data will be greater than the median, and half will be less than it.

Mode = most frequently occurring data value.

There are two ways to find the median (but they amount to the same thing):

Either: find the $\left(\frac{n+1}{2}\right)$th value in the ordered list.

If ½(n+1) isn't a whole number, take the average of the terms either side.

Or: (i) if $\frac{n}{2}$ is a whole number (i.e. n is even), then the median is the average of this term and the one above.

(ii) if $\frac{n}{2}$ is not a whole number (i.e. n is odd), just round the number up to find the position of the median.

EXAMPLE Find the mean, median and mode of the following list of data: 2, 3, 6, 2, 5, 9, 3, 8, 7, 2

Put in order first: 2, 2, 2, 3, 3, 5, 6, 7, 8, 9

Mode = 2

Mean = $\frac{2+2+2+3+3+5+6+7+8+9}{10}$ = **4.7**

Median = average of 5th and 6th values = **4**

Use a Table when there are a lot of Numbers

EXAMPLE The number of letters received one day in 100 houses was recorded. Find the mean, median and mode of the number of letters.

Number of letters	Number of houses
0	11
1	25
2	27
3	21
4	9
5	7

The number of letters received by each house is a **discrete** quantity (e.g. 3 letters). There isn't a **continuous** set of possible values between getting 3 and 4 letters (e.g. 3.45 letters).

The first thing to do is make a table like this one:

Number of letters x	Number of houses f		fx
0	11	(11)	0
1	25	(36)	25
2	27	(63)	54
3	21		63
4	9		36
5	7		35
totals	100		213

Multiply x by f to get this column.

Put the running total in brackets — it's handy when you're finding the median. (But you can stop when you get past halfway.)

$\Sigma f = 100$ $\Sigma fx = 213$

① The mean is easy — just divide the total of the fx-column (sum of all the data values) by the total of the f-column (= n, the total number of data values).

Mean = $\frac{213}{100}$ = **2.13 letters**

② To find the position of the median, add 1 to the total frequency (= $\Sigma f = n$) and then divide by 2.
Here the median is in position: (100 + 1) ÷ 2 = 50.5.

So the median is halfway between the 50th and 51st data values.

Using your running total of f, you can see that the data values in positions 37 to 63 are all 2s. This means the data values at positions 50 and 51 are both 2 — so **Median = 2 letters**

③ The highest frequency is for 2 letters — so **Mode = 2 letters**

Location: Mean, Median and Mode

Finding the mean is all very well when you have exact data values. But sometimes your data is grouped (e.g. a group might contain 'all the readings between 10 and 20', and so on). You need to know what to do.

If the data's Grouped you'll have to Estimate the Mean

If the data's grouped, you can only estimate the mean and median.

EXAMPLE The height of a number of trees was recorded. The data collected is shown in this table:

Height of tree to nearest m	0 - 5	6 - 10	11 - 15	16 - 20
Number of trees	26	17	11	6

There are no precise readings here — each reading's been put into one of these groups.

Find an estimate of the mean height of the trees, and state which class contains the median.

To estimate the mean, you assume that every reading in a class takes the mid-class value (which you find by adding the lower class boundary to the upper class boundary and dividing by 2).

It's best to make another table...

Height of tree to nearest m	Mid-class value x	Number of trees f	fx
0 - 5	2.75	26 (26)	71.5
6 - 10	8	17 (43)	136
11 - 15	13	11	143
16 - 20	18	6	108
	Totals	60 ($= \Sigma f$)	458.5($= \Sigma fx$)

Lower class boundary = 0.
Upper class boundary = 5.5.
So the mid-class value = (0 + 5.5) ÷ 2 = 2.75.

Estimated mean $= \frac{\sum fx}{\sum f} = \frac{458.5}{60} = 7.64$ **m**

The **median** is in position (60 + 1) ÷ 2 = 30.5 — the running total tells you this must be in the class **6 - 10 m.**

The Mean, Median and Mode are useful for Different Kinds of Data

These three different averages are useful for different kinds of data.

Mean:
- The mean's a good average because you use all your data in working it out.
- But it can be heavily affected by extreme values / outliers.
- And it can only be used with quantitative data (i.e. numbers).

Median: The median is not affected by extreme values, so this is a good average to use when you have outliers.

See page 108 for more about outliers.

Mode:
- The mode can be used even with non-numerical data.
- But some data sets can have more than one mode (and if every value in a data set occurs just once, then the mode isn't very helpful at all).

I can't deny it — these pages really are 'about average'...

If you have large amounts of grouped data ($n > 100$, say), it's usually okay to use the value in position $\frac{n}{2}$ (rather than $\frac{n+1}{2}$) as the median. With grouped data, you can only estimate the median anyway, and if you have a lot of data, that extra 'half a place' doesn't really make much difference. But if in any doubt, use the value in position $\frac{n+1}{2}$ — that'll always be okay.

Dispersion: Range and Interquartile Range

'Dispersion' (or 'variation') means how spread out your data is. There are different ways to measure it — and in certain circumstances, some of these measures might be better than others.

The Range is a Measure of Dispersion

The range is about the simplest measure of dispersion you could imagine.

Range = highest value – lowest value

But the range is heavily affected by extreme values, so it isn't really the most useful way to measure dispersion.

Quartiles divide the data into Four

You've seen how the median divides a data set into two halves. Well, the quartiles divide the data into four parts — with 25% of the data less than the lower quartile, and 75% of the data less than the upper quartile.

There are various ways you can find the quartiles, and they sometimes give different results. But if you use the method below, you'll be fine.

1. To find the lower quartile (Q_1), first work out $\frac{n}{4}$.
   (i) if $\frac{n}{4}$ is a whole number, then the lower quartile is the average of this term and the one above.
   (ii) if $\frac{n}{4}$ is not a whole number, just round the number up to find the position of the lower quartile.

2. To find the upper quartile (Q_3), first work out $\frac{3n}{4}$.
   (i) if $\frac{3n}{4}$ is a whole number, then the upper quartile is the average of this term and the one above.
   (ii) if $\frac{3n}{4}$ is not a whole number, just round the number up to find the position of the upper quartile.

EXAMPLE Find the median and quartiles of the following data: 2, 5, 3, 11, 6, 8, 3, 8, 1, 6, 2, 23, 9, 11, 18, 19, 22, 7.

First put the list in order: 1, 2, 2, 3, 3, 5, 6, 6, 7, 8, 8, 9, 11, 11, 18, 19, 22, 23

You need to find Q_1, Q_2 and Q_3, so work out $\frac{n}{4} = \frac{18}{4}$, $\frac{n}{2} = \frac{18}{2}$, and $\frac{3n}{4} = \frac{54}{4}$.

The median is also known as Q_2.

1) $\frac{n}{4}$ is not a whole number (= 4.5), so round up and take the 5th term: $Q_1 = 3$

2) $\frac{n}{2}$ is a whole number (= 9), so find the average of the 9th and 10th terms: $Q_2 = \frac{7+8}{2} = 7.5$

3) $\frac{3n}{4}$ is not a whole number (= 13.5), so round up and take the 14th term: $Q_3 = 11$

The Interquartile Range is Another Measure of Dispersion

Interquartile range (IQR) = upper quartile (Q_3) – lower quartile (Q_1)

The IQR shows the range of the 'middle 50%' of the data.

EXAMPLE Find the interquartile range of the data in the previous example.

$Q_1 = 3$ and $Q_3 = 11$, so the interquartile range = $Q_3 - Q_1 = 11 - 3 = 8$

The interquartile range is less heavily affected by extreme values than the range.

Sing-a-long-a-stats — "Home, home on the interquartile range..."

Right then... the range and the interquartile range are both measures of how spread out your data is. The range is pretty crude, though — one freakily high or low value in your dataset and it can become completely misleading. The interquartile range is much better, and is easy to work out (easyish, anyway). Make sure all this is clear in your head before moving on.

Dispersion: Standard Deviation

Standard deviation and variance both measure how spread out the data is from the mean — the bigger the variance, the more spread out your readings are.

The Formulas look pretty Tricky

$$\textbf{Variance} = \frac{\sum(x-\bar{x})^2}{n} = \frac{\sum x^2}{n} - \bar{x}^2 \quad \text{or} \quad \textbf{Variance} = \frac{\sum fx^2}{\sum f} - \bar{x}^2$$

$$\textbf{Standard deviation} = \sqrt{\text{variance}}$$

The formula is easier to use in these forms.

The formulas on this page are for working out the variance and standard deviation of data for an entire group of people/items (the population — see p134). There's information about the standard deviation of a sample on p135.

The x-values are the data, $\bar{x}$ is the mean, f is the frequency of each x, and n $(= \sum f)$ is the number of data values.

EXAMPLE Find the mean and standard deviation of the following numbers: 2, 3, 4, 4, 6, 11, 12

1) Find the total of the numbers first: $\sum x = 2 + 3 + 4 + 4 + 6 + 11 + 12 = 42$
2) Then the mean is easy: Mean $= \bar{x} = \frac{\sum x}{n} = \frac{42}{7} = 6$
3) Next find the sum of the squares: $\sum x^2 = 4 + 9 + 16 + 16 + 36 + 121 + 144 = 346$
4) Use this to find the variance: Variance $= \frac{\sum x^2}{n} - \bar{x}^2 = \frac{346}{7} - 6^2 = \frac{346 - 252}{7} = \frac{94}{7}$
5) And take the square root to find the standard deviation: Standard deviation $= \sqrt{\frac{94}{7}} = 3.66$ to 3 sig. fig.

Questions about Standard Deviation can look a bit Weird

They can ask questions about standard deviation in different ways. But you just need to use the same old formulas.

EXAMPLE The mean of 10 boys' heights is 180 cm, and the standard deviation is 10 cm. The mean for 9 girls is 165 cm, and the standard deviation is 8 cm. Find the mean and standard deviation of the whole group of 19 girls and boys.

① Let the boys' heights be x and the girls' heights be y.

Write down the formula for the mean and put the numbers in for the boys: $\bar{x} = \frac{\sum x}{n} \Rightarrow 180 = \frac{\sum x}{10} \Rightarrow \sum x = 1800$

Do the same for the girls: $165 = \frac{\sum y}{9} \Rightarrow \sum y = 1485$

So the sum of the heights for the boys and the girls = $\sum x + \sum y = 1800 + 1485 = 3285$

And the mean height of the boys and the girls is: $\frac{3285}{19} = 172.9$ cm

Round the fraction to 1 d.p. to give your answer. But if you need to use the mean in more calculations, use the fraction (or your calculator's memory) so you don't lose accuracy.

② Now the variance — boys first: Variance $= \frac{\sum x^2}{n} - \bar{x}^2 \Rightarrow 10^2 = \frac{\sum x^2}{10} - 180^2 \Rightarrow \sum x^2 = 10 \times (100 + 32\,400) = 325\,000$

Do the same for the girls: Variance $= \frac{\sum y^2}{n} - \bar{y}^2 \Rightarrow 8^2 = \frac{\sum y^2}{9} - 165^2 \Rightarrow \sum y^2 = 9 \times (64 + 27\,225) = 245\,601$

Okay, so the sum of the squares of the heights of the boys and the girls is: $\sum x^2 + \sum y^2 = 325\,000 + 245\,601 = 570\,601$

So for all the heights, the variance is: Variance $= \frac{570\,601}{19} - \left(\frac{3285}{19}\right)^2 = 139.0$ cm^2

Don't use the rounded mean (172.9) — you'll lose accuracy.

And finally the standard deviation of the boys and the girls is: standard deviation $= \sqrt{139.0} = 11.8$ cm

Phew.

People who enjoy this stuff are standard deviants...

The formula for the variance looks pretty scary, what with the x^2's and $\bar{x}$'s floating about. But it comes down to 'the mean of the squares minus the square of the mean'. That's how I remember it anyway — and my memory's rubbish.

Dispersion and Outliers

As on p105, if you've got grouped data, then everything's just that little bit more awkward.

Use Mid-Class Values if your data's Grouped

With grouped data, assume every reading takes the mid-class value. Then use the frequencies to find $\sum fx$ and $\sum fx^2$.

EXAMPLE The heights of sunflowers in a garden were measured and recorded in the table below. Estimate the mean height and the standard deviation.

Height of sunflower, h (cm)	$150 \le h < 170$	$170 \le h < 190$	$190 \le h < 210$	$210 \le h < 230$
Number of sunflowers	5	10	12	3

Draw up another table, and include columns for the mid-class values x, as well as fx and fx^2:

Height of sunflower (cm)	Mid-class value, x	x^2	f	fx	fx^2
$150 \le h < 170$	160	25600	5	800	128000
$170 \le h < 190$	180	32400	10	1800	324000
$190 \le h < 210$	200	40000	12	2400	480000
$210 \le h < 230$	220	48400	3	660	145200
		Totals	30 $(= \Sigma f)$	5660 $(= \Sigma fx)$	1077200 $(= \Sigma fx^2)$

fx^2 means $f \times (x^2)$ — not $(fx)^2$.

With the totals in the table, you can calculate the mean, variance, and standard deviation:

$$\text{Mean} = \bar{x} = \frac{\sum fx}{\sum f} = \frac{5660}{30} = 189 \text{ cm to 3 sig. fig.}$$

$$\text{Variance} = \frac{\sum fx^2}{\sum f} - \bar{x}^2 = \frac{1077\,200}{30} - \left(\frac{5660}{30}\right)^2 = 312 \text{ to 3 sig. fig.}$$

$$\text{Standard deviation} = \sqrt{\text{variance}} = 17.7 \text{ cm to 3 sig. fig.}$$

Outliers Affect what Measure of Dispersion is Best to Use

An outlier is a freak piece of data that lies a long way from the rest of the readings.

1) Outliers affect whether the variance and standard deviation are good measures of dispersion.
2) Outliers can make the variance and standard deviation much larger than they would otherwise be — which means these freak pieces of data are having more influence than they deserve.
3) If a data set contains outliers, then a better measure of dispersion is the interquartile range.

Outliers also affect the range — see p106.

EXAMPLE This table summarises the marks obtained in Maths 'calculator' and 'non-calculator' papers. Compare the location and dispersion of the distributions.

Calculator Paper		Non-calculator paper
28	Minimum	12
78	Maximum	82
40	Lower quartile, Q_1	35
58	Median, Q_2	42
70	Upper quartile, Q_3	56
55	Mean	46.1
21.2	Standard deviation	17.8

Location: The mean, the median and the quartiles are all higher for the calculator paper. This means that scores were generally higher on the calculator paper.

Although the maximum mark on the non-calculator paper was higher than the maximum mark on the calculator paper, this doesn't say anything about the results generally.

Dispersion: The range for the calculator paper is $78 - 28 = 50$.
The range for the non-calculator paper is $82 - 12 = 70$.
The interquartile range (IQR) for the calculator paper is $Q_3 - Q_1 = 70 - 40 = 30$.
The interquartile range (IQR) for the non-calculator paper is $Q_3 - Q_1 = 56 - 35 = 21$.

So the IQR and the standard deviation are both higher for the calculator paper, so it looks like the scores on the calculator paper are more spread out than for the non-calculator paper.
Although the range is higher for the non-calculator paper, this might not be a reliable guide to the dispersion since the data may well contain outliers.

'Outlier' is the name I give to something that my theory can't explain...

Measures of location and dispersion are supposed to capture the essential characteristics of a data set in just one or two numbers. So don't choose a measure that's heavily affected by freaky, far-flung outliers — it won't be much good.

Linear Scaling

Linear Scaling can make the *Numbers* much *Easier*

Linear scaling means doing something to every reading (like adding or multiplying by a number) to make life easier.

Finding the mean of 1001, 1002 and 1006 looks hard(ish). But take 1000 off each number and finding the mean of what's left (1, 2 and 6) is much easier — it's 3. So the mean of the original numbers must be 1003.

You usually change your original variable, x, to an easier one to work with, y (so here, if $x = 1001$, then $y = 1$).

Write down a formula connecting the two variables: e.g. $y = \frac{x-b}{a}$.

Then $\bar{y} = \frac{\bar{x}-b}{a}$ where $\bar{x}$ and $\bar{y}$ are the means of variables x and y.

Also $\text{standard deviation of } y\text{'s} = \frac{\text{standard deviation of } x\text{'s}}{a}$

You can add/subtract a number, and multiply/divide by one as well — it all depends on what will make life easiest.

Note that if you don't multiply or divide your readings by anything (i.e. if $a = 1$), then the standard deviation isn't changed.

EXAMPLE Find the mean and standard deviation of: 1 000 020, 1 000 040, 1 000 010 and 1 000 050.

The obvious thing to do is subtract a million from every reading to leave 20, 40, 10 and 50.
Then make life even simpler by dividing by 10 — giving 2, 4, 1 and 5.

1. So use: $y = \frac{x - 1000\,000}{10}$. Then $\bar{y} = \frac{\bar{x} - 1000\,000}{10}$ and $\text{s.d. of } y = \frac{\text{s.d. of } x}{10}$.

   s.d. = 'standard deviation'

2. Find the mean and standard deviation of the y values: $\bar{y} = \frac{2+4+1+5}{4} = 3$

   $\text{s.d. of } y = \sqrt{\frac{2^2+4^2+1^2+5^2}{4} - 3^2} = \sqrt{\frac{46}{4} - 9} = \sqrt{2.5} = 1.58$ to 3 sig. fig.

3. Then use the formulas to find the mean and standard deviation of the original values:

   $\bar{x} = 10\bar{y} + 1000\,000 = (10 \times 3) + 1000\,000 = 1000\,030$

   $\text{s.d. of } x = 10 \times \text{s.d. of } y = 10 \times 1.58 = 15.8$

And so variance of the x's = (s.d. of x's)^2 = (10 × s.d. of y's)^2 = 10^2 × (s.d. of y's)^2 = 10^2 × variance of the y's.

You can use these linear-scaling tricks with *Summarised Data*

This kind of question looks tricky at first — but use the same old formulas and it's a piece of cake.

EXAMPLE A set of 10 numbers (x-values) can be summarised as shown:
Find the mean and standard deviation of the numbers. $\sum(x-10) = 15$ and $\sum(x-10)^2 = 100$

1. Okay, the obvious first thing to try is: $y = x - 10$ ⟹ That means: $\sum y = 15$ and $\sum y^2 = 100$

2. Work out $\bar{y}$ and the standard deviation of the y's using the normal formulas: $\bar{y} = \frac{\sum y}{n} = \frac{15}{10} = 1.5$

   $\text{Variance of } y = \frac{\sum y^2}{n} - \bar{y}^2 = \frac{100}{10} - 1.5^2 = 10 - 2.25 = 7.75$

   so standard deviation of $y = \sqrt{7.75} = 2.78$ to 3 sig. fig.

3. Then finding the mean and standard deviation of the x-values is easy: $\bar{x} = \bar{y} + 10 = 1.5 + 10 = 11.5$

   The standard deviation of x is the same as the standard deviation of y since you've only subtracted 10 from every number. ⟹ s.d. of x = s.d. of y = 2.78 to 3 sig. fig.

Linear scaling isn't the biggest mountain you'll need to climb in Stats...

Linear scaling isn't hard, but it does need to be done carefully. And remember... adding/subtracting a number from every reading won't change the spread (the variance or standard deviation), but multiplying/dividing readings by something will.

S1 Section 1 — Practice Questions

It's important to make sure you know this stuff like the back of your hand — but unlike the back of your hand you won't have this book in the exam. So here are some practice questions to help you remember it all. We'll start off gently...

Warm-up Questions

1) Calculate the mean, median and mode of the data in this table. ⟹

x	0	1	2	3	4
f	5	4	4	2	1

2) The speeds of 60 cars travelling in a 40 mph speed limit area were measured to the nearest mph. The data is summarised in the table. Estimate the mean, and state which class contains the median.

Speed (mph)	30 - 34	35 - 39	40 - 44	45 - 50
Frequency	12	37	9	2

3) Find the median and quartiles of the data below.
Amount of pocket money (in £) received per week by twenty 15-year-olds:
10, 5, 20, 50, 5, 1, 6, 5, 15, 20, 5, 7, 5, 10, 12, 4, 8, 6, 7, 30.

4) Find the mean and standard deviation of the following numbers: 11, 12, 14, 17, 21, 23, 27.

5) The scores in an IQ test for 50 people are recorded in the table.

Score	100 - 106	107 - 113	114 - 120	121 - 127	128 - 134
Frequency	6	11	22	9	2

Estimate the mean and variance of the distribution.

6) A survey was carried out on the number of cars owned by 19 households. The results are shown in the table.

Number of cars owned	1	2	3	4	5	6	7
Frequency	8	7	2	0	0	0	2

a) Find the mean, median and mode of the data.

b) Find the following four measures of dispersion for the data: range, interquartile range, variance and standard deviation.

7) For a set of data, $n = 100$, $\sum(x - 20) = 125$, and $\sum(x - 20)^2 = 221$.
Find the mean and standard deviation of x.

8) The time taken (to the nearest minute) for a commuter to travel to work on 20 consecutive work days is recorded in the table. Use linear scaling to estimate the mean and standard deviation of the times.

Time to nearest minute	30 - 33	34 - 37	38 - 41	42 - 45
Frequency	3	6	7	4

S1 Section 1 — Practice Questions

Well wasn't that lovely. I bet you're now ready and raring to test yourself with some proper exam-style questions. Here are some I made earlier, lucky you.

Exam Questions

1 A group of 19 people played a game. The scores, x, that the people achieved are summarised by:

$$\sum(x-30) = 228 \text{ and } \sum(x-30)^2 = 3040$$

a) Calculate the mean and the standard deviation of the 19 scores. *(3 marks)*

b) Show that $\sum x = 798$ and $\sum x^2 = 33\,820$. *(3 marks)*

c) Another student played the game. Her score was 32.
Find the new mean and standard deviation of all 20 scores. *(4 marks)*

2 Two workers iron clothes. Each irons 10 items, and records the time it takes for each, to the nearest minute:

Worker A: 3 5 2 7 10 4 5 5 4 12
Worker B: 3 4 8 6 7 8 9 10 11 9

a) For both workers, find:
(i) the median *(2 marks)*
(ii) the lower and upper quartiles *(4 marks)*

b) Make one statement comparing the two sets of data. *(1 mark)*

c) Which worker would be better to employ? Give a reason for your answer. *(1 mark)*

3 In a supermarket two types of chocolate drops were compared.
The weights, a grams, of 20 chocolate drops of brand A are summarised by:

$$\Sigma a = 60.3\text{ g} \qquad \Sigma a^2 = 219\text{ g}^2$$

The mean weight of 30 chocolate drops of brand B was 2.95 g, and the standard deviation was 1 g.

a) Find the mean weight of a brand A chocolate drop. *(1 mark)*

b) Find the standard deviation of the weight of the brand A chocolate drops. *(3 marks)*

c) Compare brands A and B. *(2 marks)*

d) Find the standard deviation of the weight of all 50 chocolate drops. *(4 marks)*

4 The table shows the number of hits received by people at a paint ball party.

No. of Hits, x	12	13	14	15	16	17	18	19	20	21	22	23	24	25
Frequency, f	1	2	3	2	1	1	0	0	0	0	3	0	0	1

a) Find the mean, median and mode number of hits. *(5 marks)*

b) State which of these three averages you consider to be **least** appropriate for summarising the number of hits. Give a reason for your answer. *(2 marks)*

Random Events and Probabilities

Random events happen by chance. Probability is a measure of how likely they are. It can be a chancy business.

A Random Event has Various Outcomes

1) In a trial (or experiment) the things that can happen are called outcomes (so if I time how long it takes to eat my dinner, 63 seconds is a possible outcome).
2) Events are 'groups' of one or more outcomes (so an event might be 'it takes me less than a minute to eat my dinner every day one week').
3) When all outcomes are equally likely, you can work out the probability of an event by counting the outcomes:

$$\text{P(event)} = \frac{\text{Number of outcomes where event happens}}{\text{Total number of possible outcomes}}$$

EXAMPLE Suppose I've got a bag with 15 balls in — 5 red, 6 blue and 4 green.

If I take a ball out without looking, then any ball is equally likely — there are 15 possible outcomes. Of these 15 outcomes, 5 are red, 6 are blue and 4 are green. And so...

$\text{P(red ball)} = \frac{5}{15} = \frac{1}{3}$ $\text{P(blue ball)} = \frac{6}{15} = \frac{2}{5}$ $\text{P(red or green ball)} = \frac{5+4}{15} = \frac{9}{15} = \frac{3}{5}$

If I then do 90 trials (i.e. I pick a ball out 90 times, replacing the ball each time), then I would expect to pick:

a red ball $\frac{1}{3} \times 90 = 30$ times   a blue ball $\frac{2}{5} \times 90 = 36$ times   either a red or a green ball $\frac{3}{5} \times 90 = 54$ times

You can also use relative frequencies to assign probabilities — you use the results of trials you've already carried out.

$$\text{P(event)} = \frac{\text{Number of trials where event happened}}{\text{Total number of trials carried out}}$$

Two-Way Tables can help you work out Numbers of Outcomes

EXAMPLE The numbers of different types of pepper for sale in a greengrocer's are shown in the table below. If I select one pepper at random, find the probability that it is:
a) small, b) green, c) large and red, d) yellow or small

	Red	Yellow	Green	Total
Small	12	14	18	**44**
Large	19	16	21	**56**
Total	**31**	**30**	**39**	**100**

a) There are 44 small peppers out of a total of 100. So the probability that a randomly selected pepper will be small is: $\text{P(small)} = \frac{44}{100} = \frac{11}{25}$

Similarly...

b) $\text{P(green)} = \frac{39}{100}$ c) $\text{P(large and red)} = \frac{19}{100}$

d) $\text{P(yellow or small)} = \frac{30+44-14}{100} = \frac{60}{100} = \frac{3}{5}$

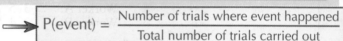

Find the number of peppers that are either yellow or small by adding the totals for 'yellow' and 'small' — but you then need to subtract 14 else you've counted the 14 small yellow peppers twice.

Venn Diagrams can show which Outcomes correspond to which Events

Say you've got 2 events, A and B — a Venn diagram can show which outcomes satisfy event A, which satisfy B, which satisfy both, and which satisfy neither.

(i) All outcomes satisfying event A go in one part of the diagram, and all outcomes satisfying event B go in another bit.

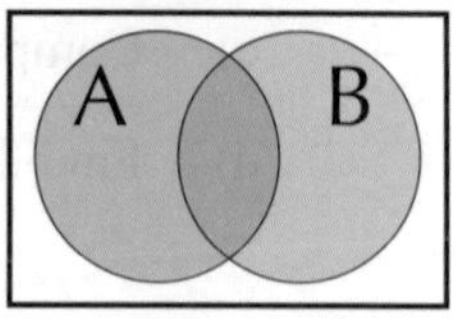

(ii) If they satisfy 'both A and B', they go in the dark green middle bit, written $A \cap B$ (and called the intersection of A and B).

(iii) The whole of the green area is written $A \cup B$ — it means 'either A or B' (and is called the union of A and B).

Again, you can work out probabilities of events by counting outcomes and using the formula above. You can also get a nice formula linking $P(A \cap B)$ and $P(A \cup B)$.

$$P(A \cup B) = P(A) + P(B) - P(A \cap B)$$

If you just add up the outcomes in A and B, you end up counting $A \cap B$ twice — that's why you have to subtract it.

EXAMPLE If you roll a dice, event A could be 'I get an even number', and B 'I get a number bigger than 4'. The Venn diagram would be:

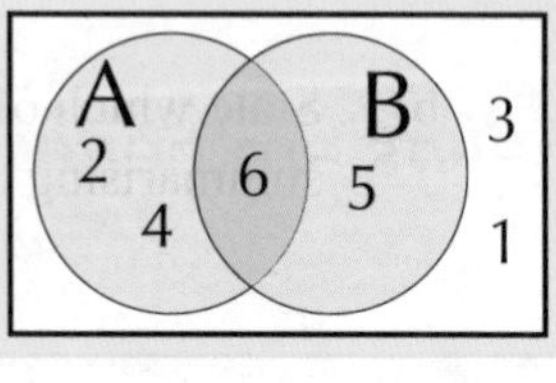

$P(A) = \frac{3}{6} = \frac{1}{2}$ $P(B) = \frac{2}{6} = \frac{1}{3}$ $P(A \cap B) = \frac{1}{6}$ $P(A \cup B) = \frac{4}{6} = \frac{2}{3}$

Here, I've just counted outcomes — but I could have used the formula.

Random Events and Probabilities

Venn diagrams can also show probabilities.

EXAMPLE A survey was carried out to find what pets people like.

The probability they like dogs is 0.6. The probability they like cats is 0.5. The probability they like gerbils is 0.4. The probability they like dogs and cats is 0.4. The probability they like cats and gerbils is 0.1, and the probability they like gerbils and dogs is 0.2. Finally, the probability they like all three kinds of animal is 0.1.

You can draw all this in a Venn diagram. (Here I've used C for 'likes cats', D for 'likes dogs' and G for 'likes gerbils'.)

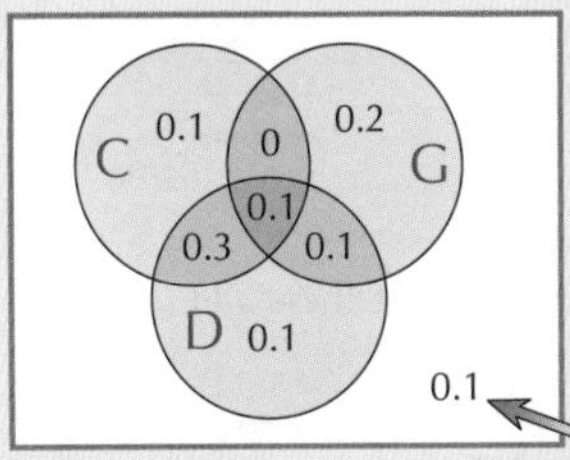

1) Stick in the middle one first — 'likes all 3 animals' (i.e. $C \cap D \cap G$).
2) Then do the 'likes 2 animals' probabilities by taking 0.1 from each of the given 'likes 2 animals' probabilities. (If they like 3 animals, they'll also be in the 'likes 2 animals' bits.)
3) Then do the 'likes 1 kind of animal' probabilities, by making sure the total probability in each circle adds up to the probability in the question.
4) Finally, subtract all the probabilities so far from 1 to find 'likes none of these animals'.

① From the Venn diagram, the probability that someone likes either dogs or cats is 0.7.

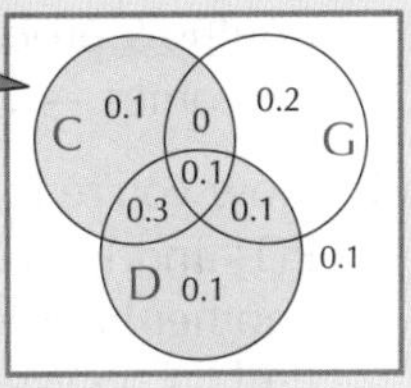

② The probability that someone likes gerbils but not dogs is 0.2.

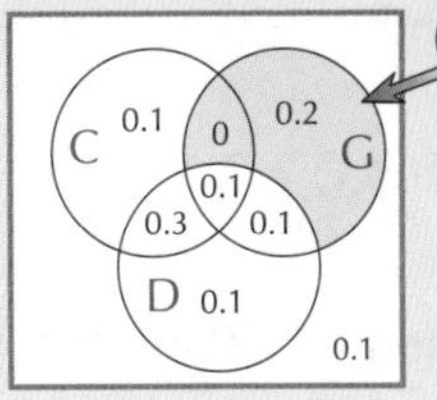

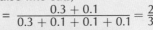

③ You can work out the probability that a dog-lover also like cats by ignoring everything outside the 'dogs' circle.

P(dog-lover also like cats)

$$= \frac{0.3 + 0.1}{0.3 + 0.1 + 0.1 + 0.1} = \frac{2}{3}$$

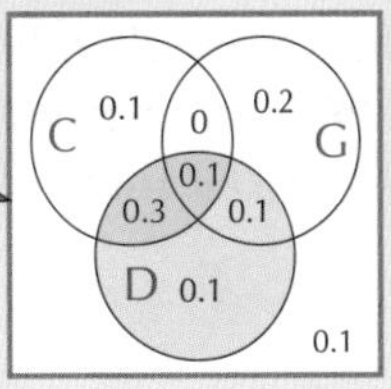

The Complement of 'Event A' is 'Not Event A'

An event A will either happen or not happen. The event 'A doesn't happen' is called the complement of A (or A').
On a Venn diagram, it would look like the picture on the right.
At least one of A and A' has to happen, so...

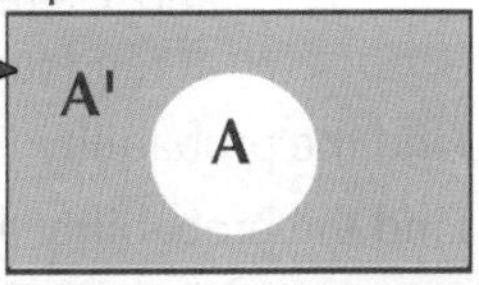

$$P(A) + P(A') = 1 \quad \text{or} \quad P(A') = 1 - P(A)$$

EXAMPLE A teacher keeps socks loose in a box. One morning, he picks out a sock. He calculates that the probability of then picking out a matching sock is 0.56. What is the probability of him not picking a matching sock?

Call event A 'picks a matching sock'. Then A' is 'doesn't pick a matching sock'. Now A and A' are complementary events (and P(A) = 0.56), so $P(A) + P(A') = 1$, and therefore $P(A') = 1 - 0.56 = 0.44$

Mutually Exclusive Events Have No Overlap

If two events can't both happen at the same time (i.e. $P(A \cap B) = 0$) they're called mutually exclusive (or just 'exclusive').
If A and B are exclusive, then the probability of A or B is: $P(A \cup B) = P(A) + P(B)$.

Use the formula from page 112, but put $P(A \cap B) = 0$.

More generally, For n exclusive events (i.e. only one of them can happen at a time):

$$P(A_1 \cup A_2 \cup ... \cup A_n) = P(A_1) + P(A_2) + ... + P(A_n)$$

EXAMPLE Find the probability that a card pulled at random from a standard pack of cards (no jokers) is either a picture card (a Jack, Queen or King) or the 7, 8 or 9 of clubs.

Call event A — 'I get a picture card', and event B — 'I get the 7, 8 or 9 of clubs'.

Events A and B are mutually exclusive — they can't both happen. Also, $P(A) = \frac{12}{52} = \frac{3}{13}$ and $P(B) = \frac{3}{52}$.

So the probability of either A or B is: $P(A \cup B) = P(A) + P(B) = \frac{12}{52} + \frac{3}{52} = \frac{15}{52}$

Two heads are better than one — though only half as likely using two coins...

You'll never actually be asked to draw a Venn diagram, but they're really useful for answering loads of questions. And one other good thing is that Venn diagrams look, well, nice somehow. More importantly, when you're filling one in, the thing to remember is that you usually need to 'start from the inside and work out'. I hope that's all clear.

Tree Diagrams

Tree Diagrams — they blossom from a tiny question-acorn into a beautiful tree of possibility. Inspiring and useful.

Tree Diagrams Show Probabilities for *Two or More* Events

Each 'chunk' of a tree diagram is a trial, and each branch of that chunk is a possible outcome. Multiplying probabilities along the branches gives you the probability of a series of outcomes.

EXAMPLE If Susan plays tennis one day, the probability that she'll play the next day is 0.2. If she doesn't play tennis, the probability that she'll play the next day is 0.6. She plays tennis on Monday. What is the probability she plays tennis:

(i) on both the Tuesday and Wednesday of that week?
(ii) on the Wednesday of the same week?

Let T mean 'plays tennis' (and then T' means 'doesn't play tennis').

(i) Then the probability that she plays on Tuesday and Wednesday is P(T and T) = 0.2 × 0.2 = 0.04 (multiply probabilities since you need a series of outcomes — T and then T).

(ii) Now you're interested in either P(T and T) or P(T' and T). To find the probability of one event or another happening, you have to add probabilities:
P(plays on Wednesday) = 0.04 + 0.48 = 0.52.

Notice that these add up to 1.

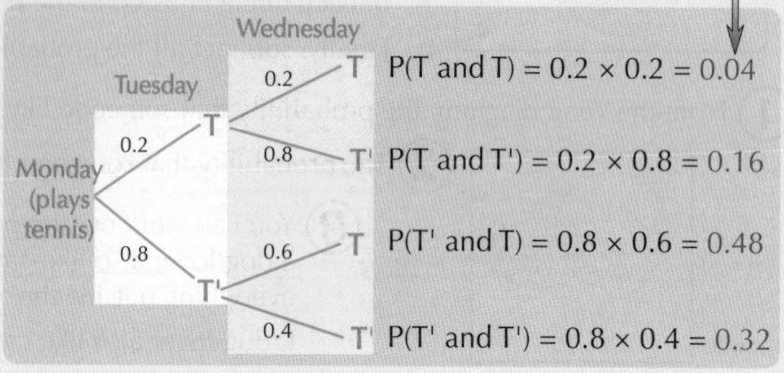

Sometimes a Branch is *Missing*

EXAMPLE A box of biscuits contains 5 chocolate biscuits and 1 lemon biscuit. George takes out 3 biscuits at random, one at a time, and eats them.

a) Find the probability that he eats 3 chocolate biscuits.
b) Find the probability that the last biscuit is chocolate.

Let C mean 'picks a chocolate biscuit' and L mean 'picks the lemon biscuit'.

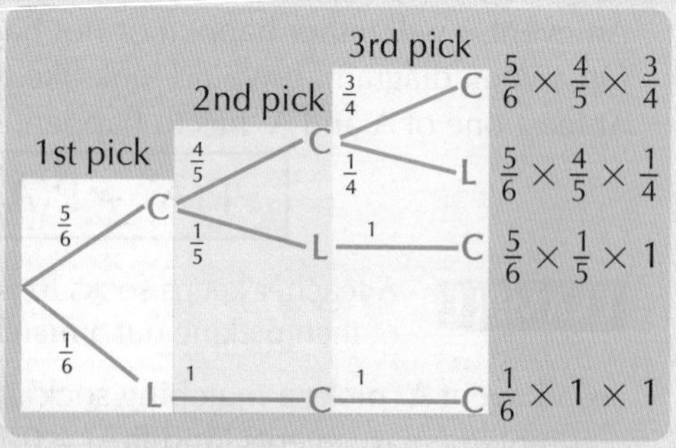

After the lemon biscuit there are only chocolate biscuits left, so the tree diagram doesn't 'branch' after an 'L'.

a) Three chocolate biscuits is shown by only one 'path' along the branches.

$$P(C \text{ and } C \text{ and } C) = \frac{5}{6} \times \frac{4}{5} \times \frac{3}{4} = \frac{60}{120} = \frac{1}{2}$$

b) The third biscuit being chocolate is shown by 3 'paths' along the branches — so you can add up the probabilities:

$$P(\text{third biscuit is chocolate}) = \left(\frac{5}{6} \times \frac{4}{5} \times \frac{3}{4}\right) + \left(\frac{5}{6} \times \frac{1}{5} \times 1\right) + \left(\frac{1}{6} \times 1 \times 1\right) = \frac{1}{2} + \frac{1}{6} + \frac{1}{6} = \frac{5}{6}$$

There's a quicker way to do this, since there's only one outcome where the chocolate isn't picked last:

$$P(\text{third biscuit is not chocolate}) = \frac{5}{6} \times \frac{4}{5} \times \frac{1}{4} = \frac{1}{6}, \text{ so } P(\text{third biscuit is chocolate}) = 1 - \frac{1}{6} = \frac{5}{6}$$

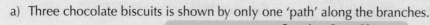

You won't be asked to draw a tree diagram — but they'll help you answer some of your questions if you use them.

Working out the probability of the complement of the event you're interested in is sometimes easier.

Sampling *with replacement* — the probabilities stay the same

In the above example, each time George takes a biscuit he eats it before taking the next one (i.e. he doesn't replace it) — this is sampling without replacement. Suppose instead that each time he takes a biscuit he puts it back in the box before taking the next one — this is sampling with replacement. All this means is that the probability of choosing a particular item remains the same for each pick.

So part a) above becomes:

$$P(C \text{ and } C \text{ and } C) = \frac{5}{6} \times \frac{5}{6} \times \frac{5}{6} = \frac{125}{216} > \frac{1}{2}$$

So the probability that George picks 3 chocolate biscuits is slightly greater when sampling is done with replacement. This makes sense because now there are, on average, more chocolate biscuits available for his 2nd and 3rd picks, so he is more likely to choose one.

Conditional Probability

After the first set of branches, tree diagrams actually show <u>conditional probabilities</u>. Read on...

P(B|A) means Probability of B, given that A has Already Happened

Conditional probability means the probability of something, given that something else has already happened. For example, P(B|A) means the probability of B, given that A has already happened. Back to tree diagrams...

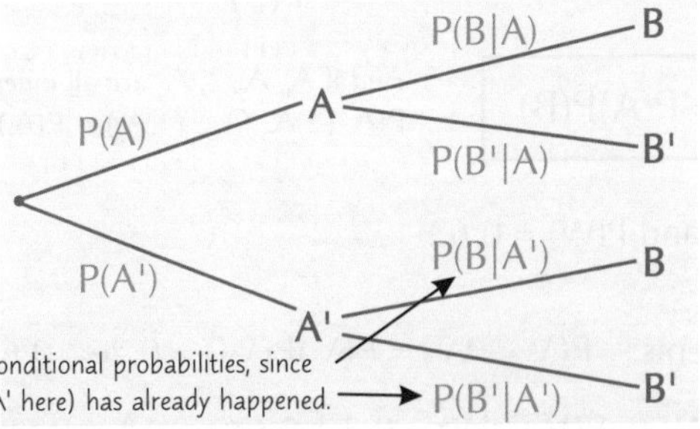

If you multiply probabilities along the branches, you get:

i.e. P(A and B) ⇒ $P(A \cap B) = P(A) \times P(B \mid A)$

You can rewrite this as:

$$P(B \mid A) = \frac{P(A \cap B)}{P(A)}$$

EXAMPLE Horace either walks (W) or runs (R) to the bus stop. If he walks he catches (C) the bus with a probability of 0.3. If he runs he catches it with a probability of 0.7. He walks to the bus stop with a probability of 0.4. Find the probability that Horace catches the bus.

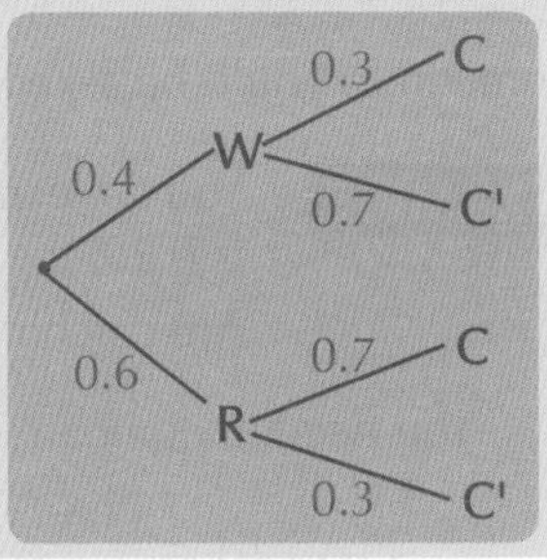

$P(C) = P(C \cap W) + P(C \cap R)$

$= P(W)\,P(C \mid W) + P(R)\,P(C \mid R)$

$= (0.4 \times 0.3) + (0.6 \times 0.7) = 0.12 + 0.42 = \underline{0.54}$

This is easier to follow if you match each part of this working to the probabilities in the tree diagram.

If B is Conditional on A then A is Conditional on B

If B depends on A then A depends on B — and it doesn't matter which event happens first.

EXAMPLE Horace turns up at school either late (L) or on time (L'). He is then either shouted at (S) or not (S'). The probability that he turns up late is 0.4. If he turns up late the probability that he is shouted at is 0.7. If he turns up on time the probability that he is shouted at is 0.2.

If you hear Horace being shouted at, what is the probability that he turned up late?

1) The probability you want is $P(L \mid S)$. Get this the right way round — he's <u>already</u> being shouted at.

2) Use the conditional probability formula: $P(L \mid S) = \dfrac{P(L \cap S)}{P(S)}$

3) The best way to find $P(L \cap S)$ and $P(S)$ is with a tree diagram.

Be careful with questions like this — the information in the question tells you what you need to know to draw the tree diagram with L (or L') considered first. But you need P(L|S) — where S is considered first. So don't rush in.

$P(L \cap S) = 0.4 \times 0.7 = 0.28$

$P(S) = P(L \cap S) + P(L' \cap S) = 0.28 + 0.12 = 0.40$

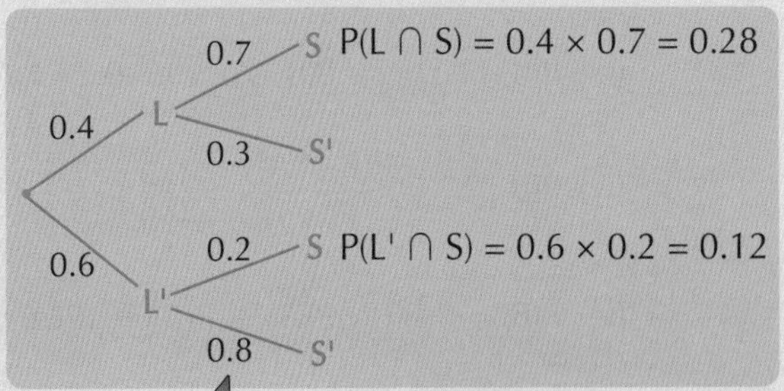

These are all conditional probabilities — e.g. $P(S' \mid L') = 0.8$.

4) Put these in your conditional probability formula to get:

$$P(L \mid S) = \frac{0.28}{0.4} = 0.7$$

There's a Formula for Working this Out — but it's Easier to Use the Tree Diagram

This formula will be on the formula sheet in your exam — it's pretty unwieldy. ⇒ $P(A_j \mid B) = \dfrac{P(A_j) \times P(B \mid A_j)}{\sum_r P(A_r) \times P(B \mid A_r)}$

Using it in the above question gives (if you replace B with S, then imagine A_1 is L, and A_2 is L'):

$$P(L \mid S) = \frac{P(L) \times P(S \mid L)}{P(L) \times P(S \mid L) + P(L') \times P(S \mid L')} = \frac{0.4 \times 0.7}{(0.4 \times 0.7) + (0.6 \times 0.2)} = \frac{0.28}{0.4} = 0.7$$

Basically the same working as with the tree diagram.

Independent Events

Independent Events Have No Effect on Each Other

If the probability of B happening doesn't depend on whether or not A has happened, then A and B are independent.

1) If A and B are independent, $P(A|B) = P(A)$.
2) If you put this in the conditional probability formula, you get: $P(A \mid B) = P(A) = \dfrac{P(A \cap B)}{P(B)}$
   Or, to put that another way:

For independent events: $P(A \cap B) = P(A)P(B)$

And if $A_1, A_2, ..., A_n$ are all independent, then $P(A_1 \cap A_2 \cap ... \cap A_n) = P(A_1)P(A_2)...P(A_n)$.

EXAMPLE V and W are independent events, where $P(V) = 0.2$ and $P(W) = 0.6$.
a) Find $P(V \cap W)$. b) Find $P(V \cup W)$.

a) Just put the numbers into the formula for independent events: $P(V \cap W) = P(V)P(W) = 0.2 \times 0.6 = 0.12$

b) Using the formula on page 112: $P(V \cup W) = P(V) + P(W) - P(V \cap W) = 0.2 + 0.6 - 0.12 = 0.68$

Sometimes you'll be asked if two events are independent or not. Here's how you work it out...

EXAMPLE You are exposed to two infectious diseases — one after the other. The probability you catch the first (A) is 0.25, the probability you catch the second (B) is 0.5, and the probability you catch both of them is 0.2. Are catching the two diseases independent events?

You need to compare $P(A|B)$ and $P(A)$ — if they're different, the events aren't independent.

$P(A \mid B) = \dfrac{P(A \cap B)}{P(B)} = \dfrac{0.2}{0.5} = 0.4$ $P(A) = 0.25$ $P(A|B)$ and $P(A)$ are different, so they're not independent.

Take Your Time with Tough Probability Questions

EXAMPLE A and B are two events, with $P(A) = 0.4$, $P(B \mid A) = 0.25$, and $P(A' \cap B) = 0.2$.
a) Find: (i) $P(A \cap B)$, (ii) $P(A')$, (iii) $P(B' \mid A)$, (iv) $P(B \mid A')$, (v) $P(B)$, (vi) $P(A \mid B)$.
b) Say whether or not A and B are independent.

a) (i) $P(B \mid A) = \dfrac{P(A \cap B)}{P(A)} = 0.25$, so $P(A \cap B) = 0.25 \times P(A) = 0.25 \times 0.4 = 0.1$

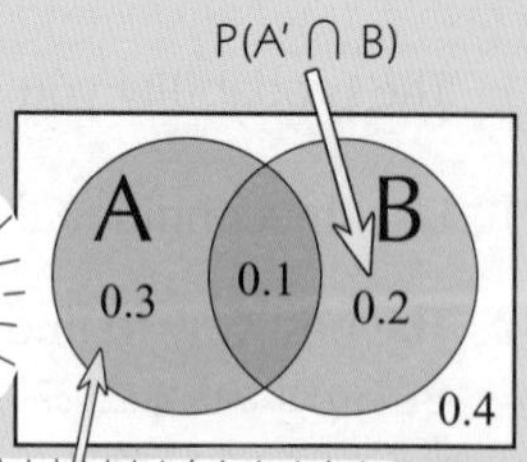

(ii) $P(A') = 1 - P(A) = 1 - 0.4 = 0.6$

A Venn diagram sometimes makes it easier to see what's going on. Fill it in as you find anything out.

(iii) $P(B' \mid A) = 1 - P(B \mid A) = 1 - 0.25 = 0.75$

Since $P(B' \mid A) + P(B \mid A) = 1$.

(iv) $P(B \mid A') = \dfrac{P(B \cap A')}{P(A')} = \dfrac{0.2}{0.6} = \dfrac{1}{3}$

$P(B \cap A') = P(A' \cap B)$

From (i) $P(A \cap B) = 0.1$, so you know this must be 0.3.

(v) $P(B) = P(B \mid A)P(A) + P(B \mid A')P(A') = (0.25 \times 0.4) + \left(\frac{1}{3} \times 0.6\right) = 0.3$

Or use the Venn diagram.

(vi) $P(A \mid B) = \dfrac{P(A \cap B)}{P(B)} = \dfrac{0.1}{0.3} = \dfrac{1}{3}$

Or you could say that $P(A \cap B) = 0.1$, while $P(A)P(B) = 0.4 \times 0.3 = 0.12$ — they're different, which shows A and B are not independent.

b) If $P(B \mid A) = P(B)$, then A and B are independent.
But $P(B \mid A) = 0.25$, while $P(B) = 0.3$, so A and B are not independent.

Statisticians say: P(Having cake ∩ Eating it) = 0...

Probability questions can be tough. For tricky questions like the last one, try drawing a Venn diagram or a tree diagram, even if the question doesn't tell you to — they're really useful for getting your head round things and understanding what on earth is going on. And don't forget the tests for independent events — you're likely to get asked a question on those.

Order of Events

If, like me, you enjoy counting things, then you're going to love this page.

Count the **Number** of **Different Ways** things could happen...

You'll often need to calculate the probability of several things happening. You need to consider how many different orders those events could have happened in — as well as the events' individual probabilities.

EXAMPLE I toss a fair coin three times. Find the probability that:
a) a person correctly guesses the outcome of the first two tosses, but guesses the third incorrectly,
b) a person correctly guesses the outcome of any two of the three tosses.

a) It's a fair coin, so: P(correct on 1st toss) = P(correct on 2nd toss) = P(incorrect on 3rd toss) = 0.5.
This means P(correct on 1st and 2nd tosses but incorrect on 3rd toss) = $0.5 \times 0.5 \times 0.5 = 0.5^3 = 0.125$

b) This time, there are three different ways to guess two out of three correctly (since you could be wrong either the first time, the second time or the third time).
Each of these three has a probability of 0.125. (For each of them, the probability will be $0.5^3 = 0.125$.)
So the probability of any of them happening must be P(guess any two correctly) = $3 \times 0.125 = 0.375$

(i) by asking "How many positions for the **Odd-One-Out**?"

EXAMPLE On any particular day, the probability that Sarah will cycle to school is always 0.6.
Find the probability that, in a normal 5-day week, she cycles to school on exactly 4 days.

You don't know on which 4 days she cycles, so you have to count how many ways it could happen.
There must be 5 different ways — since her "non-cycling day" could be any of the 5 days.

The probability of any one arrangement of 4 "cycling days" and 1 "non-cycling day" is $0.6^4 \times 0.4 = 0.05184$. (P(Sarah doesn't cycle) = 1 − 0.6 = 0.4)
So the probability that over 5 days she cycles to school exactly 4 times must be $5 \times 0.05184 = 0.2592$

(ii) by **Counting Possibilities** at each stage

EXAMPLE There are 10 CDs on a shelf: 5 jazz, 3 polka and 2 opera. Jake picks 3 CDs off the shelf at random, without replacing them. What is the probability that he picks one of each type of music?

There are two ways to approach this.

① Without a picture.
Suppose he picked the jazz CD first, then the polka CD, and then the opera CD.

- The probability of this is: $\frac{5}{10} \times \frac{3}{9} \times \frac{2}{8} = \frac{30}{720} = \frac{1}{24}$.
- But there are various ways he could have picked three CDs of different types.
  - There are 3 possibilities for the first CD.
  - Then there are 2 possibilities for the second CD.
  - That leaves just 1 possibility for the third CD.
  - So there must be $3 \times 2 \times 1 = 6$ ways to pick the CDs.
- So overall, the probability of him picking one of each type of CD must be $6 \times \frac{1}{24} = \frac{1}{4}$

② If you draw a tree diagram, you get the same answer.
Any set of branches containing one of each type of CD has a probability of $\frac{5 \times 3 \times 2}{10 \times 9 \times 8} = \frac{1}{24}$ — and there are six possible sets (the coloured ones).

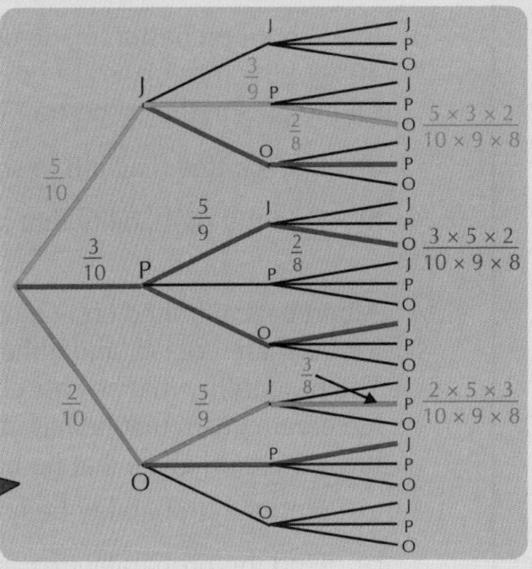

This could easily be in your exam — count yourself warned...

This is important. It often seems to appear in exam questions without drawing much attention to itself — so it's easy to forget to allow for things happening in different orders. But if you do remember it, it's easy to do — so that's good news.

S1 Section 2 — Practice Questions

Gosh. A whole page of warm-up questions. By the time you've finished these, you'll be warmer than a wolf in woollen mittens. Oh, and you'll probably be awesome at probability questions too.

Warm-up Questions

1) A standard dice and a coin are thrown and the outcomes recorded. If a head is thrown, the score on the dice is doubled. If a tail is thrown, 4 is added to the score on the dice.
   a) What is the probability that you score more than 5?
   b) If you throw a tail, what is the probability that you get an even score?

2) The hot-beverage choices of a company's workers are shown in the table. If one worker is selected at random, what is the probability that he or she:
   a) drinks coffee?
   b) drinks milky tea without sugar?
   c) either drinks tea or takes only sugar?

	Milk or sugar			
Drink	Only milk	Only sugar	Both	Neither
Tea	7	4	6	1
Coffee	5	3	2	2

3) Half the students in a sixth-form college eat sausages for dinner and 20% eat chips. 10% of those who eat chips also eat sausages. By use of a Venn diagram or otherwise, find:
   a) the percentage of students who eat both chips and sausages,
   b) the percentage of students who eat chips but not sausages,
   c) the percentage of students who eat either chips or sausages but not both.

4) There are 36 people in a room. 18 of the people have surnames beginning with the letter A, 8 have surnames beginning with B, and the rest have surnames beginning with C.
   a) Two random people leave the room. What is the probability that both people's names begin with C?
   b) Those two people come back into the room, meaning the original 36 people are present again. But then, all of a sudden, three people leave the room. What is the probability that these three people all have names beginning with a different letter?

5) In a school orchestra (made up of pupils in either the upper or lower school), 40% of the musicians are boys. Of the boys, 30% are in the upper school. Of the girls in the orchestra, 50% are in the upper school.
   a) Draw a tree diagram to show the various probabilities.
   b) Find the probability that a musician chosen at random is in the upper school.

6) Albert eats a limited choice of lunch. He eats either chicken or beef for his main course, and either chocolate pudding or ice cream for dessert. The probability that he eats chicken is 1/3, the probability that he eats ice cream given that he has chicken is 2/5, and the probability that he has ice cream given that he has beef is 3/4.
   a) Find the probability he has either chicken or ice cream — but not both.
   b) Find the probability that he eats ice cream.
   c) Find the probability that he had chicken given that you see him eating ice cream.
   d) Are the events 'Albert eats chicken' and 'Albert eats ice cream' independent? Explain your answer.

S1 Section 2 — Practice Questions

Boop. Boop. Boop. Exam simulation has begun. Repeat, exam simulation has begun. Please ensure your safety goggles are firmly attached. Emergency exits can be found on the right- and left-hand sides of the page.

Exam Questions

1 A soap company asked 120 people about the types of soap (from Brands A, B and C) they bought. Brand A was bought by 40 people, Brand B by 30 people and Brand C by 25. Both Brands A and B (and possibly C as well) were bought by 8 people, B and C (and possibly A) were bought by 10 people, and A and C (and possibly B) by 7 people. All three brands were bought by 3 people.

If a person is selected at random, find the probability that:

a) they buy at least one of the soaps. *(2 marks)*

b) they buy at least two of the soaps. *(2 marks)*

c) they buy soap B, given that they buy only one type of soap. *(3 marks)*

2 A jar contains counters of 3 different colours. There are 3 red counters, 4 white counters and 5 green counters. Two random counters are removed from the jar one at a time. Once removed, the colour of the counter is noted. The first counter is not replaced before the second one is drawn.

a) Find the probability that the second counter is green. *(2 marks)*

b) Find the probability that both the counters are red. *(2 marks)*

c) Find the probability that the two counters are not both the same colour. *(3 marks)*

3 Event J and Event K are independent events, where P(J) = 0.7 and P(K) = 0.1.

a) Find:

(i) $P(J \cap K)$, *(1 mark)*

(ii) $P(J \cup K)$. *(2 marks)*

b) If L is the event that neither J or K occurs, find $P(L|K')$. *(3 marks)*

4 For a particular biased dice, the event 'throw a 6' is called event B. P(B) = 0.2.
This biased dice and a fair dice are rolled together. Find the probability that:

a) the biased dice doesn't show a 6, *(1 mark)*

b) at least one of the dice shows a 6, *(2 marks)*

c) exactly one of the dice shows a 6, given that at least one of them shows a 6. *(3 marks)*

Discrete Random Variables

This stuff isn't hard. But learn it properly, because it's used over and over again in the rest of the section.

Getting your head round this **Basic Stuff** is Important

This first bit isn't particularly interesting. But understanding the difference between X and x (bear with me) might make the later stuff a bit less confusing. Might.

1) X (upper case) is just the name of a random variable. So X could be 'score on a dice' — it's just a name.
2) A random variable doesn't have a fixed value. Like with a dice score — the value on any 'roll' is all down to chance.
3) x (lower case) is a particular value that X can take. So for one roll of a dice, x could be 1, 2, 3, 4, 5 or 6.
4) Discrete random variables only have a certain number of possible values. Often these values are whole numbers, but they don't have to be. Usually there are only a few possible values (e.g. the possible scores with one roll of a dice).
5) A probability distribution is basically a list of the possible values of x, plus a way to find the probability for each one.
6) A probability function is a formula that generates the probabilities for different values of x.

All the Probabilities **Add up to 1**

For a discrete random variable X:

$$\sum_{\text{all}\,x} P(X=x) = 1$$

This says that if you add up the probabilities of all the possible values of X, you get 1.

EXAMPLE The random variable X can take values 1, 2, 3,..., n. If $P(X \geq 2) = k$, find $P(X < 2)$ in terms of k.

You know that $P(X = 1) + P(X = 2) + P(X = 3) + ... + P(X = n) = 1$

Or to put that another way: $P(X < 2) + P(X \geq 2) = 1$ or $P(X < 2) + k = 1$. This means: $P(X < 2) = 1 - k$

The **Cumulative Distribution Function** is a Running Total of Probabilities

Cumulative Distribution Function (c.d.f.)

A cumulative distribution function (c.d.f.) gives the probability that X will be less than or equal to a particular value. So a c.d.f. shows: $\mathbf{P}(X \leq x_0) = \sum_{x \leq x_0} p(x)$

$p(x) = P(X = x)$

EXAMPLE The random variable X can take values 1, 2, 3,..., n.
If the value of the cumulative distribution function at $x = 4$ is 0.85, find $P(X > 4)$.

At $x = 4$, the cumulative distribution function has the value 0.85 — this means $P(X \leq 4) = 0.85$.

So $P(X > 4) = 1 - P(X \leq 4) = 1 - 0.85 = 0.15$

Discrete Random Variables have a **Mean E(X)** and **Variance Var(X)**

1) You can work out the mean (or expected value) $E(X)$ for a discrete random variable X.
2) $E(X)$ is a kind of 'theoretical mean' — it's what you'd expect the mean of X to be if you took loads of readings.
3) In practice, the mean of your results is unlikely to match the theoretical mean exactly, but it should be pretty near.
4) You can also find the variance ($\text{Var}(X)$) of a random variable X. It's the 'expected variance' of a large number of readings.
5) And the standard deviation of a random variable is just the square root of its variance.

standard deviation $= \sqrt{\text{variance}} = \sqrt{\text{Var}(X)}$

Fact: the probability of this coming up in the exam is less than or equal to one...

The cumulative distribution function of a random variable is an idea you're going to need to get really comfortable with, so just have another look at that box above and let it soak into your brain properly. The same goes for the mean and variance — the idea of a 'theoretical' mean and variance is a bit strange at first, so make sure you understand it before moving on.

The Binomial Probability Function

These next few pages are basically all about success and failure.

Binomial Coefficients count Possible *Arrangements* of Successes and Failures

1) Suppose you carry out a test n times, where each time the possible outcomes are either 'success' or 'failure'.
2) If you knew the results were r successes and $(n-r)$ failures, then you could quickly work out in how many different orders those outcomes could have happened using binomial coefficients.

Binomial Coefficients

$$\binom{n}{r} = {}^nC_r = \frac{n!}{r!(n-r)!}$$

where $n! = n \times (n-1) \times (n-2) \times \ldots \times 2 \times 1$

${}^nC_r$ and $\binom{n}{r}$ both mean $\frac{n!}{r!(n-r)!}$

$n!$ = "n factorial"

$0! = 1$

EXAMPLE I toss a fair coin 5 times. How many ways are there to arrange the heads and tails if I toss:
a) 0 heads? b) 1 head?

a) Think of heads as 'success' and tails as 'failure'.

Then there's $\binom{5}{0} = \frac{5!}{0! \times (5-0)!} = 1$ way to arrange 0 heads and 5 tails.

I need to throw: tails, tails, tails, tails, tails

b) There are $\binom{5}{1} = \frac{5!}{1! \times (5-1)!} = \frac{5!}{1! \times 4!} = 5$ ways to arrange 1 head and 4 tails.

heads, tails, tails, tails, tails
tails, heads, tails, tails, tails
tails, tails, heads, tails, tails
tails, tails, tails, heads, tails
tails, tails, tails, tails, heads

The *Binomial Probability Function* gives P(r successes out of n trials)

An event's probability depends on how many different ways there are for it to happen (see p112).
That's why this next formula contains a binomial coefficient.

Binomial Probability Function

$$\text{P}(r \text{ successes in } n \text{ trials}) = \binom{n}{r} \times [\text{P(success)}]^r \times [\text{P(failure)}]^{n-r}$$

This is the probability function for a binomial probability distribution — see next page for more info.

1) Remember... if p = P(something happens), then $1 - p$ = P(that thing doesn't happen).
2) So if P(success) $= p$, then P(failure) $= 1 - p$.

EXAMPLE I roll a fair dice 5 times. Find the probability of rolling: a) 2 sixes, b) 3 sixes, c) 4 numbers less than 3.

First note that each roll of a dice is independent of the other rolls (see next page for more info).

a) For this part, call "roll a 6" a success, and "roll anything other than a 6" a failure.

Then $\text{P(roll 2 sixes)} = \binom{5}{2} \times \left(\frac{1}{6}\right)^2 \times \left(\frac{5}{6}\right)^3 = \frac{5!}{2!3!} \times \frac{1}{36} \times \frac{125}{216} = 0.161$ (to 3 d.p.).

b) Again, call "roll a 6" a success, and "roll anything other than a 6" a failure.

Then $\text{P(roll 3 sixes)} = \binom{5}{3} \times \left(\frac{1}{6}\right)^3 \times \left(\frac{5}{6}\right)^2 = \frac{5!}{3!2!} \times \frac{1}{216} \times \frac{25}{36} = 0.032$ (to 3 d.p.).

Notice how $\binom{5}{2} = \binom{5}{3}$.
In fact, $\binom{n}{r} = \binom{n}{n-r}$.

c) This time, success means "roll a 1 or a 2", while failure is now "roll a 3, 4, 5 or 6".

Then $\text{P(roll 4 numbers less than 3)} = \binom{5}{4} \times \left(\frac{1}{3}\right)^4 \times \left(\frac{2}{3}\right) = \frac{5!}{4!1!} \times \frac{1}{81} \times \frac{2}{3} = 0.041$ (to 3 d.p.).

Let this formula for success go to your head — and then keep it there...

This page is all about finding the probabilities of different numbers of successes in n trials. Now then... if you carry out n trials, there are $n + 1$ possibilities for the number of successes $(0, 1, 2, \ldots, n)$. This 'family' of possible results along with their probabilities is sounding suspiciously like a probability distribution. Oh rats... I've given away what's on the next page.

The Binomial Distribution

This page is about random variables following a binomial distribution (a particular type of probability distribution). If you liked the probability function you saw on the previous page, then you'll be happy to learn it's on this page too.

There are 5 Conditions for a Binomial Distribution

Binomial Distribution: B(n, p)

A random variable X follows a Binomial Distribution as long as these 5 conditions are satisfied:

1) There is a fixed number (n) of trials.
2) Each trial involves either "success" or "failure".
3) All the trials are independent.
4) The probability of "success" (p) is the same in each trial.
5) The variable is the total number of successes in the n trials.

Binomial random variables are discrete, since they only take values 0, 1, 2... n.

n and p are the two parameters of the binomial distribution. (Or n is sometimes called the 'index'.)

In this case, $P(X = x) = \binom{n}{x} \times p^x \times (1-p)^{n-x}$ for $x = 0, 1, 2, \ldots, n$, and you can write $X \sim B(n, p)$.

EXAMPLE: Which of the random variables described below would follow a binomial distribution? For those that do, state the distribution's parameters.

a) **The number of faulty items (T) produced in a factory per day, if the probability of each item being faulty is 0.01 and there are 10 000 items produced every day.**
Binomial — there's a fixed number (10 000) of trials with two possible results ('faulty' or 'not faulty'), a constant probability of 'success', and T is the total number of 'faulty' items.
So (as long as faulty items occur independently) $T \sim B(10\,000, 0.01)$.

b) **The number of red cards (R) drawn from a standard 52-card deck in 10 picks, not replacing the cards each time.**
Not binomial, since the probability of 'success' changes each time (as I'm not replacing the cards).

c) **The number of red cards (R) drawn from a standard 52-card deck in 10 picks, replacing the cards each time.**
Binomial — there's a fixed number (10) of independent trials with two possible results ('red' or 'black/not red'), a constant probability of success (I'm replacing the cards), and R is the number of red cards drawn. $R \sim B(10, 0.5)$.

d) **The number of times (T) I have to toss a coin before I get heads.**
Not binomial, since the number of trials isn't fixed.

e) **The number of left-handed people (L) in a sample of 500 randomly chosen people, if the fraction of left-handed people in the population as a whole is 0.13.**
Binomial — there's a fixed number (500) of independent trials with two possible results ('left-handed' or 'not left-handed'), a constant probability of success (0.13), and L is the number of left-handers. $L \sim B(500, 0.13)$.

EXAMPLE: When I toss a grape in the air and try to catch it in my mouth, my probability of success is always 0.8. The number of grapes I catch in 10 throws is described by the discrete random variable X.

a) How is X distributed? Name the type of distribution, and give the values of any parameters.
b) Find the probability of me catching at least 9 grapes.

a) There's a fixed number (10) of independent trials with two possible results ('catch' and 'not catch'), a constant probability of success (0.8), and X is the total number of catches.
Therefore X follows a binomial distribution, $X \sim B(10, 0.8)$.

b) P(at least 9 catches)
$$= P(9 \text{ catches}) + P(10 \text{ catches})$$
$$= \left\{\binom{10}{9} \times 0.8^9 \times 0.2^1\right\} + \left\{\binom{10}{10} \times 0.8^{10} \times 0.2^0\right\}$$
$$= 0.268435... + 0.107374... = 0.376 \text{ (to 3 d.p.)}.$$

Binomial distributions come with 5 strings attached...

There's a big, boring box at the top of the page with a list of 5 conditions in — and you do need to know it, unfortunately. There's only one way to learn it — keep trying to write down the 5 conditions until you can do it in your sleep.

Using Binomial Tables

Your life is just about to be made a whole lot <u>easier</u>. So smile sweetly and admit that statistics isn't <u>all</u> bad.

Look up Probabilities in **Binomial Tables**

EXAMPLE I have an unfair coin. When I toss this coin, the probability of getting heads is 0.35.
Find the probability that it will land on heads fewer than 3 times when I toss it 12 times in total.

If the random variable X represents the number of heads I get in 12 tosses, then $X \sim B(12, 0.35)$.
You need to find $P(X \leq 2)$.

① You <u>could</u> work this out 'manually'...

$$P(0 \text{ heads}) + P(1 \text{ head}) + P(2 \text{ heads}) = \left\{\binom{12}{0} \times 0.35^0 \times 0.65^{12}\right\} + \left\{\binom{12}{1} \times 0.35^1 \times 0.65^{11}\right\} + \left\{\binom{12}{2} \times 0.35^2 \times 0.65^{10}\right\}$$

$$= 0.0057 + 0.0368 + 0.1088 = 0.1513$$

See p154 for more binomial tables.

② But it's much quicker to use tables of the <u>binomial cumulative distribution function</u> (c.d.f.).
These show $P(X \leq x)$, for $X \sim B(n, p)$.

- First find the table for the <u>correct values of n and p</u>. Then the table gives you a value for $P(X \leq x)$.
- Here: $n = 12$ and $p = 0.35$.
- You need $P(X \leq 2)$. The table tells you this is 0.1513.

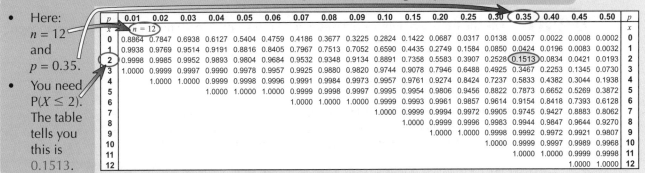

p / x	0.01	0.02	0.03	0.04	0.05	0.06	0.07	0.08	0.09	0.10	0.15	0.20	0.25	0.30	0.35	0.40	0.45	0.50	p / x
	n = 12																		
0	0.8864	0.7847	0.6938	0.6127	0.5404	0.4759	0.4186	0.3677	0.3225	0.2824	0.1422	0.0687	0.0317	0.0138	0.0057	0.0022	0.0008	0.0002	0
1	0.9938	0.9769	0.9514	0.9191	0.8816	0.8405	0.7967	0.7513	0.7052	0.6590	0.4435	0.2749	0.1584	0.0850	0.0424	0.0196	0.0083	0.0032	1
2	0.9998	0.9985	0.9952	0.9893	0.9804	0.9684	0.9532	0.9348	0.9134	0.8891	0.7358	0.5583	0.3907	0.2528	0.1513	0.0834	0.0421	0.0193	2
3	1.0000	0.9999	0.9997	0.9990	0.9978	0.9957	0.9925	0.9880	0.9820	0.9744	0.9078	0.7946	0.6488	0.4925	0.3467	0.2253	0.1345	0.0730	3
4		1.0000	1.0000	0.9999	0.9998	0.9996	0.9991	0.9984	0.9973	0.9957	0.9761	0.9274	0.8424	0.7237	0.5833	0.4382	0.3044	0.1938	4
5				1.0000	1.0000	1.0000	0.9999	0.9998	0.9997	0.9995	0.9954	0.9806	0.9456	0.8822	0.7873	0.6652	0.5269	0.3872	5
6							1.0000	1.0000	1.0000	0.9999	0.9993	0.9961	0.9857	0.9614	0.9154	0.8418	0.7393	0.6128	6
7										1.0000	0.9999	0.9994	0.9972	0.9905	0.9745	0.9427	0.8883	0.8062	7
8											1.0000	0.9999	0.9996	0.9983	0.9944	0.9847	0.9644	0.9270	8
9												1.0000	1.0000	0.9998	0.9992	0.9972	0.9921	0.9807	9
10														1.0000	0.9999	0.9997	0.9989	0.9968	10
11															1.0000	1.0000	0.9999	0.9998	11
12																	1.0000	1.0000	12

Binomial Tables Tell You **More** Than You Might Think

With a bit of <u>cunning</u>, you can get binomial tables to tell you <u>anything</u> you want to know...

EXAMPLE I have a different unfair coin. When I toss this coin, the probability of getting tails is 0.6.
The random variable X represents the number of tails in 12 tosses, so $X \sim B(12, 0.6)$.

If I toss this coin 12 times, use the table above to find the probability that:

a) it will land on tails more than 8 times,
b) it will land on heads at least 6 times,
c) it will land on heads exactly 9 times,
d) it will land on heads more than 3 but fewer than 6 times.

a) The above table only goes up to $p = 0.5$.
So <u>switch things round</u>... if P(tails) = 0.6, then P(heads) = 1 – 0.6 = <u>0.4</u> (and $p = 0.4$ is in the table).
So define a <u>new</u> random variable, Y, representing the number of <u>heads</u> in 12 throws — then $Y \sim B(12, 0.4)$.
This means $P(X > 8) = P(Y \leq 3) = 0.2253$

b) $P(Y \geq 6) = 1 - P(Y < 6) = 1 - P(Y \leq 5) = 1 - 0.6652 = 0.3348$

1) P(event happens) = 1 – P(event doesn't happen),
2) $P(Y < 6) = P(Y \leq 5)$, as Y takes whole number values.

c) $P(Y = 9) = P(Y \leq 9) - P(Y \leq 8) = 0.9972 - 0.9847 = 0.0125$

Use P(A or B) = P(A) + P(B) with the mutually exclusive events "$Y \leq 8$" and "$Y = 9$" to get $P(Y \leq 9) = P(Y \leq 8) + P(Y = 9)$.
Or you can think of it as "subtracting $P(Y \leq 8)$ from $P(Y \leq 9)$ leaves just $P(Y = 9)$".

d) $P(3 < Y < 6) = P(Y \leq 5) - P(Y \leq 3) = 0.6652 - 0.2253 = 0.4399$

Statistical tables are the original labour-saving device...

...as long as you know what you're doing. Careful, though — it's easy to trip yourself up. Basically, as long as you can find the right value of n and p (or $1 - p$, if necessary) in a table, you can use those tables to work out <u>anything</u> you might need.

Mean and Variance of B(n, p)

You know from page 120 about the mean (or expected value) and variance of a random variable.
And you also know about the binomial distribution. Put those things together, and you get this page.

For a Binomial Distribution: Mean = np

This formula will be in your formula booklet, but it's worth committing to memory anyway.

Mean of a Binomial Distribution

If $X \sim B(n, p)$, then:

Mean (or Expected Value) = $\mu = E(X) = np$

Greek letters (e.g. μ) often show something based purely on theory rather than experimental results.

Remember... the expected value is the value you'd expect the random variable to take on average if you took loads and loads of readings. It's a "theoretical mean" — the mean of experimental results is unlikely to match it exactly.

EXAMPLE If $X \sim B(20, 0.2)$, what is $E(X)$?

Just use the formula: $E(X) = np = 20 \times 0.2 = 4$

EXAMPLE What's the expected number of sixes when I roll a fair dice 30 times? Interpret your answer.

If the random variable X represents the number of sixes in 30 rolls, then $X \sim B(30, \frac{1}{6})$.

So the expected value of X is $E(X) = 30 \times \frac{1}{6} = 5$

If I were to repeatedly throw the dice 30 times, and find the average number of sixes in each set of 30 throws, then I would expect it to end up pretty close to 5.
And the more sets of 30 throws I did, the closer to 5 I'd expect the average to be.

Notice that the probability of getting exactly 5 sixes on my next set of 30 throws = $\binom{30}{5} \times \left(\frac{1}{6}\right)^5 \times \left(\frac{5}{6}\right)^{25} = 0.192$
So I'm much more likely not to get exactly 5 sixes (= 1 – 0.192 = 0.808).
This is why it only makes sense to talk about the mean as a "long-term average", and not as "what I expect to happen next".

For a Binomial Distribution: Variance = npq

Variance of a Binomial Distribution

If $X \sim B(n, p)$, then:

Variance = $\text{Var}(X) = \sigma^2 = np(1 - p) = npq$

Standard Deviation = $\sigma = \sqrt{np(1-p)} = \sqrt{npq}$

For a binomial distribution, P(success) is usually called p, and P(failure) is sometimes called q (= $1 - p$).

EXAMPLE If $X \sim B(20, 0.2)$, what is $\text{Var}(X)$?

Just use the formula: $\text{Var}(X) = np(1 - p) = 20 \times 0.2 \times 0.8 = 3.2$

EXAMPLE If $X \sim B(25, 0.2)$, find: a) $P(X \le \mu)$, b) $P(X \le \mu - \sigma)$, c) $P(X \le \mu - 2\sigma)$

$E(X) = \mu = 25 \times 0.2 = 5$, and $\text{Var}(X) = \sigma^2 = 25 \times 0.2 \times (1 - 0.2) = 4$, which gives $\sigma = 2$.

So, using tables (for $n = 25$ and $p = 0.2$):
a) $P(X \le \mu) = P(X \le 5) = 0.6167$
b) $P(X \le \mu - \sigma) = P(X \le 3) = 0.2340$
c) $P(X \le \mu - 2\sigma) = P(X \le 1) = 0.0274$

See page 154.

For B(n, p) — the variance is always less than the mean...

Nothing too fancy there really. A couple of easy-to-remember formulas, and some stuff about how to interpret these figures which you've seen before anyway. So learn the formulas, put the kettle on, and have a cup of tea while the going's good.

Binomial Distribution Problems

That's everything you need to know about binomial distributions.
So it's time to put it all together and have a look at the kind of thing you might get asked in the exam.

EXAMPLE 1: Selling Double Glazing

A double-glazing salesman is handing out leaflets in a busy shopping centre. He knows that the probability of each passing person taking a leaflet is always 0.3. During a randomly chosen one-minute interval, 30 people passed him.

a) Suggest a suitable model to describe the number of people (X) who take a leaflet.
b) What is the probability that more than 10 people take a leaflet?
c) How many people would the salesman expect to take a leaflet?
d) Find the variance and standard deviation of X.

a) During this one-minute interval, there's a fixed number (30) of independent trials with two possible results ("take a leaflet" and "do not take a leaflet"), a constant probability of success (0.3), and X is the total number of people taking leaflets. So $X \sim B(30, 0.3)$.

b) $P(X > 10) = 1 - P(X \leq 10) = 1 - 0.7304 = 0.2696$

Use binomial tables for this — see p154.

c) The number of people the salesman could expect to take a leaflet is $E(X) = np = 30 \times 0.3 = 9$

d) Variance $= np(1-p) = 30 \times 0.3 \times (1 - 0.3) = 6.3$ Standard deviation $= \sqrt{6.3} = 2.51$ (to 2 d.p.)

EXAMPLE 2: Multiple-Choice Guessing

A student has to take a 50-question multiple-choice exam, where each question has five possible answers, of which only one is correct. He believes he can pass the exam by guessing answers at random.

a) How many questions could the student be expected to guess correctly?
b) If the pass mark is 15, what is the probability that the student will pass the exam?
c) The examiner decides to set the pass mark so that it is at least 3 standard deviations above the expected number of correct guesses. What should the minimum pass mark be?

Let X be the number of correct guesses over the 50 questions. Then $X \sim B(50, 0.2)$.

Define your random variable first, and say how it will be distributed.

a) $E(X) = np = 50 \times 0.2 = 10$

b) $P(X \geq 15) = 1 - P(X < 15) = 1 - P(X \leq 14) = 1 - 0.9393 = 0.0607$

c) $\text{Var}(X) = np(1-p) = 50 \times 0.2 \times 0.8 = 8$ — so the standard deviation $= \sqrt{8} = 2.828$ (to 3 d.p.).
So the pass mark needs to be at least $10 + (3 \times 2.828) \approx 18.5$ — i.e. the minimum pass mark should be 19.

EXAMPLE 3: An Unfair Coin

I am spinning a coin that I know is three times as likely to land on heads as it is on tails.

a) What is the probability that it lands on tails for the first time on the third spin?
b) What is the probability that in 10 spins, it lands on heads at least 7 times?

You know that P(heads) = 3 × P(tails), and that P(heads) + P(tails) = 1.
This means that P(heads) = 0.75 and P(tails) = 0.25.

Careful... this doesn't need you to use one of the binomial formulas.

a) P(lands on tails for the first time on the third spin) $= 0.75 \times 0.75 \times 0.25 = 0.141$ (to 3 d.p.).

b) If X represents the number of heads in 10 spins, then $X \sim B(10, 0.75)$.
This means the number of tails in 10 spins can be described by the random variable Y, where $Y \sim B(10, 0.25)$.
$P(X \geq 7) = P(Y \leq 3) = 0.7759$

p = 0.75 isn't in your tables, so define a new binomial random variable Y with probability of success p = 0.25.

Proof that you shouldn't send a monkey to take your multi-choice exams...

You can see now how useful a working knowledge of statistics is. Ever since you first started using CGP books, I've been banging on about how hard it is to pass an exam without revising. Well, now you can prove I was correct using a bit of knowledge and binomial tables. Yup... statistics can help out with some of those tricky situations you face in life.

S1 Section 3 — Practice Questions

Hopefully, everything you've just read will already be stuck in your brain. But if you need a bit of help to wedge it in place, then try these questions. Actually... I reckon you'd best try them anyway — a little suffering is good for the soul.

Warm-up Questions

1) In how many different orders can the following be arranged?

a) 6 heads and 4 tails, if I toss a coin a total of 10 times.

b) 4 even numbers and 3 odd numbers, if I roll a dice 7 times.

2) What is the probability of the following?

a) Getting exactly 5 heads when you spin a fair coin 10 times.

b) Getting exactly 9 heads when you spin a fair coin 10 times.

3) Which of the following would follow a binomial distribution? Explain your answers.

a) The number of prime numbers you throw in 30 throws of a standard dice.

b) The number of people in a particular class at a school who get 'heads' when they flip a coin.

c) The number of aces in a 7-card hand dealt from a standard deck of 52 cards.

d) The number of shots I have to take before I score from the free-throw line in basketball.

4) What is the probability of the following?

a) Getting at least 5 heads when you spin a fair coin 10 times.

b) Getting at least 9 heads when you spin a fair coin 10 times.

5) If $X \sim B(14, 0.27)$, find:

a) $P(X = 4)$ b) $P(X < 2)$ c) $P(5 < X \leq 8)$

6) If $X \sim B(25, 0.15)$ and $Y \sim B(15, 0.65)$ find:

a) $P(X \leq 3)$ b) $P(X \leq 7)$

c) $P(X \leq 15)$ d) $P(Y \leq 3)$

e) $P(Y \leq 7)$ f) $P(Y \leq 15)$

7) Find the required probability for each of the following binomial distributions.

a) $P(X \leq 15)$ if $X \sim B(20, 0.4)$ b) $P(X < 4)$ if $X \sim B(40, 0.15)$

c) $P(X > 7)$ if $X \sim B(25, 0.45)$ d) $P(X \geq 40)$ if $X \sim B(50, 0.8)$

e) $P(X = 20)$ if $X \sim B(30, 0.7)$ f) $P(X = 7)$ if $X \sim B(10, 0.75)$

8) Find the mean and variance of the following random variables.

a) $X \sim B(20, 0.4)$ b) $X \sim B(40, 0.15)$

c) $X \sim B(25, 0.45)$ d) $X \sim B(50, 0.8)$

e) $X \sim B(30, 0.7)$ f) $X \sim B(45, 0.012)$

S1 Section 3 — Practice Questions

Right then... you're nearly at the end of the section, and with any luck your tail is up, the wind is in your sails and the going is good. But the real test of whether you're ready for the exam is some exam questions. And as luck would have it, there are some right here. So give them a go and test your mettle, see if you can walk the walk... and so on.

Exam Questions

1 a) The random variable X follows the binomial distribution B(12, 0.6). Find:

(i) $P(X < 8)$,

(2 marks)

(ii) $P(X = 5)$,

(2 marks)

(iii) $P(3 < X \leq 7)$.

(3 marks)

b) If $Y \sim B(11, 0.8)$, find:

(i) $P(Y = 4)$,

(2 marks)

(ii) $E(Y)$,

(1 mark)

(iii) $Var(Y)$.

(1 mark)

2 The probability of an apple containing a maggot is 0.15.

a) Find the probability that in a random sample of 40 apples there are:

(i) fewer than 6 apples containing maggots,

(2 marks)

(ii) more than 2 apples containing maggots,

(2 marks)

(iii) exactly 12 apples containing maggots.

(2 marks)

b) These apples are sold in crates of 40. Ed buys 3 crates.
Find the probability that more than 1 crate contains more than 2 apples with maggots.

(3 marks)

3 Simon tries to solve the crossword puzzle in his newspaper every day for two weeks.
He either succeeds in solving the puzzle, or he fails to solve it.

a) Simon believes that this situation can be modelled by a random variable following a binomial distribution.

(i) State two conditions needed for a binomial distribution to arise here.

(2 marks)

(ii) State which quantity would follow a binomial distribution (assuming the above conditions are satisfied).

(1 mark)

b) Simon believes a random variable X follows the distribution $B(18, p)$.
If $P(X = 4) = P(X = 5)$, find p.

(5 marks)

Normal Distributions

The normal distribution is a continuous probability distribution — this means there are 'no gaps' between its possible values. It's also one of the most useful probability distributions in all of statistics.

The Normal Distribution is 'Bell-Shaped'

1) Loads of things in real life are most likely to fall 'somewhere in the middle', and are much less likely to take extremely high or extremely low values. In this kind of situation, you often get a normal distribution.
2) If you were to draw a graph showing how likely different values are, you'd end up with a graph that looks a bit like a bell. There's a peak in the middle at the mean (or expected value). And the graph is symmetrical — so values the same distance above and below the mean are equally likely.

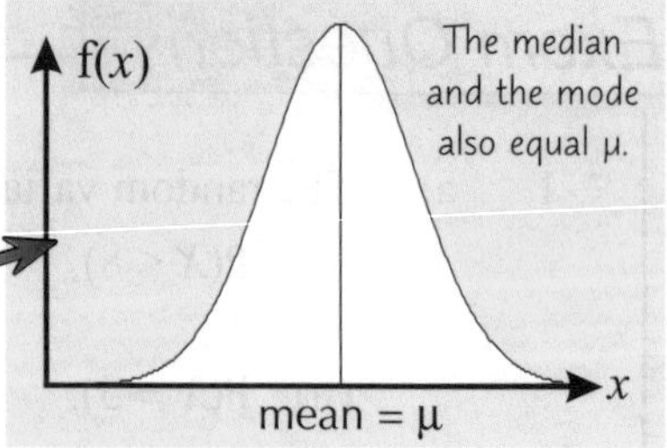

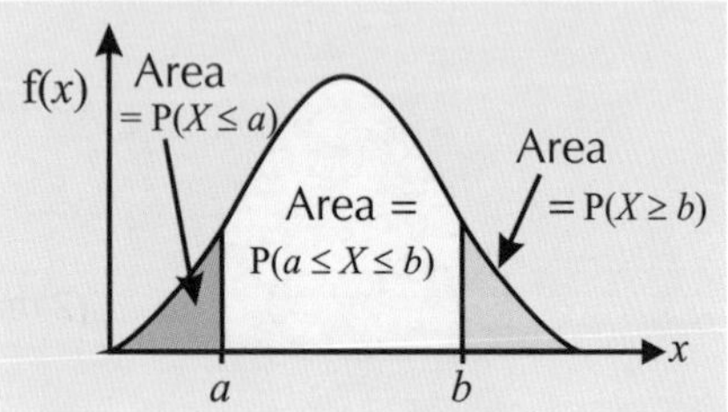

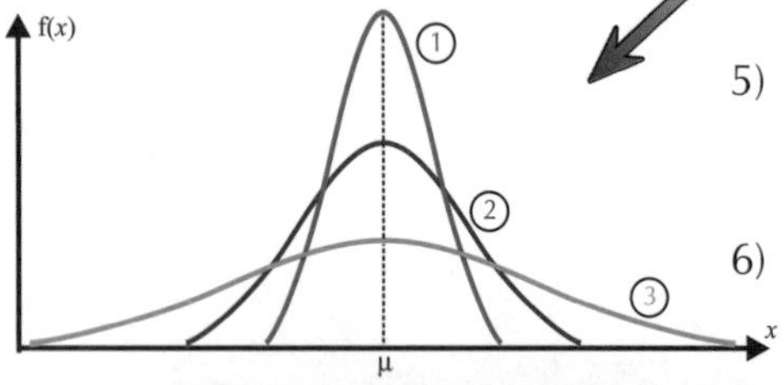

3) With a graph of a normal distribution, the probability of the random variable taking a value between two limits is the area under the graph between those limits. And since the total probability is 1, the total area under the graph must also be 1.
4) These three graphs all show normal distributions with the same mean (μ), but different variances (σ^2). Graph 1 has a small variance, and graph 3 has a larger variance — but the total area under all three curves is the same (= 1).
5) In all three graphs, about two-thirds of the readings would fall within one standard deviation (σ) of the mean (μ) (i.e. they'd be between $\mu - \sigma$ and $\mu + \sigma$).
6) The most important normal distribution is the standard normal distribution, or Z — this has a mean of zero and a variance of 1.

Normal Distribution $N(\mu, \sigma^2)$

- If X is normally distributed with **mean** μ and **variance** σ^2, it's written $X \sim N(\mu, \sigma^2)$.
- The standard normal distribution Z has **mean** 0 and **variance** 1, i.e. $Z \sim N(0, 1)$.

Use Tables to Work out Probabilities of Z

Working out the area under a normal distribution curve is usually hard. But for Z, there are tables you can use (see p153). You look up a value of z and these tables (labelled $\Phi(z)$) tell you the probability that $Z \leq z$.

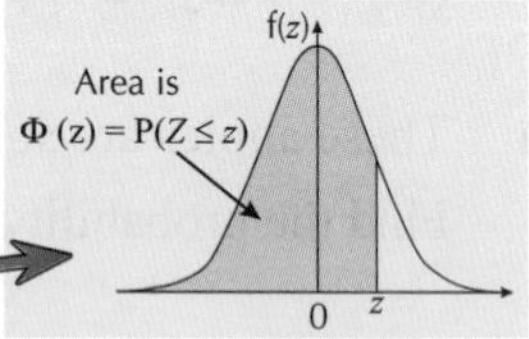

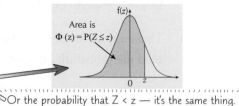

Or the probability that $Z < z$ — it's the same thing.

EXAMPLE Find the probability that:

a) $Z < 0.1$, b) $Z \leq 0.64$, c) $Z > 0.23$, d) $Z \geq -0.42$, e) $Z \leq -1.94$, f) $0.12 < Z \leq 0.82$, g) $Z = 1$

Tables only tell you the probability of Z being less than a particular value — use a sketch to work out anything else.

a) $P(Z < 0.1) = 0.53983$ ← Just look up z = 0.1 in the tables.

b) $P(Z \leq 0.64) = 0.73891$ ← For continuous distributions like Z (i.e. with 'no gaps' between its possible values): $P(Z \leq 0.64) = P(Z < 0.64)$.

c) $P(Z > 0.23) = 1 - P(Z \leq 0.23) = 1 - 0.59095 = 0.40905$

d) $P(Z \geq -0.42) = P(Z \leq 0.42) = 0.66276$ ← Use the symmetry of the graph:

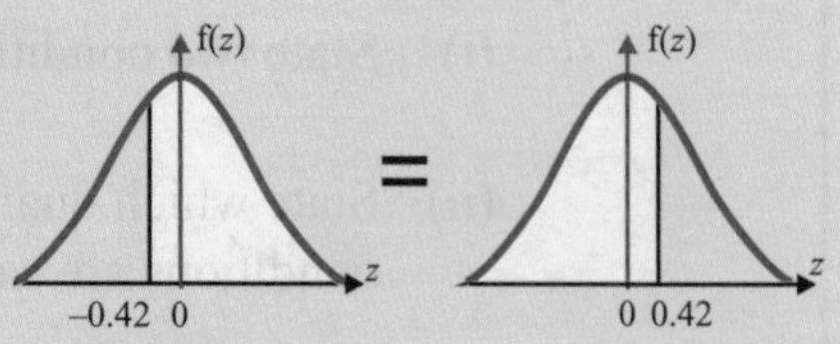

e) $P(Z \leq -1.94) = P(Z \geq 1.94) = 1 - P(Z < 1.94) = 1 - 0.97381 = 0.02619$ ← Again, draw a graph and use the symmetry:

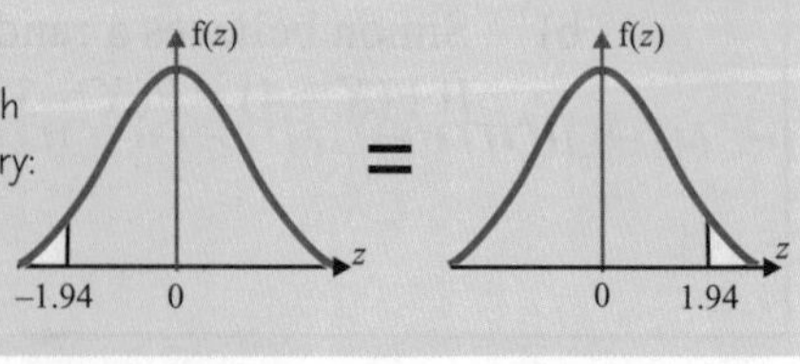

f) $P(0.12 < Z \leq 0.82) = P(Z \leq 0.82) - P(Z \leq 0.12)$
$= 0.79389 - 0.54776 = 0.24613$

g) $P(Z = 1) = 0$ ← $P(Z = k) = 0$ for any k (since the area under f(z) at a single point is zero).

The Standard Normal Distribution, Z

You can use that big table of the normal distribution function ($\Phi(z)$) 'the other way round' as well — by starting with a probability and finding a value for z.

Use **Tables** to Find a **Value** of z if You're Given a **Probability**

EXAMPLE If $P(Z < z) = 0.95543$, then what is the value of z?

Using the table for $\Phi(z)$ (the normal cumulative distribution function), $z = 1.70$.

All the probabilities in the table of $\Phi(z)$ are greater than 0.5, but you can still use the tables with values less than this. You'll most likely need to subtract the probability from 1, and then use a sketch.

EXAMPLE If $P(Z < z) = 0.26109$, then what is the value of z?

① Subtract from 1 to get a probability greater than 0.5: $1 - 0.26109 = 0.73891$

② If $P(Z < z) = 0.73891$, then from the table, $z = 0.64$.

③ So if $P(Z < z) = 0.26109$, then, $z = -0.64$.

If $P(Z < z) = 0.26109$, then z must be negative.

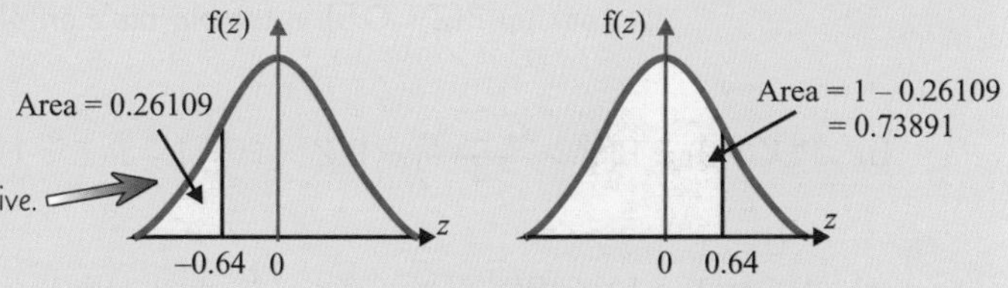

The **Percentage Points** Table also Tells You z if You're Given a **Probability**

You use the percentage points table in a similar way — i.e. you start with a probability, and look up a value for z.

EXAMPLE If $P(Z < z) = 0.85$, then what is the value of z?

Using the percentage points table, $z = 1.0364$.

You might need to use a bit of imagination and a sketch to get the most out of the percentage points table.

EXAMPLE Find z if: a) $P(Z > z) = 0.95$, b) $P(Z > z) = 0.1$, c) $P(Z \leq z) = 0.99$

a) Using the percentage points table, if $P(Z < z) = 0.95$, then $z = 1.6449$.
So if $P(Z > z) = 0.95$, then z must equal -1.6449.

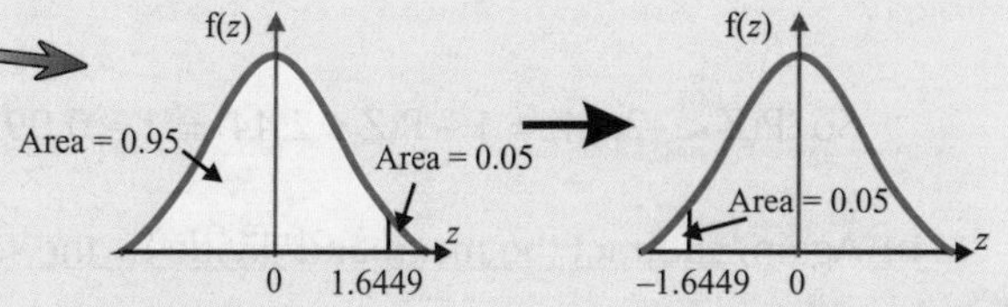

b) If $P(Z > z) = 0.1$, then $P(Z < z) = 0.9$.
Using the percentage points table, $z = 1.2816$.

c) If $P(Z \leq z) = 0.99$, then $z = 2.3263$.

Again, you can treat the "less than or equals sign" as a "less than sign", since for a continuous distribution, $P(Z \leq z) = P(Z < z)$.

The medium of a random variable follows a paranormal distribution...

It's definitely worth sketching the graph when you're finding probabilities using a normal distribution — you're much less likely to make a daft mistake. So even if the question looks a simple one, draw a quick sketch — it's probably worth it. And make sure you know exactly how to use both of the normal tables — the one for $\Phi(z)$ and the percentage points table.

Normal Distributions and Z-Tables

<u>All</u> normally-distributed variables can be transformed to Z — which is a marvellous thing.

Transform to Z by **Subtracting** μ, then **dividing by** σ

1) You can convert <u>any</u> normally-distributed variable to Z by:
   i) <u>subtracting the mean</u>, and then
   ii) <u>dividing by the standard deviation</u>.

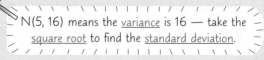

This means that if you subtract μ from any numbers in the question and then divide by σ — you can use your tables for Z.

$$\text{If } X \sim N(\mu, \sigma^2)\text{, then } \frac{X-\mu}{\sigma} = Z\text{, where } Z \sim N(0,1)$$

2) Once you've transformed a variable like this, you can use the <u>Z-tables</u>.

EXAMPLE If $X \sim N(5, 16)$ find: a) $P(X < 7)$, b) $P(X > 9)$, c) $P(5 < X < 11)$

Subtract μ (= 5) from any numbers and divide by σ (= $\sqrt{16}$ = 4) — then you'll have a value for $Z \sim N(0, 1)$.

N(5, 16) means the <u>variance</u> is 16 — take the <u>square root</u> to find the <u>standard deviation</u>.

a) $P(X < 7) = P\left(Z < \frac{7-5}{4}\right) = P(Z < 0.5) = 0.69146$

Look up $P(Z < 0.5)$ in the big table on p153.

b) $P(X > 9) = P\left(Z > \frac{9-5}{4}\right) = P(Z > 1) = 1 - P(Z < 1) = 1 - 0.84134 = 0.15866$

c) $P(5 < X < 11) = P\left(\frac{5-5}{4} < Z < \frac{11-5}{4}\right) = P(0 < Z < 1.5)$

$= P(Z < 1.5) - P(Z < 0) = 0.93319 - 0.5 = 0.43319$

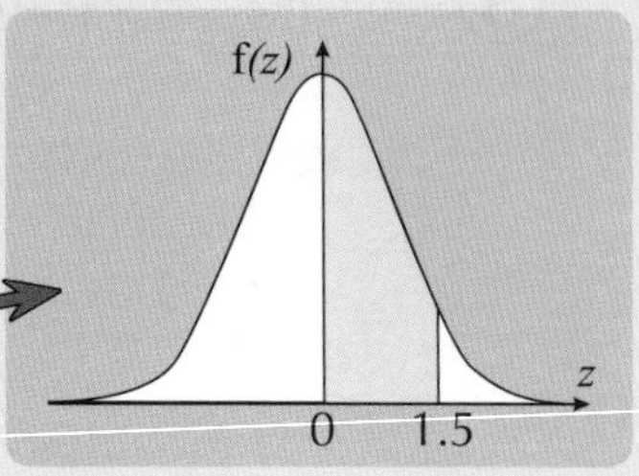

Find the area to the left of 1.5 and subtract the area to the left of 0.

The **Z-Distribution** Can be Used in **Real-Life** Situations

EXAMPLE The times taken by a group of people to complete an assault course are normally distributed with a mean of 600 seconds and a variance of 105 seconds. Find the probability that a randomly selected person took:
a) fewer than 575 seconds, b) more than 620 seconds.

If X represents the time taken in seconds, then $X \sim N(600, 105)$.
It's a normal distribution — so your first thought should be to try and '<u>standardise</u>' it by converting it to Z.

a) <u>Subtract the mean</u> and <u>divide by the standard deviation</u>: $P(X < 575) = P\left(\frac{X-600}{\sqrt{105}} < \frac{575-600}{\sqrt{105}}\right)$
$= P(Z < -2.44)$

So $P(Z < -2.44) = 1 - P(Z \leq 2.44) = 1 - 0.99266 = 0.00734$.

b) Again, <u>subtract the mean</u> and <u>divide by the standard deviation</u>: $P(X > 620) = P\left(\frac{X-600}{\sqrt{105}} > \frac{620-600}{\sqrt{105}}\right)$
$= P(Z > 1.95)$
$= 1 - P(Z \leq 1.95)$
$= 1 - 0.97441 = 0.02559$

Transform to Z and use Z-tables — I repeat: transform to Z and use Z-tables...

The basic idea is always the same — transform your normally-distributed variable to Z, and then use the Z-tables. Statisticians call this the "normal two-step". Well... some of them probably do, anyway. I admit, this stuff is all a bit weird and confusing at first. But as always, work through a few examples and it'll start to click. So get some practice.

Normal Distributions and Z-Tables

You might be given some probabilities and asked to find μ and σ. Just use the same old ideas...

Find μ and σ by First *Transforming* to Z

EXAMPLE $X \sim N(\mu, 2^2)$ and $P(X < 23) = 0.90147$. Find μ.

This is a normal distribution — so your first thought should be to convert it to Z.

① $P(X < 23) = P[\frac{X-\mu}{2} < \frac{23-\mu}{2}] = P[Z < \frac{23-\mu}{2}] = 0.90147$

Substitute Z for $\frac{X-\mu}{\sigma}$.

But 0.90147 isn't in the percentage points table, so you'll need to look it up in the 'big' table instead (the one for $\Phi(z)$).

② If $P(Z < z) = 0.90147$, then $z = 1.29$

③ So $\frac{23-\mu}{2} = 1.29$ — now solve this to find $\mu = 23 - (2 \times 1.29) = 20.42$

EXAMPLE $X \sim N(53, \sigma^2)$ and $P(X < 50) = 0.2$. Find σ.

Again, this is a normal distribution — so you need to use that lovely standardising equation again.

① $P(X < 50) = P[Z < \frac{50-53}{\sigma}] = P[Z < -\frac{3}{\sigma}] = 0.2$

Ideally, you'd look up 0.2 in the percentage points table to find $-\frac{3}{\sigma}$. Unfortunately, it isn't there, so you have to think a bit...

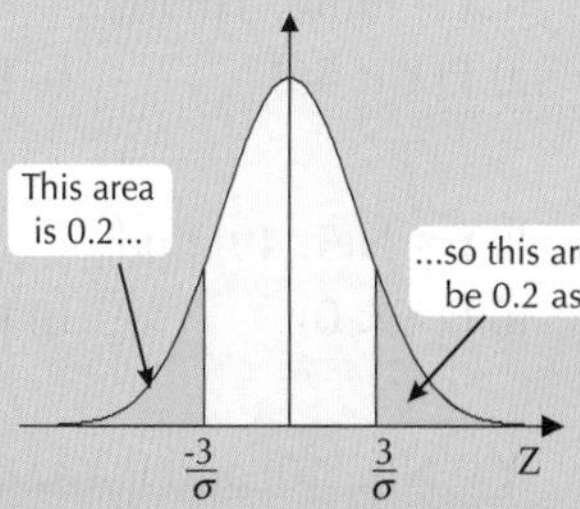

This area is 0.2...
...so this area must be 0.2 as well.
-3/σ
3/σ
Z

② $P[Z < -\frac{3}{\sigma}]$ is 0.2, so from the symmetry of the graph, $P[Z < \frac{3}{\sigma}]$ must be 0.8.

So look up 0.8 in the 'percentage points' table to find that an area of 0.8 is to the left of $z = 0.8416$.

This tells you that $\frac{3}{\sigma} = 0.8416$, or $\sigma = 3.56$ (to 3 sig. fig.)

If You Have to Find μ and σ, You'll Need to Solve *Simultaneous Equations*

EXAMPLE The random variable $X \sim N(\mu, \sigma^2)$. If $P(X < 9) = 0.55962$ and $P(X > 14) = 0.03216$, then find μ and σ.

① $P(X < 9) = P[Z < \frac{9-\mu}{\sigma}] = 0.55962$.

Using the table for $\Phi(z)$, this tells you that $\frac{9-\mu}{\sigma} = 0.15$, or $9 - \mu = 0.15\sigma$.

② $P(X > 14) = P[Z > \frac{14-\mu}{\sigma}] = 0.03216$, which means that $P[Z < \frac{14-\mu}{\sigma}] = 1 - 0.03216 = 0.96784$.

Using the table for $\Phi(z)$, this tells you that $\frac{14-\mu}{\sigma} = 1.85$, or $14 - \mu = 1.85\sigma$.

③ Subtract the equations: $(14 - \mu) - (9 - \mu) = 1.85\sigma - 0.15\sigma$, or $5 = 1.7\sigma$. This gives $\sigma = 5 \div 1.7 = 2.94$.

Now use one of the other equations to find μ: $\mu = 9 - (0.15 \times 2.94) = 8.56$.

The Norman Distribution — came to England in 1066...

It's always the same — you always need to do the normal two-step of subtracting the mean and dividing by the standard deviation. Just make sure you don't use the variance by mistake — remember, in $N(\mu, \sigma^2)$, the second number always shows the variance. I'm sure you wouldn't make a mistake... it's just that I'm such a worrier and I do so want you to do well.

S1 Section 4 — Practice Questions

Normal distributions... there's nothing terribly hard about them. They just get fiddly and annoying.
But you need to know how to answer normal distribution questions, so here are some to practise with.

Warm-up Questions

1) Find the probability that:
   a) $Z < 0.84$, b) $Z < 2.95$,
   c) $Z > 0.68$, d) $Z \geq 1.55$,
   e) $Z < -2.10$, f) $Z \leq -0.01$,
   g) $Z > 0.10$, h) $Z \leq 0.64$,
   i) $Z > 0.23$, j) $0.10 < Z \leq 0.50$,
   k) $-0.62 \leq Z < 1.10$, l) $-0.99 < Z \leq -0.74$.

2) Find the value of z if:
   a) $P(Z < z) = 0.91309$, b) $P(Z < z) = 0.58706$,
   c) $P(Z > z) = 0.03593$, d) $P(Z < z) = 0.99$,
   e) $P(Z \leq z) = 0.40129$, f) $P(Z \geq z) = 0.995$.

3) If $X \sim N(50, 16)$ find:
   a) $P(X < 55)$, b) $P(X < 42)$, c) $P(X > 56)$ d) $P(47 < X < 57)$.

4) If $X \sim N(5, 49)$ find:
   a) $P(X < 0)$, b) $P(X < 1)$, c) $P(X > 7)$ d) $P(2 < X < 4)$.

5) $X \sim N(\mu, 10)$ and $P(X < 8) = 0.89251$. Find μ.

6) $X \sim N(\mu, 8^2)$ and $P(X > 221) = 0.30854$. Find μ.

7) $X \sim N(11, \sigma^2)$ and $P(X < 13) = 0.6$. Find σ.

8) $X \sim N(108, \sigma^2)$ and $P(X \leq 110) = 0.96784$. Find σ.

9) The random variable $X \sim N(\mu, \sigma^2)$.
   If $P(X < 15.2) = 0.97831$ and $P(X > 14.8) = 0.10565$, then find μ and σ.

10) The mass of items produced by a factory is normally distributed with a mean of 55 grams and a standard deviation of 4.4 grams. Find the probability of a randomly chosen item having a mass of:
    a) less than 55 grams, b) less than 50 grams, c) more than 60 grams.

11) The mass of eggs laid by an ostrich is normally distributed with a mean of 1.4 kg and a standard deviation of 300 g. Find the probability of a randomly chosen egg from this bird having a mass of:
    a) less than 1 kg, b) more than 1.5 kg, c) between 1300 and 1600 g.

12) $X \sim N(80, 15)$
    a) If $P(X > a) = 0.01$, find a.
    b) If $P(|X - 80| < b) = 0.8$, find b.

S1 Section 4 — Practice Questions

You're nearly at the end of another section, so it's time to celebrate. And what better way to celebrate than with a page of questions that look a lot like the ones you'll face on your exam day.

Exam Questions

1 The exam marks for 1000 candidates are normally distributed with mean 50 marks and standard deviation 30 marks.

a) The pass mark is 41. Estimate the number of candidates who passed the exam. *(3 marks)*

b) Find the mark needed for a distinction if the top 10% of the candidates achieved a distinction. *(3 marks)*

2 The lifetimes of a particular type of battery are normally distributed with mean μ and standard deviation σ. A student using these batteries finds that 40% last less than 20 hours and 80% last less than 30 hours. Find μ and σ. *(7 marks)*

3 The random variable X has a normal distribution with mean 120 and standard deviation 25.

a) Find $P(X > 145)$. *(3 marks)*

b) Find the value of j such that $P(120 < X < j) = 0.46407$. *(4 marks)*

4 The diameters of the pizza bases made at a restaurant are normally distributed. The mean diameter is 12 inches, and 5% of the bases measure more than 13 inches.

a) Write down the median diameter of the pizza bases. *(1 mark)*

b) Find the standard deviation of the diameters of the pizza bases. *(4 marks)*

Any pizza base with a diameter of less than 10.8 inches is considered too small and is discarded.

c) If 100 pizza bases are made in an evening, approximately how many would you expect to be discarded due to being too small? *(3 marks)*

Three pizza bases are selected at random.

d) Find the probability that at least one of these bases is too small. *(3 marks)*

5 A garden centre sells bags of compost. The volume of compost in the bags is normally distributed with a mean of 50 litres.

a) If the standard deviation of the volume is 0.4 litres, find the probability that a randomly selected bag will contain less than 49 litres of compost. *(3 marks)*

b) If 1000 of these bags of compost are bought, find the expected number of bags containing more than 50.5 litres of compost. *(5 marks)*

A different garden centre sells bags of similar compost. The volume of compost, in litres, in these bags is described by the random variable Y, where $Y \sim N(75, \sigma^2)$. It is found that 10% of the bags from this garden centre contain less than 74 litres of compost.

c) Find σ. *(3 marks)*

Sampling and Statistics

No time for any small talk, I'm afraid. It's straight on with the business of populations and how to find out about them.

A Population is a Group of People or Items

In any statistical investigation, there'll be a group of people or items you want to find out about.
This group is called the population, and could be:

- All the students in a maths class
- All the penguins in Antarctica
- All the chocolate puddings produced by a company in a year

You could find out about a population by collecting information from every single member.
But that's usually difficult, expensive and time-consuming — so it's more common to use a sample instead.

A Sample needs to be Representative of its Population

1) If collecting information from every member of a population is impossible or impractical, you can find out about a population by questioning or examining just a selection of people or items. This selected group is called a sample.
2) Data collected from a sample is used to draw conclusions about the whole population. So it's vital that the sample is as much like the population as possible. A biased sample is one which doesn't fairly represent the population.

To Avoid Sampling Bias:

① Select from the correct population and make sure none of the population is excluded. This usually involves making an accurate list of everyone/everything in the population. E.g. if you want to find out the views of residents from a particular street, your sample should only include residents from that street and should be chosen from a full list of the residents.

② Select your sample at random — see below. Non-random sampling methods include things like the sampler just asking their friends (who might all give similar answers), or asking for volunteers (meaning they only get people with strong views on a subject).

③ Make sure all your sample members respond — otherwise the results could be biased. E.g. if some of your sampled residents are out when you go to interview them, it's important that you go back and get their views another time.

3) Using a random sample means there will be no systematic bias in your sample. A 'simple random sample of size n' is one where n members of the population are chosen at random from a full list of the population.

In a Simple Random Sample...

- Every person or item in the population has an equal chance of being in the sample.
- Each selection is independent of every other selection.

This means every single possible sample is equally likely.

Statistics are Calculated Using Only the Values from your Sample

You're nearly at the end of the AS Statistics sections now. So it must be time to say what a statistic actually is.

Statistics

A statistic is a quantity calculated only from the known observations in a sample.
A statistic is a random variable — it takes different values for different samples.

EXAMPLE If $X_1, \ldots, X_{10}$ form a random sample from a population with unknown mean, μ, then:

- $X_{10} - X_1$ and $\frac{\sum X_i}{n}$ are both statistics.
- $\sum X_i^2 - \mu$ isn't a statistic (since it depends on the unknown value of μ).

$\frac{\sum X}{n}$ is the sample mean, usually written $\overline{X}$.

1) If you took a sample of observations and calculated a particular statistic, then took another sample and calculated the same statistic, then took another sample, etc., you'd end up with lots of values of the same statistic.
2) The probability distribution of a statistic is called a sampling distribution.

A sample should be a miniature version of the population...

Simple random sampling is important, since it reduces the risk of sampling bias and increases the chances of reliable results — which is the whole point. Just because there aren't any sums to do doesn't mean you can skim over this stuff, by the way.

Using Statistics

The point of finding statistics isn't really to find out about your sample — but for something far more important. You actually calculate statistics to find out something about the whole population that the sample came from.

Statistics are used to Estimate Population Parameters

1) The characteristics of a population (e.g. its mean or variance) can be described by quantities called parameters.
2) Statistics are used to estimate these parameters. For example, you can use the sample mean ($\overline{X} = \frac{\Sigma X}{n}$) to estimate the population mean (μ). The sample mean is called an estimator of the population mean.
3) The sample mean is actually an unbiased estimator of the population mean — this means the expected value of the sample mean is the same as the population mean (a good thing).

Greek letters (e.g. σ) are often used for parameters, and Latin letters (e.g. s) are used for statistics.

It's easy to end up with a Biased estimate of Population Variance

Not all estimators are unbiased. In fact, the formula for population variance from page 107 is a biased estimator of population variance when used with a sample.

1) If you have a sample $X_1, ..., X_n$ taken from a population, then $\frac{\sum(X_i - \overline{X})^2}{n} = \frac{\sum X_i^2}{n} - \left(\frac{\sum X_i}{n}\right)^2$ will give a biased estimate of the population variance (σ^2).
2) For an unbiased estimate of the population variance, there's a different formula...

This is the formula from p107.

Unbiased Estimators of Population Variance and Standard Deviation

These are unbiased estimators of the population variance and population standard deviation:

Sample variance: $S^2 = \frac{n}{n-1}\left[\frac{\sum X_i^2}{n} - \left(\frac{\sum X_i}{n}\right)^2\right] = \frac{\sum(X_i - \overline{X})^2}{n-1}$

Sample standard deviation: $S = \sqrt{\text{sample variance}}$

This one's in your formula booklet, but the one in red is most useful: "the mean of the squares minus the square of the mean, all multiplied by $\frac{n}{n-1}$".

EXAMPLE A random sample was taken from a population whose mean (μ) and variance (σ^2) are unknown. If the observed values were 5.6, 5.8, 5.5, 5.8 and 5.7, find unbiased estimates of μ and σ^2.

The sample mean ($\bar{x}$) will give an unbiased estimate of the population mean (μ). $\Rightarrow \sum x = 28.4$, so $\bar{x} = \frac{\sum x}{n} = \frac{28.4}{5} = 5.68$

Use the above formula to find an unbiased estimate of the population variance (σ^2). $\Rightarrow \sum x^2 = 161.38$, so $s^2 = \frac{n}{n-1}\left[\frac{\sum x^2}{n} - \left(\frac{\sum x}{n}\right)^2\right] = \frac{5}{4}\left[\frac{161.38}{5} - \left(\frac{28.4}{5}\right)^2\right] = 0.017$

If your data's grouped, you need to use mid-class values (just like on pages 105 and 108).

EXAMPLE The table shows how many pets a random sample of 95 students from a school have. Calculate unbiased estimates of the mean and variance for the whole school.

Number of pets	0	1	2	3	4	5–6	7-8	9-10
Frequency, f	4	13	24	17	15	11	5	6

Add some extra rows to the table and work out the following totals.

$\sum f = 95 = n, \sum fx = 327$ and $\sum fx^2 = 1657.5$

Mid-class value, x	0	1	2	3	4	5.5	7.5	9.5
fx	0	13	48	51	60	60.5	37.5	57
x^2	0	1	4	9	16	30.25	56.25	90.25
fx^2	0	13	96	153	240	332.75	281.25	541.5

This gives: $\bar{x} = \frac{\sum fx}{\sum f} = \frac{327}{95} = 3.44$ (to 3 sig.fig.),

and $s^2 = \frac{n}{n-1}\left[\frac{\sum fx^2}{n} - \left(\frac{\sum fx}{n}\right)^2\right] = \frac{95}{94}\left[\frac{1657.5}{95} - \left(\frac{327}{95}\right)^2\right] = 5.7$ (to 1 d.p.)

'Statisticiness'* — when cool-sounding stats are used to convince people of a lie...

Remember... you use the variance formula on p107 when you have values for the whole population. You use the one on this page when you just have a sample of values and are estimating the population variance. I know... it's annoying.

*Not actually a real word... but I'm sure you know what I mean.

Standard Error and the Central Limit Theorem

The Central Limit Theorem is pretty much the best thing ever. No, I mean it... it really is quite cool.

The Standard Error of $\overline{X}$ is: σ ÷ the Square Root of the Sample Size

1) Suppose you've got a random variable X, with mean μ and variance σ^2.
2) When you take a sample of n readings from the distribution of X, you can work out its sample mean $\overline{X}$. If you now keep taking samples of size n from that distribution and working out the sample means, then you get a collection of sample means, drawn from the sampling distribution of $\overline{X}$.
3) This sampling distribution also has mean μ, but a variance of $\frac{\sigma^2}{n}$ (and so a standard deviation of $\frac{\sigma}{\sqrt{n}}$). This standard deviation is called the standard error of the sample mean.

> The standard error of $\overline{X}$ is $\frac{\sigma}{\sqrt{n}}$.

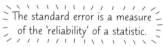

The standard error is a measure of the 'reliability' of a statistic.

If X is Normally Distributed, $\overline{X} \sim N(\mu, \sigma^2/n)$

1) From above, you know that for any random variable X with mean μ and variance σ^2, the sampling distribution of the sample mean $\overline{X}$ has mean μ, but a variance of $\frac{\sigma^2}{n}$.
2) But... if X follows a normal distribution, then the sampling distribution of $\overline{X}$ will also be normal. This is important.

> If $X \sim N(\mu, \sigma^2)$, then $\overline{X} \sim N\left(\mu, \frac{\sigma^2}{n}\right)$

For Large n, even when X Isn't Normally Distributed, $\overline{X} \sim N(\mu, \sigma^2/n)$ (approximately)

The Central Limit Theorem tells you something about $\overline{X}$ — even if you know nothing at all about the distribution of X.

The Central Limit Theorem

Suppose you take a sample of n readings from any distribution with mean μ and variance σ^2.

- For large n, the distribution of the sample mean, $\overline{X}$, is approximately normal: $\overline{X} \sim N\left(\mu, \frac{\sigma^2}{n}\right)$.
- The bigger n is, the better the approximation will be. (For $n > 30$ it's pretty good.)

EXAMPLE A sample of size 50 is taken from a population with mean 20 and variance 10. Find the probability that the sample mean is less than 19.

① Since n (= 50) is quite large, you can use the Central Limit Theorem. Here $\overline{X} \sim N\left(20, \frac{10}{50}\right) = N(20, 0.2)$ (approx).

② You need $P(\overline{X} < 19)$. Since $\overline{X}$ has an (approximately) normal distribution, you can standardise it and use tables for Z (see page 130). So subtract the mean from all the numbers and divide by the standard deviation (= $\sqrt{0.2}$):

$$P(\overline{X} < 19) \approx P\left(Z < \frac{19-20}{\sqrt{0.2}}\right) = P(Z < -2.24)$$

③ Now you can use your normal-distribution tables (see p153).

$$P(Z < -2.24) = P(Z > 2.24) = 1 - P(Z \le 2.24) = 1 - 0.98745 = 0.01255$$

As always... if you need to, draw a sketch.

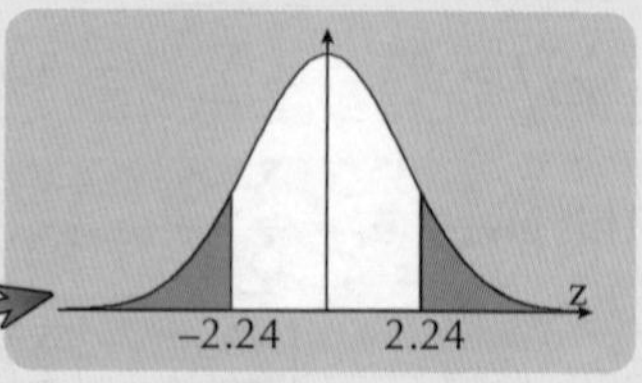

The Central Limit Theorem — the best thing since those sheets of bread...

Personally, I can't help but think the Central Limit Theorem's pretty amazing — you can take samples from any distribution you like and say something meaningful (excuse pun) about the mean. This is why the normal distribution is probably the most important distribution in all of statistics. This is a tricky page, I admit, so learn it carefully — every last detail.

Confidence Intervals for Normal Distributions

You can be pretty confident (ho ho) the stuff on this page will come up in your exam.

A *Confidence Interval* is a Type of *Interval Estimation*

A confidence interval for the mean of a population is a range of values (usually symmetrical about your sample mean) that you're 'fairly sure' contains the true population mean. Look at this example:

"(11.2, 14.4) is a 95% confidence interval for μ"

Remember... the value of μ is fixed — it's the interval that will change depending on your sample.

This means that there's a 95% probability that the interval (11.2, 14.4) includes the true value of μ.

A Confidence Interval Extends "z × Standard Error" from the Sample Mean

For a population with a normal distribution and known variance, it's easy to find a confidence interval for the mean.

Confidence Intervals for μ — Normal Distribution and Known Variance

A confidence interval for the population mean is: $\left(\overline{X} - z\frac{\sigma}{\sqrt{n}}, \overline{X} + z\frac{\sigma}{\sqrt{n}}\right)$

The value of z depends on the 'level of confidence' you need — for example:

- for a 95% confidence interval, choose z so that $P(-z < Z < z) = 0.95$,
- for a 98% confidence interval, choose z so that $P(-z < Z < z) = 0.98$, and so on.

This relies on $\overline{X}$ being normally distributed. From the previous page, you know that if X is normally distributed, then so is $\overline{X}$ — which is handy. In fact, the Central Limit Theorem tells you that if you use a big enough sample size, $\overline{X}$ is always (approximately) normally distributed — even better.

EXAMPLE A machine makes chocolate bars with weights that are normally distributed with a standard deviation of 8.2 grams. A random sample of 10 bars is chosen, with weights (in grams) as follows:
251.3 257.2 249.5 254.3 252.6 256.1 254.7 255.3 254.8 252.6.

a) Calculate a 95% confidence interval for the mean of all the bars made by the machine.

b) A manager says that the mean weight of all the bars made is 256 g. Comment on this claim.

a) • Find $\overline{X}$, the sample mean.

$\overline{X} = (251.3 + 257.2 + 249.5 + 254.3 + 252.6 + 256.1 + 254.7 + 255.3 + 254.8 + 252.6) \div 10 = 253.84$

• Using the normal 'percentage points' table (p153), find z so that $P(-z < Z < z) = 0.95$. This means looking up $p = 0.975$ — you find that $z = 1.96$.

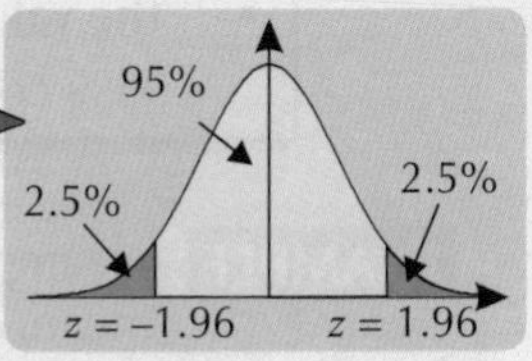

• So your confidence interval is:

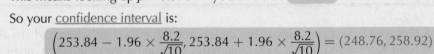

$\left(253.84 - 1.96 \times \frac{8.2}{\sqrt{10}}, 253.84 + 1.96 \times \frac{8.2}{\sqrt{10}}\right) = (248.76, 258.92)$

b) The manager's claim is within the confidence interval — so these results provide no evidence to doubt the claim.

If the manager had claimed the mean weight was 260 grams, the results would not support this claim.

EXAMPLE A sample of size 100 is taken from a population with a standard deviation of 16 but an unknown mean. If the sample mean is 44, find a 99% confidence interval for the population mean.

① Since n (= 100) is quite large, you can use the Central Limit Theorem to say that $\overline{X} \sim N\left(\mu, \frac{16^2}{100}\right)$ (approximately).

② So the confidence interval is $\left(\overline{X} - z\frac{\sigma}{\sqrt{n}}, \overline{X} + z\frac{\sigma}{\sqrt{n}}\right)$, where I choose z so that $P(-z < Z < z) = 0.99$.
Using $p = 0.995$ in the percentage points table gives $z = 2.5758$.

③ This gives a confidence interval of $\left(44 - 2.5758 \times \frac{16}{\sqrt{100}}, 44 + 2.5758 \times \frac{16}{\sqrt{100}}\right) = (39.9, 48.1)$

Useful Quote: "Equations are the devil's sentences" *

There are a few handy values of z that it's probably worth remembering — e.g. for a 90% confidence interval, $z = 1.6449$; for a 95% confidence interval, $z = 1.96$, and for a 99% confidence interval, $z = 2.5758$. You'll be able to look them up in the exam if you don't remember them though, so don't worry too much if your brain's already full.

* Stephen Colbert

Confidence Intervals for Large n

To find a confidence interval for the mean, you need to know the sample mean's standard error. To find that, you need to know the population standard deviation (σ). But sometimes you don't know that. Never mind.

Estimate the Standard Error $\sigma/\sqrt{n}$ using the Estimator $S/\sqrt{n}$

1) You can estimate the population's standard deviation using the unbiased estimator S (see p135). For large n ($n > 30$ or so), this estimate should be pretty good.
2) You can then make an unbiased estimate of the sample mean's standard error ($\frac{\sigma}{\sqrt{n}}$), using $\frac{s}{\sqrt{n}}$.

EXAMPLE A random sample of 80 newts were weighed, and the summary statistics below calculated. Estimate the sample mean's standard error.

$$\sum x = 7238, \sum x^2 = 684\,688$$

$$s^2 = \frac{n}{n-1}\left[\frac{\sum x^2}{n} - \left(\frac{\sum x}{n}\right)^2\right] = \frac{80}{79}\left[\frac{684\,688}{80} - \left(\frac{7238}{80}\right)^2\right] = 377.594...$$

So an estimate of the mean's standard error is: $\frac{s}{\sqrt{n}} = \frac{\sqrt{377.594...}}{\sqrt{80}} = 2.17$ (to 3 sig.fig.)

And because n is quite large, this should be a pretty good estimate.

If n is Large, you can Always find a Confidence Interval for the Mean

When you're working out confidence intervals for a population mean, a large sample size (n) is handy for two reasons:

1) If you use a sample of more than about 30 observations, the Central Limit Theorem tells you that $\overline{X}$ will be approximately normally distributed — so you can use the confidence interval formula on page 137.
2) And even if you don't know the population standard deviation (σ), you can estimate it pretty well using s. This gives you the formula below (which is the same as the one on p137, only with S instead of σ).

Confidence Intervals for μ — Large n but Unknown Variance

A confidence interval for the population mean where the population variance is unknown is given by:

$$\left(\overline{X} - z\frac{S}{\sqrt{n}}, \overline{X} + z\frac{S}{\sqrt{n}}\right)$$

- Here, S is an unbiased estimator of the population standard deviation.
- The value of z depends on the 'level of confidence' you need (e.g. for a 95% confidence interval, choose z so that $P(-z < Z < z) = 0.95$, and so on).

EXAMPLE A sample of size 50 is taken from a population with an unknown standard deviation and an unknown mean. If $\sum x = 80$ and $\sum x^2 = 192$, find a 95% confidence interval for the population mean.

① n (= 50) is quite large, so the Central Limit Theorem says $\overline{X}$ will be (approximately) normally distributed.

② The sample mean is $\overline{x} = \frac{\sum x}{n} = \frac{80}{50} = 1.6$.

③ An unbiased estimate of the population variance is:

$$s^2 = \frac{n}{n-1}\left[\frac{\sum x^2}{n} - \left(\frac{\sum x}{n}\right)^2\right] = \frac{50}{49}\left[\frac{192}{50} - \left(\frac{80}{50}\right)^2\right] = \frac{1}{49}[192 - 128] = \frac{64}{49}$$

So an unbiased estimate of the population standard deviation is: $s = \sqrt{\frac{64}{49}} = \frac{8}{7}$

Because n is quite large, this should be fairly reliable.

④ This gives you a confidence interval of:

$$\left(\overline{x} - z\frac{s}{\sqrt{n}}, \overline{x} + z\frac{s}{\sqrt{n}}\right) = \left(1.6 - 1.96 \times \frac{8}{7\sqrt{50}}, 1.6 + 1.96 \times \frac{8}{7\sqrt{50}}\right) = (1.28, 1.92) \text{ (to 2 d.p.)}$$

For large n, you can be confident of being able to find a confidence interval...

Well... this page might be tricky, or it might not be — I'm none too sure. In a way, it's fairly easy because it just relies on things from earlier in the section. But make sure you understand that tricky bit about why a large sample size is so useful.

S1 Section 5 — Practice Questions

That was quite a section... starting off with some fairly okayish sampling bits and bobs and then finishing up with some pretty awkward calculations. Try these questions to see what stuck in your head and what didn't.

Warm-up Questions

1) The manager of a tennis club wants to know if members are happy with the facilities provided. She decides to ask the views of a sample of members.
   a) Identify the population the manager is interested in.
   b) Explain how the manager could avoid systematic bias in her sample.

2) What is a simple random sample?

3) The weights of a population of jars of pickled onions have unknown mean μ and standard deviation σ. A random sample of 50 weights (X_1, ..., X_{50}) are recorded. For each of the following, say whether or not it's a statistic:
   a) $\frac{X_{25} + X_{26}}{2}$ b) $\sum X_i - \sigma$ c) $\sum X_i^2 + \mu$ d) $\frac{\sum X_i}{50}$

4) A random sample was taken from a population whose mean (μ) and variance (σ^2) are unknown. Find unbiased estimates of μ and σ^2 if the sample values were:
   8.4, 8.6, 7.2, 6.5, 9.1, 7.7, 8.1, 8.4, 8.5, 8.0.

5) A random sample of size 15 (X_1,..., X_{15}) is taken from a normally-distributed population with mean 7.2 and standard deviation 5.
   a) Calculate the standard error of the sample mean.
   b) Write down the sampling distribution of the sample mean.
   c) Find $P(\overline{X} < 7.5)$.

6) A random sample of size 80 is taken from a population with mean 18 and standard deviation 4.
   a) What can you say about the sampling distribution of the sample mean? Explain your answer fully.
   b) Find $P(\overline{X} > 19)$.

7) A random sample of size 25 is taken from a normally-distributed population with standard deviation 0.4. Find a 99% confidence interval for the population mean if the sample mean is 18.2.

8) A random sample of size 120 is taken from a population with standard deviation 3. Find a 95% confidence interval for the population mean if the sample mean is 33.8. State clearly any assumptions you make.

9) A random sample of size 100 was taken from a population. The following statistics were then calculated: $\sum x = 104$ and $\sum x^2 = 122.1$
   Find an unbiased estimate of the standard error of the sample mean.

10) A sample of size 80 was taken from a population with an unknown mean. The following statistics were then calculated: $\sum x = 24.2$ and $\sum x^2 = 41.3$
   Find a 95% confidence interval for the population mean.
   State clearly any assumptions you make.

S1 Section 5 — Practice Questions

This is real statistics now — the kind that's too tricky for most newspapers to report accurately.
But if you can answer these questions, then you'll probably be able to explain where they're going wrong.

Exam Questions

1 A sample of 50 sunflowers is selected. Their heights are measured.
The results are summarised in the table below.

Height, h (cm)	0-40	40-80	80-100	100-140	140-160
Frequency, f	5	8	11	19	7

Calculate an unbiased estimate of:

a) the mean height of the population, *(1 mark)*

b) the standard deviation of the population heights. *(3 marks)*

2 The weights of a population of marsh frogs are normally distributed with a standard deviation of 8.5 grams. The weights of a random sample of 30 marsh frogs are measured.
The total weight of all 30 frogs is found to be 3.84 kg.

a) Calculate the mean weight in grams of the frogs in this sample. *(1 mark)*

b) Calculate the standard error in grams of the sample mean weight. *(2 marks)*

c) Construct a 95% confidence interval for the mean weight of this population of marsh frogs. *(3 marks)*

d) Within what range would you expect 99% of individual marsh frogs' weights to fall? *(3 marks)*

3 A sample of 60 racing snails sold by a business was selected at random.
The distance covered (x metres) in 10 minutes by each of the 60 snails was measured.
The following statistics were then calculated: $\sum x = 184$ and $\sum x^2 = 620$

a) Find an unbiased estimate of the sample mean's standard error. *(4 marks)*

b) Construct a 98% confidence interval for the population mean.
State clearly any assumptions you make. *(5 marks)*

c) The manager of the business claims that, on average, its snails can cover over 3 metres in 10 minutes. Comment on this claim. *(2 marks)*

4 The number of minutes (X) taken by Ravi to travel to work each day may be modelled by a normal distribution with unknown mean μ and standard deviation σ.
The value of X was measured on 40 days. The results are summarised below.

$$\sum x = 1676 \text{ and } \sum (x - \bar{x})^2 = 1382$$

a) Find an unbiased estimate of the population variance. *(1 mark)*

b) Construct a 99% confidence interval for the population mean. *(4 marks)*

c) Ravi claims it takes him, on average, 42 minutes to travel to work.
His manager believes the mean time is actually over 46 minutes.
Comment on each of these claims. *(2 marks)*

Correlation

Correlation is all about how closely two quantities are linked. And it can involve a fairly hefty formula.

Draw a Scatter Diagram to see Patterns in Data

Sometimes variables are measured in pairs — maybe because you want to find out how closely they're linked.
These pairs of variables might be things like: — 'my age' and 'length of my feet', or
— 'temperature' and 'number of accidents on a stretch of road'.

You can plot readings from a pair of variables on a scatter diagram — this'll tell you something about the data.

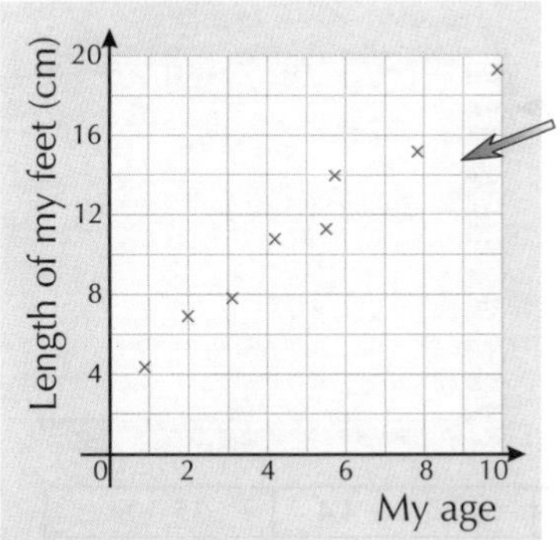

The variables 'my age' and 'length of my feet' seem linked — all the points lie close to a line. As I got older, my feet got bigger and bigger (though I stopped measuring when I was 10).

It's a lot harder to see any connection between the variables 'temperature' and 'number of accidents' — the data seems scattered pretty much everywhere.

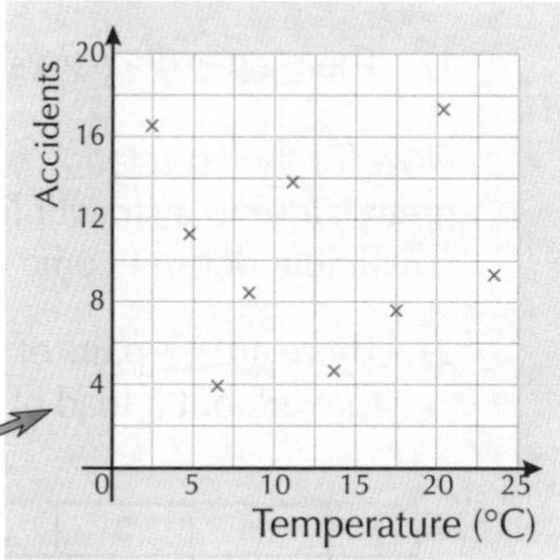

Correlation is a measure of How Closely variables are Linked

1) Sometimes, as one variable gets bigger, the other one also gets bigger — then the scatter diagram might look like the one on the right. Here, a line of best fit would have a positive gradient. The two variables are positively correlated (or there's a positive correlation between them).

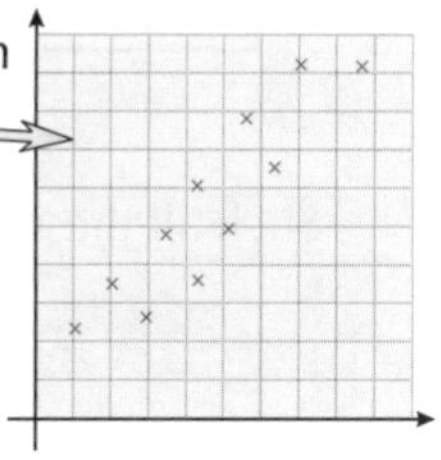

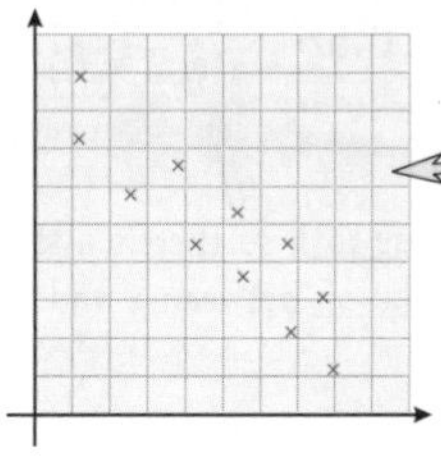

2) But if one variable gets smaller as the other one gets bigger, then the scatter diagram might look like this one — and the line of best fit would have a negative gradient. The two variables are negatively correlated (or there's a negative correlation between them).

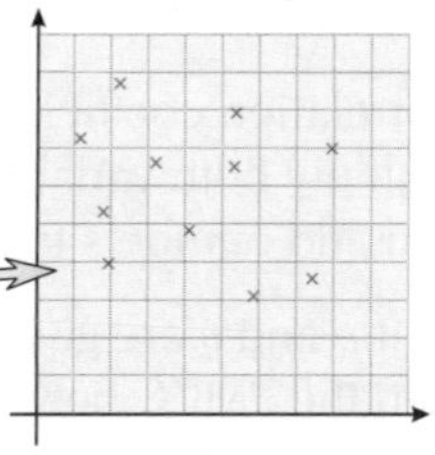

3) And if the two variables aren't linked at all, you'd expect a random scattering of points — it's hard to say where the line of best fit would be. The variables aren't correlated (or there's no correlation).

The Product-Moment Correlation Coefficient (r) measures Correlation

1) The Product-Moment Correlation Coefficient (PMCC, or r, for short) measures how close to a straight line the points on a scatter graph lie.
2) The PMCC is always between +1 and –1.
If all your points lie exactly on a straight line with a positive gradient (perfect positive correlation), $r = +1$.
If all your points lie exactly on a straight line with a negative gradient (perfect negative correlation), $r = -1$.
(In reality, you'd never expect to get a PMCC of +1 or –1 — your scatter graph points might lie pretty close to a straight line, but it's unlikely they'd all be on it.)
3) If $r = 0$ (or more likely, pretty close to 0), that would mean the variables aren't correlated.
4) The formula for the PMCC is a real stinker. But some calculators can work it out if you type in the pairs of readings, which makes life easier. Otherwise, just take it nice and slow.

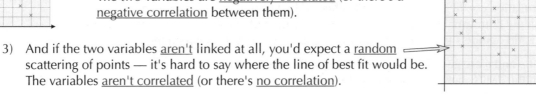

$$r = \frac{S_{xy}}{\sqrt{S_{xx}S_{yy}}} = \frac{\sum[x-\bar{x}][y-\bar{y}]}{\sqrt{(\sum[x-\bar{x}]^2)(\sum[y-\bar{y}]^2)}} = \frac{\sum xy - \frac{[\sum x][\sum y]}{n}}{\sqrt{\left(\sum x^2 - \frac{[\sum x]^2}{n}\right)\left(\sum y^2 - \frac{[\sum y]^2}{n}\right)}}$$

This is the easiest one to use, but it's still a bit hefty. Fortunately, it'll be on your formula sheet.

See page 143 for more about S_{xy} and S_{xx}.

Correlation

Don't rush questions on correlation. In fact, take your time and draw yourself a nice table.

EXAMPLE Illustrate the following data with a scatter diagram, and find the product-moment correlation coefficient (r) between the variables x and y.

If $p = 4x - 3$ and $q = 9y + 17$, what is the PMCC between p and q?

x	1.6	2.0	2.1	2.1	2.5	2.8	2.9	3.3	3.4	3.8	4.1	4.4
y	11.4	11.8	11.5	12.2	12.5	12.0	12.9	13.4	12.8	13.4	14.2	14.3

1) The scatter diagram's the easy bit — just plot the points.

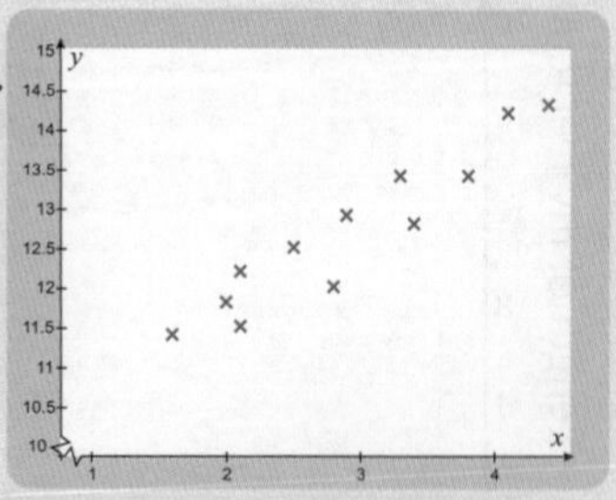

Now for the correlation coefficient. From the scatter diagram, the points lie pretty close to a straight line with a positive gradient — so if the correlation coefficient doesn't come out pretty close to +1, we'd need to worry...

2) There are 12 pairs of readings, so $n = 12$. That bit's easy — now you have to work out a load of sums. It's best to add a few extra rows to your table...

x	1.6	2	2.1	2.1	2.5	2.8	2.9	3.3	3.4	3.8	4.1	4.4	$35 = \Sigma x$
y	11.4	11.8	11.5	12.2	12.5	12	12.9	13.4	12.8	13.4	14.2	14.3	$152.4 = \Sigma y$
x^2	2.56	4	4.41	4.41	6.25	7.84	8.41	10.89	11.56	14.44	16.81	19.36	$110.94 = \Sigma x^2$
y^2	129.96	139.24	132.25	148.84	156.25	144	166.41	179.56	163.84	179.56	201.64	204.49	$1946.04 = \Sigma y^2$
xy	18.24	23.6	24.15	25.62	31.25	33.6	37.41	44.22	43.52	50.92	58.22	62.92	$453.67 = \Sigma xy$

Stick all these in the formula to get: $$r = \frac{\left[453.67 - \frac{35 \times 152.4}{12}\right]}{\sqrt{\left[110.94 - \frac{35^2}{12}\right] \times \left[1946.04 - \frac{152.4^2}{12}\right]}} = \frac{9.17}{\sqrt{8.857 \times 10.56}} = 0.948 \text{ (to 3 s.f.)}$$

This is pretty close to 1, so there's a strong positive correlation between x and y.

3) Correlation coefficients aren't greatly affected by linear scaling — you can multiply both variables by positive numbers (or multiply them both by negative numbers), and add numbers to them, and you won't change the PMCC between them.

So if p and q are given by $p = 4x - 3$ and $q = 9y + 17$, then the PMCC between p and q is also 0.948.

If you multiply one of the variables by a negative number, the PMCC changes sign.

Don't make **Sweeping Statements** using Statistics

1) A high correlation coefficient doesn't necessarily mean that one factor causes the other.

EXAMPLE The number of televisions sold in Japan and the number of cars sold in America may well be correlated, but that doesn't mean that high TV sales in Japan cause high car sales in the US.

2) The PMCC is only a measure of a linear relationship between two variables (i.e. how close they'd be to a line if you plotted a scatter diagram).

EXAMPLE In the diagram on the right, the PMCC would be pretty low, but the two variables definitely look linked. It looks like the points lie on a parabola (the shape of an x^2 curve) — not a line.

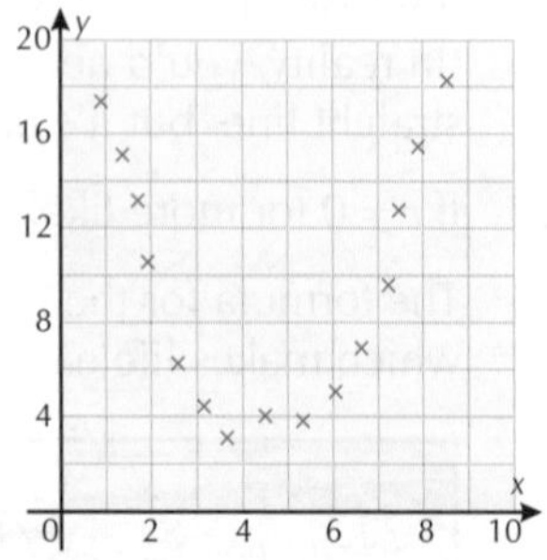

What's a statistician's favourite soap — Correlation Street... (Boom boom)

It's worth remembering that the PMCC assumes that both variables are normally distributed — chances are you won't get asked a question about that, but there's always the possibility that you might, so learn it.

Linear Regression

Linear regression is just fancy stats-speak for 'finding lines of best fit'. Not so scary now, eh...

Decide which is the *Independent Variable* and which is the *Dependent*

EXAMPLE The data below shows the load on a lorry, x (in tonnes), and the fuel efficiency, y (in km per litre).

x	5.1	5.6	5.9	6.3	6.8	7.4	7.8	8.5	9.1	9.8
y	9.6	9.5	8.6	8.0	7.8	6.8	6.7	6	5.4	5.4

1) The variable along the x-axis is the explanatory or independent variable — it's the variable you can control, or the one that you think is affecting the other. The variable 'load' goes along the x-axis here.
2) The variable up the y-axis is the response or dependent variable — it's the variable you think is being affected. In this example, this is the fuel efficiency.

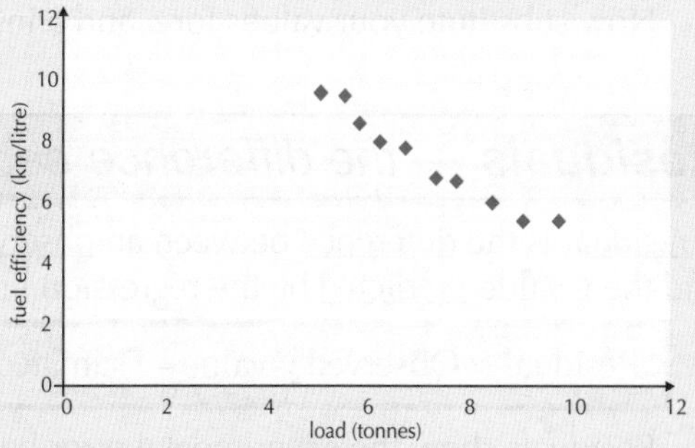

The *Regression Line* (Line of Best Fit) is in the form *y = a + bx*

To find the line of best fit for the above data you need to work out some sums.
Then it's quite easy to work out the equation of the line. If your line of best fit is $y = a + bx$, this is what you do...

① First work out these four sums — a table is probably the best way: $\sum x$, $\sum y$, $\sum x^2$, $\sum xy$.

x	5.1	5.6	5.9	6.3	6.8	7.4	7.8	8.5	9.1	9.8	$72.3 = \sum x$
y	9.6	9.5	8.6	8	7.8	6.8	6.7	6	5.4	5.4	$73.8 = \sum y$
x^2	26.01	31.36	34.81	39.69	46.24	54.76	60.84	72.25	82.81	96.04	$544.81 = \sum x^2$
xy	48.96	53.2	50.74	50.4	53.04	50.32	52.26	51	49.14	52.92	$511.98 = \sum xy$

② Then work out S_{xy}, given by: $S_{xy} = \sum(x-\bar{x})(y-\bar{y}) = \sum xy - \frac{(\sum x)(\sum y)}{n}$

and S_{xx}, given by: $S_{xx} = \sum(x-\bar{x})^2 = \sum x^2 - \frac{(\sum x)^2}{n}$

These are the same as the terms used to work out the PMCC (see p141).

③ The gradient (b) of your regression line is given by: $b = \frac{S_{xy}}{S_{xx}}$

④ And the intercept (a) is given by: $a = \bar{y} - b\bar{x}$.

⑤ Then the regression line is just: $y = a + bx$.

Loads of calculators will work out regression lines for you — but you still need to know this method, since they might give you just the sums from Step 1.

EXAMPLE Find the equation of the regression line of y on x for the data above.

The 'regression line of y on x' means that x is the independent variable, and y is the dependent variable.

1) Work out the sums: $\sum x = 72.3$, $\sum y = 73.8$, $\sum x^2 = 544.81$, $\sum xy = 511.98$.
2) Then work out S_{xy} and S_{xx}: $S_{xy} = 511.98 - \frac{72.3 \times 73.8}{10} = -21.594$, $S_{xx} = 544.81 - \frac{72.3^2}{10} = 22.081$
3) So the gradient of the regression line is: $b = \frac{-21.594}{22.081} = -0.978$ (to 3 sig. fig.)
4) And the intercept is: $a = \frac{\sum y}{n} - b\frac{\sum x}{n} = \frac{73.8}{10} - (-0.978) \times \frac{72.3}{10} = 14.451 = 14.5$ (to 3 sig. fig.)
5) This all means that your regression line is: $y = 14.5 - 0.978x$

Remember: $\bar{x} = \frac{\sum x}{n}$

The regression line always goes through the point $(\bar{x}, \bar{y})$.

This tells you: (i) for every extra tonne carried, you'd expect the lorry's fuel efficiency to fall by 0.978 km per litre, and (ii) with no load ($x = 0$), you'd expect the lorry to do 14.5 km per litre of fuel. Assuming the trend continues down to $x = 0$.

Linear Regression

Regression with **Linear Scaling** involves an **Extra Stage**

Examiners might ask you to find a regression line for data that's undergone linear scaling. You could be asked to find the regression line of T on S, say, where $S = x + 15$ and $T = \frac{y}{10}$. Looks confusing, but you know that $y = a + bx$ is the regression line for y on x, and you've already found values for a and b above. So just carry out a couple of extra steps...

1) Find expressions for x and y in terms of S and T. Here $x = S - 15$ and $y = 10T$. Substitute for x and y in $y = a + bx$ to find an expression for T. $10T = a + b(S - 15) \Rightarrow T = \frac{a - 15b}{10} + \frac{b}{10}S$
2) Now substitute your values for a and b into the expression $T = \frac{a - 15b}{10} + \frac{b}{10}S$ to find the regression line of T on S.

Residuals — the difference between **Practice** and **Theory**

A residual is the difference between an observed y-value and the y-value predicted by the regression line.

Residual = Observed y-value – Estimated y-value

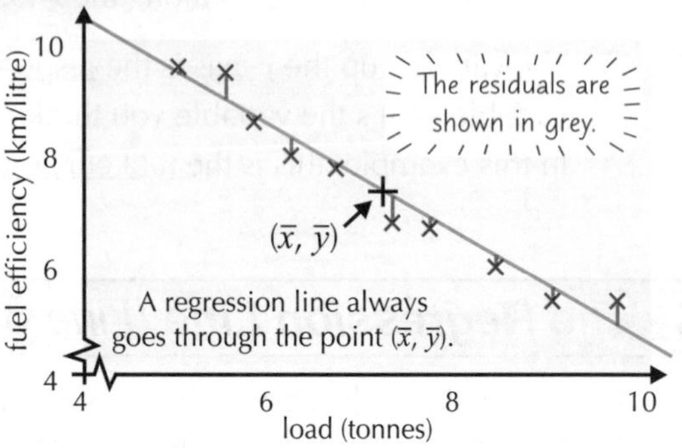

fuel efficiency (km/litre)
10
8
6
4
The residuals are shown in grey.
$(\bar{x}, \bar{y})$
A regression line always goes through the point $(\bar{x}, \bar{y})$.
4
6
8
10
load (tonnes)

1) Residuals show the experimental error between the y-value that's observed and the y-value your regression line says it should be.
2) Residuals are shown by a vertical line from the actual point to the regression line.
3) Ideally, you'd like your residuals to be small — this would show your regression line fits the data well. If they're large (i.e. a high percentage of the dependent variable), then that could mean your model won't be a very reliable one.

EXAMPLE For the fuel efficiency example on the last page, calculate the residuals for: (i) $x = 5.6$, (ii) $x = 7.4$.

(i) When $x = 5.6$, the residual $= 9.5 - (-0.978 \times 5.6 + 14.451) = 0.526$ (to 3 sig. fig.)
(ii) When $x = 7.4$, the residual $= 6.8 - (-0.978 \times 7.4 + 14.451) = -0.414$ (to 3 sig. fig.)

A positive residual means the regression line is too low for that value of x.
A negative residual means the regression line is too high.

This kind of regression is called Least Squares Regression, because you're finding the equation of the line which minimises the sum of the squares of the residuals (i.e. $\sum e_k^2$ is as small as possible, where the e_k are the residuals).

Use Regression Lines **With Care**

You can use a regression line to predict values of y, but it's best not to use x-values outside the range of your original data.

EXAMPLE Use your regression equation to estimate the value of y when: (i) $x = 7.6$, (ii) $x = 12.6$

(i) When $x = 7.6$, $y = -0.978 \times 7.6 + 14.451 = 7.02$ (to 3 sig. fig.). This should be a pretty reliable guess, since $x = 7.6$ falls in the range of x we already have readings for — this is called interpolation.

(ii) When $x = 12.6$, $y = -0.978 \times 12.6 + 14.451 = 2.13$ (to 3 sig. fig.). This may well be unreliable since $x = 12.6$ is bigger than the biggest x-value we already have — this is called extrapolation.

Outliers can also be a problem — they can have a big effect on the regression line's equation, and drag it far away from the rest of the data values.
Here, the circled data value is an outlier.

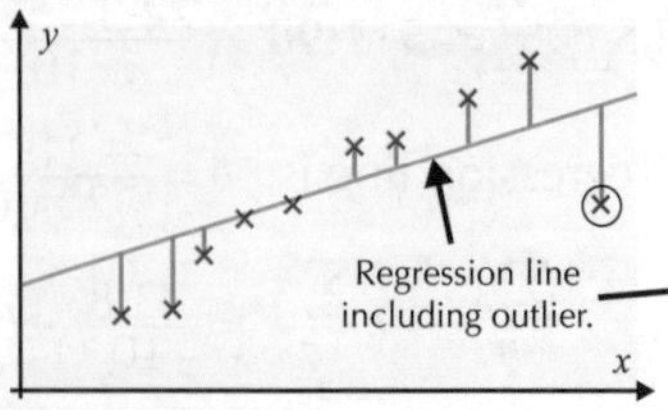

y
x
Regression line including outlier.

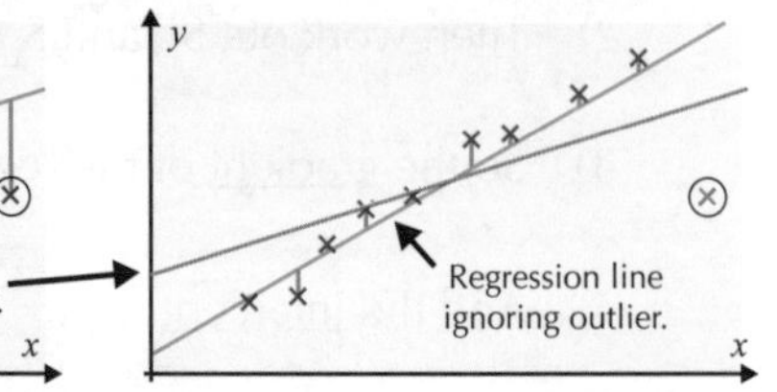

y
x
Regression line ignoring outlier.

99% of all statisticians make sweeping statements...

Be careful with that extrapolation business — it's like me saying that because I grew at an average rate of 10 cm a year for the first few years of my life, by the time I'm 50 I should be 5 metres tall. (There's still time, but I can't see that happening.)

S1 Section 6 — Practice Questions

That was a short section, but chock-full of fiddly terms and hefty equations. The only way to learn all those details is by using them — so stretch your maths muscles and take a jog around this obstacle course of practice questions.

Warm-up Questions

1) The table below shows the results of some measurements concerning alcoholic cocktails. Here, x = total volume in ml, and y = percentage alcohol concentration by volume.

x	90	100	100	150	160	200	240	250	290	300
y	40	35	25	30	25	25	20	25	15	7

a) Draw a scatter diagram representing this information.

b) Calculate the product-moment correlation coefficient (PMCC) of these values.

c) What does the PMCC tell you about these results?

2) For each pair of variables below, state which would be the dependent variable and which would be the independent variable.

a) • the annual number of volleyball-related injuries
• the annual number of sunny days

b) • the annual number of rainy days
• the annual number of Monopoly-related injuries

c) • a person's disposable income
• a person's spending on luxuries

d) • the number of trips to the loo per day
• the number of cups of tea drunk per day

e) • the number of festival tickets sold
• the number of pairs of Wellington boots bought

3) The radius in mm, r, and the weight in grams, w, of 10 randomly selected blueberry pancakes are given in the table below.

r	48.0	51.0	52.0	54.5	55.1	53.6	50.0	52.6	49.4	51.2
w	100	105	108	120	125	118	100	115	98	110

a) Find: (i) $S_{rr} = \sum r^2 - \frac{(\sum r)^2}{n}$, (ii) $S_{rw} = \sum rw - \frac{(\sum r)(\sum w)}{n}$

The regression line of w on r has equation $w = a + br$.

b) Find b, the gradient of the regression line.

c) Find a, the intercept of the regression line on the w-axis.

d) Write down the equation of the regression line of w on r.

e) Use your regression line to estimate the weight of a blueberry pancake of radius 60 mm.

f) Comment on the reliability of your estimate, giving a reason for your answer.

4) Mmm, blueberry pancakes... must have another question about blueberry pancakes...

The variables P and Q are defined as $P = r - 5$ and $Q = \frac{w}{8}$, where r and w are as used in Question 3.

Find the regression line of Q on P.

S1 Section 6 — Practice Questions

Land ahoy, ye lily-livered yellow-bellies.
Just a quick heave-ho through these exam questions, then drop anchor and row ashore. Yaarrr, freedom...

Exam Questions

1 Values of two variables x and y are recorded in the table below.

x	1	2	3	4	5	6	7	8
y	0.50	0.70	0.10	0.82	0.50	0.36	0.16	0.80

a) Represent this data on a scatter diagram. *(2 marks)*

b) Calculate the product-moment correlation coefficient (PMCC) between the two variables. *(4 marks)*

c) What does this value of the PMCC tell you about these variables? *(1 mark)*

2 The following times (in seconds) were taken by eight different runners to complete distances of 20 metres and 60 metres.

Runner	A	B	C	D	E	F	G	H
20-metre time (x)	3.39	3.20	3.09	3.32	3.33	3.27	3.44	3.08
60-metre time (y)	8.78	7.73	8.28	8.25	8.91	8.59	8.90	8.05

a) Plot a scatter diagram to represent the data. *(2 marks)*

b) Find the equation of the regression line of y on x, and plot it on your scatter diagram. *(8 marks)*

c) Use the equation of the regression line to estimate the value of y when:
(i) $x = 3.15$, (ii) $x = 3.88$.
Comment on the reliability of your estimates. *(4 marks)*

d) Find the residuals for:
(i) $x = 3.32$ (ii) $x = 3.27$.
Illustrate them on your scatter diagram. *(4 marks)*

3 A journalist at British Biking Monthly recorded the distance in miles, x, cycled by 20 different cyclists in the morning and the number of calories, y, eaten at lunch.
The following summary statistics were provided:

$$S_{xx} = 310\,880 \qquad S_{yy} = 788.95 \qquad S_{xy} = 12\,666$$

a) Use these values to calculate the product-moment correlation coefficient. *(2 marks)*

b) Give an interpretation of your answer to part (a). *(1 mark)*

A Swedish cycling magazine calculated the product-moment correlation coefficient of the data after converting the distances to km.

c) State the value of the product-moment correlation coefficient in this case. *(1 mark)*

4 The equation of the line of regression for a set of data is $y = 211.599 + 9.602x$.

a) Use the equation of the regression line to estimate the value of y when:
(i) $x = 12.5$ (ii) $x = 14.7$. *(2 marks)*

b) Calculate the residuals if the respective observed y-values were $y = 332.5$ and $y = 352.1$. *(2 marks)*

General Certificate of Education
Advanced Subsidiary (AS) and Advanced Level

Statistics S1 — Practice Exam One

Time Allowed: 1 hour 30 min

Graphical calculators may be used for this exam.

Give any non-exact numerical answers to three significant figures.

Statistical tables can be found on page 153.

There are 75 marks available for this paper.

1 The events A and B are mutually exclusive.
P(A) = 0.3 and P(B) = 0.4.

a) Write down $P(A \cap B)$. *(1 mark)*

b) Find $P(A \cup B)$. *(2 marks)*

c) Hence find the probability that neither event happens. *(2 marks)*

d) Write down $P(A \mid B)$. Explain your answer. *(2 marks)*

2 A group of 10 friends play a round of minigolf.
Their total score ($\sum x$) is 500 and $\sum x^2 = 25\,622$.

a) Find the mean, μ, and the standard deviation, σ, for this data. *(3 marks)*

b) Another friend wants to incorporate his score of 50. Without further calculation and giving reasons, explain the effect of adding this score on:

(i) the mean, *(2 marks)*

(ii) the standard deviation. *(2 marks)*

3 The temperature, T °C, reached by water in a kettle before the kettle switches off is normally distributed, with a mean of 93 °C. On 20% of occasions, the water reaches 95 °C before switching off.

a) Write down the median temperature reached. Explain your answer. *(2 marks)*

b) Find the standard deviation of the switch-off temperature's distribution. *(4 marks)*

c) In a restaurant, tea made with water at less than 88 °C will attract complaints.
Calculate the probability that a customer will complain if this kettle is used. *(2 marks)*

d) Find d such that $P(93 - d \leq T \leq 93 + d) = 0.99$. *(4 marks)*

4 Each student in a class has a standard pack of 52 cards, where each pack is made up of the same number of red and black cards and contains 12 picture cards in total.

a) One student picks 3 cards at random from her pack, without replacing the selected cards before the next pick. The random variable X represents the number of picture cards picked. Explain why X does not follow a binomial distribution. *(1 mark)*

b) Another student chooses 3 cards at random, but replaces each of the selected cards before the next pick. The random variable Y represents the number of picture cards picked. Calculate:

(i) the probability that the student chooses exactly two picture cards, *(3 marks)*

(ii) the mean of Y, *(1 mark)*

(ii) the variance of Y. *(1 mark)*

c) All 20 students in the group now choose 4 cards at random from their pack, replacing their selected cards each time. The random variable Q represents the number of students that choose exactly 3 red cards. Find the probability that Q is at least 2 but no greater than 8. *(4 marks)*

5 A survey to find the time taken to service a sample of 10 cars was carried out. The results are shown in the table below.

Number of days (x)	1	2	3	4	5	6	7	8
Frequency (f)	5	3	1	0	0	0	0	1

a) For this data, find the:

(i) mean, *(2 marks)*

(ii) median. *(2 marks)*

b) Sam wants to use the range as a summary of the variation in the data.

Do you consider this to be an appropriate measure to use? Give a reason for your answer. *(2 marks)*

6 A farmer wants to know the lengths of his crop of cucumbers. He measures the lengths (x cm) of 80 randomly selected cucumbers, and obtains the following results:

$$\sum x = 2246.3 \qquad \sum x^2 = 63\,885.4$$

a) Find unbiased estimates of:

(i) the population mean, *(1 mark)*

(ii) the population variance. *(2 marks)*

b) Construct a 95% confidence interval for the mean length of the farmer's cucumbers. *(4 marks)*

c) The farmer claims these results show the average length of his crop of cucumbers is greater than 27 cm. Comment on this claim. *(2 marks)*

7 Of 30 drivers interviewed, 9 have been involved in a car crash at some time.
Of those who have been involved in a crash, 5 wear glasses.
The probability of wearing glasses, given that the driver has not had a car accident, is $\frac{1}{3}$.

a) What is the probability that a person chosen at random:

(i) has not been involved in a crash? *(1 mark)*

(ii) wears glasses? *(3 marks)*

b) What is the probability that a glasses wearer has been a crash victim? *(3 marks)*

c) Out of the original 30 drivers, 13 were thirty-five years of age or older.
Four people are selected at random from this group of 30.
Calculate the probability that only one of these people is less than 35 years of age. *(3 marks)*

8 A teacher collects the following data showing students' marks in an examination (y) and the amount of revision undertaken in hours (x).

x	12	10	9	5	14	11	12	6
y	88	72	65	59	92	75	80	69

($\Sigma x = 79$, $\Sigma y = 600$, $\Sigma x^2 = 847$, $\Sigma y^2 = 45\,884$ and $\Sigma xy = 6143$.)

a) Calculate S_{xx}, S_{yy} and S_{xy}. *(3 marks)*

b) Calculate the product-moment correlation coefficient for x and y. *(2 marks)*

c) Which of x and y is the explanatory variable? Explain your answer. *(2 marks)*

d) The teacher believes she can fit a linear regression line $y = a + bx$ to the data.
Give one reason to support her conclusion. *(1 mark)*

e) Find the equation of the regression line of y on x in the form $y = a + bx$. *(4 marks)*

f) The teacher must estimate the examination mark of a student who did not take the examination.
The teacher knows this student to have done 8 hours of revision.
Estimate the likely mark for this student. *(1 mark)*

g) Comment on the reliability of your estimate in f). *(1 mark)*

General Certificate of Education
Advanced Subsidiary (AS) and Advanced Level

Statistics S1 — Practice Exam Two

Time Allowed: 1 hour 30 min

Graphical calculators may be used for this exam.

Give any non-exact numerical answers to three significant figures.

Statistical tables can be found on page 153.

There are 75 marks available for this paper.

1 A box of chocolates contains a total of 40 individual sweets.
Each sweet is made with either plain, milk or white chocolate, and has either a hard or soft centre.
The number of sweets of each type is shown in the table below.

	Plain chocolate	Milk chocolate	White chocolate
Hard centre	8	11	5
Soft centre	6	8	2

a) A chocolate is selected at random. Find the probability of it:

(i) having a soft centre, *(1 mark)*

(ii) not being made with white chocolate, *(2 marks)*

(iii) **either** having a hard centre **or** being made with milk chocolate, *(2 marks)*

(iv) having a hard centre, given that it's made with plain chocolate. *(2 marks)*

b) Three chocolates are selected at random, without replacement.
Find the probability that one is made with plain chocolate, one with milk chocolate and the other with white chocolate. *(4 marks)*

2 The number of seconds (T) that trained divers can hold their breath for may be modelled by a normal distribution with a mean of 132 seconds and a standard deviation of 40 seconds.
Find:

a) $P(T > 160)$, *(3 marks)*

b) $P(60 \leq T \leq 90)$, *(4 marks)*

c) P(T equals exactly 135). *(1 mark)*

3 The sales figures of a gift shop for a 12-week period are shown below.

Week	1	2	3	4	5	6	7	8	9	10	11	12
Sales (£'000s)	5.5	4.2	5.8	9.1	3.8	4.6	6.4	6.2	4.9	5.9	6.0	4.1

(You may use $\Sigma x = 66.5$, and $\Sigma x^2 = 390.97$.)

a) Find the mean and variance of the weekly sales. *(4 marks)*

b) Find the median and quartiles of the sales data. *(3 marks)*

A different gift shop collects weekly sales figures in a different format.
The table below shows in how many weeks sales fell into various categories.

Sales (£'000s)	$4 \le y < 5$	$5 \le y < 6$	$6 \le y < 7$	$7 \le y < 8$	$8 \le y < 9$	$9 \le y < 10$
Number of weeks, f	1	2	2	3	4	1

c) For this shop's weekly sales, calculate an estimate of:

(i) the mean, *(2 marks)*

(ii) the variance. *(3 marks)*

4 A construction company measures the length (y metres) of a cable when put under different amounts of tension (x, measured in kilonewtons, kN). The results of its tests are shown below.

x (kN)	1	2	3	5	8	10	15	20
y (metres)	3.05	3.1	3.13	3.15	3.27	3.4	3.5	3.6

a) Draw a scatter graph to show these results. *(2 marks)*

b) Calculate S_{xx} and S_{xy}.
(You may use $\Sigma x^2 = 828$ and $\Sigma xy = 219.05$.) *(2 marks)*

An engineer believes a linear regression line of the form $y = a + bx$ could be found to accurately describe the results.

c) Find the equation of this regression line. *(4 marks)*

d) Explain what your values of a and b represent. *(2 marks)*

e) Use your regression line to predict the length of the cable when put under a tension of 30 kilonewtons. *(2 marks)*

f) Comment on the reliability of the estimate in e). *(1 mark)*

5 A manager in an aircraft-maintenance business wants to know how long it takes the company's engineers to change a fuel filter. He knows that the times taken are normally distributed with a variance of 16 minutes. Thirty-six randomly selected engineers were timed as they changed a fuel filter. The following summary statistics were obtained, where x is the number of minutes taken:

$$\sum x = 1344 \qquad \sum x^2 = 51983$$

a) Find an unbiased estimate of the mean time needed by one of the company's engineers to change a fuel filter. *(1 mark)*

b) For this sample size, calculate an unbiased estimate of the standard error of the sample mean. *(5 marks)*

c) Find a 99% confidence interval for the mean time needed for an engineer to change a fuel filter. *(2 marks)*

6 The duration in minutes, X, of a car wash is a little erratic, but it is normally distributed with a mean of 8 minutes and a variance of 1.2. Find:

a) $P(X < 7.5)$, *(3 marks)*

b) the probability that the duration deviates from the mean by more than 1 minute, *(4 marks)*

c) the duration in minutes, d, such that there is no more than a 1% probability that the car wash will take longer than this duration. *(4 marks)*

7 An average of 10% of chocolate bars made by a particular manufacturer contain a 'golden ticket'. A student buys 5 chocolate bars every week for 4 weeks. The number of golden tickets he finds is represented by the random variable X.

a) Assuming that $X \sim B(20, 0.10)$, find:

(i) $P(X > 1)$, *(2 marks)*

(ii) $P(2 < X \leq 6)$ *(3 marks)*

b) Two other students also buy a total of 20 chocolate bars each.
Find the probability that the three students find a total of exactly 3 golden tickets between them. *(3 marks)*

c) A number of other students at this school each bought 10 bars of chocolate.

(i) Calculate the mean and variance for the number of golden tickets, T, that each of these students would expect to find. *(2 marks)*

(ii) In fact, the number of golden tickets each student found had a mean of 0.2 and a variance of 0.19.

Comment on the claim that the students were not buying chocolate bars at random. *(2 marks)*

AQA S1 — STATISTICAL TABLES

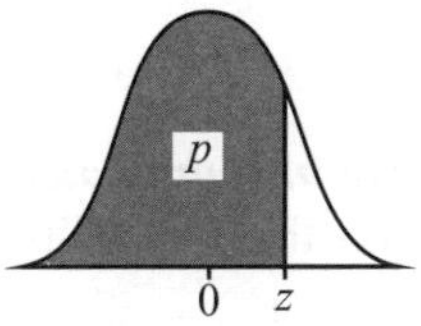

The normal distribution function

This table gives the probability, p, that the random variable $Z \sim N(0, 1)$ is less than or equal to z.

z	0.00	0.01	0.02	0.03	0.04	0.05	0.06	0.07	0.08	0.09	z
0.0	0.50000	0.50399	0.50798	0.51197	0.51595	0.51994	0.52392	0.52790	0.53188	0.53586	**0.0**
0.1	0.53983	0.54380	0.54776	0.55172	0.55567	0.55962	0.56356	0.56749	0.57142	0.57535	**0.1**
0.2	0.57926	0.58317	0.58706	0.59095	0.59483	0.59871	0.60257	0.60642	0.61026	0.61409	**0.2**
0.3	0.61791	0.62172	0.62552	0.62930	0.63307	0.63683	0.64058	0.64431	0.64803	0.65173	**0.3**
0.4	0.65542	0.65910	0.66276	0.66640	0.67003	0.67364	0.67724	0.68082	0.68439	0.68793	**0.4**
0.5	0.69146	0.69497	0.69847	0.70194	0.70540	0.70884	0.71226	0.71566	0.71904	0.72240	**0.5**
0.6	0.72575	0.72907	0.73237	0.73565	0.73891	0.74215	0.74537	0.74857	0.75175	0.75490	**0.6**
0.7	0.75804	0.76115	0.76424	0.76730	0.77035	0.77337	0.77637	0.77935	0.78230	0.78524	**0.7**
0.8	0.78814	0.79103	0.79389	0.79673	0.79955	0.80234	0.80511	0.80785	0.81057	0.81327	**0.8**
0.9	0.81594	0.81859	0.82121	0.82381	0.82639	0.82894	0.83147	0.83398	0.83646	0.83891	**0.9**
1.0	0.84134	0.84375	0.84614	0.84849	0.85083	0.85314	0.85543	0.85769	0.85993	0.86214	**1.0**
1.1	0.86433	0.86650	0.86864	0.87076	0.87286	0.87493	0.87698	0.87900	0.88100	0.88298	**1.1**
1.2	0.88493	0.88686	0.88877	0.89065	0.89251	0.89435	0.89617	0.89796	0.89973	0.90147	**1.2**
1.3	0.90320	0.90490	0.90658	0.90824	0.90988	0.91149	0.91309	0.91466	0.91621	0.91774	**1.3**
1.4	0.91924	0.92073	0.92220	0.92364	0.92507	0.92647	0.92785	0.92922	0.93056	0.93189	**1.4**
1.5	0.93319	0.93448	0.93574	0.93699	0.93822	0.93943	0.94062	0.94179	0.94295	0.94408	**1.5**
1.6	0.94520	0.94630	0.94738	0.94845	0.94950	0.95053	0.95154	0.95254	0.95352	0.95449	**1.6**
1.7	0.95543	0.95637	0.95728	0.95818	0.95907	0.95994	0.96080	0.96164	0.96246	0.96327	**1.7**
1.8	0.96407	0.96485	0.96562	0.96638	0.96712	0.96784	0.96856	0.96926	0.96995	0.97062	**1.8**
1.9	0.97128	0.97193	0.97257	0.97320	0.97381	0.97441	0.97500	0.97558	0.97615	0.97670	**1.9**
2.0	0.97725	0.97778	0.97831	0.97882	0.97932	0.97982	0.98030	0.98077	0.98124	0.98169	**2.0**
2.1	0.98214	0.98257	0.98300	0.98341	0.98382	0.98422	0.98461	0.98500	0.98537	0.98574	**2.1**
2.2	0.98610	0.98645	0.98679	0.98713	0.98745	0.98778	0.98809	0.98840	0.98870	0.98899	**2.2**
2.3	0.98928	0.98956	0.98983	0.99010	0.99036	0.99061	0.99086	0.99111	0.99134	0.99158	**2.3**
2.4	0.99180	0.99202	0.99224	0.99245	0.99266	0.99286	0.99305	0.99324	0.99343	0.99361	**2.4**
2.5	0.99379	0.99396	0.99413	0.99430	0.99446	0.99461	0.99477	0.99492	0.99506	0.99520	**2.5**
2.6	0.99534	0.99547	0.99560	0.99573	0.99585	0.99598	0.99609	0.99621	0.99632	0.99643	**2.6**
2.7	0.99653	0.99664	0.99674	0.99683	0.99693	0.99702	0.99711	0.99720	0.99728	0.99736	**2.7**
2.8	0.99744	0.99752	0.99760	0.99767	0.99774	0.99781	0.99788	0.99795	0.99801	0.99807	**2.8**
2.9	0.99813	0.99819	0.99825	0.99831	0.99836	0.99841	0.99846	0.99851	0.99856	0.99861	**2.9**
3.0	0.99865	0.99869	0.99874	0.99878	0.99882	0.99886	0.99889	0.99893	0.99896	0.99900	**3.0**
3.1	0.99903	0.99906	0.99910	0.99913	0.99916	0.99918	0.99921	0.99924	0.99926	0.99929	**3.1**
3.2	0.99931	0.99934	0.99936	0.99938	0.99940	0.99942	0.99944	0.99946	0.99948	0.99950	**3.2**
3.3	0.99952	0.99953	0.99955	0.99957	0.99958	0.99960	0.99961	0.99962	0.99964	0.99965	**3.3**
3.4	0.99966	0.99968	0.99969	0.99970	0.99971	0.99972	0.99973	0.99974	0.99975	0.99976	**3.4**
3.5	0.99977	0.99978	0.99978	0.99979	0.99980	0.99981	0.99981	0.99982	0.99983	0.99983	**3.5**
3.6	0.99984	0.99985	0.99985	0.99986	0.99986	0.99987	0.99987	0.99988	0.99988	0.99989	**3.6**
3.7	0.99989	0.99990	0.99990	0.99990	0.99991	0.99991	0.99992	0.99992	0.99992	0.99992	**3.7**
3.8	0.99993	0.99993	0.99993	0.99994	0.99994	0.99994	0.99994	0.99995	0.99995	0.99995	**3.8**
3.9	0.99995	0.99995	0.99996	0.99996	0.99996	0.99996	0.99996	0.99996	0.99997	0.99997	**3.9**

Percentage points of the normal distribution

The table gives the values of z satisfying $P(Z \le z) = p$, where $Z \sim N(0, 1)$.

p	0.00	0.01	0.02	0.03	0.04	0.05	0.06	0.07	0.08	0.09	p
0.5	0.0000	0.0251	0.0502	0.0753	0.1004	0.1257	0.1510	0.1764	0.2019	0.2275	**0.5**
0.6	0.2533	0.2793	0.3055	0.3319	0.3585	0.3853	0.4125	0.4399	0.4677	0.4959	**0.6**
0.7	0.5244	0.5534	0.5828	0.6128	0.6433	0.6745	0.7063	0.7388	0.7722	0.8064	**0.7**
0.8	0.8416	0.8779	0.9154	0.9542	0.9945	1.0364	1.0803	1.1264	1.1750	1.2265	**0.8**
0.9	1.2816	1.3408	1.4051	1.4758	1.5548	1.6449	1.7507	1.8808	2.0537	2.3263	**0.9**
p	**0.000**	**0.001**	**0.002**	**0.003**	**0.004**	**0.005**	**0.006**	**0.007**	**0.008**	**0.009**	p
0.95	1.6449	1.6546	1.6646	1.6747	1.6849	1.6954	1.7060	1.7169	1.7279	1.7392	**0.95**
0.96	1.7507	1.7624	1.7744	1.7866	1.7991	1.8119	1.8250	1.8384	1.8522	1.8663	**0.96**
0.97	1.8808	1.8957	1.9110	1.9268	1.9431	1.9600	1.9774	1.9954	2.0141	2.0335	**0.97**
0.98	2.0537	2.0749	2.0969	2.1201	2.1444	2.1701	2.1973	2.2262	2.2571	2.2904	**0.98**
0.99	2.3263	2.3656	2.4089	2.4573	2.5121	2.5758	2.6521	2.7478	2.8782	3.0902	**0.99**

AQA S1 — STATISTICAL TABLES

The cumulative binomial distribution function

This table gives the probability $P(X \leq x)$, where the random variable $X \sim B(n, p)$.

p	0.01	0.02	0.03	0.04	0.05	0.06	0.07	0.08	0.09	0.10	0.15	0.20	0.25	0.30	0.35	0.40	0.45	0.50	p
x	$n =$ 2																		x
0	0.9801	0.9604	0.9409	0.9216	0.9025	0.8836	0.8649	0.8464	0.8281	0.8100	0.7225	0.6400	0.5625	0.4900	0.4225	0.3600	0.3025	0.2500	0
1	0.9999	0.9996	0.9991	0.9984	0.9975	0.9964	0.9951	0.9936	0.9919	0.9900	0.9775	0.9600	0.9375	0.9100	0.8775	0.8400	0.7975	0.7500	1
2	1.0000	1.0000	1.0000	1.0000	1.0000	1.0000	1.0000	1.0000	1.0000	1.0000	1.0000	1.0000	1.0000	1.0000	1.0000	1.0000	1.0000	1.0000	2
x	$n =$ 3																		x
0	0.9703	0.9412	0.9127	0.8847	0.8574	0.8306	0.8044	0.7787	0.7536	0.7290	0.6141	0.5120	0.4219	0.3430	0.2746	0.2160	0.1664	0.1250	0
1	0.9997	0.9988	0.9974	0.9953	0.9928	0.9896	0.9860	0.9818	0.9772	0.9720	0.9393	0.8960	0.8438	0.7840	0.7183	0.6480	0.5748	0.5000	1
2	1.0000	1.0000	1.0000	0.9999	0.9999	0.9998	0.9997	0.9995	0.9993	0.9990	0.9966	0.9920	0.9844	0.9730	0.9571	0.9360	0.9089	0.8750	2
3				1.0000	1.0000	1.0000	1.0000	1.0000	1.0000	1.0000	1.0000	1.0000	1.0000	1.0000	1.0000	1.0000	1.0000	1.0000	3
x	$n =$ 4																		x
0	0.9606	0.9224	0.8853	0.8493	0.8145	0.7807	0.7481	0.7164	0.6857	0.6561	0.5220	0.4096	0.3164	0.2401	0.1785	0.1296	0.0915	0.0625	0
1	0.9994	0.9977	0.9948	0.9909	0.9860	0.9801	0.9733	0.9656	0.9570	0.9477	0.8905	0.8192	0.7383	0.6517	0.5630	0.4752	0.3910	0.3125	1
2	1.0000	1.0000	0.9999	0.9998	0.9995	0.9992	0.9987	0.9981	0.9973	0.9963	0.9880	0.9728	0.9492	0.9163	0.8735	0.8208	0.7585	0.6875	2
3			1.0000	1.0000	1.0000	1.0000	1.0000	1.0000	0.9999	0.9999	0.9995	0.9984	0.9961	0.9919	0.9850	0.9744	0.9590	0.9375	3
4									1.0000	1.0000	1.0000	1.0000	1.0000	1.0000	1.0000	1.0000	1.0000	1.0000	4
x	$n =$ 5																		x
0	0.9510	0.9039	0.8587	0.8154	0.7738	0.7339	0.6957	0.6591	0.6240	0.5905	0.4437	0.3277	0.2373	0.1681	0.1160	0.0778	0.0503	0.0313	0
1	0.9990	0.9962	0.9915	0.9852	0.9774	0.9681	0.9575	0.9456	0.9326	0.9185	0.8352	0.7373	0.6328	0.5282	0.4284	0.3370	0.2562	0.1875	1
2	1.0000	0.9999	0.9997	0.9994	0.9988	0.9980	0.9969	0.9955	0.9937	0.9914	0.9734	0.9421	0.8965	0.8369	0.7648	0.6826	0.5931	0.5000	2
3		1.0000	1.0000	1.0000	1.0000	0.9999	0.9999	0.9998	0.9997	0.9995	0.9978	0.9933	0.9844	0.9692	0.9460	0.9130	0.8688	0.8125	3
4						1.0000	1.0000	1.0000	1.0000	1.0000	0.9999	0.9997	0.9990	0.9976	0.9947	0.9898	0.9815	0.9688	4
5											1.0000	1.0000	1.0000	1.0000	1.0000	1.0000	1.0000	1.0000	5
x	$n =$ 6																		x
0	0.9415	0.8858	0.8330	0.7828	0.7351	0.6899	0.6470	0.6064	0.5679	0.5314	0.3771	0.2621	0.1780	0.1176	0.0754	0.0467	0.0277	0.0156	0
1	0.9985	0.9943	0.9875	0.9784	0.9672	0.9541	0.9392	0.9227	0.9048	0.8857	0.7765	0.6554	0.5339	0.4202	0.3191	0.2333	0.1636	0.1094	1
2	1.0000	0.9998	0.9995	0.9988	0.9978	0.9962	0.9942	0.9915	0.9882	0.9842	0.9527	0.9011	0.8306	0.7443	0.6471	0.5443	0.4415	0.3438	2
3		1.0000	1.0000	1.0000	0.9999	0.9998	0.9997	0.9995	0.9992	0.9987	0.9941	0.9830	0.9624	0.9295	0.8826	0.8208	0.7447	0.6563	3
4					1.0000	1.0000	1.0000	1.0000	1.0000	0.9999	0.9996	0.9984	0.9954	0.9891	0.9777	0.9590	0.9308	0.8906	4
5										1.0000	1.0000	0.9999	0.9998	0.9993	0.9982	0.9959	0.9917	0.9844	5
6												1.0000	1.0000	1.0000	1.0000	1.0000	1.0000	1.0000	6
x	$n =$ 7																		x
0	0.9321	0.8681	0.8080	0.7514	0.6983	0.6485	0.6017	0.5578	0.5168	0.4783	0.3206	0.2097	0.1335	0.0824	0.0490	0.0280	0.0152	0.0078	0
1	0.9980	0.9921	0.9829	0.9706	0.9556	0.9382	0.9187	0.8974	0.8745	0.8503	0.7166	0.5767	0.4449	0.3294	0.2338	0.1586	0.1024	0.0625	1
2	1.0000	0.9997	0.9991	0.9980	0.9962	0.9937	0.9903	0.9860	0.9807	0.9743	0.9262	0.8520	0.7564	0.6471	0.5323	0.4199	0.3164	0.2266	2
3		1.0000	1.0000	0.9999	0.9998	0.9996	0.9993	0.9988	0.9982	0.9973	0.9879	0.9667	0.9294	0.8740	0.8002	0.7102	0.6083	0.5000	3
4				1.0000	1.0000	1.0000	1.0000	0.9999	0.9999	0.9998	0.9988	0.9953	0.9871	0.9712	0.9444	0.9037	0.8471	0.7734	4
5								1.0000	1.0000	1.0000	0.9999	0.9996	0.9987	0.9962	0.9910	0.9812	0.9643	0.9375	5
6											1.0000	1.0000	0.9999	0.9998	0.9994	0.9984	0.9963	0.9922	6
7													1.0000	1.0000	1.0000	1.0000	1.0000	1.0000	7
x	$n =$ 8																		x
0	0.9227	0.8508	0.7837	0.7214	0.6634	0.6096	0.5596	0.5132	0.4703	0.4305	0.2725	0.1678	0.1001	0.0576	0.0319	0.0168	0.0084	0.0039	0
1	0.9973	0.9897	0.9777	0.9619	0.9428	0.9208	0.8965	0.8702	0.8423	0.8131	0.6572	0.5033	0.3671	0.2553	0.1691	0.1064	0.0632	0.0352	1
2	0.9999	0.9996	0.9987	0.9969	0.9942	0.9904	0.9853	0.9789	0.9711	0.9619	0.8948	0.7969	0.6785	0.5518	0.4278	0.3154	0.2201	0.1445	2
3	1.0000	1.0000	0.9999	0.9998	0.9996	0.9993	0.9987	0.9978	0.9966	0.9950	0.9786	0.9437	0.8862	0.8059	0.7064	0.5941	0.4770	0.3633	3
4			1.0000	1.0000	1.0000	1.0000	0.9999	0.9999	0.9997	0.9996	0.9971	0.9896	0.9727	0.9420	0.8939	0.8263	0.7396	0.6367	4
5							1.0000	1.0000	1.0000	1.0000	0.9998	0.9988	0.9958	0.9887	0.9747	0.9502	0.9115	0.8555	5
6											1.0000	0.9999	0.9996	0.9987	0.9964	0.9915	0.9819	0.9648	6
7												1.0000	1.0000	0.9999	0.9998	0.9993	0.9983	0.9961	7
8														1.0000	1.0000	1.0000	1.0000	1.0000	8
x	$n =$ 9																		x
0	0.9135	0.8337	0.7602	0.6925	0.6302	0.5730	0.5204	0.4722	0.4279	0.3874	0.2316	0.1342	0.0751	0.0404	0.0207	0.0101	0.0046	0.0020	0
1	0.9966	0.9869	0.9718	0.9522	0.9288	0.9022	0.8729	0.8417	0.8088	0.7748	0.5995	0.4362	0.3003	0.1960	0.1211	0.0705	0.0385	0.0195	1
2	0.9999	0.9994	0.9980	0.9955	0.9916	0.9862	0.9791	0.9702	0.9595	0.9470	0.8591	0.7382	0.6007	0.4628	0.3373	0.2318	0.1495	0.0898	2
3	1.0000	1.0000	0.9999	0.9997	0.9994	0.9987	0.9977	0.9963	0.9943	0.9917	0.9661	0.9144	0.8343	0.7297	0.6089	0.4826	0.3614	0.2539	3
4			1.0000	1.0000	1.0000	0.9999	0.9998	0.9997	0.9995	0.9991	0.9944	0.9804	0.9511	0.9012	0.8283	0.7334	0.6214	0.5000	4
5						1.0000	1.0000	1.0000	1.0000	0.9999	0.9994	0.9969	0.9900	0.9747	0.9464	0.9006	0.8342	0.7461	5
6										1.0000	1.0000	0.9997	0.9987	0.9957	0.9888	0.9750	0.9502	0.9102	6
7												1.0000	0.9999	0.9996	0.9986	0.9962	0.9909	0.9805	7
8													1.0000	1.0000	0.9999	0.9997	0.9992	0.9980	8
9															1.0000	1.0000	1.0000	1.0000	9
x	$n =$ 10																		x
0	0.9044	0.8171	0.7374	0.6648	0.5987	0.5386	0.4840	0.4344	0.3894	0.3487	0.1969	0.1074	0.0563	0.0282	0.0135	0.0060	0.0025	0.0010	0
1	0.9957	0.9838	0.9655	0.9418	0.9139	0.8824	0.8483	0.8121	0.7746	0.7361	0.5443	0.3758	0.2440	0.1493	0.0860	0.0464	0.0233	0.0107	1
2	0.9999	0.9991	0.9972	0.9938	0.9885	0.9812	0.9717	0.9599	0.9460	0.9298	0.8202	0.6778	0.5256	0.3828	0.2616	0.1673	0.0996	0.0547	2
3	1.0000	1.0000	0.9999	0.9996	0.9990	0.9980	0.9964	0.9942	0.9912	0.9872	0.9500	0.8791	0.7759	0.6496	0.5138	0.3823	0.2660	0.1719	3
4			1.0000	1.0000	0.9999	0.9998	0.9997	0.9994	0.9990	0.9984	0.9901	0.9672	0.9219	0.8497	0.7515	0.6331	0.5044	0.3770	4
5					1.0000	1.0000	1.0000	1.0000	0.9999	0.9999	0.9986	0.9936	0.9803	0.9527	0.9051	0.8338	0.7384	0.6230	5
6									1.0000	1.0000	0.9999	0.9991	0.9965	0.9894	0.9740	0.9452	0.8980	0.8281	6
7											1.0000	0.9999	0.9996	0.9984	0.9952	0.9877	0.9726	0.9453	7
8												1.0000	1.0000	0.9999	0.9995	0.9983	0.9955	0.9893	8
9														1.0000	1.0000	0.9999	0.9997	0.9990	9
10																1.0000	1.0000	1.0000	10

AQA S1 — STATISTICAL TABLES

The binomial cumulative distribution function (continued)

p	0.01	0.02	0.03	0.04	0.05	0.06	0.07	0.08	0.09	0.10	0.15	0.20	0.25	0.30	0.35	0.40	0.45	0.50	p
x	$n=$ 11																		x
0	0.8953	0.8007	0.7153	0.6382	0.5688	0.5063	0.4501	0.3996	0.3544	0.3138	0.1673	0.0859	0.0422	0.0198	0.0088	0.0036	0.0014	0.0005	0
1	0.9948	0.9805	0.9587	0.9308	0.8981	0.8618	0.8228	0.7819	0.7399	0.6974	0.4922	0.3221	0.1971	0.1130	0.0606	0.0302	0.0139	0.0059	1
2	0.9998	0.9988	0.9963	0.9917	0.9848	0.9752	0.9630	0.9481	0.9305	0.9104	0.7788	0.6174	0.4552	0.3127	0.2001	0.1189	0.0652	0.0327	2
3	1.0000	1.0000	0.9998	0.9993	0.9984	0.9970	0.9947	0.9915	0.9871	0.9815	0.9306	0.8389	0.7133	0.5696	0.4256	0.2963	0.1911	0.1133	3
4			1.0000	1.0000	0.9999	0.9997	0.9995	0.9990	0.9983	0.9972	0.9841	0.9496	0.8854	0.7897	0.6683	0.5328	0.3971	0.2744	4
5					1.0000	1.0000	1.0000	0.9999	0.9998	0.9997	0.9973	0.9883	0.9657	0.9218	0.8513	0.7535	0.6331	0.5000	5
6								1.0000	1.0000	1.0000	0.9997	0.9980	0.9924	0.9784	0.9499	0.9006	0.8262	0.7256	6
7											1.0000	0.9998	0.9988	0.9957	0.9878	0.9707	0.9390	0.8867	7
8												1.0000	0.9999	0.9994	0.9980	0.9941	0.9852	0.9673	8
9													1.0000	1.0000	0.9998	0.9993	0.9978	0.9941	9
10															1.0000	1.0000	0.9998	0.9995	10
11																	1.0000	1.0000	11
x	$n=$ 12																		x
0	0.8864	0.7847	0.6938	0.6127	0.5404	0.4759	0.4186	0.3677	0.3225	0.2824	0.1422	0.0687	0.0317	0.0138	0.0057	0.0022	0.0008	0.0002	0
1	0.9938	0.9769	0.9514	0.9191	0.8816	0.8405	0.7967	0.7513	0.7052	0.6590	0.4435	0.2749	0.1584	0.0850	0.0424	0.0196	0.0083	0.0032	1
2	0.9998	0.9985	0.9952	0.9893	0.9804	0.9684	0.9532	0.9348	0.9134	0.8891	0.7358	0.5583	0.3907	0.2528	0.1513	0.0834	0.0421	0.0193	2
3	1.0000	0.9999	0.9997	0.9990	0.9978	0.9957	0.9925	0.9880	0.9820	0.9744	0.9078	0.7946	0.6488	0.4925	0.3467	0.2253	0.1345	0.0730	3
4		1.0000	1.0000	0.9999	0.9998	0.9996	0.9991	0.9984	0.9973	0.9957	0.9761	0.9274	0.8424	0.7237	0.5833	0.4382	0.3044	0.1938	4
5				1.0000	1.0000	1.0000	0.9999	0.9998	0.9997	0.9995	0.9954	0.9806	0.9456	0.8822	0.7873	0.6652	0.5269	0.3872	5
6							1.0000	1.0000	1.0000	0.9999	0.9993	0.9961	0.9857	0.9614	0.9154	0.8418	0.7393	0.6128	6
7										1.0000	0.9999	0.9994	0.9972	0.9905	0.9745	0.9427	0.8883	0.8062	7
8											1.0000	0.9999	0.9996	0.9983	0.9944	0.9847	0.9644	0.9270	8
9												1.0000	1.0000	0.9998	0.9992	0.9972	0.9921	0.9807	9
10														1.0000	0.9999	0.9997	0.9989	0.9968	10
11															1.0000	1.0000	0.9999	0.9998	11
12																	1.0000	1.0000	12
x	$n=$ 13																		x
0	0.8775	0.7690	0.6730	0.5882	0.5133	0.4474	0.3893	0.3383	0.2935	0.2542	0.1209	0.0550	0.0238	0.0097	0.0037	0.0013	0.0004	0.0001	0
1	0.9928	0.9730	0.9436	0.9068	0.8646	0.8186	0.7702	0.7206	0.6707	0.6213	0.3983	0.2336	0.1267	0.0637	0.0296	0.0126	0.0049	0.0017	1
2	0.9997	0.9980	0.9938	0.9865	0.9755	0.9608	0.9422	0.9201	0.8946	0.8661	0.6920	0.5017	0.3326	0.2025	0.1132	0.0579	0.0269	0.0112	2
3	1.0000	0.9999	0.9995	0.9986	0.9969	0.9940	0.9897	0.9837	0.9758	0.9658	0.8820	0.7473	0.5843	0.4206	0.2783	0.1686	0.0929	0.0461	3
4		1.0000	1.0000	0.9999	0.9997	0.9993	0.9987	0.9976	0.9959	0.9935	0.9658	0.9009	0.7940	0.6543	0.5005	0.3530	0.2279	0.1334	4
5				1.0000	1.0000	0.9999	0.9999	0.9997	0.9995	0.9991	0.9925	0.9700	0.9198	0.8346	0.7159	0.5744	0.4268	0.2905	5
6						1.0000	1.0000	1.0000	0.9999	0.9999	0.9987	0.9930	0.9757	0.9376	0.8705	0.7712	0.6437	0.5000	6
7									1.0000	1.0000	0.9998	0.9988	0.9944	0.9818	0.9538	0.9023	0.8212	0.7095	7
8											1.0000	0.9998	0.9990	0.9960	0.9874	0.9679	0.9302	0.8666	8
9												1.0000	0.9999	0.9993	0.9975	0.9922	0.9797	0.9539	9
10													1.0000	0.9999	0.9997	0.9987	0.9959	0.9888	10
11														1.0000	1.0000	0.9999	0.9995	0.9983	11
12																1.0000	1.0000	0.9999	12
13																		1.0000	13
x	$n=$ 14																		x
0	0.8687	0.7536	0.6528	0.5647	0.4877	0.4205	0.3620	0.3112	0.2670	0.2288	0.1028	0.0440	0.0178	0.0068	0.0024	0.0008	0.0002	0.0001	0
1	0.9916	0.9690	0.9355	0.8941	0.8470	0.7963	0.7436	0.6900	0.6368	0.5846	0.3567	0.1979	0.1010	0.0475	0.0205	0.0081	0.0029	0.0009	1
2	0.9997	0.9975	0.9923	0.9833	0.9699	0.9522	0.9302	0.9042	0.8745	0.8416	0.6479	0.4481	0.2811	0.1608	0.0839	0.0398	0.0170	0.0065	2
3	1.0000	0.9999	0.9994	0.9981	0.9958	0.9920	0.9864	0.9786	0.9685	0.9559	0.8535	0.6982	0.5213	0.3552	0.2205	0.1243	0.0632	0.0287	3
4		1.0000	1.0000	0.9998	0.9996	0.9990	0.9980	0.9965	0.9941	0.9908	0.9533	0.8702	0.7415	0.5842	0.4227	0.2793	0.1672	0.0898	4
5				1.0000	1.0000	0.9999	0.9998	0.9996	0.9992	0.9985	0.9885	0.9561	0.8883	0.7805	0.6405	0.4859	0.3373	0.2120	5
6						1.0000	1.0000	1.0000	0.9999	0.9998	0.9978	0.9884	0.9617	0.9067	0.8164	0.6925	0.5461	0.3953	6
7									1.0000	1.0000	0.9997	0.9976	0.9897	0.9685	0.9247	0.8499	0.7414	0.6047	7
8											1.0000	0.9996	0.9978	0.9917	0.9757	0.9417	0.8811	0.7880	8
9												1.0000	0.9997	0.9983	0.9940	0.9825	0.9574	0.9102	9
10													1.0000	0.9998	0.9989	0.9961	0.9886	0.9713	10
11														1.0000	0.9999	0.9994	0.9978	0.9935	11
12															1.0000	0.9999	0.9997	0.9991	12
13																1.0000	1.0000	0.9999	13
14																		1.0000	14
x	$n=$ 15																		x
0	0.8601	0.7386	0.6333	0.5421	0.4633	0.3953	0.3367	0.2863	0.2430	0.2059	0.0874	0.0352	0.0134	0.0047	0.0016	0.0005	0.0001	0.0000	0
1	0.9904	0.9647	0.9270	0.8809	0.8290	0.7738	0.7168	0.6597	0.6035	0.5490	0.3186	0.1671	0.0802	0.0353	0.0142	0.0052	0.0017	0.0005	1
2	0.9996	0.9970	0.9906	0.9797	0.9638	0.9429	0.9171	0.8870	0.8531	0.8159	0.6042	0.3980	0.2361	0.1268	0.0617	0.0271	0.0107	0.0037	2
3	1.0000	0.9998	0.9992	0.9976	0.9945	0.9896	0.9825	0.9727	0.9601	0.9444	0.8227	0.6482	0.4613	0.2969	0.1727	0.0905	0.0424	0.0176	3
4		1.0000	0.9999	0.9998	0.9994	0.9986	0.9972	0.9950	0.9918	0.9873	0.9383	0.8358	0.6865	0.5155	0.3519	0.2173	0.1204	0.0592	4
5			1.0000	1.0000	0.9999	0.9999	0.9997	0.9993	0.9987	0.9978	0.9832	0.9389	0.8516	0.7216	0.5643	0.4032	0.2608	0.1509	5
6					1.0000	1.0000	1.0000	0.9999	0.9998	0.9997	0.9964	0.9819	0.9434	0.8689	0.7548	0.6098	0.4522	0.3036	6
7								1.0000	1.0000	1.0000	0.9994	0.9958	0.9827	0.9500	0.8868	0.7869	0.6535	0.5000	7
8											0.9999	0.9992	0.9958	0.9848	0.9578	0.9050	0.8182	0.6964	8
9											1.0000	0.9999	0.9992	0.9963	0.9876	0.9662	0.9231	0.8491	9
10												1.0000	0.9999	0.9993	0.9972	0.9907	0.9745	0.9408	10
11													1.0000	0.9999	0.9995	0.9981	0.9937	0.9824	11
12														1.0000	0.9999	0.9997	0.9989	0.9963	12
13															1.0000	1.0000	0.9999	0.9995	13
14																	1.0000	1.0000	14

The binomial cumulative distribution function (continued)

p	0.01	0.02	0.03	0.04	0.05	0.06	0.07	0.08	0.09	0.10	0.15	0.20	0.25	0.30	0.35	0.40	0.45	0.50	p
x	$n =$ 20																		x
0	0.8179	0.6676	0.5438	0.4420	0.3585	0.2901	0.2342	0.1887	0.1516	0.1216	0.0388	0.0115	0.0032	0.0008	0.0002	0.0000	0.0000	0.0000	**0**
1	0.9831	0.9401	0.8802	0.8103	0.7358	0.6605	0.5869	0.5169	0.4516	0.3917	0.1756	0.0692	0.0243	0.0076	0.0021	0.0005	0.0001	0.0000	**1**
2	0.9990	0.9929	0.9790	0.9561	0.9245	0.8850	0.8390	0.7879	0.7334	0.6769	0.4049	0.2061	0.0913	0.0355	0.0121	0.0036	0.0009	0.0002	**2**
3	1.0000	0.9994	0.9973	0.9926	0.9841	0.9710	0.9529	0.9294	0.9007	0.8670	0.6477	0.4114	0.2252	0.1071	0.0444	0.0160	0.0049	0.0013	**3**
4		1.0000	0.9997	0.9990	0.9974	0.9944	0.9893	0.9817	0.9710	0.9568	0.8298	0.6296	0.4148	0.2375	0.1182	0.0510	0.0189	0.0059	**4**
5			1.0000	0.9999	0.9997	0.9991	0.9981	0.9962	0.9932	0.9887	0.9327	0.8042	0.6172	0.4164	0.2454	0.1256	0.0553	0.0207	**5**
6				1.0000	1.0000	0.9999	0.9997	0.9994	0.9987	0.9976	0.9781	0.9133	0.7858	0.6080	0.4166	0.2500	0.1299	0.0577	**6**
7						1.0000	1.0000	0.9999	0.9998	0.9996	0.9941	0.9679	0.8982	0.7723	0.6010	0.4159	0.2520	0.1316	**7**
8								1.0000	1.0000	0.9999	0.9987	0.9900	0.9591	0.8867	0.7624	0.5956	0.4143	0.2517	**8**
9										1.0000	0.9998	0.9974	0.9861	0.9520	0.8782	0.7553	0.5914	0.4119	**9**
10											1.0000	0.9994	0.9961	0.9829	0.9468	0.8725	0.7507	0.5881	**10**
11												0.9999	0.9991	0.9949	0.9804	0.9435	0.8692	0.7483	**11**
12												1.0000	0.9998	0.9987	0.9940	0.9790	0.9420	0.8684	**12**
13													1.0000	0.9997	0.9985	0.9935	0.9786	0.9423	**13**
14														1.0000	0.9997	0.9984	0.9936	0.9793	**14**
15															1.0000	0.9997	0.9985	0.9941	**15**
16																1.0000	0.9997	0.9987	**16**
17																	1.0000	0.9998	**17**
18																		1.0000	**18**
x	$n =$ 25																		x
0	0.7778	0.6035	0.4670	0.3604	0.2774	0.2129	0.1630	0.1244	0.0946	0.0718	0.0172	0.0038	0.0008	0.0001	0.0000	0.0000	0.0000	0.0000	**0**
1	0.9742	0.9114	0.8280	0.7358	0.6424	0.5527	0.4696	0.3947	0.3286	0.2712	0.0931	0.0274	0.0070	0.0016	0.0003	0.0001	0.0000	0.0000	**1**
2	0.9980	0.9868	0.9620	0.9235	0.8729	0.8129	0.7466	0.6768	0.6063	0.5371	0.2537	0.0982	0.0321	0.0090	0.0021	0.0004	0.0001	0.0000	**2**
3	0.9999	0.9986	0.9938	0.9835	0.9659	0.9402	0.9064	0.8649	0.8169	0.7636	0.4711	0.2340	0.0962	0.0332	0.0097	0.0024	0.0005	0.0001	**3**
4	1.0000	0.9999	0.9992	0.9972	0.9928	0.9850	0.9726	0.9549	0.9314	0.9020	0.6821	0.4207	0.2137	0.0905	0.0320	0.0095	0.0023	0.0005	**4**
5		1.0000	0.9999	0.9996	0.9988	0.9969	0.9935	0.9877	0.9790	0.9666	0.8385	0.6167	0.3783	0.1935	0.0826	0.0294	0.0086	0.0020	**5**
6			1.0000	1.0000	0.9998	0.9995	0.9987	0.9972	0.9946	0.9905	0.9305	0.7800	0.5611	0.3407	0.1734	0.0736	0.0258	0.0073	**6**
7					1.0000	0.9999	0.9998	0.9995	0.9989	0.9977	0.9745	0.8909	0.7265	0.5118	0.3061	0.1536	0.0639	0.0216	**7**
8						1.0000	1.0000	0.9999	0.9998	0.9995	0.9920	0.9532	0.8506	0.6769	0.4668	0.2735	0.1340	0.0539	**8**
9								1.0000	1.0000	0.9999	0.9979	0.9827	0.9287	0.8106	0.6303	0.4246	0.2424	0.1148	**9**
10										1.0000	0.9995	0.9944	0.9703	0.9022	0.7712	0.5858	0.3843	0.2122	**10**
11											0.9999	0.9985	0.9893	0.9558	0.8746	0.7323	0.5426	0.3450	**11**
12											1.0000	0.9996	0.9966	0.9825	0.9396	0.8462	0.6937	0.5000	**12**
13												0.9999	0.9991	0.9940	0.9745	0.9222	0.8173	0.6550	**13**
14												1.0000	0.9998	0.9982	0.9907	0.9656	0.9040	0.7878	**14**
15													1.0000	0.9995	0.9971	0.9868	0.9560	0.8852	**15**
16														0.9999	0.9992	0.9957	0.9826	0.9461	**16**
17														1.0000	0.9998	0.9988	0.9942	0.9784	**17**
18															1.0000	0.9997	0.9984	0.9927	**18**
19																0.9999	0.9996	0.9980	**19**
20																1.0000	0.9999	0.9995	**20**
21																	1.0000	0.9999	**21**
22																		1.0000	**22**
x	$n =$ 30																		x
0	0.7397	0.5455	0.4010	0.2939	0.2146	0.1563	0.1134	0.0820	0.0591	0.0424	0.0076	0.0012	0.0002	0.0000	0.0000	0.0000	0.0000	0.0000	**0**
1	0.9639	0.8795	0.7731	0.6612	0.5535	0.4555	0.3694	0.2958	0.2343	0.1837	0.0480	0.0105	0.0020	0.0003	0.0000	0.0000	0.0000	0.0000	**1**
2	0.9967	0.9783	0.9399	0.8831	0.8122	0.7324	0.6487	0.5654	0.4855	0.4114	0.1514	0.0442	0.0106	0.0021	0.0003	0.0000	0.0000	0.0000	**2**
3	0.9998	0.9971	0.9881	0.9694	0.9392	0.8974	0.8450	0.7842	0.7175	0.6474	0.3217	0.1227	0.0374	0.0093	0.0019	0.0003	0.0000	0.0000	**3**
4	1.0000	0.9997	0.9982	0.9937	0.9844	0.9685	0.9447	0.9126	0.8723	0.8245	0.5245	0.2552	0.0979	0.0302	0.0075	0.0015	0.0002	0.0000	**4**
5		1.0000	0.9998	0.9989	0.9967	0.9921	0.9838	0.9707	0.9519	0.9268	0.7106	0.4275	0.2026	0.0766	0.0233	0.0057	0.0011	0.0002	**5**
6			1.0000	0.9999	0.9994	0.9983	0.9960	0.9918	0.9848	0.9742	0.8474	0.6070	0.3481	0.1595	0.0586	0.0172	0.0040	0.0007	**6**
7				1.0000	0.9999	0.9997	0.9992	0.9980	0.9959	0.9922	0.9302	0.7608	0.5143	0.2814	0.1238	0.0435	0.0121	0.0026	**7**
8					1.0000	1.0000	0.9999	0.9996	0.9990	0.9980	0.9722	0.8713	0.6736	0.4315	0.2247	0.0940	0.0312	0.0081	**8**
9							1.0000	0.9999	0.9998	0.9995	0.9903	0.9389	0.8034	0.5888	0.3575	0.1763	0.0694	0.0214	**9**
10								1.0000	1.0000	0.9999	0.9971	0.9744	0.8943	0.7304	0.5078	0.2915	0.1350	0.0494	**10**
11										1.0000	0.9992	0.9905	0.9493	0.8407	0.6548	0.4311	0.2327	0.1002	**11**
12											0.9998	0.9969	0.9784	0.9155	0.7802	0.5785	0.3592	0.1808	**12**
13											1.0000	0.9991	0.9918	0.9599	0.8737	0.7145	0.5025	0.2923	**13**
14												0.9998	0.9973	0.9831	0.9348	0.8246	0.6448	0.4278	**14**
15												0.9999	0.9992	0.9936	0.9699	0.9029	0.7691	0.5722	**15**
16												1.0000	0.9998	0.9979	0.9876	0.9519	0.8644	0.7077	**16**
17													0.9999	0.9994	0.9955	0.9788	0.9286	0.8192	**17**
18													1.0000	0.9998	0.9986	0.9917	0.9666	0.8998	**18**
19														1.0000	0.9996	0.9971	0.9862	0.9506	**19**
20															0.9999	0.9991	0.9950	0.9786	**20**
21															1.0000	0.9998	0.9984	0.9919	**21**
22																1.0000	0.9996	0.9974	**22**
23																	0.9999	0.9993	**23**
24																	1.0000	0.9998	**24**
25																		1.0000	**25**

AQA S1 — STATISTICAL TABLES

The binomial cumulative distribution function (continued)

p	0.01	0.02	0.03	0.04	0.05	0.06	0.07	0.08	0.09	0.10	0.15	0.20	0.25	0.30	0.35	0.40	0.45	0.50	p
x	$n=$ 40																		x
0	0.6690	0.4457	0.2957	0.1954	0.1285	0.0842	0.0549	0.0356	0.0230	0.0148	0.0015	0.0001	0.0000	0.0000	0.0000	0.0000	0.0000	0.0000	0
1	0.9393	0.8095	0.6615	0.5210	0.3991	0.2990	0.2201	0.1594	0.1140	0.0805	0.0121	0.0015	0.0001	0.0000	0.0000	0.0000	0.0000	0.0000	1
2	0.9925	0.9543	0.8822	0.7855	0.6767	0.5665	0.4625	0.3694	0.2894	0.2228	0.0486	0.0079	0.0010	0.0001	0.0000	0.0000	0.0000	0.0000	2
3	0.9993	0.9918	0.9686	0.9252	0.8619	0.7827	0.6937	0.6007	0.5092	0.4231	0.1302	0.0285	0.0047	0.0006	0.0001	0.0000	0.0000	0.0000	3
4	1.0000	0.9988	0.9933	0.9790	0.9520	0.9104	0.8546	0.7868	0.7103	0.6290	0.2633	0.0759	0.0160	0.0026	0.0003	0.0000	0.0000	0.0000	4
5		0.9999	0.9988	0.9951	0.9861	0.9691	0.9419	0.9033	0.8535	0.7937	0.4325	0.1613	0.0433	0.0086	0.0013	0.0001	0.0000	0.0000	5
6		1.0000	0.9998	0.9990	0.9966	0.9909	0.9801	0.9624	0.9361	0.9005	0.6067	0.2859	0.0962	0.0238	0.0044	0.0006	0.0001	0.0000	6
7			1.0000	0.9998	0.9993	0.9977	0.9942	0.9873	0.9758	0.9581	0.7559	0.4371	0.1820	0.0553	0.0124	0.0021	0.0002	0.0000	7
8				1.0000	0.9999	0.9995	0.9985	0.9963	0.9919	0.9845	0.8646	0.5931	0.2998	0.1110	0.0303	0.0061	0.0009	0.0001	8
9					1.0000	0.9999	0.9997	0.9990	0.9976	0.9949	0.9328	0.7318	0.4395	0.1959	0.0644	0.0156	0.0027	0.0003	9
10						1.0000	0.9999	0.9998	0.9994	0.9985	0.9701	0.8392	0.5839	0.3087	0.1215	0.0352	0.0074	0.0011	10
11							1.0000	1.0000	0.9999	0.9996	0.9880	0.9125	0.7151	0.4406	0.2053	0.0709	0.0179	0.0032	11
12									1.0000	0.9999	0.9957	0.9568	0.8209	0.5772	0.3143	0.1285	0.0386	0.0083	12
13										1.0000	0.9986	0.9806	0.8968	0.7032	0.4408	0.2112	0.0751	0.0192	13
14											0.9996	0.9921	0.9456	0.8074	0.5721	0.3174	0.1326	0.0403	14
15											0.9999	0.9971	0.9738	0.8849	0.6946	0.4402	0.2142	0.0769	15
16											1.0000	0.9990	0.9884	0.9367	0.7978	0.5681	0.3185	0.1341	16
17												0.9997	0.9953	0.9680	0.8761	0.6885	0.4391	0.2148	17
18												0.9999	0.9983	0.9852	0.9301	0.7911	0.5651	0.3179	18
19												1.0000	0.9994	0.9937	0.9637	0.8702	0.6844	0.4373	19
20													0.9998	0.9976	0.9827	0.9256	0.7870	0.5627	20
21													1.0000	0.9991	0.9925	0.9608	0.8669	0.6821	21
22														0.9997	0.9970	0.9811	0.9233	0.7852	22
23														0.9999	0.9989	0.9917	0.9595	0.8659	23
24														1.0000	0.9996	0.9966	0.9804	0.9231	24
25															0.9999	0.9988	0.9914	0.9597	25
26															1.0000	0.9996	0.9966	0.9808	26
27																0.9999	0.9988	0.9917	27
28																1.0000	0.9996	0.9968	28
29																	0.9999	0.9989	29
30																	1.0000	0.9997	30
31																		0.9999	31
32																		1.0000	32
x	$n=$ 50																		x
0	0.6050	0.3642	0.2181	0.1299	0.0769	0.0453	0.0266	0.0155	0.0090	0.0052	0.0003	0.0000	0.0000	0.0000	0.0000	0.0000	0.0000	0.0000	0
1	0.9106	0.7358	0.5553	0.4005	0.2794	0.1900	0.1265	0.0827	0.0532	0.0338	0.0029	0.0002	0.0000	0.0000	0.0000	0.0000	0.0000	0.0000	1
2	0.9862	0.9216	0.8108	0.6767	0.5405	0.4162	0.3108	0.2260	0.1605	0.1117	0.0142	0.0013	0.0001	0.0000	0.0000	0.0000	0.0000	0.0000	2
3	0.9984	0.9822	0.9372	0.8609	0.7604	0.6473	0.5327	0.4253	0.3303	0.2503	0.0460	0.0057	0.0005	0.0000	0.0000	0.0000	0.0000	0.0000	3
4	0.9999	0.9968	0.9832	0.9510	0.8964	0.8206	0.7290	0.6290	0.5277	0.4312	0.1121	0.0185	0.0021	0.0002	0.0000	0.0000	0.0000	0.0000	4
5	1.0000	0.9995	0.9963	0.9856	0.9622	0.9224	0.8650	0.7919	0.7072	0.6161	0.2194	0.0480	0.0070	0.0007	0.0001	0.0000	0.0000	0.0000	5
6		0.9999	0.9993	0.9964	0.9882	0.9711	0.9417	0.8981	0.8404	0.7702	0.3613	0.1034	0.0194	0.0025	0.0002	0.0000	0.0000	0.0000	6
7		1.0000	0.9999	0.9992	0.9968	0.9906	0.9780	0.9562	0.9232	0.8779	0.5188	0.1904	0.0453	0.0073	0.0008	0.0001	0.0000	0.0000	7
8			1.0000	0.9999	0.9992	0.9973	0.9927	0.9833	0.9672	0.9421	0.6681	0.3073	0.0916	0.0183	0.0025	0.0002	0.0000	0.0000	8
9				1.0000	0.9998	0.9993	0.9978	0.9944	0.9875	0.9755	0.7911	0.4437	0.1637	0.0402	0.0067	0.0008	0.0001	0.0000	9
10					1.0000	0.9998	0.9994	0.9983	0.9957	0.9906	0.8801	0.5836	0.2622	0.0789	0.0160	0.0022	0.0002	0.0000	10
11						1.0000	0.9999	0.9995	0.9987	0.9968	0.9372	0.7107	0.3816	0.1390	0.0342	0.0057	0.0006	0.0000	11
12							1.0000	0.9999	0.9996	0.9990	0.9699	0.8139	0.5110	0.2229	0.0661	0.0133	0.0018	0.0002	12
13								1.0000	0.9999	0.9997	0.9868	0.8894	0.6370	0.3279	0.1163	0.0280	0.0045	0.0005	13
14									1.0000	0.9999	0.9947	0.9393	0.7481	0.4468	0.1878	0.0540	0.0104	0.0013	14
15										1.0000	0.9981	0.9692	0.8369	0.5692	0.2801	0.0955	0.0220	0.0033	15
16											0.9993	0.9856	0.9017	0.6839	0.3889	0.1561	0.0427	0.0077	16
17											0.9998	0.9937	0.9449	0.7822	0.5060	0.2369	0.0765	0.0164	17
18											0.9999	0.9975	0.9713	0.8594	0.6216	0.3356	0.1273	0.0325	18
19											1.0000	0.9991	0.9861	0.9152	0.7264	0.4465	0.1974	0.0595	19
20												0.9997	0.9937	0.9522	0.8139	0.5610	0.2862	0.1013	20
21												0.9999	0.9974	0.9749	0.8813	0.6701	0.3900	0.1611	21
22												1.0000	0.9990	0.9877	0.9290	0.7660	0.5019	0.2399	22
23													0.9996	0.9944	0.9604	0.8438	0.6134	0.3359	23
24													0.9999	0.9976	0.9793	0.9022	0.7160	0.4439	24
25													1.0000	0.9991	0.9900	0.9427	0.8034	0.5561	25
26														0.9997	0.9955	0.9686	0.8721	0.6641	26
27														0.9999	0.9981	0.9840	0.9220	0.7601	27
28														1.0000	0.9993	0.9924	0.9556	0.8389	28
29															0.9997	0.9966	0.9765	0.8987	29
30															0.9999	0.9986	0.9884	0.9405	30
31															1.0000	0.9995	0.9947	0.9675	31
32																0.9998	0.9978	0.9836	32
33																0.9999	0.9991	0.9923	33
34																1.0000	0.9997	0.9967	34
35																	0.9999	0.9987	35
36																	1.0000	0.9995	36
37																		0.9998	37
38																		1.0000	38

Constant Acceleration Equations

Welcome to the technicolour world of Mechanics 1. Fashions may change, but there will always be M1 questions that involve objects travelling in a straight line. It's just a case of picking the right equations to solve the problem.

There are Five Constant Acceleration Equations

Examiners call these "uvast" questions (pronounced ewe-vast, like a large sheep) because of the five variables involved:

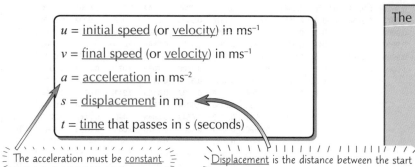

u = initial speed (or velocity) in ms^{-1}

v = final speed (or velocity) in ms^{-1}

a = acceleration in ms^{-2}

s = displacement in m

t = time that passes in s (seconds)

The acceleration must be constant.

Displacement is the distance between the start and end points of motion. It's not always the same as the total distance travelled.

The constant acceleration equations are:

$$v = u + at$$

$$s = ut + \tfrac{1}{2}at^2$$

$$s = \tfrac{1}{2}(u + v)t$$

$$v^2 = u^2 + 2as$$

$$s = vt - \tfrac{1}{2}at^2$$

None of those equations are in the formula book, so you're going to have to learn them. Questions will usually give you three variables — your job is to choose the equation that will find you the missing fourth variable.

EXAMPLE

A jet ski travels in a straight line along a river. It passes under two bridges 200 m apart and is observed to be travelling at 5 ms^{-1} under the first bridge and at 9 ms^{-1} under the second bridge. Calculate its acceleration (assuming it is constant).

List the variables ("uvast"):

$u = 5$

$v = 9$

$a = a$ — You have to work out a.

$s = 200$

$t = t$ — You're not told about the time taken.

Check you're using the right units — m, s, ms^{-1} and ms^{-2}.

Choose the equation with u, v, s and a in it: $v^2 = u^2 + 2as$

Substitute values: $9^2 = 5^2 + (2 \times a \times 200)$

Simplify: $81 = 25 + 400a$

Rearrange: $400a = 81 - 25 = 56$

Then solve: $a = \frac{56}{400} = 0.14\ ms^{-2}$

Motion under Gravity just means taking a = g...

Don't be put off by questions involving objects moving freely under gravity — they're just telling you the acceleration is g.

Use the value of g given on the front of the paper or in the question. If you don't, you risk losing a mark because your answer won't match the examiners' answer.

EXAMPLE

A pebble is dropped into a well 18 m deep and moves freely under gravity until it hits the bottom. Calculate the time it takes to reach the bottom. (Take g = 9.8 ms^{-2}.)

First, list the variables:

$u = 0$ — Because the pebble was dropped, not thrown.

$v = v$

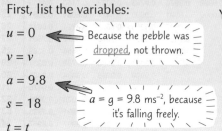

$a = 9.8$ — $a = g = 9.8\ ms^{-2}$, because it's falling freely.

$s = 18$

$t = t$

You need the equation with u, a, s and t in it: $s = ut + \tfrac{1}{2}at^2$

Substitute values: $18 = (0 \times t) + (\tfrac{1}{2} \times 9.8 \times t^2)$

Simplify: $18 = 4.9t^2$

Rearrange to give t^2: $t^2 = \frac{18}{4.9} = 3.67...$

Solve by square-rooting: $t = \sqrt{3.67...} = 1.92$ s

Watch out for tricky questions like this — at first it looks like they've only given you one variable.
You have to spot that the pebble was dropped (so it started with no velocity) and that it's moving freely under gravity.

Constant Acceleration Equations

...or a = –g

EXAMPLE

A ball is projected vertically upwards at 3 ms^{-1} from a point 1.5 m above the ground. How long does it take to reach its maximum height?
How fast will the ball be travelling when it hits the ground? (Take $g = 9.8$ ms^{-2}.)

First, list the variables, taking up as the positive direction:

$u = 3$

$v = 0$ — When projected objects reach the top of their motion, they stop momentarily.

$a = -9.8$ — Because g always acts downwards and up was taken as positive, a is negative.

$s = s$

$t = ?$

Use the equation with u, v, a and t in it: $v = u + at$

Substitute values: $0 = 3 + (-9.8 \times t)$

Simplify: $0 = 3 - 9.8t$

Rearrange and solve to find t: $t = \frac{3}{9.8} = 0.306$ s

To find the speed of the ball when it hits the ground consider the complete path of the ball.

$s = -1.5$ as it's the displacement from the ball's original position, not total distance travelled.

s is negative as the ground is below the point of projection.

Using $v^2 = u^2 + 2as$ where $u = 3$, $v = ?$, $a = -9.8$, $s = -1.5$:

$v^2 = u^2 + 2as = 3^2 + 2(-9.8 \times -1.5) = 38.4$, so $v = \sqrt{38.4} = 6.20$ ms^{-1} (to 3 s.f.)

Sometimes there's More Than One Object Moving at the Same Time

For these questions, t is often the same (or connected as in this example) because time ticks along for both objects at the same rate. The distance travelled might also be connected.

EXAMPLE

A car, A, travelling along a straight road at a constant 30 ms^{-1} passes point R at $t = 0$. Exactly 2 seconds later, a second car, B, travelling at 25 ms^{-1}, moves in the same direction from point R. Car B accelerates at a constant 2 ms^{-2}. Show that the two cars are level when $\mathbf{t^2 - 9t - 46 = 0}$, where t is the time taken by car A.

For each car, there are different "uvast" equations, so you write separate lists and separate equations.

CAR A

$u_A = 30$ $v_A = 30$

$a_A = 0$ $s_A = s$

$t_A = t$

Constant speed so $a_A = 0$

The two cars are level, so choose an equation with s in it:

$s = ut + \frac{1}{2}at^2$

Substitute values: $s = 30t + (\frac{1}{2} \times 0 \times t^2)$

Simplify: $\mathbf{s = 30t}$

CAR B

$u_B = 25$ $v_B = v$

$a_B = 2$ $s_B = s$

$t_B = (t - 2)$

s is the same for both cars because they're level.

B starts moving 2 seconds after A passes point R.

Use the same equation for car B: $s = ut + \frac{1}{2}at^2$

Substitute values: $s = 25(t - 2) + (\frac{1}{2} \times 2 \times (t - 2)^2)$

Simplify: $s = 25t - 50 + (t - 2)(t - 2)$

$s = 25t - 50 + (t^2 - 4t + 4)$

$\mathbf{s = t^2 + 21t - 46}$

The distance travelled by both cars is equal, so put the expressions for s equal to each other:

$30t = t^2 + 21t - 46$

$t^2 - 9t - 46 = 0$ — That's the result you were asked to find.

Constant acceleration questions involve modelling assumptions (simplifications to real life so you can use the equations):

1) The object is a particle — this just means it's very small and so isn't affected by air resistance as cars or stones would be in real life.
2) Acceleration is constant — without it, the equations couldn't be used.

As Socrates once said, "The unexamined life is not worth living"... *but what did he know...*

Make sure you:
1) Make a list of the uvast variables EVERY time you get one of these questions.
2) Look out for "hidden" values — e.g. "particle initially at rest..." means $u = 0$.
3) Choose and solve the equation that goes with the variables you've got.

Motion Graphs

You can use displacement-time (x/t), velocity-time (v/t) and acceleration-time (a/t) graphs to represent all sorts of motion.

Displacement-time Graphs: Height = Displacement and Gradient = Velocity

The steeper the line, the greater the velocity. A horizontal line has a zero gradient, so that means the object isn't moving.

EXAMPLE A cyclist's journey is shown on this x/t graph. Describe the motion.

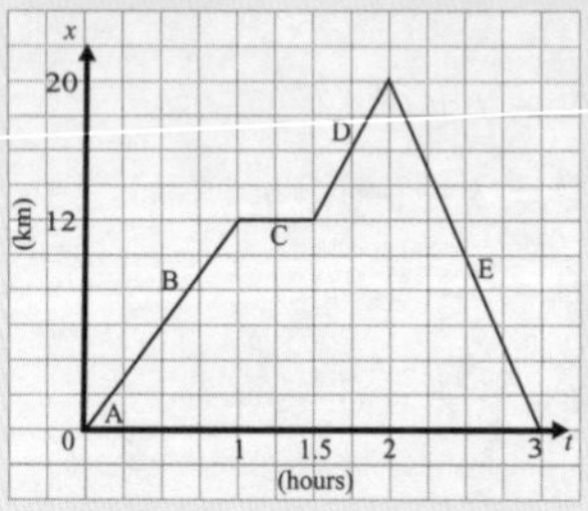

A: Starts from rest (when $t = 0$, $x = 0$)

B: Travels 12 km in 1 hour at a velocity of 12 kmh^{-1}

C: Rests for ½ hour ($v = 0$)

D: Cycles 8 km in ½ hour at a velocity of 16 kmh^{-1}

E: Returns to starting position, cycling 20 km in 1 hour at a velocity of –20 kmh^{-1} (i.e. 20 kmh^{-1} in the opposite direction)

You may also get a distance-time graph. These show total distance travelled rather than displacement. The gradient gives the speed rather than the velocity, so will always be positive (or zero).

EXAMPLE A girl jogs 2 km in 15 minutes and a boy runs 1.5 km in 6 min, rests for 1 min then walks the last 0.5 km in 8 min. Show the two journeys on an x/t graph.

Girl: constant velocity, so there's just one straight line for her journey from (0, 0) to (15, 2)

Boy: three parts to the journey, so there are three straight lines: A - run, B - rest, C - walk

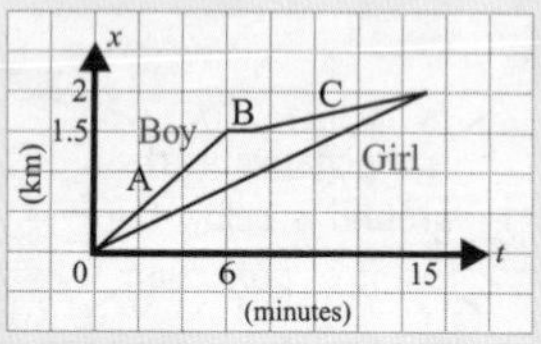

Velocity-time Graphs: Area = Distance and Gradient = Acceleration

The area under the graph can be calculated by splitting the area into rectangles, triangles or trapeziums. Work out the areas separately, then add them all up at the end.

EXAMPLE A train journey is shown on the v/t graph on the right. Find the distance travelled and the rate of deceleration as the train comes to a stop.

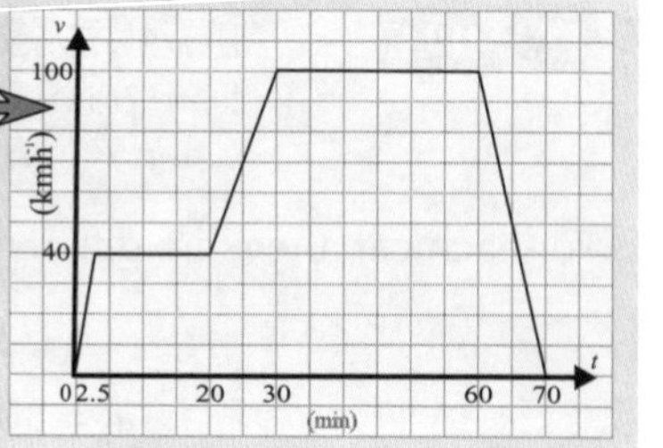

The time is given in minutes and the velocity as kilometres per hour, so divide the time in minutes by 60 to get the time in hours.

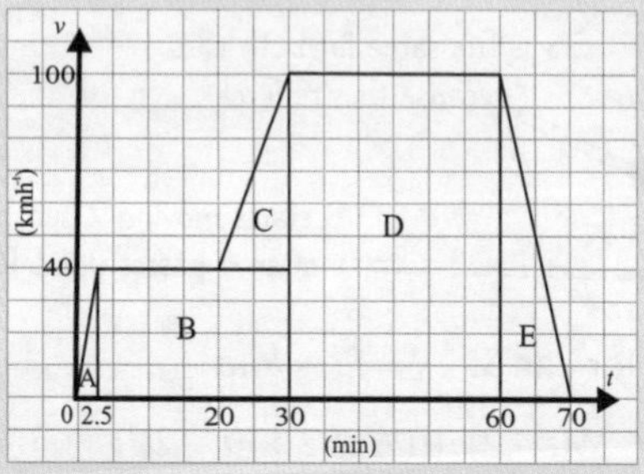

Area of A: $(2.5 \div 60 \times 40) \div 2 = 0.833...$

Area of B: $27.5 \div 60 \times 40 = 18.33...$

Area of C: $(10 \div 60 \times 60) \div 2 = 5$

Area of D: $30 \div 60 \times 100 = 50$

Area of E: $(10 \div 60 \times 100) \div 2 = 8.33...$

Total area = 82.5 so distance is 82.5 km

You might get a speed-time graph instead of a velocity-time graph — they're pretty much the same, except speeds are always positive, whereas you can have negative velocities.

The gradient of the graph at the end of the journey is $-100\ \text{kmh}^{-1} \div (10 \div 60)$ hours $= -600\ \text{kmh}^{-2}$
So the train decelerates at 600 kmh^{-2}.

Acceleration-time Graphs: Area = Velocity

EXAMPLE The acceleration of a parachutist who jumps from a plane is modelled by the a/t graph on the right. Describe the motion of the parachutist and find her velocity when she is no longer accelerating.

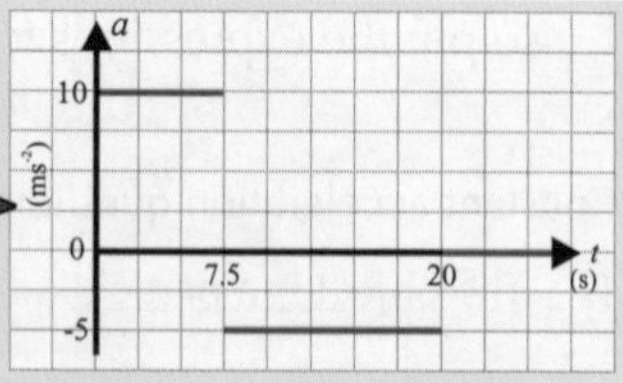

She falls with acceleration due to gravity of 10 ms^{-2} for 7.5 s. The parachute opens and the resultant acceleration due to gravity and the air resistance of the parachute is 5 ms^{-2} acting upwards for 12.5 s. After 20 s, the acceleration is zero and so she falls with constant velocity. This velocity is the area under the graph:

Area A: $10 \times 7.5 = 75\ \text{ms}^{-1}$ Area B: $5 \times 12.5 = 62.5\ \text{ms}^{-1}$

Area B is under the horizontal axis, so subtract area B from area A:

Velocity $= 75\ \text{ms}^{-1} - 62.5\ \text{ms}^{-1} = 12.5\ \text{ms}^{-1}$

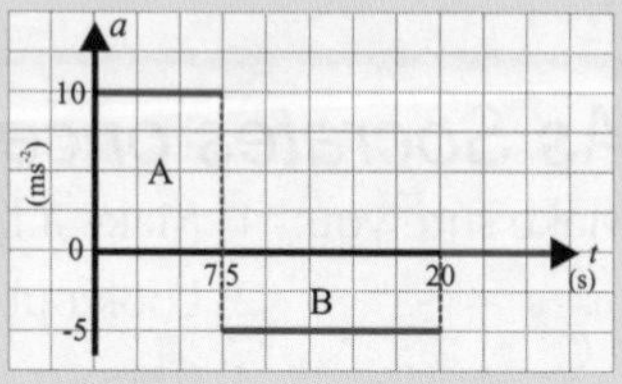

Motion Graphs

Graphs can be used to *Solve Complicated Problems*

As well as working out distance, velocity and acceleration from graphs, you can also solve more complicated problems. These might involve working out information not shown directly on the graph.

EXAMPLE

A jogger and a cyclist set off at the same time. The jogger runs with a constant velocity. The cyclist accelerates from rest, reaching a velocity of 5 ms^{-1} after 6 s and then continues at this velocity. The cyclist overtakes the jogger after 15 s.

Draw a graph of the motion and find the velocity of the jogger.

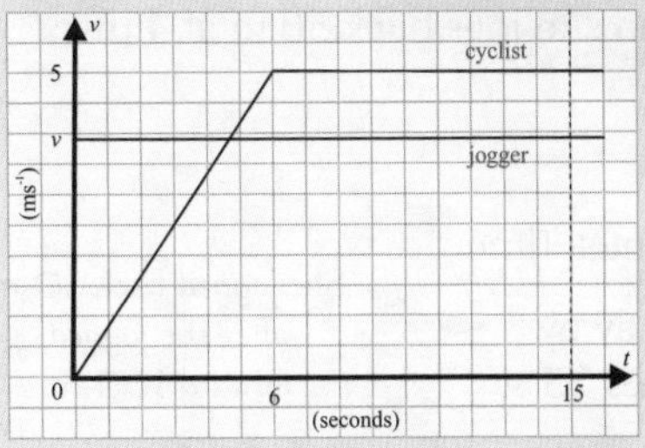

Call the velocity of the jogger v.

After 15 s the distance each has travelled is the same, so you can work out the area under the two graphs to get the distances:

Jogger: Area = distance = $15v$

Cyclist: Area = distance = $(6 \times 5) \div 2 + (9 \times 5) = 60$

So $15v = 60$
$v = 4$ ms^{-1}

area of triangle + area of rectangle

EXAMPLE

A man throws a pebble vertically upwards from ground level with a speed of u ms^{-1}. The pebble takes 2.6 s to return to ground level and reaches a maximum height of 9.4 m.

Draw a graph of the motion and find the value of u.

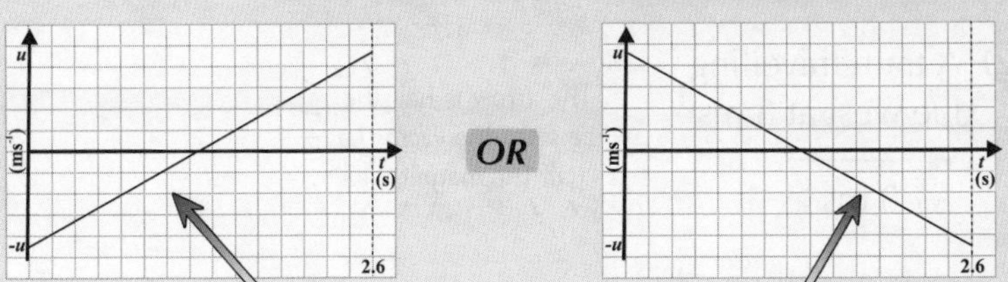

The pebble's velocity is negative for half of its motion. As you can choose up or down to be the positive direction, there are two possible graphs that describe the motion. The graph on the left takes 'down' as positive, while the one on the right takes 'up' as positive.

Using $s = \frac{1}{2}(u + v)t$ until pebble reaches maximum height, where $v = 0$ ms^{-1}:

$9.4 = \frac{1}{2}(u + 0)1.3$

so $u = \frac{9.4}{0.65} = 14.5$ ms^{-1}

The pebble reaches maximum height in half the time it takes to fall to the ground.

EXAMPLE

A bus is travelling at V ms^{-1}. When it reaches point A it accelerates uniformly for 4 s, reaching a speed of 21 ms^{-1} as it passes point B. At point B, the driver brakes uniformly until the bus comes to a halt 7 s later. The rate of deceleration is twice the rate of acceleration.

Draw a speed-time graph of the motion and find the value of V.

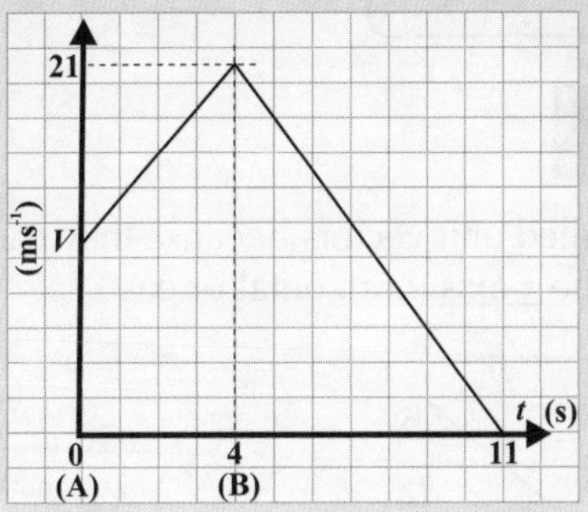

To find the deceleration use $a = \frac{(v - u)}{t}$:

$a = \frac{(0 - 21)}{7} = -3$ ms^{-2}

Just v = u + at rearranged.

Rate of deceleration is twice the rate of acceleration, so rate of acceleration = 1.5 ms^{-2}.

Finding V using $u = v - at$:

$V = 21 - (1.5 \times 4)$

So, $V = 15$ ms^{-1}

Random tongue-twister #1 — I wish to wash my Irish wristwatch...

If a picture can tell a thousand words then a graph can tell... um... a thousand and one. Make sure you know what type of graph you're using and learn what the gradient and the area under each type of graph tells you.

Vectors

'Vector' might sound like a really dull Bond villain, but... well, it's not. Vectors have both size (or magnitude) and direction. If a measurement just has size but not direction, it's called a scalar.

- So, Vector, you expect me to talk?
- No, Mr Bond, I expect you to die! Ak ak ak!

Vectors have *Magnitude* and *Direction* — *Scalars* only have *Magnitude*

Vectors: velocity, displacement, acceleration, force. E.g. a train heading due east at 16 ms^{-1}.

Scalars: speed, distance. E.g. a car travelling at 4 ms^{-1}.

A really important thing to remember is that an object's speed and velocity aren't always the same:

EXAMPLE A runner sprints 100 m along a track at a speed of 8 ms^{-1} and then she jogs back 50 m at 4 ms^{-1}.

Average Speed

Speed = Distance ÷ Time

The runner takes (100 ÷ 8) + (50 ÷ 4) = 25 s to travel **150 m**.

So the average speed is 150 ÷ 25 = 6 ms^{-1}

Average Velocity

Velocity = Change in displacement ÷ Time

In total, the runner ends up **50 m** away from her start point and it takes 25 s.

So the average velocity is 50 ÷ 25 = 2 ms^{-1} in the direction of the sprint.

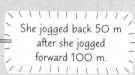

She jogged back 50 m after she jogged forward 100 m.

The *Length* of the *Arrow* shows the *Magnitude of a Vector*

You can draw vectors as arrows where the length shows the magnitude:

① A train travelling due east at 16 ms^{-1}

16 ms^{-1}

② A train travelling due west at 8 ms^{-1}

8 ms^{-1}

The arrow is half the size since the vector has half the magnitude.

You can add vectors together by drawing the arrows nose to tail.
The single vector that goes from the start to the end of the vectors is called the resultant vector.

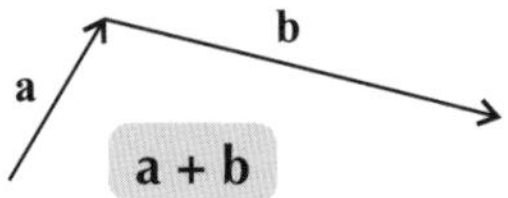

a + b

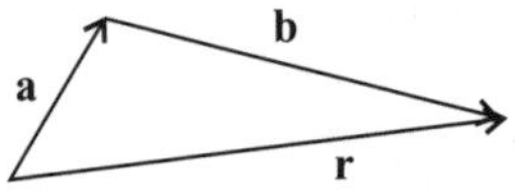

Resultant: **r = a + b**

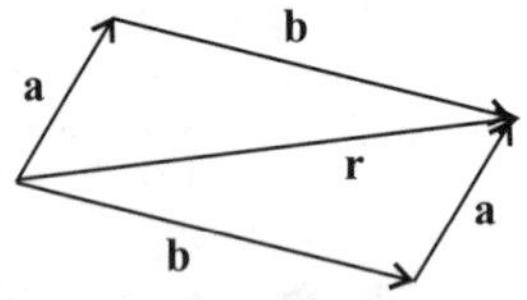

a + b = b + a

You can also multiply a vector by a scalar (just a number): the length changes but the direction stays the same.

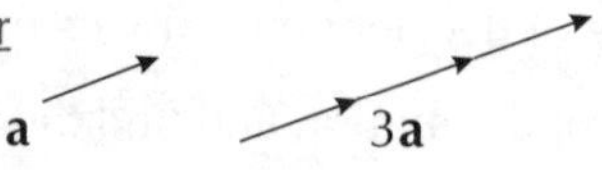

Vectors can be described using *i + j Units*

You can describe vectors using the unit vectors **i** + **j**. They're called unit vectors because they each have a magnitude of 1 and a direction. **i** is in the direction of the *x*-axis (horizontal or east), and **j** is in the direction of the *y*-axis (vertical or north).

E.g. $\overrightarrow{AB}$ = 5**i** +4**j** :

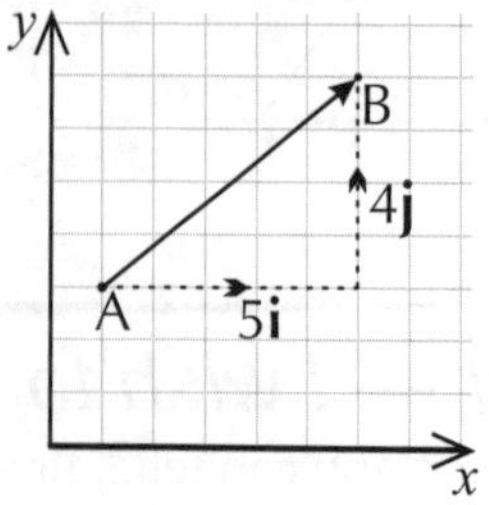

$\overrightarrow{AB}$ is just another way of writing a vector — it means 'the displacement of B from A'.

Vectors

Resolving means writing a vector as Component Vectors

Splitting a vector up like this means you can work things out with one component at a time. When adding vectors to get a resultant vector, it's easier to add the horizontal and vertical components separately. So you split the vector into components first — this is resolving. If the vectors are in **i** + **j** notation already, you don't need to split them up — just add the **i** and **j**'s:

EXAMPLE $\overrightarrow{AB} = 3\mathbf{i} + 2\mathbf{j}$ and $\overrightarrow{BC} = 5\mathbf{i} - 3\mathbf{j}$. Work out $\overrightarrow{AC}$.

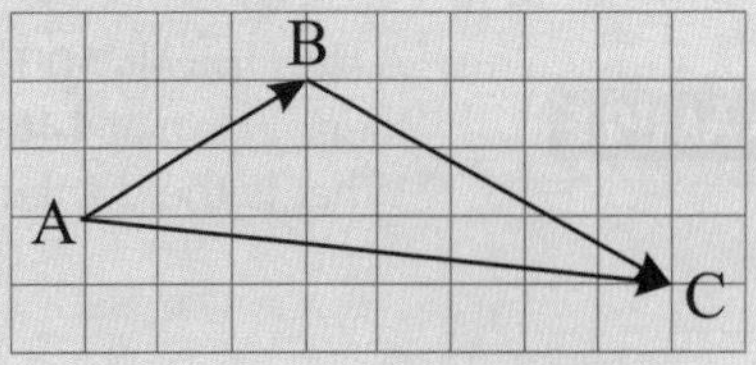

Add the horizontal and vertical components separately.

$\overrightarrow{AC} = \overrightarrow{AB} + \overrightarrow{BC} = (3\mathbf{i} + 2\mathbf{j}) + (5\mathbf{i} - 3\mathbf{j}) = 8\mathbf{i} - \mathbf{j}$

Use Trig and Pythagoras to Change a vector into Component Form

EXAMPLE The speed of a ball is 5 ms^{-1} at an angle of 30° to the horizontal. Find the horizontal and vertical components of the ball's velocity, **v**.

First, draw a diagram and make a right-angle triangle:

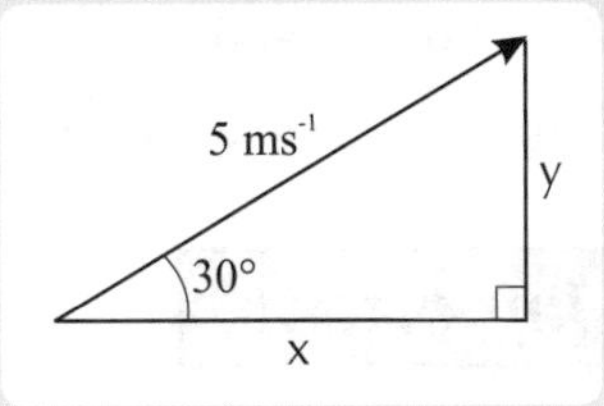

Using trigonometry, we can find x and y:

$\cos 30° = \frac{x}{5}$ so $x = 5\cos 30°$

$\sin 30° = \frac{y}{5}$ so $y = 5\sin 30°$

So $\boldsymbol{v} = (5\cos 30°\mathbf{i} + 5\sin 30°\mathbf{j})$ ms^{-1}

EXAMPLE The acceleration of a body is given by the vector $\mathbf{a} = 6\mathbf{i} - 2\mathbf{j}$. Find the magnitude and direction of the acceleration.

Start with a diagram again. Remember, the y-component "–2" means "down 2".

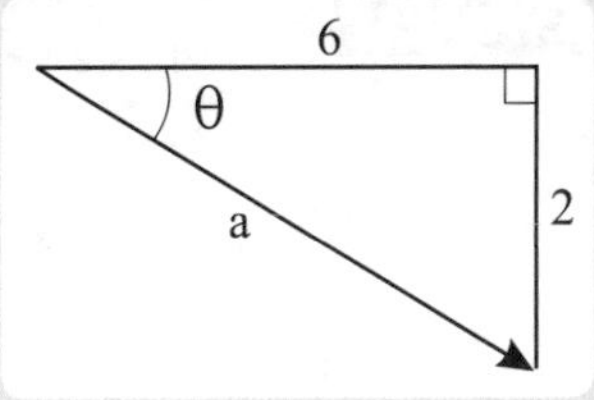

Using Pythagoras' theorem, you can work out the magnitude of **a**:

$a^2 = 6^2 + (-2)^2 = 40$

so $a = \sqrt{40} = 6.32$ (to 3 s.f.)

Use trigonometry to work out the angle:

$\tan\theta = \frac{2}{6} \Rightarrow \theta = \tan^{-1}\left(\frac{2}{6}\right) = 18.4°$

So vector **a** has magnitude 6.32 and direction 18.4° below the horizontal.

The angle θ is usually measured between the vector and the horizontal.

In general, a vector **r** with magnitude r and direction θ can be written as $r\cos\theta\mathbf{i} + r\sin\theta\mathbf{j}$

The vector $\mathbf{r} = x\mathbf{i} + y\mathbf{j}$ has magnitude $r = |\mathbf{r}| = \sqrt{x^2 + y^2}$ and direction $\theta = \tan^{-1}\left(\frac{y}{x}\right)$

The magnitude of a vector **a** is denoted |**a**| or a.

You can Resolve in any two Perpendicular Directions — not just x and y

EXAMPLE Find the resultant of the forces shown in the diagram.

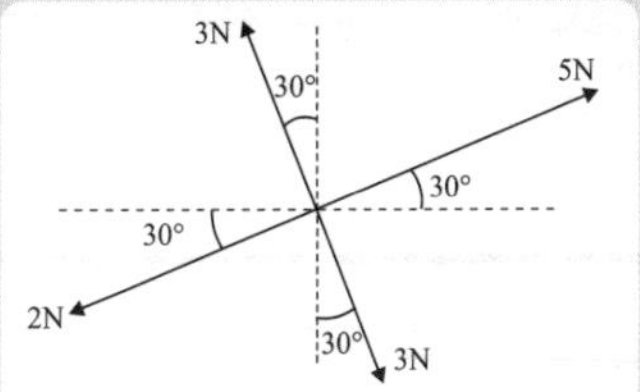

Resolving in ↖ direction: 3 N – 3 N = 0

Resolving in ↗ direction: 5 N – 2 N = 3 N

So the resultant force is 3 N in the direction of the 5 N force.

The two forces of 3 N are acting in opposite directions, so you take one away from the other — balancing them out.

Vectors

Vectors are much more than just a pretty face (or arrow). Once you deal with all the waffle in the question, all sorts of problems involving displacement, velocity, acceleration and forces can be solved using vectors.

***Draw a Diagram** if there are **Lots of Vectors** floating around*

In fact, draw a diagram when there are only a couple of vectors. But it's vital when there are lots of the little beggars.

EXAMPLE

A ship travels 100 km at a bearing of 025°, then 75 km at 140° before going 125 km at 215°. What is the displacement of the ship from its starting point?

Remember that bearings are always measured clockwise starting from north.

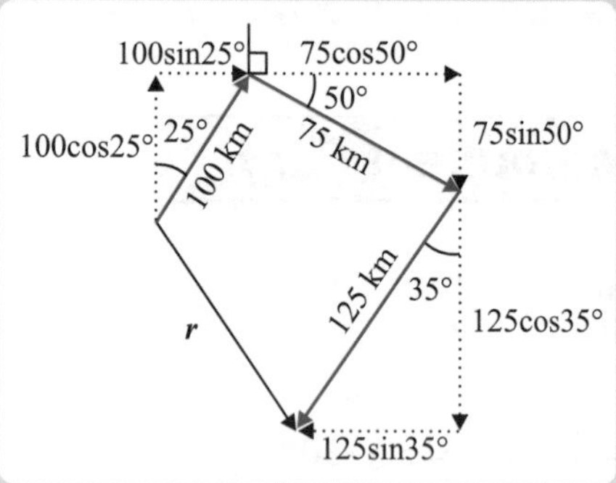

Resolve East: $100\sin25° + 75\cos50° - 125\sin35° = 18.8$ km

Resolve North: $100\cos25° - 75\sin50° - 125\cos35° = -69.2$ km

Magnitude of $\mathbf{r} = \sqrt{18.8^2 + (-69.2)^2} = 71.7$ km

direction $\theta = \tan^{-1}\left(\frac{69.2}{18.8}\right) = 74.8°$

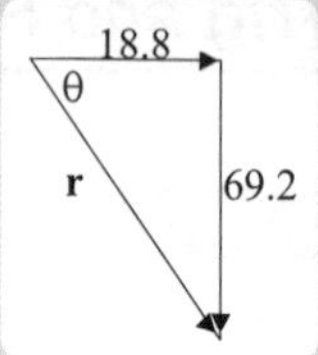

Bearing is $90° + 74.8° = 164.8°$

So the displacement is 71.7 km on a bearing of 164.8°

*The **Direction** part of a vector is **Really Important***

...and that means that you've got to make sure your diagram is spot on. These two problems look similar, and the final answers are pretty similar too. But look closely at the diagrams and you will see they are a bit different.

EXAMPLE

A canoe is paddled at 4 ms^{-1} in a direction perpendicular to the seashore. The sea current has a velocity of 1 ms^{-1} parallel to the shore. Find the resultant velocity **r** of the canoe.

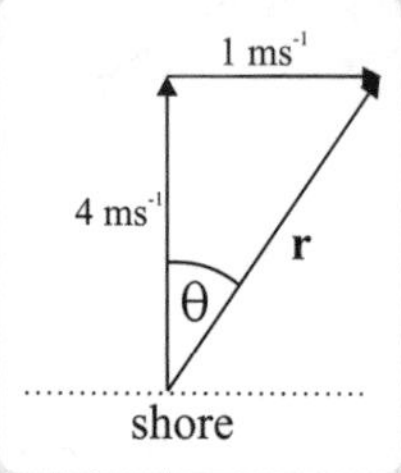

The resultant velocity **r** is the hypotenuse of the right-angled triangle.

Magnitude of $\mathbf{r} = \sqrt{4^2 + 1^2} = 4.12$ ms^{-1}

Direction: $\theta = \tan^{-1}\left(\frac{1}{4}\right) = 14.0°$ clockwise from the direction of paddling.

EXAMPLE

A canoe can be paddled at 4 ms^{-1} in still water. The sea current has a velocity of 1 ms^{-1} parallel to the shore. Find the angle at which the canoe must be paddled in order to travel in a direction perpendicular to the shore and the magnitude of the resultant velocity.

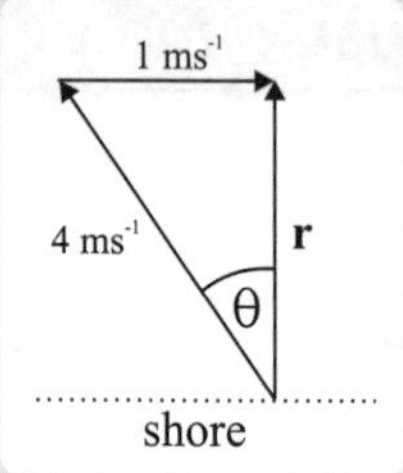

The resultant velocity **r** isn't the hypotenuse this time.

Magnitude of $\mathbf{r} = \sqrt{4^2 - 1^2} = 3.87$ ms^{-1}

Direction: $\theta = \sin^{-1}\left(\frac{1}{4}\right) = 14.5°$ anticlockwise from the perpendicular to the shore.

Bet you can't say 'perpendicular' 10 times fast...

Next time you're baffled by vectors, just start resolving and Bob's your mother's brother.

i + j Vectors

The only way to get your head around **i** + **j** vectors is to practise, then practise a bit more. I'm sure you get the idea.

For particles travelling at Constant Velocity, s = vt

If a particle is travelling at a constant velocity, v, then its displacement, s, after time, t, can be found by $s = vt$. This is handy for questions involving position vectors — these are vectors which describe the position of something relative to an origin.

EXAMPLE At $t = 0$ a particle has position vector $(6\mathbf{i} + 8\mathbf{j})$ m relative to a fixed origin O. The particle is travelling at constant velocity $(2\mathbf{i} - 6\mathbf{j})$ ms^{-1}. Find its position vector at $t = 4$ s.

First find its displacement using $\mathbf{s} = \mathbf{v}t$: $\mathbf{s} = 4(2\mathbf{i} - 6\mathbf{j}) = (8\mathbf{i} - 24\mathbf{j})$ m

Then add this to its original position vector:

$(6\mathbf{i} + 8\mathbf{j}) + (8\mathbf{i} - 24\mathbf{j}) = (14\mathbf{i} - 16\mathbf{j})$ m

You might need to use the Constant Acceleration Equations

If a particle is accelerating, you'll need to use the constant acceleration equations from page 158.

EXAMPLE Find the velocity and speed of a particle after 3 s if its initial velocity is $(6\mathbf{i} + 4\mathbf{j})$ ms^{-1} and acceleration is $(0.3\mathbf{i} + 0.5\mathbf{j})$ ms^{-2}.

Using $\mathbf{v} = \mathbf{u} + \mathbf{a}t$: $\mathbf{v} = (6\mathbf{i} + 4\mathbf{j}) + 3(0.3\mathbf{i} + 0.5\mathbf{j}) = (6\mathbf{i} + 4\mathbf{j}) + (0.9\mathbf{i} + 1.5\mathbf{j}) = (6.9\mathbf{i} + 5.5\mathbf{j})$ ms^{-1}

Speed = magnitude of $\mathbf{v} = \sqrt{6.9^2 + 5.5^2} = 8.82$ ms^{-1}

EXAMPLE At $t = 0$, a particle, P, has position vector $(3\mathbf{i} + \mathbf{j})$ m relative to a fixed origin, and velocity $(2\mathbf{i} - 5\mathbf{j})$ ms^{-1}, where **i** and **j** are the horizontal and vertical unit vectors respectively.

i) Given that P accelerates at a rate of $(-2\mathbf{i} + \mathbf{j})$ ms^{-2}, find its position vector at $t = 6$ s.

ii) Find the average velocity of the particle during this time.

iii) Find the velocity of the particle and the value of t when it is travelling parallel to **i**.

i) Use $\mathbf{s} = \mathbf{u}t + \frac{1}{2}\mathbf{a}t^2$:

$\mathbf{s} = 6(2\mathbf{i} - 5\mathbf{j}) + \frac{1}{2}(6^2)(-2\mathbf{i} + \mathbf{j}) = (12\mathbf{i} - 30\mathbf{j}) + (-36\mathbf{i} + 18\mathbf{j}) = (-24\mathbf{i} - 12\mathbf{j})$ m

New position vector $= (3\mathbf{i} + \mathbf{j}) + (-24\mathbf{i} - 12\mathbf{j}) = (-21\mathbf{i} - 11\mathbf{j})$ m

ii) Use $\mathbf{v} = \frac{\mathbf{s}}{t}$ where **s** is the difference between the final and initial position vectors (calculated in part i)): $\mathbf{v}_{\text{average}} = \frac{(-24\mathbf{i} - 12\mathbf{j})}{6} = (-4\mathbf{i} - 2\mathbf{j})$ ms^{-1}.

iii) Consider the horizontal and vertical components of motion separately:

Vertically:

When the particle is travelling parallel to **i**, the component of its velocity in the direction of **j** is zero. So:

$u_v = -5$, $v_v = 0$, $a_v = 1$

$v_v = u_v + a_v t \Rightarrow 0 = -5 + t \Rightarrow t = 5$

u_v, v_v and a_v are the vertical components of **u**, **v** and **a** respectively.

Horizontally:

$u_h = 2$, $a_h = -2$, $t = 5$.

$v_h = u_h + a_h t$

$\Rightarrow v_h = 2 + (-2 \times 5) = -8$

v_h is non-zero, so you know the particle is definitely moving.

So the velocity of the particle when it's travelling parallel to **i** is $-8\mathbf{i}$ ms^{-1}, and this occurs at time $t = 5$ s.

M1 Section 1 — Practice Questions

Time for the first set of M1 questions — just a few to get you up to speed gently.
A bit of advice — don't panic, take it step by step, and above all ~~don't get hurt~~ keep practising until it's second nature.

Warm-up Questions

Take $g = 9.8$ ms^{-2} in each of these questions.

1) A motorcyclist accelerates uniformly from 3 ms^{-1} to 9 ms^{-1} in 2 seconds.
What is the distance travelled by the motorcyclist during this acceleration?

2) A runner starts from rest and accelerates at 0.5 ms^{-2} for 5 seconds. She maintains a constant velocity for 20 seconds then decelerates to a stop at 0.25 ms^{-2}.
Draw a *v/t* graph to show the motion and find the distance the runner travelled.

3) The start of a journey is shown on the *a/t* graph to the right.
Find the velocity when: a) $t = 3$ b) $t = 5$ c) $t = 6$

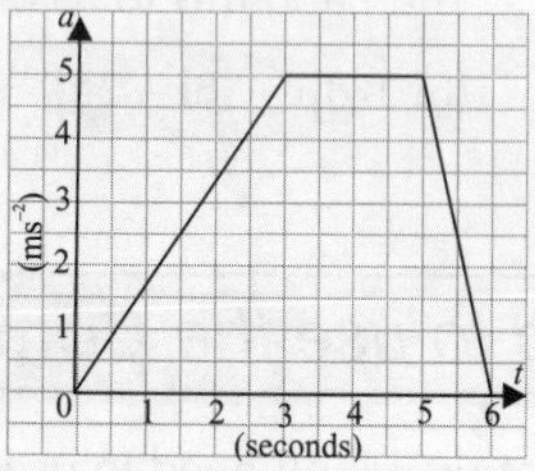

4) Find the average velocity of a cyclist who cycles at 15 kmh^{-1} north for 15 minutes and then cycles south at 10 kmh^{-1} for 45 minutes.

5) Find $\mathbf{a} + 2\mathbf{b} - 3\mathbf{c}$ where $\mathbf{a} = 3\mathbf{i} + 7\mathbf{j}$; $\mathbf{b} = -2\mathbf{i} + 2\mathbf{j}$; $\mathbf{c} = \mathbf{i} - 3\mathbf{j}$.

6) A plane flies 40 miles due south, then 60 miles southeast before going 70 miles on a bearing of 020°. Find the distance and bearing on which the plane must fly to return to its starting point.

Hopefully those questions above were no trouble, so it's time to have a go at the kind of questions you're likely to see in the exam. You'll find them below, in a different font and everything...

Exam Questions

Take $g = 9.8$ ms^{-2} in each of these questions.

1 A window cleaner of a block of flats accidentally drops his sandwich, which falls freely to the ground. The speed of the sandwich as it passes a high floor is u. After a further 1.2 seconds the sandwich is moving at a speed of 17 ms^{-1}. The distance between the consecutive floors of the building is h.

a) Find the value of u.

(3 marks)

b) It takes the sandwich another 2.1 seconds to fall the remaining 14 floors to the ground. Find h.

(4 marks)

2 A train starts from rest at station A and travels with constant acceleration for 20 s. It then travels with constant speed V ms^{-1} for 2 minutes. It then decelerates with constant deceleration for 40 s before coming to rest at station B. The total distance between stations A and B is 2.1 km.

a) Sketch a speed-time graph for the motion of the train between stations A and B. *(3 marks)*

b) Hence or otherwise find the value of V. *(3 marks)*

c) Calculate the distance travelled by the train while decelerating. *(2 marks)*

d) Sketch an acceleration-time graph for this motion. *(3 marks)*

M1 Section 1 — Practice Questions

There's more where that came from. Oh yes indeed...

3 A rocket is projected vertically upwards from a point 8 m above the ground at a speed of u ms^{-1} and travels freely under gravity. The rocket hits the ground at 20 ms^{-1}. Find:

a) the value of u, *(3 marks)*

b) how long it takes to hit the ground. *(3 marks)*

4 A girl can swim at 2 ms^{-1} in still water. She is swimming across a river in a direction perpendicular to the riverbank. The river is flowing at 3 ms^{-1} and so it carries the girl downstream.

Find the magnitude of the resultant velocity of the girl and the angle it makes with the riverbank. *(3 marks)*

5 A particle is initially at point O, with position vector $\mathbf{s}_O = (\mathbf{i} + 2\mathbf{j})$ m. It travels with constant velocity $(3\mathbf{i} + \mathbf{j})$ ms^{-1}. After 8 s it reaches point A. A second particle has constant velocity $(-4\mathbf{i} + 2\mathbf{j})$ ms^{-1} and takes 5 s to travel from point A to point B.

Find $\mathbf{s}_A$ and $\mathbf{s}_B$, the position vectors of points A and B. *(4 marks)*

6 At $t = 0$, a particle P is at position vector $(\mathbf{i} + 5\mathbf{j})$ m, relative to a fixed origin, where $\mathbf{i}$ and $\mathbf{j}$ are the unit vectors directed east and north respectively. P is moving with a constant velocity of $(7\mathbf{i} - 3\mathbf{j})$ ms^{-1}. After 4 s the velocity of P changes to $(a\mathbf{i} + b\mathbf{j})$ ms^{-1}. After a further 3.5 seconds P reaches position vector $(15\mathbf{i})$ m. Find:

a) the speed of P at $t = 0$, *(2 marks)*

b) the bearing of P at $t = 0$, *(3 marks)*

c) the values of a and b. *(4 marks)*

7 A particle moves on a smooth horizontal plane. It is initially at point P, with position vector $(\mathbf{i} + \mathbf{j})$ m relative to a fixed origin. At P, the particle has velocity $(3\mathbf{i} - \mathbf{j})$ ms^{-1}. The particle moves with constant acceleration for 10 seconds until it reaches point Q, with position vector $(6\mathbf{i} + 11\mathbf{j})$ m.

a) Find the particle's acceleration. *(3 marks)*

b) Show that 6 seconds after the particle leaves P, it is moving parallel to the unit vector $\mathbf{j}$. *(3 marks)*

Forces and Modelling

M1 questions talk about forces all the time, so you need to understand what each type of force is. Then you can use that information to create a model to work from.

Hint: 'modelling' in maths doesn't have anything to do with plastic aeroplane kits... or catwalks.

Types of forces

WEIGHT (W)

Due to the particle's mass, m and the force of gravity, g: $W = mg$ — weight always acts downwards.

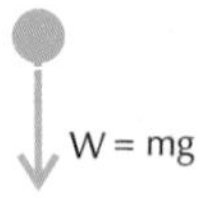

THE NORMAL REACTION (R OR N)

The reaction from a surface. Reaction is always at 90° to the surface.

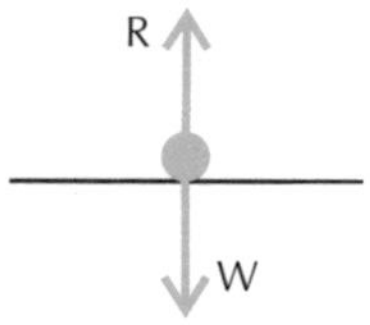

TENSION (T)

Force in a taut rope, wire or string.

FRICTION (F)

Due to the roughness between a body and a surface. Always acts against motion, or likely motion.

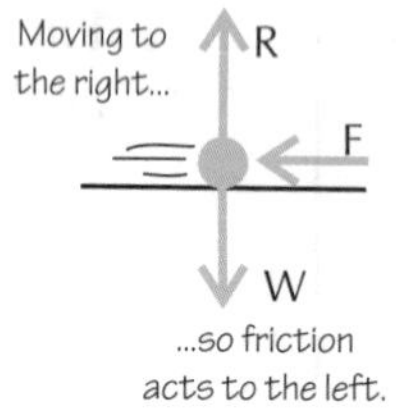

THRUST

Force in a rod (e.g. the pole of an open umbrella).

Talk the Talk

Maths questions in M1 use a lot of words that you already know, but here they're used to mean something very precise. Learn these definitions so you don't get caught out:

Particle	the body is a point so its dimensions don't matter	**Rigid**	the body does not bend
Light	the body has no mass	**Thin**	the body has no thickness
Static	not moving	**Equilibrium**	nothing's moving
Rough	the surface will oppose motion with friction / drag	**Plane**	a flat surface
Beam or Rod	a long, straight body (e.g. a broom handle)	**Inextensible**	the body can't be stretched
Uniform	the mass is evenly spread out throughout the body	**Smooth**	the surface doesn't have friction / drag opposing motion
Non-uniform	the mass is unevenly spread out		

Mathematical Modelling

You'll have to make lots of assumptions in M1. Doing this is called 'modelling', and you do it to make a sticky real-life situation simpler.

EXAMPLE **The ice hockey player**

You might have to assume:

- no friction between the skates and the ice
- no drag (air resistance)
- the skater generates a constant forward force S
- the skater is very small (a point mass)
- there is only one point of contact with the ice
- the weight acts downwards

Tom Horton — coffee shop owner

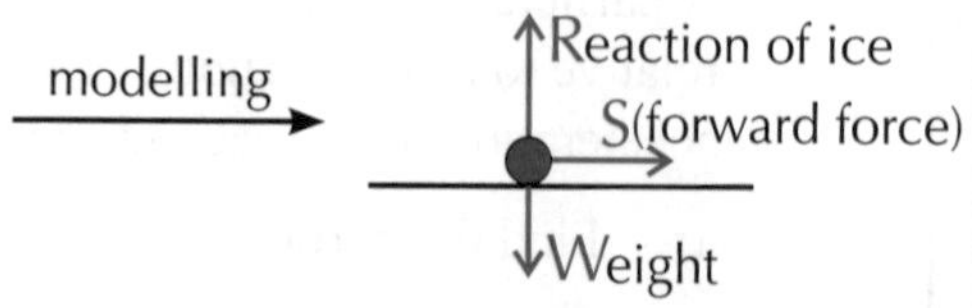

Tom as a **point mass** with forces

The simplified model you end up with can then be used as your vector diagram.

Forces and Modelling

Modelling is a *Cycle*

Having created a model you can later improve it by making more (or fewer) assumptions.
Solve the problem using the initial assumptions, compare it to real life, evaluate the model and then use that information to change the assumptions. Then keep going until you're satisfied with the model.

Always start by drawing a *Simple Diagram* of the *Model*

Here are a couple of old chestnuts that often turn up in M1 exams in one form or another.

EXAMPLE
The book on a table

A book is put flat on a table. One end of the table is slowly lifted and the angle to the horizontal is measured when the book starts to slide. What assumptions might you make?

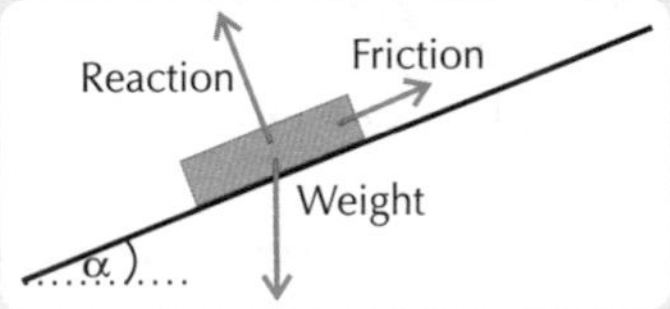

Assumptions:
The book is rigid, so it doesn't bend or open.
The book is a particle, so its dimensions don't matter.
There's no wind or other external forces involved.

EXAMPLE
The pulley

Two particles are connected by a string that passes over a fixed peg. The particles are released from rest. Draw a model of the forces. What assumptions have you made?

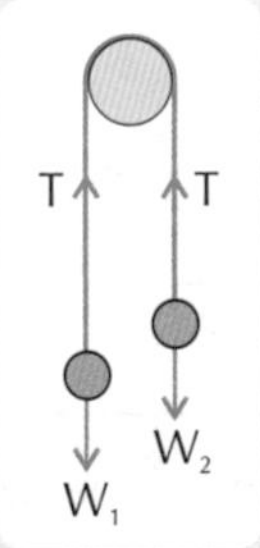

Assumptions:
The peg is smooth, so there's no friction.
The string is light, so its mass can be ignored.
The string is inextensible, so it doesn't stretch.
The tension in the string is the same either side of the pulley.

EXAMPLE
The sledge

A sledge is being steadily pulled by a small child on horizontal snow. Draw a force diagram for a model of the sledge. List your assumptions.

It's quicker and easier to use just the first letter of the force in your diagram, e.g. *F* = friction.

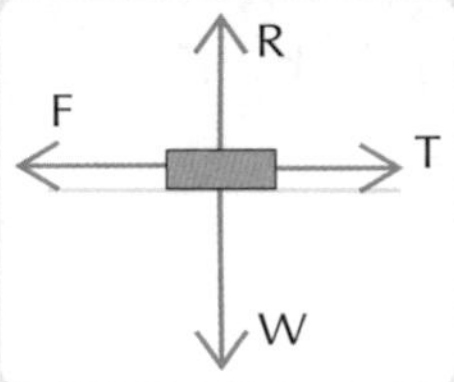

Assumptions:
Friction is too big to be ignored (i.e. it's not ice).
The string is horizontal (it's a small child).
The sledge is a small particle (so its size doesn't matter).

EXAMPLE
The mass on a string

A ball is held by two strings, A and B, at angles α and β to the vertical. Draw a diagram to model this scenario. State your assumptions.

Assumptions:
The ball is modelled as a particle (its dimensions don't matter).
The strings are light (their mass can be ignored).
The strings are inextensible (they can't stretch).

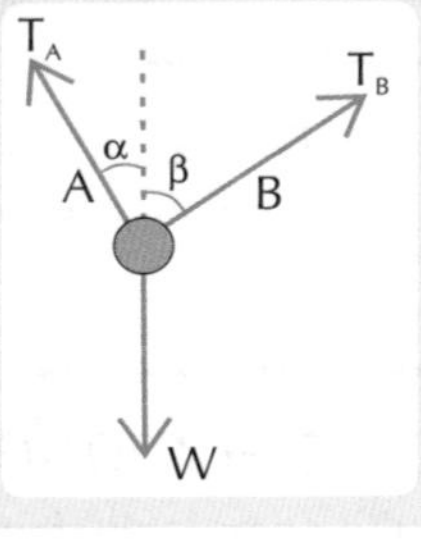

Forces and Modelling

Now for some proper examples — yeah baby.

You Might be asked to **Comment** on your **Model**

EXAMPLE

A wooden block of mass m kg is held at rest on a plane inclined at an angle of 30° to the horizontal. The block is then released from rest, and slides down the plane. A model of this situation makes the assumption that the only forces acting on the block are its weight and the normal reaction to the plane, N.

a) Draw and label a diagram showing the forces acting on the block and its motion.

The block slides down the plane with constant acceleration, moving a distance of 8 m in 4 s. The model predicts that the block will accelerate at 4.9 ms^{-2}.

b) Find the actual acceleration of the block.

c) Explain the difference between the actual acceleration and the value predicted by the model.

a)

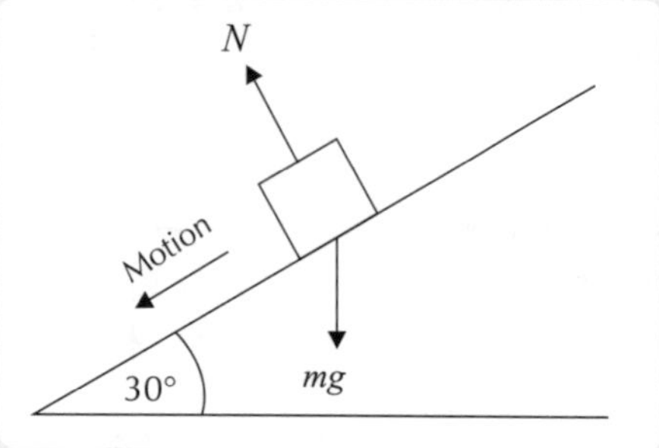

b) Taking down the slope as positive:

$u = 0,\ s = 8,\ t = 4,\ a = ?.$

$$s = ut + \frac{1}{2}at^2$$

$$\Rightarrow 8 = \frac{16}{2}a$$

$$\Rightarrow a = 1 \text{ ms}^{-2}$$

You're told in the question that acceleration is constant, so you should be on the lookout to use one of the 'uvast' equations.

c) There will be resistive forces such as friction and air resistance acting on the block which will reduce the acceleration, but these forces aren't included in the model.

EXAMPLE

a) A ball is dropped from a height of 5 m above the ground.

(i) How long does it take to reach the ground?

(ii) What assumptions have you made in modelling the ball's motion?

b) A larger ball is dropped from the same height. Using the same modelling assumptions, how would the time for this ball to reach the ground to compare to your answer to part a) (i)? Explain your answer.

a) (i) Taking down as positive:

$u = 0,\ a = 9.8,\ s = 5,\ t = ?$

$$s = ut + \frac{1}{2}at^2 \Rightarrow 5 = \frac{9.8}{2}t^2$$

$$t^2 = \frac{10}{9.8} \Rightarrow t = 1.01 \text{ s}$$

(ii)

- Acceleration due to gravity, g, is constant.
- There are no external forces (e.g. wind, air resistance) acting on the particle,
- The ball is modelled as a particle (its dimensions don't matter).

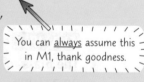

You can always assume this in M1, thank goodness.

b) Using the same modelling assumptions, the time taken for the larger ball to reach the ground would be the same as for the smaller ball. This is because both balls were modelled as particles, i.e. a mass with no size.

I used to be a model when I was younger, but then I fell apart...

Make sure you're completely familiar with the different forces and all the jargon that gets bandied about in mechanics. Keep your models as simple as possible — that will make answering the questions as simple as possible too.

Forces are Vectors

Forces have direction and magnitude, which makes them vectors — this means that all the stuff you learnt about vectors you'll need here. To help you out, we've given you some more vector examples all about forces...

Forces have ***Components***

You've done a fair amount of trigonometry already, so this should be as straightforward as watching dry paint.

EXAMPLE A particle is acted on by a force of 15 N at 30° above the horizontal. Find the horizontal and vertical components of the force.

A bit of trigonometry is all that's required:

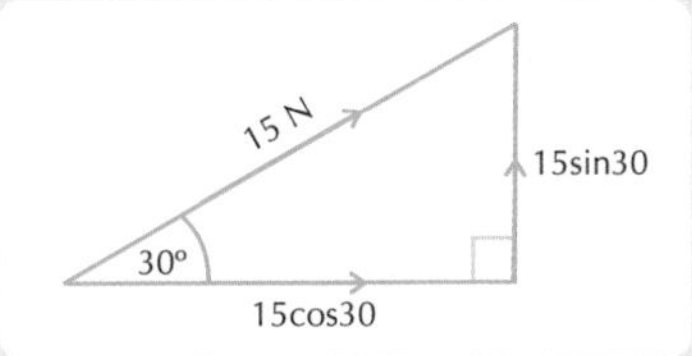

Force = 15cos30°$\mathbf{i}$ + 15sin30°$\mathbf{j}$

= (13.0 $\mathbf{i}$ + 7.5 $\mathbf{j}$) N

(i.e. 13 N horizontally and 7.5 N upwards)

Add Forces Nose to Tail *to get the* ***Resultant***

The important bit when you're drawing a diagram to find the resultant is to make sure the arrows are the right way round. Repeat after me: nose to tail, nose to tail, nose to tail.

EXAMPLE A second horizontal force of 20 N to the right is also applied to the particle in the example above. Find the resultant of these forces.

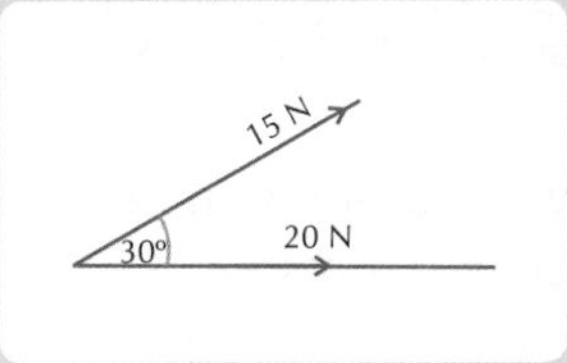

You need to put the arrows nose to tail:

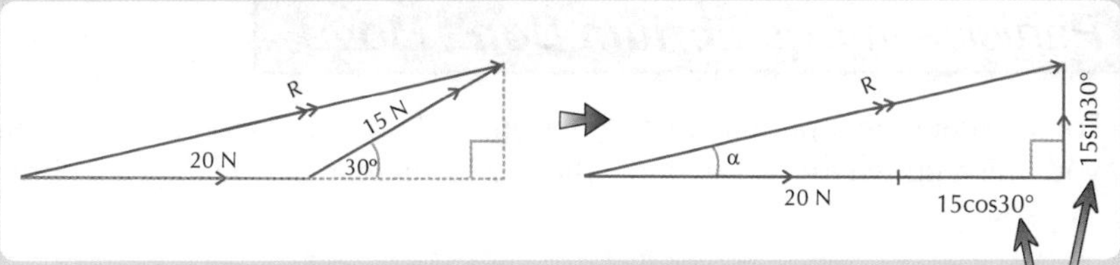

The 15 N force has been split into horizontal and vertical components.

Using Pythagoras and trigonometry:

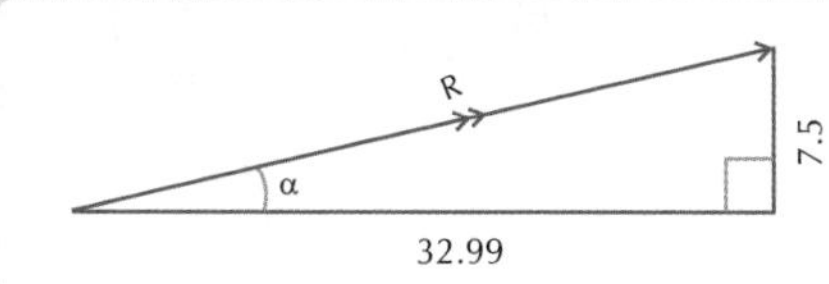

$R = \sqrt{32.99^2 + 7.5^2} = 33.8$ N

$\alpha = \tan^{-1}\left(\frac{7.5}{32.99}\right) = 12.8°$ above the horizontal

EXAMPLE Find the magnitude and direction of the resultant of the forces shown acting on the particle.

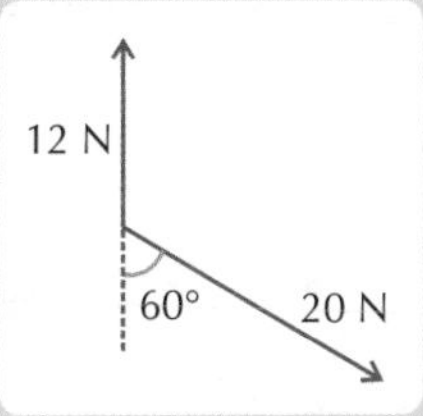

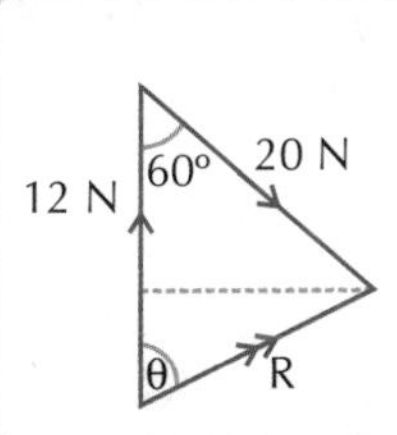

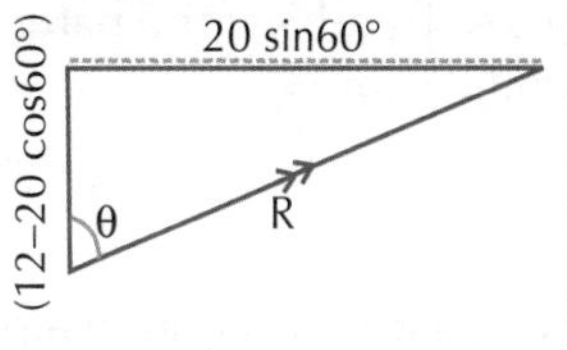

Hint: you could also use the cosine rule here to get R.

$R = \sqrt{(12 - 20\cos 60°)^2 + (20\sin 60°)^2} = 17.4$ N

$\theta = \tan^{-1}\left(\frac{20\sin 60°}{12 - 20\cos 60°}\right) = 83.4°$ to the vertical

If a particle is released it will move in the direction of the resultant.

Forces are Vectors

Resolving more than Two Forces

A question could involve more than two forces. You still work it out the same though — resolve, resolve, resolve...

EXAMPLE Three forces of magnitudes 9 N, 12 N and 13 N act on a particle P in the directions shown in the diagram.
Find the magnitude and direction of the resultant of the three forces.

One of the forces is already aligned with the y-axis, so it makes sense to start by resolving the other forces relative to this.

Along the y-axis:

Sum of components = 9sin35° + 12sin50° – 13
= **1.355 N** (4 s.f.)

Along the x-axis:

Sum of components = 12cos50° – 9cos35°
= **0.3411 N** (4 s.f.)

Magnitude of resultant = $\sqrt{1.355^2 + 0.3411^2}$
= 1.40 N (3 s.f.)

Direction of resultant:
$\theta = \tan^{-1}\frac{1.355}{0.3411}$ = 75.9° above the positive x-axis.

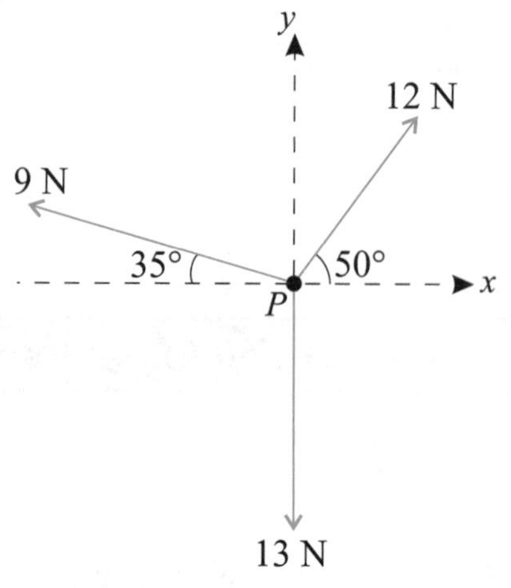

Particles in Equilibrium Don't Move

Forces acting on a particle in equilibrium add up to zero force. That means when you draw all the arrows nose to tail, you finish up where you started. That's why diagrams showing equilibrium are called 'polygons of forces'.

EXAMPLE Two perpendicular forces of magnitude 20 N act on a particle. A third force, P, acts at 45° to the horizontal, as shown. Given that the particle is in equilibrium, find the magnitude of P.

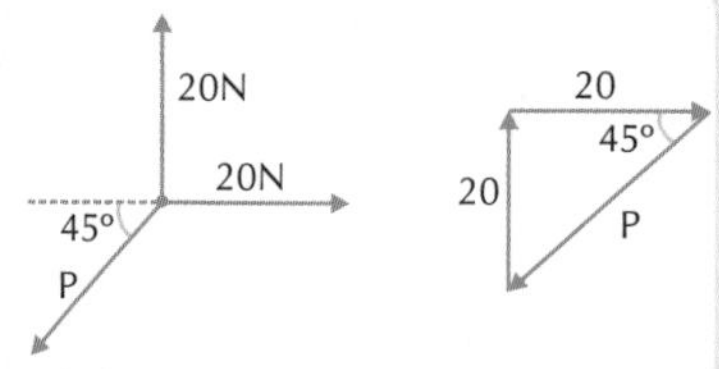

$P\cos 45° = 20$

$P = 28.3$ N

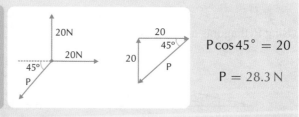

EXAMPLE A force of 50 N acts on a particle at an angle of 20° to the vertical, as shown. Find the magnitude of the two other forces, T and S, if the particle is in equilibrium.

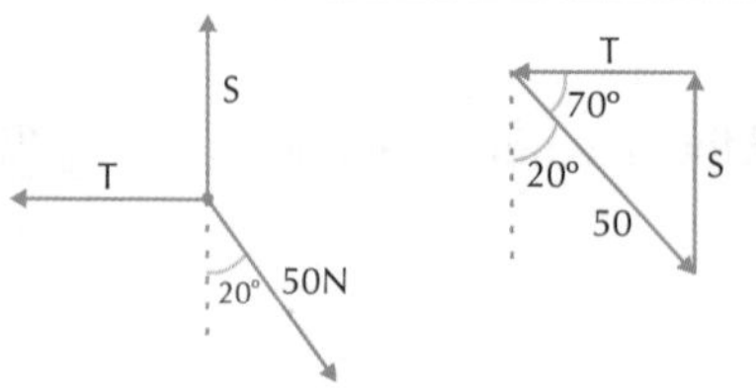

S = 50sin70° = 47.0 N
T = 50cos70° = 17.1 N

EXAMPLE Three forces act upon a particle. A force of magnitude 85 N acts horizontally, the force Q acts vertically, and the force P acts at 55° to the horizontal, as shown. The particle is in equilibrium. Find the magnitude of P and Q.

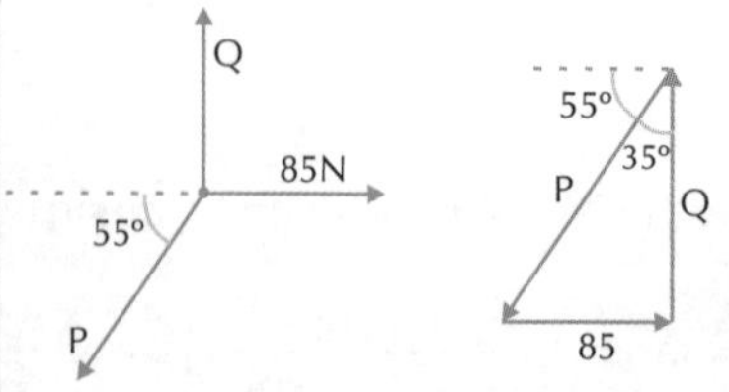

$\sin 35° = \frac{85}{P}$ so $P = 148$ N

$\tan 35° = \frac{85}{Q}$ so $Q = 121$ N

Forces are Vectors

An *Inclined Plane* is a *Sloping Surface*

EXAMPLE A sledge of weight 1000 N is being held on a rough inclined plane at an angle of 35° by a force parallel to the slope of 700 N acting up the slope and a frictional force, F acting down the slope. Find F and N, the normal contact force between the slope and the sledge.

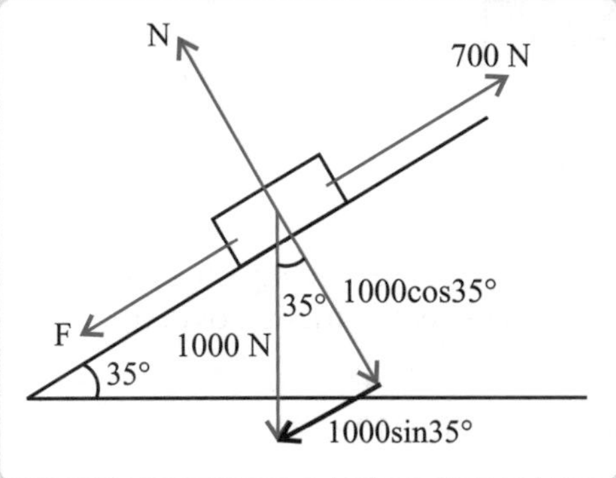

All the forces, except the weight, involved here act either parallel or perpendicular to the slope so it makes sense to resolve in these directions.

Perpendicular to the slope:

N – 1000cos35° = 0
So N = 1000cos35° = 819 N (to 3 s.f.

Parallel to the slope:

700 – 1000sin35° – F = 0
So F = 700 – 1000sin35° = 126 N (to 3 s.f.)

There's more about friction on the next page, and on pages 181-182. Gripping.

Masses on Strings *Produce Tension*

EXAMPLE A mass of 12 kg is held by two light strings, P and Q, acting at 40° and 20° to the vertical, as shown. Find T_P and T_Q, the tension in each string. Take $g = 9.8$ ms^{-2}.

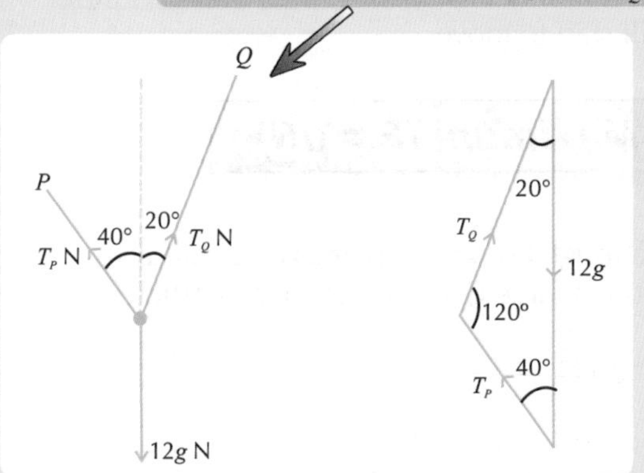

Sine rule:

$$\frac{T_P}{\sin 20°} = \frac{12g}{\sin 120°}$$

So $T_P = 46.4$ N

$$\frac{T_Q}{\sin 40°} = \frac{12g}{\sin 120°}$$

So $T_Q = 87.3$ N

Always look out for sine rule triangles in your polygons of forces.

i + j vectors can be used to Describe Forces

Treat forces described in terms of **i** + **j** just the same as any other force or vector you deal with in M1.

EXAMPLE A force acts on a particle P in a direction which is parallel to the vector (3**i** – 2**j**). Find the angle between the vector **j** and the force. (The unit vectors **i** and **j** are due east and due north respectively.)

The force is parallel to (3**i** – 2**j**), so the angle between it and **j** is the same as the angle between (3**i** – 2**j**) and **j**:

$\tan\theta = \frac{2}{3}$ and so θ = 33.7° south of east.
So, angle to **j** is 33.7 + 90 = 123.7° = 124° (to 3 s.f.)

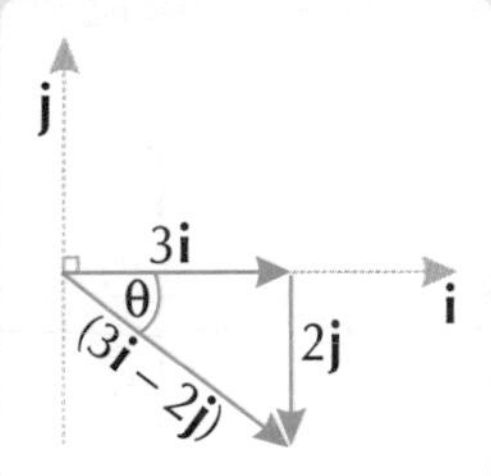

Cliché #27 — "The more things change, the more they stay the same"...

As it makes their lives easier, examiners always stick the same kinds of questions into M1 exams. So, no need to panic — just keep practising the questions and then there'll be no surprises when it comes to the exam. Simple.

Friction

Friction tries to prevent motion, but don't let it prevent you getting marks in the exam — revise this page and it won't.

Friction Tries to **Prevent Motion**

Push hard enough and a particle will move, even though there's friction opposing the motion. A friction force, F, has a maximum value. This depends on the roughness of the surface and the value of the normal reaction from the surface.

$$F \leq \mu R \quad \text{OR} \quad F \leq \mu N$$

(where R and N both stand for normal reaction)

μ has no units.
μ is pronounced as 'mu'.

μ is called the "coefficient of friction". The rougher the surface, the bigger μ gets.

EXAMPLE

What range of values can a friction force take in resisting a horizontal force P acting on a particle Q, of mass 12 kg, resting on a rough horizontal plane which has a coefficient of friction of 0.4? (Take $g = 9.8$ ms^{-2}.)

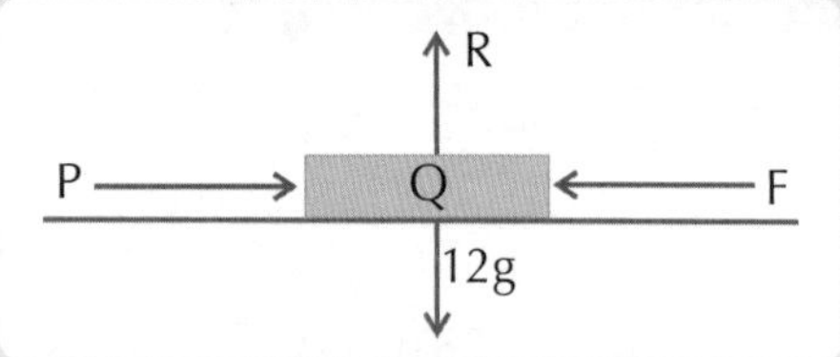

Resolving vertically: $R = 12g$

Use formula from above: $F \leq \mu R$

$F \leq 0.4(12g)$

$F \leq 47.0$ N (3 s.f.)

So friction can take any value between 0 and 47.0 N, depending on how large P is.

If $P \leq 47.0$ N then Q remains in equilibrium. If $P = 47.0$ N then Q is on the point of sliding — i.e. friction is at its limit. If $P > 47.0$ then Q will start to move.

Limiting Friction is When Friction is at *Maximum* ($F = \mu R$)

EXAMPLE

A particle of mass 6 kg is placed on a rough horizontal plane which has a coefficient of friction of 0.3. A horizontal force Q is applied to the particle. Describe what happens if Q is: a) 16 N b) 20 N
Take $g = 9.8$ ms^{-2}.

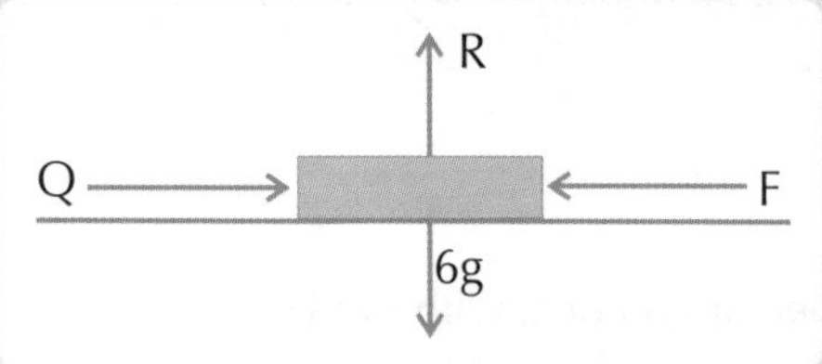

Resolving vertically: $R = 6g$

Using formula above: $F \leq \mu R$

$F \leq 0.3(6g)$

$F \leq 17.6$ N (3 s.f.)

a) Since $Q < 17.6$ it won't move.

b) Since $Q > 17.6$ it'll start moving. No probs.

EXAMPLE

A particle of mass 4 kg at rest on a rough horizontal plane is being pushed by a horizontal force of 30 N. Given that the particle is on the point of moving, find the coefficient of friction.

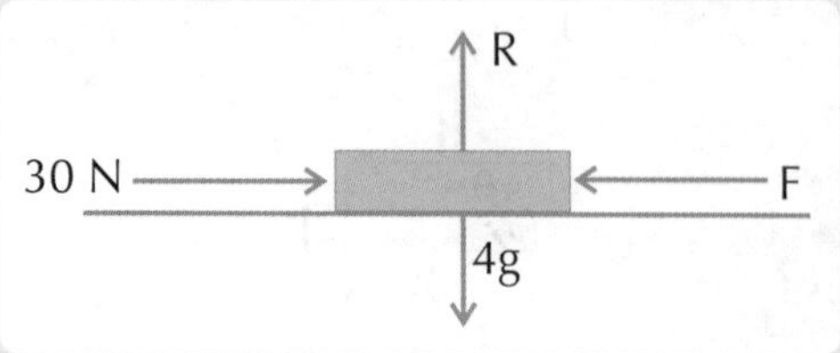

Resolving horizontally: $F = 30$

Resolving vertically: $R = 4g$

The particle's about to move, so friction is at its limit:

$F = \mu R$

$30 = \mu(4g)$

$\mu = \frac{30}{4g} = 0.765$ (3 s.f.)

Sometimes friction really rubs me up the wrong way...

Friction can be a right nuisance, but without it we'd just slide all over the place, which would be worse (I imagine).

M1 Section 2 — Practice Questions

Time to resolve the force applied to your revision along the question axis... That didn't sound as good as I hoped. Still, at least you can distract yourself from my terrible sense of humour with these excellent practice questions.

Warm-up Questions

Take $g = 9.8$ ms^{-2} in each of these questions.

1) The following items are dropped from a height of 2 m onto a cushion:

   a) a full 330 ml drinks can b) an empty drinks can c) a table tennis ball

   The time each takes to fall is measured.
   Draw a model of each situation and list any assumptions which you've made.

2) A car is travelling at 25 mph along a straight level road.
   Draw a model of the situation and list any assumptions which you've made.

3) Find the magnitudes and directions to the horizontal of the resultant force in each situation.

   a)
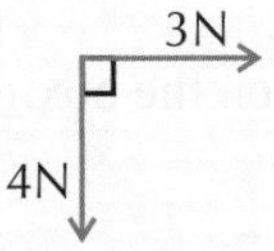

   b)
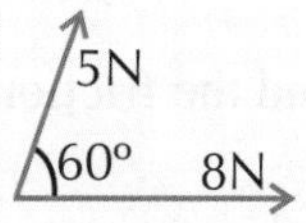

   c)
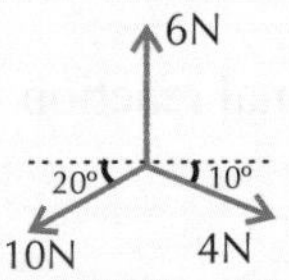

4) A mass of m kg is suspended by two light wires A and B, with angles 60° and 30° to the vertical respectively, as shown.
   The tension in A is 20 N. Find:

   a) the tension in wire B,

   b) the value of m.

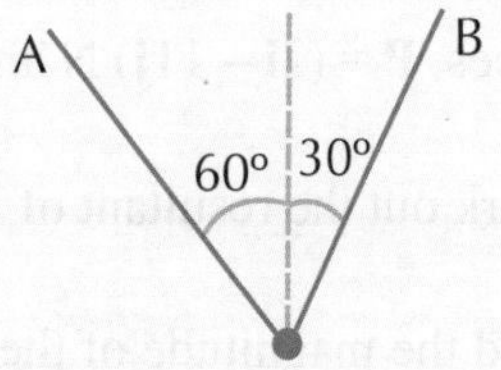

5)
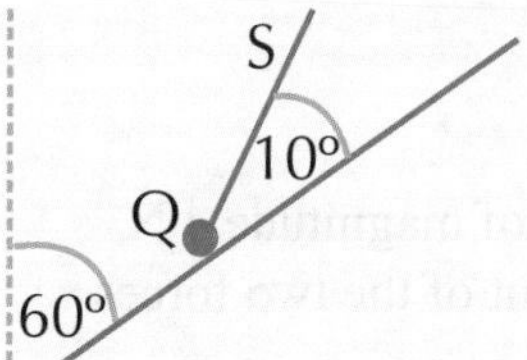

A particle, Q, of mass m kg, is in equilibrium on a smooth plane which makes an angle of 60° to the vertical. This is achieved by an attached string S, with tension 70 N, angled at 10° to the plane, as shown. Draw a force diagram and find both the mass of Q and the reaction on it from the surface.

6) A toy train of weight 25 N is pulled up a slope of 20°. The tension in the string is 25 N and a frictional force of 5 N acts on the train. Find the normal contact force and the resultant force acting on the train.

7) a) Describe the motion of a mass of 12 kg pushed by a force of 50 N parallel to the rough horizontal plane on which the mass is placed. The plane has coefficient of friction $\mu = \frac{1}{2}$.

   b) What minimum force would be needed to move the mass in part a)?

M1 Section 2 — Practice Questions

Now you can apply your vector and statics knowledge to some exam questions so that when you take the real thing there'll be no need for your pen to remain at equilibrium.

Exam Questions

Take $g = 9.8$ ms^{-2} in each of these questions.

1 The diagram shows two forces acting on a particle. Find the magnitude and direction of the resultant force.

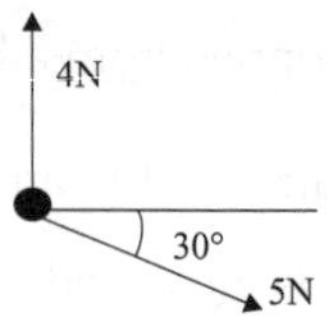

(4 marks)

2 A box of mass 39 kg is at rest on a rough horizontal surface. The box is pushed with a force of 140 N from an angle of 20° above the horizontal. The box remains stationary.

a) Draw a diagram to show the four forces acting on the box. *(2 marks)*

b) Calculate the magnitude of the normal reaction force and the frictional force on the box. *(4 marks)*

3 Two forces, $\mathbf{P} = (2\mathbf{i} - 11\mathbf{j})$ N and $\mathbf{Q} = (7\mathbf{i} + 5\mathbf{j})$ N, act on a particle.

a) Work out the resultant of **P** and **Q**. *(2 marks)*

b) Find the magnitude of the resultant force. *(2 marks)*

4 A force of magnitude 7 N acts horizontally on a particle. Another force, of magnitude 4 N, acts on the particle at an angle of 30° to the horizontal force. The resultant of the two forces has a magnitude R at an angle α above the horizontal.
Find:

a) R, *(4 marks)*

b) the angle α. *(2 marks)*

5 A particle, M, is attached to the end of a light inextensible rod fixed at point X. A force of magnitude 10 N is applied to M at 14° to the horizontal. M is held in equilibrium at an angle of 35° as shown in the diagram. Find the mass of M.

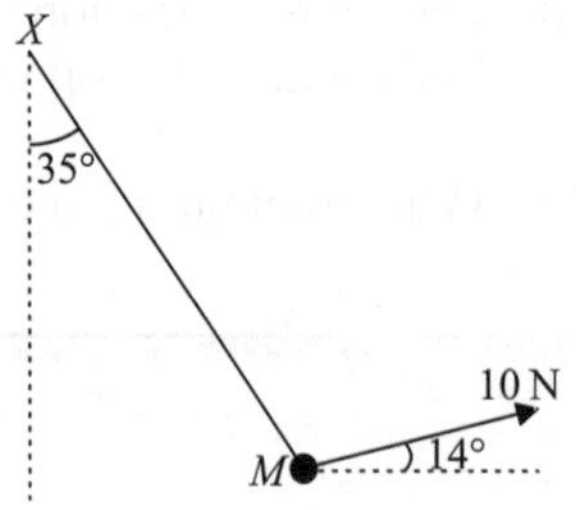

(4 marks)

M1 Section 2 — Practice Questions

Another page of questions to practise on here, so keep on truckin' (remembering that "truckin'" is a vector quantity with both magnitude *and* direction). The normal reaction to that joke is a sigh... on with the questions.

6 Three forces of magnitudes 15 N, 12 N and W act on a particle as shown. Given that the particle is in equilibrium, find:

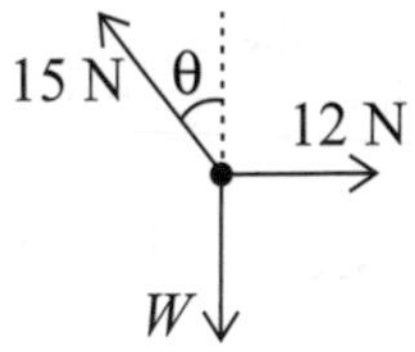

a) the value of θ *(2 marks)*

b) the force W. *(2 marks)*

The force W is now removed.

c) State the magnitude and direction of the resultant of the two remaining forces. *(2 marks)*

7 A sledge is held at rest on a smooth slope which makes an angle of 25° with the horizontal. The rope holding the sledge is at an angle of 20° above the slope. The normal reaction acting on the sledge due to contact with the surface is 80 N.

Find:

a) The tension, T, in the rope. *(3 marks)*

b) The weight of the sledge. *(2 marks)*

8 A particle of mass m kg is held in equilibrium by two light inextensible strings. One string is attached at an angle of 50° to the horizontal, as shown, and has tension T N. The other string is horizontal and has tension 58 N. Find:

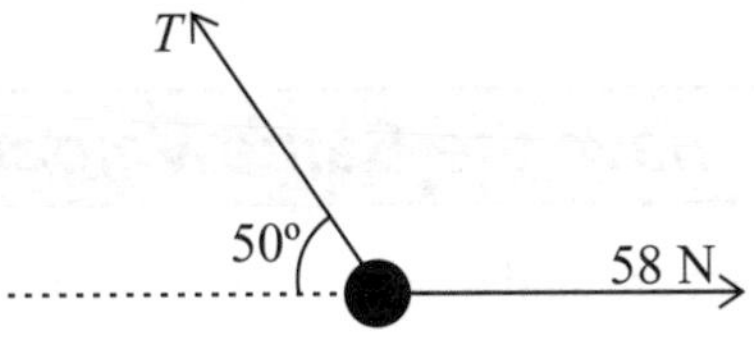

a) the magnitude of T. *(3 marks)*

b) the mass, m, of the particle. *(3 marks)*

9 A 2 kg ring threaded on a rough horizontal rod is pulled by a rope having tension S and attached at 40° to the horizontal, as shown. The coefficient of friction between the rod and the ring is $\frac{3}{10}$. Given that the ring is about to slide, find the magnitude of S.

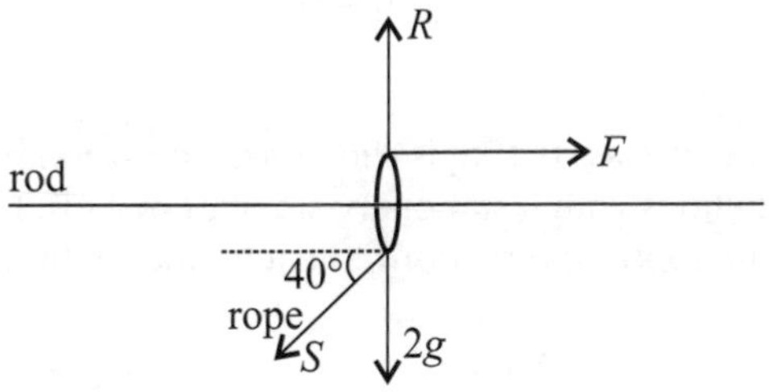

(4 marks)

Momentum

Momentum has *Magnitude and Direction*

Momentum is a measure of how much "umph" a moving object has, due to its mass and velocity.
Total momentum before a collision equals total momentum after a collision.
This idea is called "Conservation of Momentum".
Because it's a vector, the sign of the velocity in momentum is important.

Momentum = Mass × Velocity

The unit of momentum is kgms^{-1} or Ns

EXAMPLE

Particles A and B, each of mass 5 kg, move in a straight line with velocities 6 ms^{-1} and 2 ms^{-1} respectively. After colliding with B, A continues in the same direction with velocity 4.2 ms^{-1}. Find the velocity of B after impact.

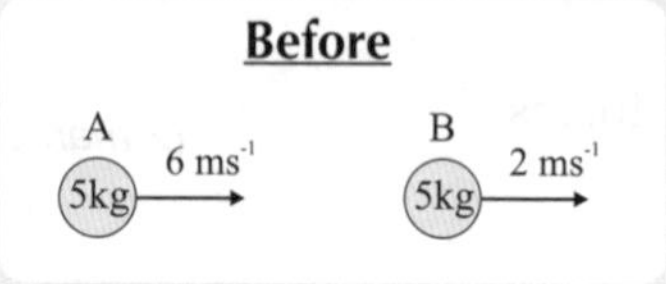

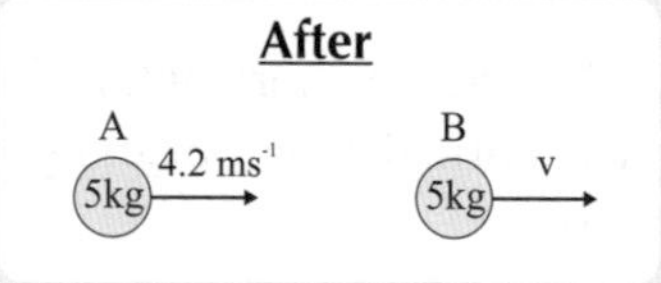

Momentum A + Momentum B = Momentum A + Momentum B

$(5 \times 6) + (5 \times 2) = (5 \times 4.2) + (5 \times v)$

$40 = 21 + 5v$

So $v = 3.8$ ms^{-1} in the same direction as before

Draw 'before' and 'after' diagrams to help you see what's going on.

Stick to saying 'same' or 'opposite' direction, rather than left or right — there's less chance of confusion.

EXAMPLE

Particles A and B of mass 6 kg and 3 kg are moving towards each other at speeds of 2 ms^{-1} and 1 ms^{-1} respectively. Given that B rebounds with speed 3 ms^{-1} in the opposite direction to its initial velocity, find the velocity of A after the collision.

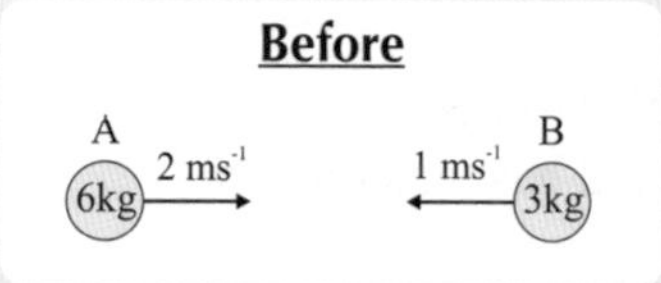

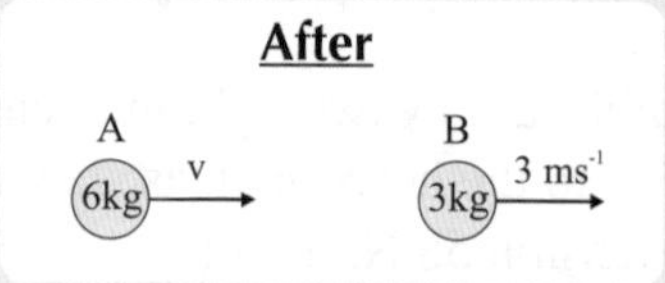

$(6 \times 2) + (3 \times -1) = (6 \times v) + (3 \times 3)$

$9 = 6v + 9$

$v = 0$

Masses *Joined Together* have the *Same Velocity*

Particles that stick together after impact are said to "coalesce". After that you can treat them as just one object.

EXAMPLE

Two particles of mass 40 g and M kg move towards each other with speeds of 6 ms^{-1} and 3 ms^{-1} respectively. Given that the particles coalesce after impact and move with a speed of 2 ms^{-1} in the same direction as that of the 40 g particle's initial velocity, find M.

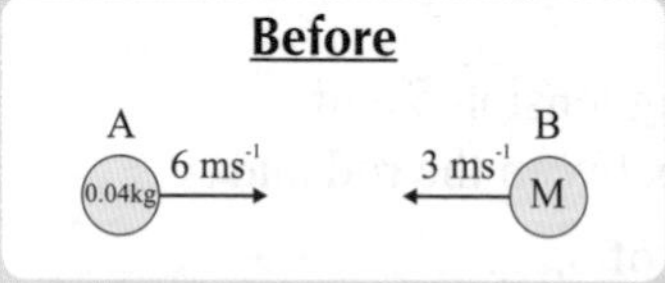

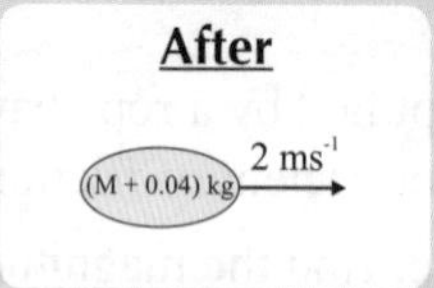

$(0.04 \times 6) + (M \times -3) = [(M + 0.04) \times 2]$

$0.24 - 3M = 2M + 0.08$

$5M = 0.16$

$M = 0.032$ kg

Don't forget to convert all masses to the same units.

EXAMPLE

A lump of ice of mass 0.1 kg is slid across the smooth surface of a frozen lake with speed 4 ms^{-1}. It collides with a stationary stone of mass 0.3 kg. The lump of ice and the stone then move in opposite directions to each other with the same speeds. Find their speed.

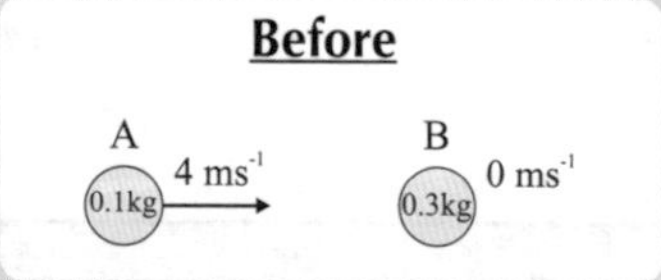

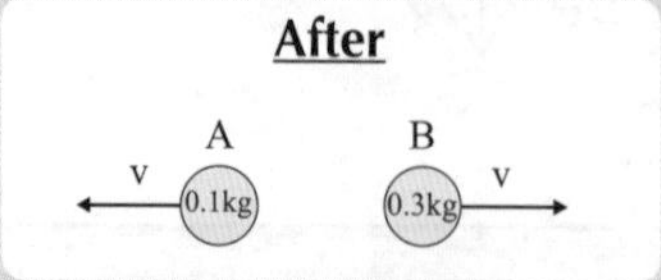

$(0.1 \times 4) + (0.3 \times 0) = (0.1 \times -v) + 0.3v$

$0.4 = -0.1v + 0.3v$

$v = 2$ ms^{-1}

Momentum

Momentum works the same in **2 Dimensions**

In your exam you could be asked about the momentum of particles moving on a plane.
Don't worry though — everything is the same as on the previous page, just with two components of velocity.
You'll usually be given the velocities as column vectors. They're just a different way of writing vectors: $\binom{x}{y} = x\mathbf{i} + y\mathbf{j}$.

EXAMPLE

Two particles, A and B, move on a smooth horizontal plane. A has mass m kg and velocity $\binom{U}{V}$ ms^{-1}. B has mass $3m$ kg and velocity $\binom{U}{4}$ ms^{-1}.
A and B collide and coalesce to form a single particle, C, with velocity $\binom{1}{5}$ ms^{-1}.
Find the values of U and V.

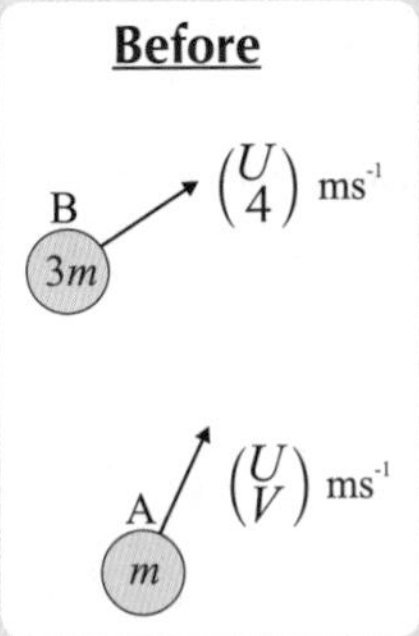

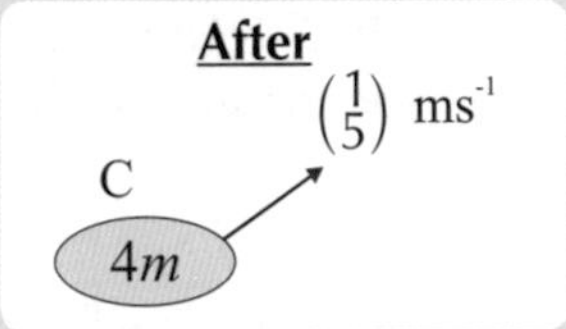

$$m\binom{U}{V} + 3m\binom{U}{4} = 4m\binom{1}{5}$$

$$\Rightarrow m\binom{U+3U}{V+12} = m\binom{4}{20}$$

Equate the tops and bottoms of the column vectors separately.

$$\Rightarrow \begin{matrix} 4U = 4 \\ V + 12 = 20 \end{matrix} \Rightarrow \begin{matrix} U = 1 \\ V = 8 \end{matrix}$$

EXAMPLE

A fly of mass 4 g and a bee of mass 6 g are involved in a collision.
Before the incident, the velocity of the fly was $\binom{5}{k}$ ms^{-1} (with $k < 0$) and the velocity of the bee was $\binom{1}{-1}$ ms^{-1}.

a) Given that the initial speed of the fly was 13 ms^{-1}, find the value of k.

Immediately following the collision, the velocity of the fly was measured to be $\binom{2}{0}$ ms^{-1}.

b) Determine the velocity of the bee immediately following the collision.

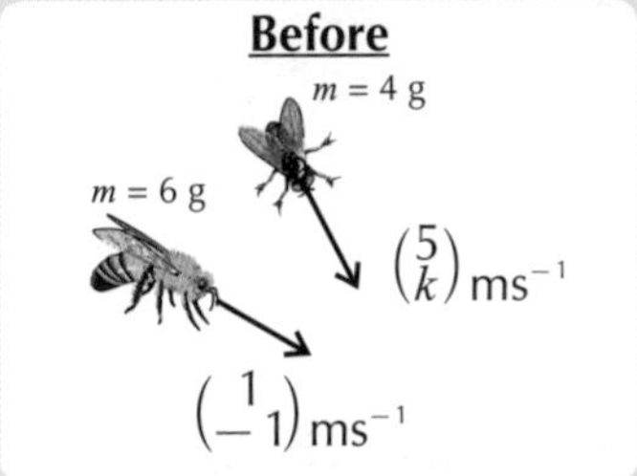

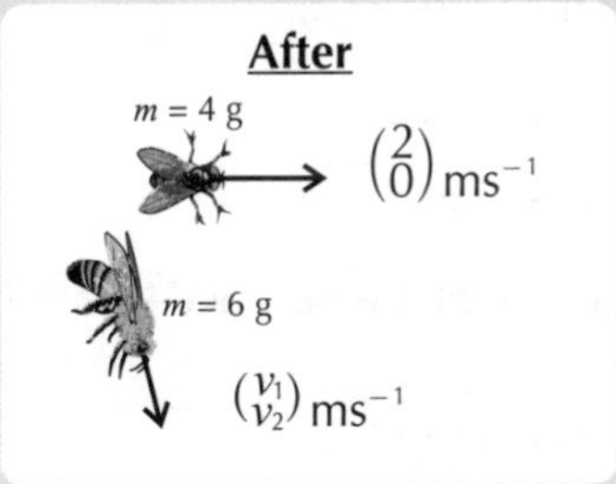

a) Use Pythagoras' Theorem:

$13 = \sqrt{5^2 + k^2} \Rightarrow 169 = 25 + k^2$
$\Rightarrow k = \pm\sqrt{144} = \pm 12$
You are given in the question that $k < 0$, so $k = -12$.

b) Use conservation of momentum:

$$4\binom{5}{-12} + 6\binom{1}{-1} = 4\binom{2}{0} + 6\binom{v_1}{v_2}$$

$$\Rightarrow 6\binom{v_1}{v_2} = \binom{20+6-8}{-48-6-0}$$

$$\Rightarrow \binom{v_1}{v_2} = \frac{1}{6}\binom{18}{-54} = \binom{3}{-9}\text{ms}^{-1}$$

It's okay to use grams for the mass here, as you're not asked to give the actual momentum in Ns. Just make sure that you use the same units in all parts of the working, otherwise you'll be in a pickle.

Ever heard of Hercules?

Well, he carried out 12 tasks. Nothing to do with momentum, but if you're feeling sorry for yourself for doing M1, think on.

Newton's Laws

That clever chap Isaac Newton established 3 laws involving motion. You need to know all of them.

Newton's Laws of Motion

Newton's First Law

A body will stay at rest or maintain a constant velocity — unless an extra force acts to change that motion.

Newton's Second Law

$F_{net} = ma$

F_{net} (the overall resultant force) is equal to the mass multiplied by the acceleration. Also, F_{net} and a are in the same direction.

Newton's Third Law

For two bodies in contact with each other, the force each applies to the other is equal in magnitude but opposite in direction.

Hint: $\boldsymbol{F_{net} = ma}$ is sometimes just written as $\boldsymbol{F = ma}$, but it means the same thing.

The resultant force acting on a body can also be found by calculating the rate of change of momentum of the body.

Resolve Forces in Perpendicular Directions

EXAMPLE A mass of 4 kg at rest on a smooth horizontal plane is acted on by a horizontal force of 5 N. Find the acceleration of the mass and the normal reaction from the plane. Take $g = 9.8$ ms^{-2}.

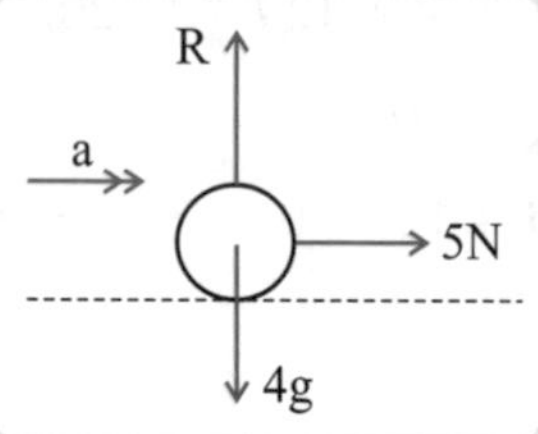

Resolve horizontally:

$F_{net} = ma$

$5 = 4a$

$a = 1.25$ ms^{-2} in the direction of the horizontal force

Always write $F_{net} = ma$ first.

Resolve vertically:

$F_{net} = ma$, so $R - 4g = 4 \times 0$

$R = 4g = 39.2$ N

EXAMPLE A particle of weight 30 N is being accelerated across a smooth horizontal plane by a force of 6 N acting at an angle of 25° to the horizontal, as shown. Given that the particle starts from rest, find:

a) its speed after 4 seconds b) the magnitude of the normal reaction with the plane.

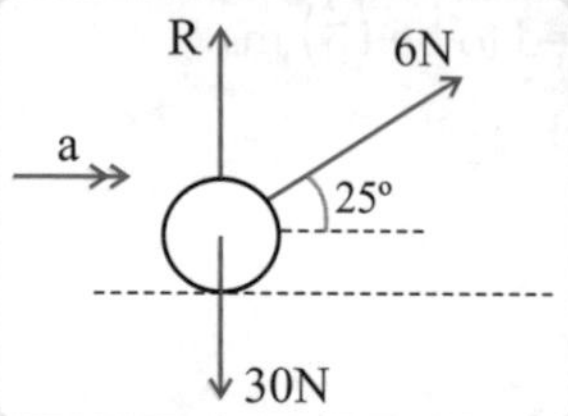

a) Resolve horizontally:

$F_{net} = ma$

$6\cos 25° = \frac{30}{g}a$ so $a = 1.776...$ ms^{-2}

$v = u + at$

$v = 0 + 1.776... \times 4 = 7.11$ ms^{-1} (to 3 s.f.)

b) Resolve vertically:

$F_{net} = ma$

$R + 6\sin 25° - 30 = \frac{30}{g} \times 0$

So $R = 30 - 6\sin 25° = 27.5$ N (to 3 s.f.)

You can apply F = ma to i + j Vectors too

EXAMPLE A particle of mass m kg is acted upon by two forces, $(6\mathbf{i} - \mathbf{j})$ N and $(2\mathbf{i} + 4\mathbf{j})$ N, resulting in an acceleration of magnitude 9 ms^{-2}. Find the value of m.

Resultant force, $\mathbf{R} = (6\mathbf{i} - \mathbf{j}) + (2\mathbf{i} + 4\mathbf{j}) = (8\mathbf{i} + 3\mathbf{j})$ N

Magnitude of $\mathbf{R} = |\mathbf{R}| = \sqrt{8^2 + 3^2} = \sqrt{73}$ N

$F_{net} = ma$, so $\sqrt{73} = 9m$

hence $m = 0.949$ kg (3 s.f.)

The force of $(2\mathbf{i} + 4\mathbf{j})$ N is removed from the particle. Calculate the magnitude of the new acceleration.

Magnitude $= \sqrt{6^2 + (-1)^2} = \sqrt{37}$ N

$a = \frac{F_{net}}{m} = \frac{\sqrt{37}}{0.9493} = 6.41$ ms^{-2} (3 s.f.)

Interesting Newton fact: Isaac Newton had a dog called Diamond...

Did you know that Isaac Newton and Stephen Hawking both held the same position at Cambridge University? And the dog fact about Newton is true — don't ask me how I know such things, just bask in my amazing knowledge of all things trivial.

Friction and Inclined Planes

Solving these problems involves careful use of $F_{net} = ma$, $F = \mu R$ and the constant acceleration equations.

Use F = ma in **Two Directions** for **Inclined Plane** questions

For inclined slope questions, it's much easier to resolve forces parallel and perpendicular to the plane's surface.

EXAMPLE A mass of 600 g is propelled up the line of greatest slope of a smooth plane inclined at 30° to the horizontal. If its velocity is 3 ms^{-1} after the propelling force has stopped, find the distance it travels before coming to rest and the magnitude of the normal reaction. Use $g = 9.8$ ms^{-2}.

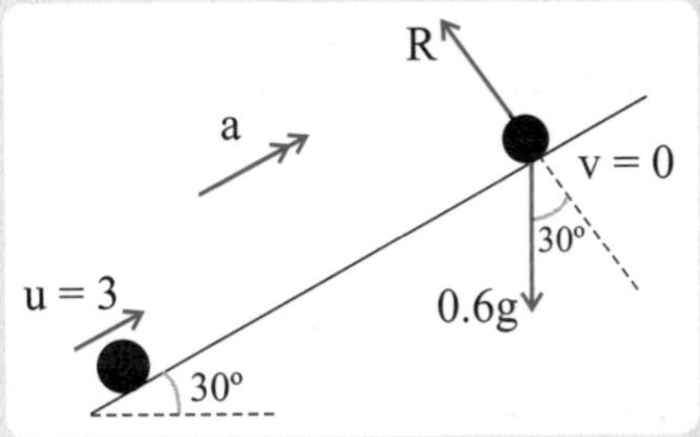

Resolve in ↖ direction:

$F_{net} = ma$

$R - 0.6g\cos 30° = 0.6 \times 0$

So $R = 5.09$ N (3 s.f.)

Resolve in ↗ direction:

$F_{net} = ma$

$-0.6g\sin 30° = 0.6a$ — Taking up the plane as +ve.

$\boldsymbol{a = -4.9}$ **ms^{-2}**

$v^2 = u^2 + 2as$

$0 = 3^2 + 2(-4.9)s$

So $s = 0.918$ m (3 s.f.)

Remember that friction always acts in the opposite direction to the motion.

EXAMPLE A small body of weight 20 N accelerates uniformly from rest and moves a distance of 5 m down a rough plane angled at 15° to the horizontal.

a) Assuming that air resistance can be ignored, draw a force diagram and find the coefficient of friction between the body and the plane given that the motion takes 6 seconds. Take $g = 9.8$ ms^{-2}.

b) In reality, air resistance does have an affect on the body. How would you expect this to change the body's acceleration?

a) $u = 0, \quad s = 5, \quad t = 6, \quad a = ?$

$s = ut + \frac{1}{2}at^2 \Rightarrow 5 = (\frac{1}{2}a \times 6^2)$

$\Rightarrow \boldsymbol{a = 0.2778}$ **ms^{-2}**

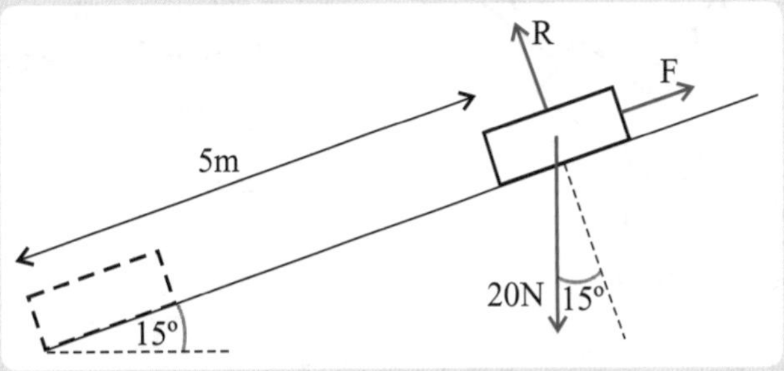

Resolving in ↙ direction:

$F_{net} = ma$

$20\sin 15° - F = \frac{20}{g} \times 0.2778$

$\boldsymbol{F = 4.609}$ **N**

Resolving in ↖ direction:

$F_{net} = ma$

$R - 20\cos 15° = \frac{20}{g} \times 0$

$R = 20\cos 15° = \mathbf{19.32}$ **N**

It's sliding, so: $F = \mu R$

$4.609 = \mu \times 19.32$

$\mu = 0.24$ (to 2 d.p.)

b) As air resistance acts against the motion of the body, it would reduce the magnitude of the resultant force acting on the body. So, because $F_{net} = ma$, and m is constant, the body's acceleration would decrease.

Friction and Inclined Planes

Friction opposes motion, so it also increases the tension in whatever's doing the pulling...

EXAMPLE

A mass of 3 kg is being pulled up a plane inclined at 20° to the horizontal by a rope parallel to the surface. Given that the mass is accelerating at 0.6 ms^{-2} and that the coefficient of friction is 0.4, find the tension in the rope. Take g = 9.8 ms^{-2}.

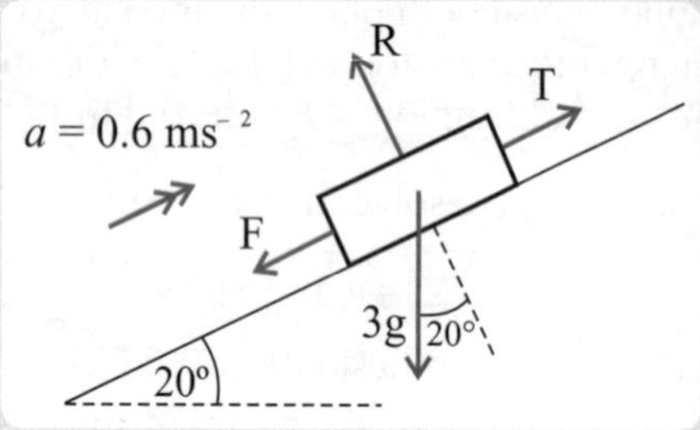

Resolving in ↖ direction:

$F_{net} = ma$

$R - 3g\cos 20° = 3 \times 0$

so: $R = 3g\cos 20° = 27.63$ N

The mass is sliding, so $F = \mu R$

$= 0.4 \times 27.63 = 11.05$ N

Resolving in ↗ direction:

$F_{net} = ma$

$T - F - 3g\sin 20° = 3 \times 0.6$

$T = 1.8 + 11.05 + 3g\sin 20° = 22.9$ N

Friction **Opposes Limiting Motion**

For a body at rest but on the point of moving down a plane, the friction force is up the plane. A body about to move up a plane is opposed by friction down the plane. Remember it well — it's about to come in handy.

EXAMPLE

A 4 kg box is placed on a 30° plane where $\mu = 0.4$. A force Q maintains equilibrium by acting up the plane parallel to the line of greatest slope. Find Q if the box is on the point of sliding:

a) up the plane b) down the plane.

A body like this, on the point of moving, is said to be in 'limiting equilibrium'.

a)

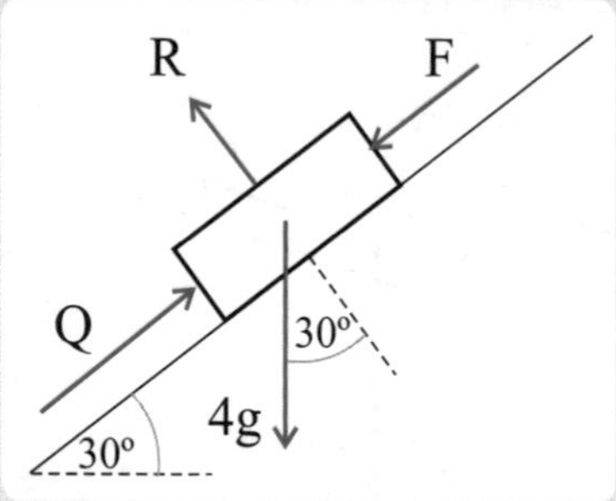

$F_{net} = ma$

Resolving in ↖ direction:

$R - 4g\cos 30° = 0$

$R = 4g\cos 30°$

$F = \mu R$

$= 0.4 \times 4g\cos 30°$

$= 1.6g\cos 30°$

Resolving in ↗ direction:

$Q - 4g\sin 30° - F = 4 \times 0$

$Q = 4g\sin 30° + 1.6g\cos 30°$

$= 33.2$ N

b)

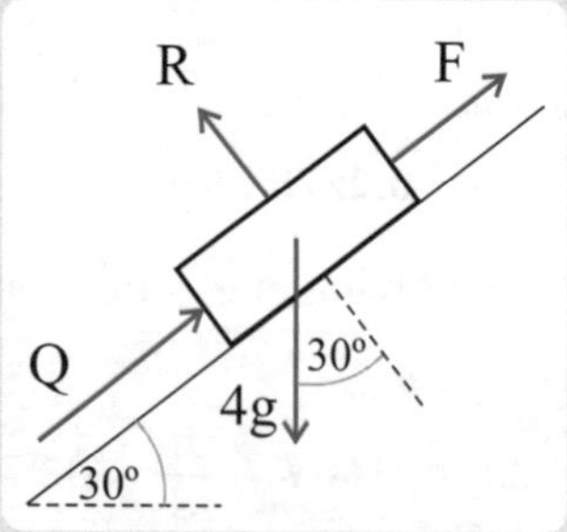

Resolving in ↖ direction:

$R = 4g\cos 30°$

$F = 1.6g\cos 30°$

Resolving in ↗ direction:

$Q - 4g\sin 30° + F = 4 \times 0$

$Q = 4g\sin 30° - 1.6g\cos 30°$

$Q = 6.02$ N

So for equilibrium $6.02 \text{ N} \leq Q \leq 33.2 \text{ N}$

Inclined planes — nothing to do with suggestible Boeing 737s...

The main thing to remember is that you can choose to resolve in any two directions as long as they're perpendicular. It makes sense to choose the directions that involve doing as little work as possible. Obviously.

Connected Particles

Like Laurel goes with Hardy and Posh goes with Becks, some particles are destined to be together...

Connected Particles act like One Mass

Particles connected together have the same speeds and magnitudes of acceleration as each other, as long as the connection holds. So, train carriages moving together have the same acceleration.

EXAMPLE

A 30 tonne locomotive engine is pulling a single 10 tonne carriage as shown. They are accelerating at 0.3 ms^{-2} due to the force P generated by the engine. It's assumed that there are no forces resistant to motion. Find P and the tension in the coupling.

Here's the pretty picture:

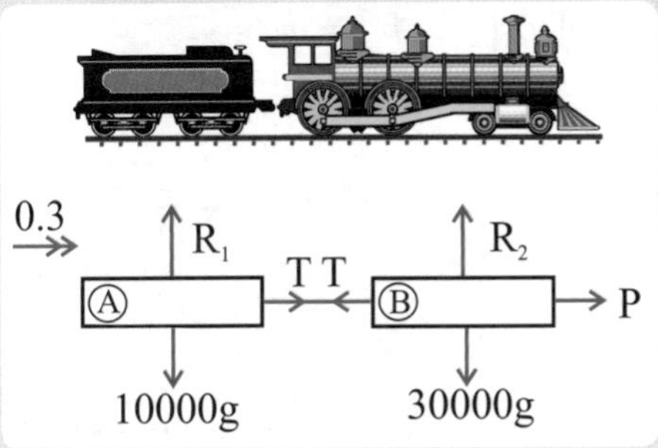

For A: $F_{net} = ma$

$T = 10\,000 \times 0.3$

$T = 3000$ N

For B: $F_{net} = ma$

$P - T = 30\,000 \times 0.3$

$P = 12000$ N

Pulleys (and 'Pegs') are always Smooth

In M1 questions, you can always assume that the tension in a string will be the same either side of a smooth pulley.

EXAMPLE

Masses of 3 kg and 5 kg are connected by an inextensible string and hang vertically either side of a smooth pulley. They are released from rest. Find their acceleration and the time it takes for each to move 40 cm. State any assumptions made in the model. Take $g = 9.8$ ms^{-2}.

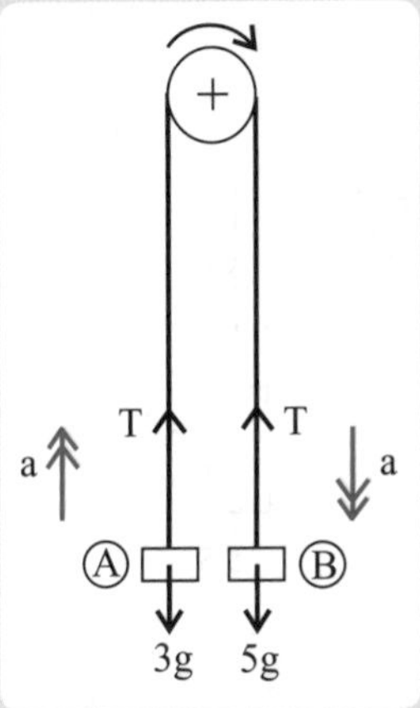

For A: $F_{net} = ma$

Resolving upwards: $T - 3g = 3a$ ①

For B: $F_{net} = ma$

Resolving downwards: $5g - T = 5a$

$T = 5g - 5a$ ②

Eliminating T from ① and ②: $(5g - 5a) - 3g = 3a$

$a = 2.45$ ms^{-2}

List variables: $u = 0$; $a = 2.45$; $s = 0.4$

Use an equation with u, a, s and t in it:

$s = ut + \frac{1}{2}at^2$

$0.4 = (0 \times t) + (\frac{1}{2} \times 2.45 \times t^2)$ So $t = \sqrt{\frac{0.8}{2.45}} = 0.57$ s

Assumptions: The 3 kg mass does not hit the pulley; the 5 kg mass does not hit the ground; there's no air resistance; the string is 'light' so has zero mass; the string doesn't break; the pulley is fixed.

Connected Particles

Use F = ma in the Direction Each Particle Moves

EXAMPLE A mass of 3 kg is placed on a smooth horizontal table. A light inextensible string connects it over a smooth peg to a 5 kg mass which hangs vertically as shown. Find the tension in the string if the system is released from rest. Take $g = 9.8\ \text{ms}^{-2}$.

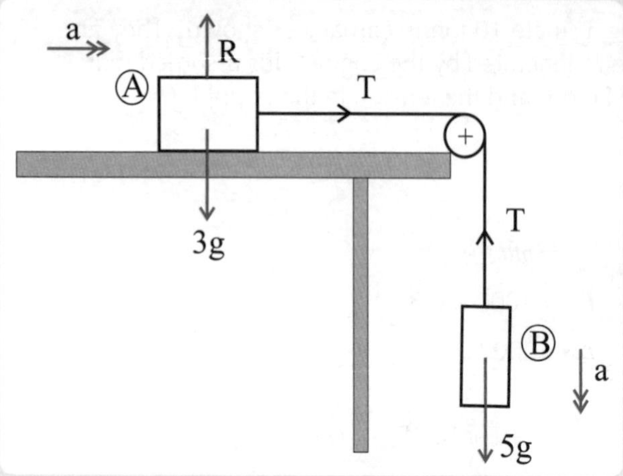

For A:
Resolve horizontally: →

$F_{net} = ma$

$T = 3a$

$a = \frac{T}{3}$ ①

For B:
Resolve vertically: ↓

$F_{net} = ma$

$5g - T = 5a$ ②

Sub ① into ②:

$5g - T = 5 \times \frac{T}{3}$

So $\frac{8}{3}T = 5g$

$T = 18.4$ N (to 3 s.f.)

Remember to use F ≤ μR on Rough Planes

More complicated pulley and peg questions have <u>friction</u> for you to enjoy too.

EXAMPLE The peg system of the example above is set up again. However, this time a constant friction force, F, acts on the 3 kg mass due to the table top now being rough, with coefficient of friction $\mu = 0.5$. Find the new tension in the string when the particles are released from rest. Take $g = 9.8\ \text{ms}^{-2}$.

For B: Resolving vertically: $5g - T = 5a$ ①

For A: Resolving horizontally: $F_{net} = ma$

$T - F = 3a$ ②

Resolving vertically: $R - 3g = 0$

$\mathbf{R = 3g}$

The particles are moving, so $F = \mu R = 0.5 \times 3g$

$= \mathbf{14.7\ N}$

Sub this into ②: $T - 14.7 = 3a$

$\mathbf{a = \frac{1}{3}(T - 14.7)}$

Sub this into ①:

$5g - T = 5 \times \frac{1}{3}(T - 14.7)$

$8T = 147 + 73.5$

$T = 27.6$ N

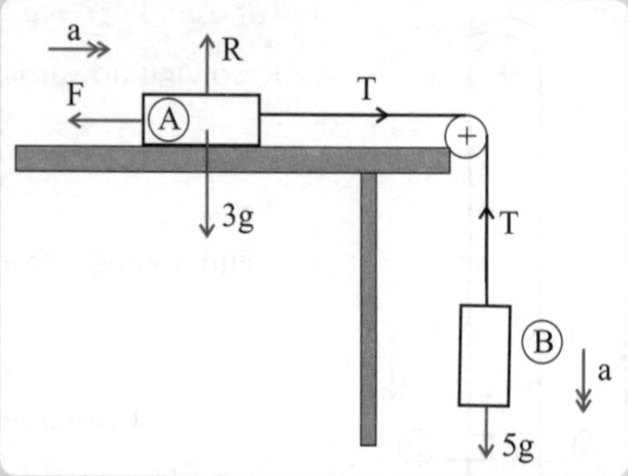

Useful if you're hanging over a Batman-style killer crocodile pit...

It makes things a lot easier when you know that connected particles have the same speed and magnitude of acceleration, and that in M1 pulleys can always be treated as smooth. Those examiners occasionally do try to make your life easier, honestly.

Connected Particles

EXAMPLE

Particles A and B of mass 4 kg and 10 kg respectively are connected by a light inextensible string over a smooth pulley as shown. A force of 15 N acts on A at an angle of 25° to a rough horizontal plane where $\mu = 0.7$. When B is released from rest it takes 2 s to fall d m to the ground. Find d. Take $g = 9.8$ ms^{-2}.

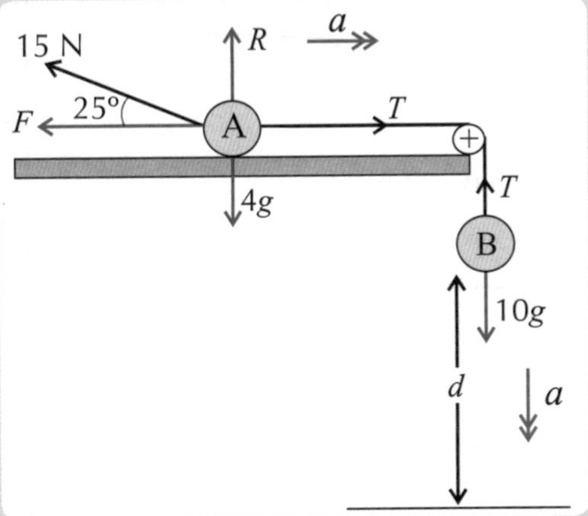

For A: Resolve vertically to find R:
$R = 4g - 15\sin25° = 32.86$ N
$F = \mu R = 0.7 \times 32.86 = 23.00$ N

$F_{net} = ma$
$T - 23.00 - 15\cos25° = 4a$
so $T = 4a + 36.59$ ①

For B: $F_{net} = ma$
$10g - T = 10a$
so $T = 98 - 10a$ ②

Substitute ① into ②:
$4a + 36.59 = 98 - 10a$
so $a = 4.39$ ms^{-2}

Using $s = ut + \frac{1}{2}at^2$: $s = d$, $u = 0$, $t = 2$, $a = 4.39$
so $d = (0 \times 2) + \frac{1}{2}(4.39 \times 2^2) = 8.8$ m (2 s.f.)

When B hits the ground A carries on moving along the plane. How long does it take A to stop after B hits the ground?

Speed of A when B hits the ground:
$v = u + at$: $u = 0$, $a = 4.39$, $t = 2$
$v = 0 + (4.39 \times 2) = 8.78$ ms^{-1}

Resolve to find new acceleration of A:
$-F - 15\cos25° = 4a$

Acceleration, $a = \frac{-23 - 15\cos 25°}{4} = -9.15$ ms^{-2} (deceleration)

Time taken to stop:
$t = \frac{(v-u)}{a}$: $u = 8.78$, $v = 0$, $a = -9.15$
so $t = \frac{0 - 8.78}{-9.15} = 0.96$ s (to 2 s.f.)

To answer this question, you have to assume that A stops before it hits the pulley.

And to finish, a 'towing' question with friction.

EXAMPLE

A truck of mass 5000 kg tows a log of mass 1500 kg across rough ground. The truck and log are connected by a cable which can be assumed to remain horizontal throughout the motion. They move with a constant acceleration of 0.9 ms^{-2}. The coefficient of friction between the log and the ground is 0.7. Find P, the driving force of the truck's engine, and T, the tension in the cable.

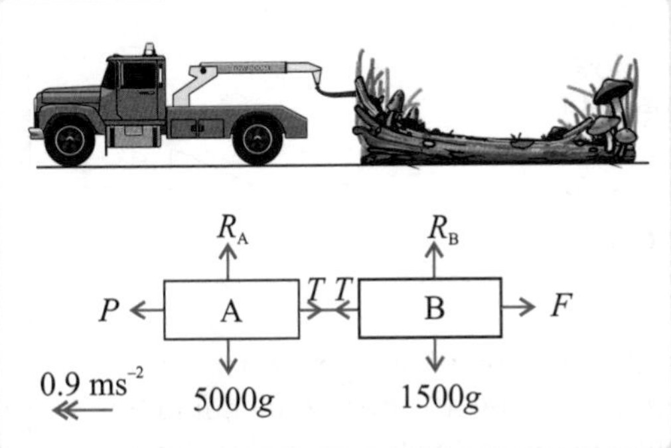

Call the truck 'A' and the log 'B'.

The truck is moving to the left, so left has been taken as positive.

For B: Use $F = \mu R_B$ to find the frictional force, F:
$F = 0.7 \times 1500 \times 9.8 = 10\,290$ N

There's no acceleration vertically, so reaction force $R_B = 1500g$.

$F_{net} = ma$
$T - F = 1500 \times 0.9$
So, $T = (1500 \times 0.9) + F = 1350 + 10\,290 = 11\,640$ N

For A: $F_{net} = ma$
$P - T = 5000 \times 0.9$
$P = (5000 \times 0.9) + T = 4500 + 11\,640 = 16\,140$ N

Suppose that the cable was not horizontal, but inclined at an angle to the horizontal, with the higher end attached to the truck. Assuming that the total resistive and driving forces remained unchanged, how would the tension in the rope differ to the answer you found above?

Call the new tension T_2. The horizontal component of T_2 would need to be equal to T, above. i.e. $T_2\cos\alpha = 11\,640$ N, where α is the angle between the cable and the horizontal. So T_2 would be greater than the value of T found above.

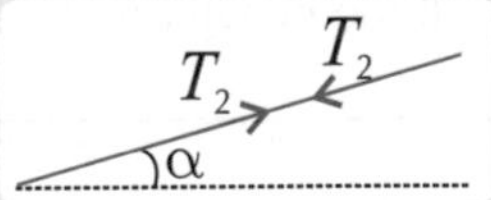

Connected particles — together forever... *isn't it beautiful?*

The key word here is rough. If a question mentions the surface is rough, then cogs should whirr and the word 'friction' should pop into your head. Take your time with force diagrams — I had a friend who rushed into drawing a diagram, and he ended up with a broken arm. But that was years later, now that I come to think of it. Still — better safe than sorry.

M1 Section 3 — Practice Questions

Find the coefficient of friction between a student's pen and a sheet of paper. Model the pen as a rod and the paper as a rough plane... or else you could just answer the questions below, which would be a better use of your time.

Warm-up Questions

Take g = 9.8 ms^{-2} in each of these questions.

1) Each diagram represents the motion of two particles moving in a straight line. Find the missing mass or velocity (all masses are in kg and all velocities are in ms^{-1}).

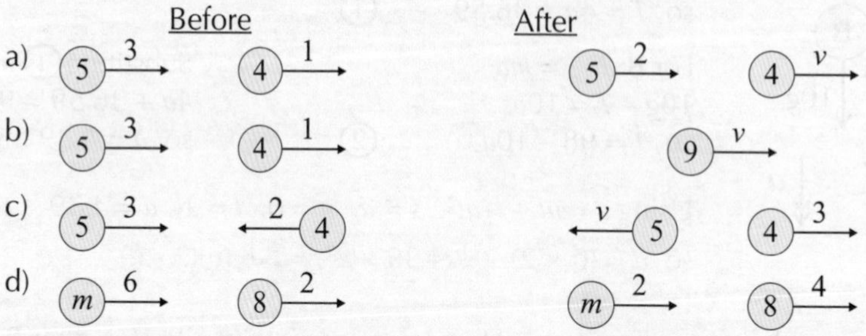

2) A horizontal force of 2 N acts on a 1.5 kg particle initially at rest on a smooth horizontal plane. Find the speed of the particle 3 seconds later.

3) Two forces act on a particle of mass 8 kg which is initially at rest on a smooth horizontal plane. The two forces are (24**i** + 18**j**) N and (6**i** + 22**j**) N (with **i** and **j** being perpendicular unit vectors in the plane). Find the magnitude and direction of the resulting acceleration of the particle and the displacement of the particle after 3 seconds.

4) A horizontal force P acting on a 2 kg mass generates an acceleration of 0.3 ms^{-2}. Given that the mass is in contact with a rough horizontal plane which resists motion with a force of 1 N, find P. Then find the coefficient of friction, μ, to 2 d.p.

5) A brick of mass 1.2 kg is sliding down a rough plane which is inclined at 25° to the horizontal. Given that its acceleration is 0.3 ms^{-2}, find the coefficient of friction between the brick and the plane. What assumptions have you made?

6) An army recruit of weight 600 N steps off a tower and accelerates down a "death slide" wire as shown. The recruit hangs from a light rope held between her hands and looped over the wire. The coefficient of friction between the rope and wire is 0.5. Given that the wire is 20 m long and makes an angle of 30° to the horizontal throughout its length, find how fast the recruit is travelling when she reaches the end of the wire.

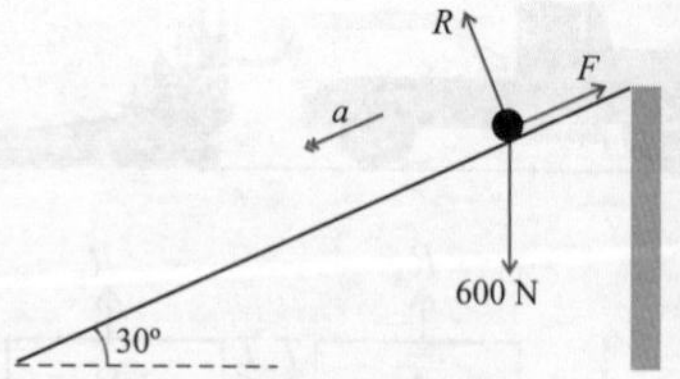

7) A 2 tonne tractor experiences a resistance force of 1000 N whilst driving along a straight horizontal road. If the tractor engine provides a forward force of 1500 N and it's pulling a 1 tonne trailer, find the resistance force acting on the trailer, and the tension in the coupling between the tractor and trailer, if they are moving with constant speed.

8) Two particles are connected by a light inextensible string, and hang in a vertical plane either side of a smooth pulley. When released from rest the particles accelerate at 1.2 ms^{-2}. If the heavier mass is 4 kg, find the weight of the other.

M1 Section 3 — Practice Questions

Right, those warm-up questions should have given you the momentum to get straight into these exam questions. May the *ma* be with you...

Exam Questions

Take $g = 9.8\ ms^{-2}$ in each of these questions.

1 A crane moves a mass of 300 kg, A, suspended by two light cables AB and AC attached to a horizontal movable beam BC. The mass is moved in the direction of the line of the supporting beam BC during which time the cables maintain a constant angle of 40° to the horizontal, as shown.

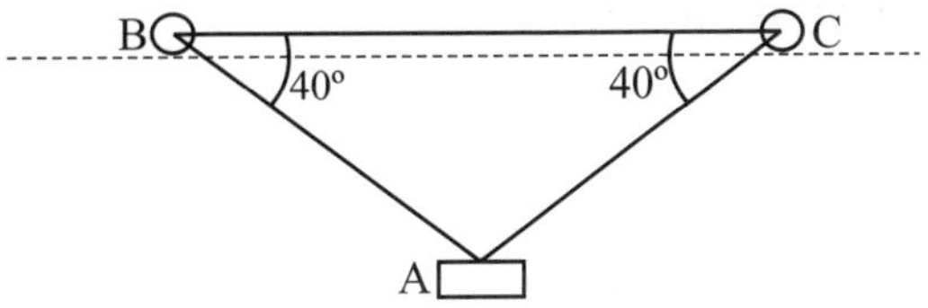

a) The mass is initially moving with constant speed. Find the tension in each cable. *(4 marks)*

b) The crane then moves the mass with a constant acceleration of $0.4\ ms^{-2}$. Find the tension in each cable. *(6 marks)*

c) What modelling assumptions have you made in part b)? *(2 marks)*

2 A horizontal force of 8 N just stops a mass of 7 kg from sliding down a plane inclined at 15° to the horizontal, as shown.

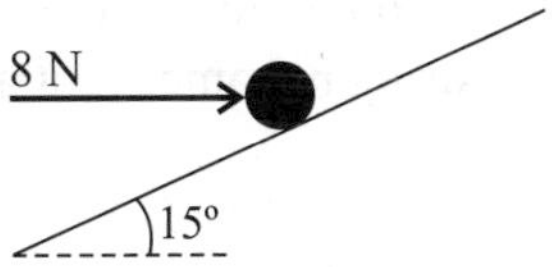

a) Calculate the coefficient of friction between the mass and the plane to 2 d.p. *(5 marks)*

b) The 8 N force is now removed. Find how long the mass takes to slide a distance of 3 m down the line of greatest slope. *(7 marks)*

3 A car of mass 1500 kg is pulling a caravan of mass 500 kg. They experience resistance forces totalling 1000 N and 200 N respectively. The forward force generated by the car's engine is 2500 N. The coupling between the two does not break.

a) Find the acceleration of the car and caravan. *(3 marks)*

b) Find the tension in the coupling. *(2 marks)*

4 Two particles of mass 0.8 kg and 1.2 kg travel in the same direction along a straight, horizontal line with speeds of $4\ ms^{-1}$ and $2\ ms^{-1}$ respectively until they collide. After the collision the 0.8 kg mass has a velocity of $2.5\ ms^{-1}$ in the same direction. The 1.2 kg mass then continues with its new velocity until it collides with a mass of m kg travelling with a speed of $4\ ms^{-1}$ in the opposite direction to it.
Given that both particles are brought to rest by this collision, find the mass m. *(4 marks)*

M1 Section 3 — Practice Questions

5 A coal wagon of mass 4 tonnes is rolling along a straight rail track at 2.5 ms^{-1}. It collides with a stationary wagon of mass 1 tonne. During the collision the wagons become coupled and move together along the track.

a) Find their speed after the collision. *(2 marks)*

b) State two assumptions made in your model. *(2 marks)*

6 Two forces, $(x\mathbf{i} + y\mathbf{j})$ N and $(5\mathbf{i} + \mathbf{j})$ N, act on a particle P of mass 2.5 kg. The resultant of the two forces is $(8\mathbf{i} - 3\mathbf{j})$ N.
Find:

a) the values of x and y, *(2 marks)*

b) the magnitude of the acceleration of P. *(3 marks)*

7 Two particles A and B are connected by a light inextensible string which passes over a smooth fixed pulley as shown. A has a mass of 7 kg and B has a mass of 3 kg. The particles are released from rest with the string taut, and A falls freely until it strikes the ground travelling at a speed of 5.9 ms^{-1}. A does not rebound after hitting the floor.

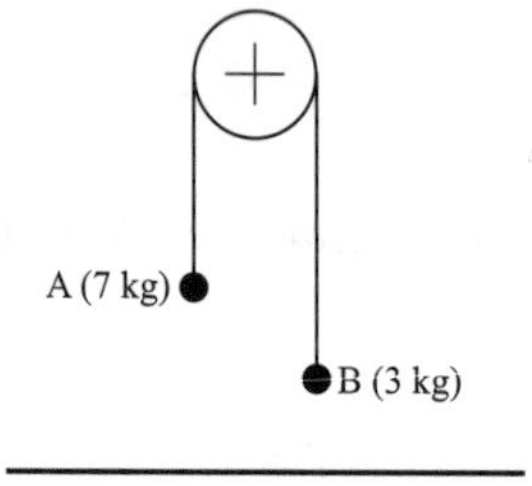

a) Find the time taken for A to hit the ground. *(4 marks)*

b) How far will B have travelled when A hits the ground? *(2 marks)*

c) Find the time (in s) from when A hits the ground until the string becomes taut again. *(4 marks)*

8 Two particles, A and B, move on a smooth horizontal plane. A has mass m kg and velocity $\binom{u}{1}$ ms^{-1}. B has mass 5 kg and velocity $\binom{0}{-5}$ ms^{-1}.

a) Find an expression for the total momentum of the system. *(1 mark)*

b) The particles collide and coalesce to form a new single particle. This particle has velocity $\binom{2}{-1}$ ms^{-1}. Find:

(i) the value of m, *(3 marks)*

(ii) the speed of A before the collision *(3 marks)*

9 Particle P, of mass m kg and velocity $\binom{8}{4}$ ms^{-1}, collides with the stationary particle Q, of mass $3m$ kg. The velocities of P and Q immediately following the collision are $\binom{v_P}{1}$ ms^{-1} and $\binom{1}{v_Q}$ ms^{-1} respectively.
Find:

a) v_P and v_Q, *(3 marks)*

b) the speeds of both P and Q immediately following the collision. *(3 marks)*

Projectiles

A 'projectile' is just any old object that's been lobbed through the air. When you're doing projectile questions you'll have to model the motion of particles in two dimensions whilst ignoring air resistance.

Split Velocity of Projection into Two Components

A particle projected with a speed u at an angle α to the horizontal has two components of initial velocity — one horizontal (parallel to the x-axis) and one vertical (parallel to the y-axis).
These are called x and y components, and they make projectile questions dead easy to deal with:

Here's the same information in a diagram:

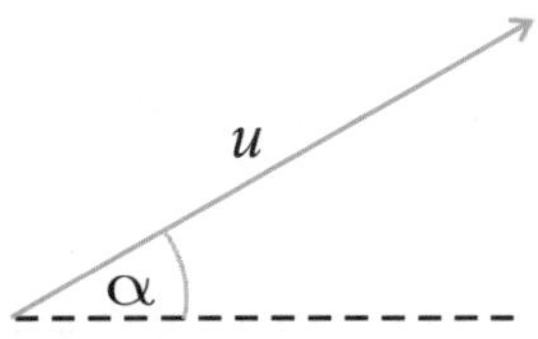

Split the velocity into its x and y components:

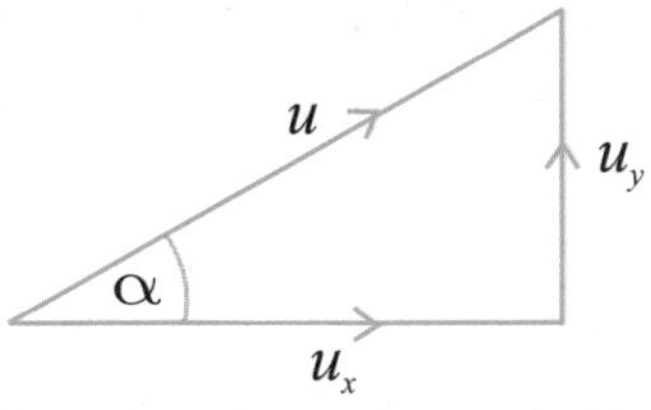

Finally, work out the values of the components using trigonometry:

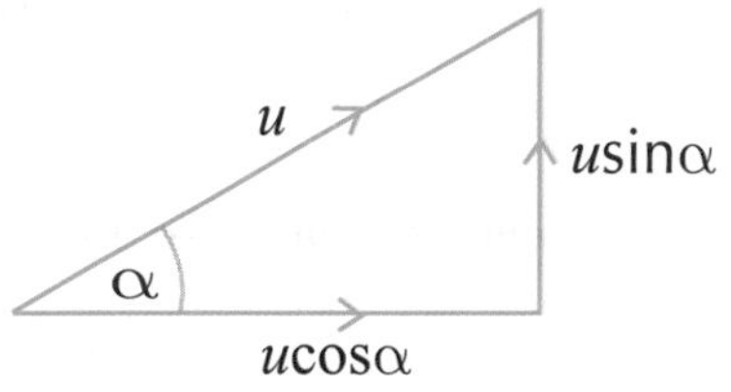

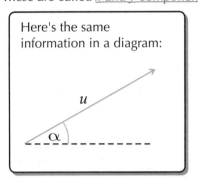

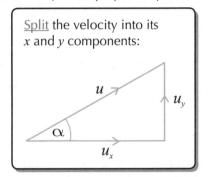

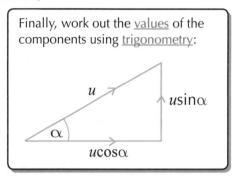

Split the Motion into Horizontal and Vertical Components too

Split everything you know about the motion into horizontal and vertical components too. Then you can deal with them separately using the 'uvast' equations. The only thing that's the same in both directions is time — so this connects the two directions. Remember that the only acceleration is due to gravity — so horizontal acceleration is zero.

EXAMPLE

A stone is thrown horizontally with speed 10 ms^{-1} from a height of 2 m above the horizontal ground. Find the time taken for the stone to hit the ground and the horizontal distance travelled before impact. Find also the stone's velocity after 0.5 s.

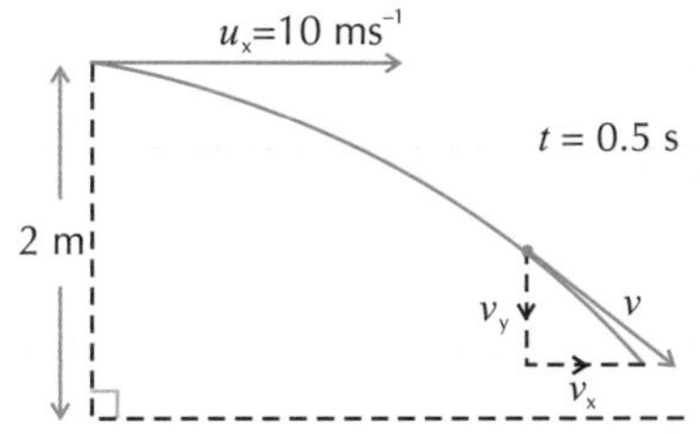

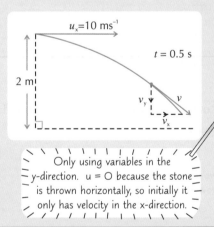

Only using variables in the y-direction. u = 0 because the stone is thrown horizontally, so initially it only has velocity in the x-direction.

Resolving vertically (take down as +ve):

$u = u_y = 0$ $s = 2$
$a = 9.8$ $t = ?$

$s = ut + \frac{1}{2}at^2$

$2 = 0 \times t + \frac{1}{2} \times 9.8 \times t^2$

$t = 0.639$ s (to 3 s.f.)
i.e. the stone lands after 0.639 seconds

Resolving horizontally (take right as +ve):

The same as for the vertical motion.

$u = u_x = 10$ $s = ?$
$a = 0$ $t = 0.639$

$s = ut + \frac{1}{2}at^2$

$= 10 \times 0.639 + \frac{1}{2} \times 0 \times 0.639^2$

$= 6.39$ m

i.e. the stone has gone 6.39 m horizontally when it lands.

Now find the velocity after 0.5 s — again, keep the vertical and horizontal bits separate.

$v = u + at$

$v_y = 0 + 9.8 \times 0.5$

$= 4.9$ ms^{-1}

$v = u + at$

$v_x = 10 + 0 \times \frac{1}{2}$

$= 10$ ms^{-1}

v_x is always equal to u_x when there's no horizontal acceleration.

Now you can find the speed and direction...

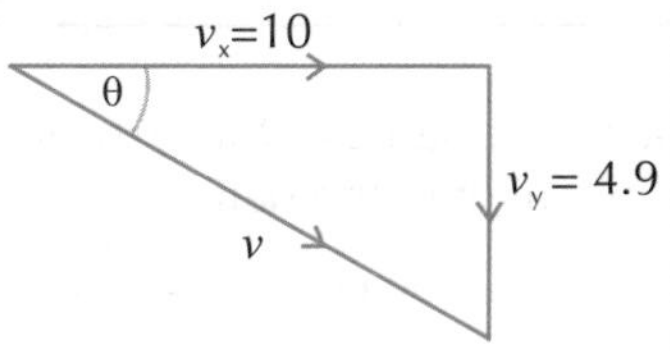

$v = \sqrt{4.9^2 + 10^2} = 11.1$ ms^{-1}

$\tan\theta = \frac{4.9}{10}$

So $\theta = 26.1°$ below horizontal

Projectiles

EXAMPLE A cricket ball is projected with a speed of 30 ms⁻¹ at an angle of 25° to the horizontal. Assume the ground is horizontal and the ball is struck from a point 1.5 m above the ground. Find:

a) the maximum height the ball reaches (h),

b) the horizontal distance travelled by the ball before it hits the ground (r),

c) the length of time the ball is at least 5 m above the ground.

d) the magnitude of the vertical component of the ball's velocity when its speed is 29 ms⁻¹.

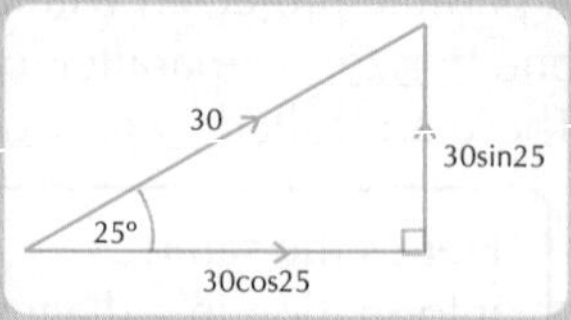

A heavier ball is projected with the same initial velocity under exactly the same conditions.

e) How would the range of the heavier ball predicted by this model compare with your answer to part b)?

a) **Resolving vertically** (take up as +ve):

$u = 30\sin25°$ $v = 0$
$a = -9.8$ $s = ?$

The ball will momentarily stop moving vertically when it reaches its maximum height.

$v^2 = u^2 + 2as$

$0 = (30\sin 25°)^2 + 2(-9.8 \times s)$

$s = 8.20\text{ m}$

$h = 8.20\text{ m} + 1.5\text{ m} = 9.70\text{ m}$

Don't forget to add the height from which the ball is hit.

---

b) **Resolving vertically** (take up as +ve):

$s = -1.5$
$a = -9.8$
$u = 30\sin25°$
$t = ?$

$s = ut + \frac{1}{2}at^2$

$-1.5 = (30\sin 25°)t - \frac{1}{2}(9.8)t^2$

$t^2 - 2.587t - 0.306 = 0$

$t = -0.11$ or $\mathbf{t = 2.70\text{ s}}$

Using the quadratic formula you get two answers, but time can't be negative, so forget about this answer.

Resolving horizontally (take right as +ve)

$s = r$ $u = 30\cos25°$
$t = 2.70$ $a = 0$

$s = ut + \frac{1}{2}at^2$

$r = 30\cos 25° \times 2.70 + \frac{1}{2} \times 0 \times 2.70^2$

$= 73.4\text{ m}$

---

c) **Resolving vertically** (take up as +ve):

$s = 3.5$
$a = -9.8$
$u = 30\sin25°$
$t = ?$

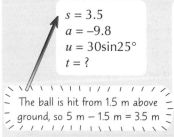

The ball is hit from 1.5 m above ground, so 5 m – 1.5 m = 3.5 m

$s = ut + \frac{1}{2}at^2$

$3.5 = (30\sin 25°)t - \frac{1}{2}(9.8)t^2$

$t^2 - 2.587t + 0.714 = 0$

$t = 0.31$ or $t = 2.27\text{ s}$

These are the two times when the ball is 5 m above the ground.

So, length of time at least 5 m above the ground:

2.27 s – 0.31 s = 1.96 s

---

d) Speed, $v = \sqrt{v_x^2 + v_y^2}$. v_x is always 30cos25°, as there is no acceleration horizontally, so:

$29 = \sqrt{(30\cos 25)^2 + v_y^2} \Rightarrow v_y^2 = 29^2 - (30\cos 25)^2 = 101.75.$

So, magnitude of v_y = 10.1 ms⁻¹ (3 s.f.).

Remember, v_x is the horizontal component and v_y the vertical component of the ball's velocity.

---

e) This model would predict the range of the heavier ball as exactly the same as the answer to b), because the mass does not feature in any of the equations used to model the flight of the ball.

Projectiles

Just one last example of projectile motion. But boy is it a beauty...

EXAMPLE

A golf ball is struck from a point *A* on a horizontal plane. The ball is modelled as a particle and is projected with initial speed u ms^{-1} at an angle α.

a) Show that x, the horizontal range of the ball, satisfies

$$x = \frac{2u^2}{g}\sin\alpha\cos\alpha$$

where g is acceleration due to gravity.

The ball is hit with initial speed $u = 60$ ms^{-1}, at an angle of $\alpha = 40°$ to the horizontal.

The hole that the golfer is aiming for is 330 m away.

b) Show that the ball will land 32 m beyond the hole.

c) Find the speed of the ball as it passes above the hole.

a) Displacement, acceleration and initial velocity are the only variables in the formula, so use these. Also use time, because that's the variable which connects the two components of motion. The formula includes motion in both directions (x and y), so consider these separately:

Resolving vertically (taking up as +ve):

$u_y = u\sin\alpha$ $\quad a = -g$

$t = t$ $\quad s = 0$

Putting $s = 0$ finds the total time from the ball being struck to when it lands.

Using $s = ut + \frac{1}{2}at^2$:

$0 = (u\sin\alpha \times t) - \frac{1}{2}gt^2$

$\Rightarrow 0 = t\left(u\sin\alpha - \frac{1}{2}gt\right)$

$\Rightarrow t = 0$ or $t = \frac{2u\sin\alpha}{g}$

Resolving horizontally (taking right as +ve):

$u_x = u\cos\alpha$ $\quad a = 0$

$s = x$ $\quad t = \frac{2u\sin\alpha}{g}$

Using $s = ut + \frac{1}{2}at^2$:

$x = u\cos\alpha \times \frac{2u\sin\alpha}{g}$

$\Rightarrow x = \frac{2u^2}{g}\sin\alpha\cos\alpha$, as required.

b) Using the result from a), and substituting $u = 60$ and $\alpha = 40°$, the horizontal range of the ball is:

$x = \frac{2 \times 60^2}{9.8} \times \sin 40° \times \cos 40° = 362$ m (3 s.f.). So the ball will land 362 – 330 = 32 m beyond the hole.

c) Again, split the motion into horizontal and vertical components:

Resolving horizontally (taking right as +ve):

$u_x = 60\cos 40°$ $\quad a = 0$

$s = 330$ $\quad t = t$

Using $s = ut + \frac{1}{2}at^2$:

$330 = 60\cos 40° \times t \Rightarrow t = 7.180$ s.

Also, $v_x = u_x = 60\cos 40° = \mathbf{45.96}$ **ms^{-1}**.

Remember — with projectiles there's no horizontal acceleration, so v_x always equals u_x.

Resolving vertically (taking up as +ve):

$u_y = 60\sin 40°$ $\quad a = -g$

$t = 7.180$ $\quad v_y = ?$

Using $v = u + at$:

$v_y = 60\sin 40° - (9.8 \times 7.180) = \mathbf{-31.80}$ **ms^{-1}**

The velocity is negative because the ball is moving downwards.

Now you can find the speed: $v = \sqrt{v_x^2 + v_y^2} = 55.9$ ms^{-1} (3 s.f.)

Projectiles — they're all about throwing up. Or across. Or slightly down...

The main thing to remember here is that <u>horizontal acceleration is zero</u> — great news because it makes half the calculations as easy as a log-falling beginner's class. Oh — and remember to keep track of which directions are positive and negative.

M1 Section 4 — Practice Questions

Well, if you've been revising M1 in order, that's your lot. Have a butchers at these warm-up questions, then it's onwards and upwards, projectile style.

Warm-up Questions

Take $g = 9.8 \text{ ms}^{-2}$ in each of these questions.

1) A particle is projected with initial velocity u ms^{-1} at an angle α to the horizontal. What is the initial velocity of the particle in the direction parallel to the horizontal in terms of u and α?
2) A rifle fires a bullet horizontally at 120 ms^{-1}. The target is hit at a horizontal distance of 60 m from the end of the rifle. Find how far the target is vertically below the end of the rifle. Take $g = 9.8 \text{ ms}^{-2}$.
3) A golf ball takes 4 seconds to land after being hit with a golf club from a point on the horizontal ground. If it leaves the club with a speed of 22 ms^{-1}, at an angle of α to the horizontal, find α. Take $g = 9.8 \text{ ms}^{-2}$.
4) A particle is projected with velocity 20 ms^{-1} at an angle of 20° above the horizontal. The particle lands at the level of projection 30 seconds later. Find the minimum speed of the particle during its flight.

And it's the final furlong — one last dash to the finish.

Exam Questions

Take $g = 9.8 \text{ ms}^{-2}$ in each of these questions.

1 A stationary football is kicked with a speed of 20 ms^{-1}, at an angle of 30° to the horizontal, towards a goal 30 m away. The crossbar is 2.5 m above the level ground. Assuming the path of the ball is not impeded, determine whether the ball passes above or below the crossbar. What assumptions does your model make?

(6 marks)

2

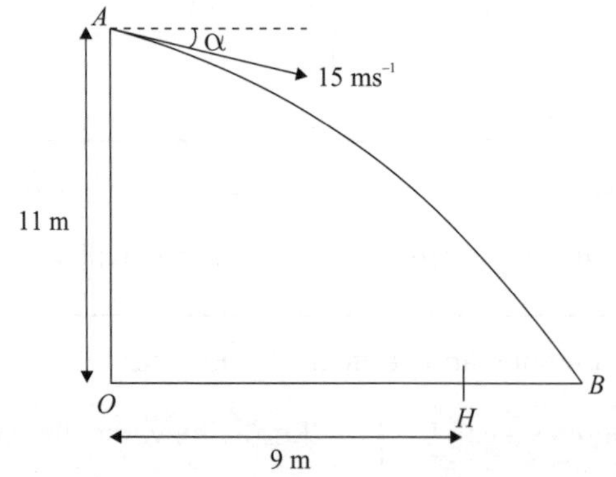

A stone is thrown from point A on the edge of a cliff, towards a point H, which is on horizontal ground. The point O is on the ground, 11 m vertically below the point of projection. The stone is thrown with speed 15 ms^{-1} at an angle α below the horizontal, where $\tan\alpha = \frac{3}{4}$.

The horizontal distance from O to H is 9 m.

The stone misses the point H and hits the ground at point B, as shown above. Find:

a) the time taken by the stone to reach the ground, *(5 marks)*

b) the horizontal distance the stone misses H by, *(3 marks)*

c) the speed of the stone when it lands at B, *(5 marks)*

d) the angle between the direction of motion and the horizontal as the ball lands at B. *(2 marks)*

General Certificate of Education
Advanced Subsidiary (AS) and Advanced Level

Mechanics M1 — Practice Exam One

Time Allowed: 1 hour 30 min

Graphical calculators may be used for this exam.

Whenever a numerical value of g is required, take $g = 9.8$ ms^{-2}.

Unless told otherwise, give any non-exact numerical answers to 3 significant figures.

There are 75 marks available for this paper.

1 A motorcyclist is travelling at 15 ms^{-1}. As he passes point A on a straight section of road, he accelerates uniformly for 4 s until he passes point B at 40 ms^{-1}. He then immediately decelerates at 2.8 ms^{-2} so that when he passes point C he is travelling at 26 ms^{-1}.
Find:

a) his acceleration between A and B, *(2 marks)*

b) the time to travel from B to C, *(2 marks)*

c) the distance from A to C. *(3 marks)*

2 A car, of mass 1600 kg, is towing a boat of mass 3000 kg along a straight, horizontal road. The car experiences a resistive force of magnitude 400 N and the boat experiences a constant resistive force of magnitude R N. A driving force of magnitude 4300 N acts on the car. The vehicles accelerate at 0.8 ms^{-2}.

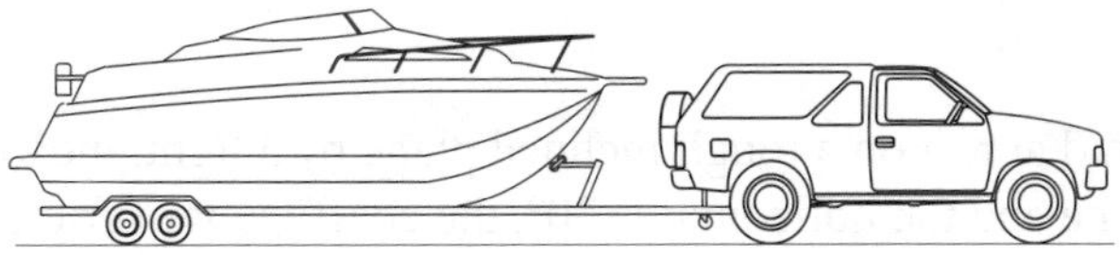

a) Assuming that the tow bar connecting the car to the boat is horizontal, find:

(i) the tension, T, in the tow bar, *(3 marks)*

(ii) the resistance force, R, acting on the boat. *(3 marks)*

b) When the car and boat are travelling at a speed of 7 ms^{-1}, the boat becomes detached from the car. Assuming that the only horizontal force now acting on the boat is the resistive force R, find:

(i) the boat's deceleration, *(2 marks)*

(ii) the distance the boat travels before it comes to rest. *(3 marks)*

3 Two railway trucks A and B, with masses 3.5 tonnes and 1.5 tonnes respectively, are travelling towards each other on straight horizontal rails. Both trucks are travelling at 3 ms^{-1} when they collide. Assume that the trucks can be modelled as particles.

a) If the trucks couple together on impact and move together as a single body on the rails, find the speed and direction of the combined trucks immediately after the collision. *(3 marks)*

b) Assume that the trucks do not couple together, and following the collision, A is moving with speed 1 ms^{-1} in the opposite direction to before the collision. Find the speed and direction of B immediately after the collision. *(3 marks)*

4 A cyclist starts from rest and accelerates at 1.5 ms^{-2} for 8 s along a straight horizontal road and then continues at a constant speed. A motorist sets off from the same point as the cyclist at the same time and accelerates at a constant rate for 6 s, reaching a maximum speed of V ms^{-1}. The car then decelerates at a constant rate until it comes to rest after a further 18 s. At the instant when the car comes to rest, it is overtaken by the cycle.

a) Calculate the greatest speed attained by the cyclist. *(2 marks)*

b) Sketch a speed-time graph for the motion of the cyclist. *(2 marks)*

c) Sketch a speed-time graph for the motion of the car. *(2 marks)*

d) Calculate the distance travelled by the cyclist before overtaking the car. *(3 marks)*

e) Calculate the value of V. *(3 marks)*

5 A block of mass 5 kg is held at rest on a rough inclined plane by a light, inextensible string. The angle between the plane and the horizontal is 30°, the coefficient of friction between the block and the plane is $\mu = 0.3$, and the tension in the string is T N. The string is parallel to the slope, and the block is on the point of moving up the slope.

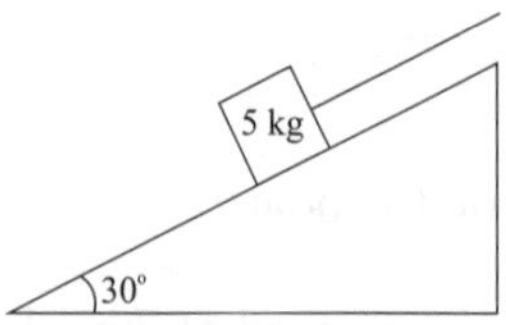

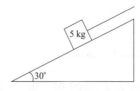

a) Find the reaction force of the plane on the block. *(2 marks)*

b) Find the value of T. *(3 marks)*

The string holding the block is cut, and the block begins to move down the slope.

c) Calculate the resultant force, F_{net}, acting on the block parallel to the slope. *(2 marks)*

d) Hence find the block's acceleration down the slope. *(2 marks)*

6 A particle of mass 0.5 kg moves under the action of two forces, $\mathbf{F}_1$ and $\mathbf{F}_2$, where $\mathbf{F}_1 = (3\mathbf{i} + 2\mathbf{j})$ N and $\mathbf{F}_2 = (2\mathbf{i} - \mathbf{j})$ N.

a) Find, in degrees correct to one decimal place, the angle which the resultant force makes with the direction of **i**. *(3 marks)*

b) Find, correct to three significant figures, the magnitude of the acceleration of the particle. *(4 marks)*

7 A helicopter *P* sets off from its base *O* on the coast and flies at constant height with initial velocity $(40\mathbf{i} + 108\mathbf{j})$ kmh^{-1} and constant acceleration $(4\mathbf{i} + 12\mathbf{j})$ kmh^{-2}, where **i** and **j** are the unit vectors in the direction of east and north respectively. Thirty minutes after *P* has left the base, the pilot informs the base that he has engine failure and is ditching into the sea.

a) Find the position vector of *P* relative to *O* at the time it experiences engine failure. *(3 marks)*

The base immediately informs another helicopter, *Q*, that *P* is in need of assistance. *Q* flies at a constant height from position $(2.5\mathbf{i} + 25.5\mathbf{j})$ km with initial velocity $(60\mathbf{i} + 120\mathbf{j})$ kmh^{-1} towards *P*'s position.

b) Given that *Q* takes 15 minutes to reach *P*, find the magnitude of the acceleration of *Q*. *(4 marks)*

The base sends out a third helicopter, *R*, on another mission. *R* leaves the base with velocity $(30\mathbf{i} - 40\mathbf{j})$ kmh^{-1} and accelerates at $(5\mathbf{i} + 8\mathbf{j})$ kmh^{-2} towards its destination.

c) Find the position vector of *R* when it is travelling due east. *(6 marks)*

8

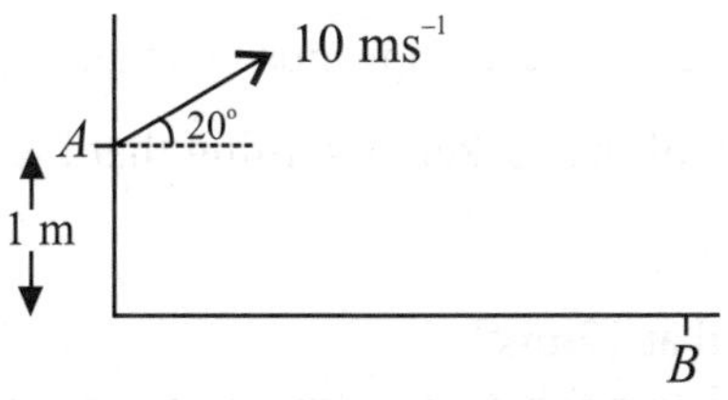

A frisbee is thrown from *A*, at a height of 1 m above the ground. The frisbee's initial velocity is 10 ms^{-1} at an angle of 20° above the horizontal. The frisbee lands on horizontal ground, at *B*. Find:

a) the maximum height the frisbee reaches during its flight, *(3 marks)*

b) the time it takes the frisbee to travel from *A* to *B*, *(4 marks)*

c) the horizontal distance between *A* and *B*. *(3 marks)*

General Certificate of Education
Advanced Subsidiary (AS) and Advanced Level

Mechanics M1 — Practice Exam Two

Time Allowed: 1 hour 30 min

Graphical calculators may be used for this exam.

Whenever a numerical value of g is required, take g = 9.8 ms^{-2}.

Unless told otherwise, give any non-exact numerical answers to 3 significant figures.

There are 75 marks available for this paper.

1 Three forces act at a point, O, as shown below.

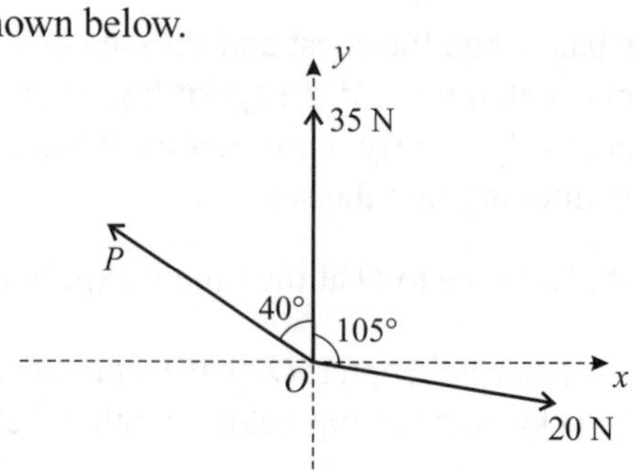

a) Given that there is no net force in the direction of the x-axis, find P. *(2 marks)*

b) Find the magnitude and direction of the resultant force, R. *(3 marks)*

2 A rocket with mass 1 400 000 kg is launched vertically upwards by engines providing a force of 34 000 000 N.

a) Calculate the expected acceleration of the rocket, assuming that the only other force acting on the rocket is its weight. *(2 marks)*

b) The actual acceleration is measured at 12 ms^{-2}.
Find the magnitude of the total resistive force, R, acting on the rocket. *(2 marks)*

c) Find the time taken to reach a height of 20 km. Give your answer to the nearest second. *(2 marks)*

d) At a height of 20 km, the engines stop firing, and the rocket moves freely under gravity (i.e. the resistive force can be ignored). Find the maximum height reached by the rocket. *(4 marks)*

e) State one modelling assumption you have made. *(1 mark)*

3 A small ring of mass m is threaded onto a rough inextensible rope held taut and horizontal. The ring is held in limiting equilibrium by a string pulling upwards at an angle of 51.3° to the horizontal. The coefficient of friction between the ring and the rope is 0.6 and the magnitude of the frictional force is 1.5 N.

a) Sketch a diagram showing the ring and the forces acting upon it. *(2 marks)*

b) Find, to 2 s.f., the tension in the string, T, and the mass of the ring, m. *(6 marks)*

4 A roller coaster moves from rest along a track with constant acceleration. The roller coaster takes 2 s to reach a speed of 6 ms^{-1}, then travels at a constant speed for 15 s. The roller coaster decelerates to rest in 1 s as it reaches the top of the hill. The roller coaster waits for 5 s before accelerating down a vertical slope for 4 s, reaching a maximum speed of 30 ms^{-1}.

a) Draw a speed-time graph representing the motion of the roller coaster. *(2 marks)*

b) Find the greatest acceleration experienced by the roller coaster. *(3 marks)*

c) What is the total distance travelled by the roller coaster? *(3 marks)*

5 A frog jumps off a rock of height 0.5 m with initial velocity U at an angle $\alpha°$ above the horizontal. The frog lands on horizontal ground at point P, as shown.

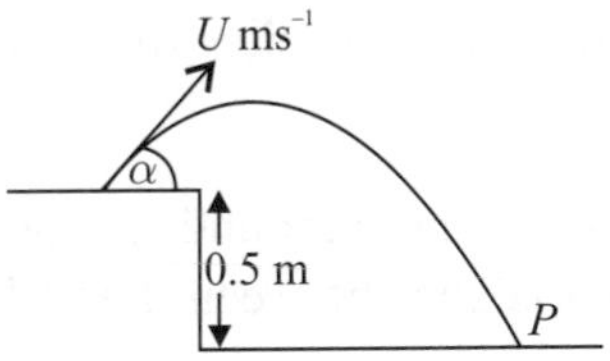

a) Show that the frog reaches maximum height $h = \dfrac{g + U^2\sin^2\alpha}{2g}$ m above the ground, where g is acceleration due to gravity. *(4 marks)*

b) Find an expression for the time, t, it takes the frog to reach this height, in terms of U, g and α. *(3 marks)*

c) Show that V, the speed of the frog when it lands at P, is $V = \sqrt{U^2 + g}$ ms^{-1}. *(5 marks)*

6 Particle P is moving on a smooth, horizontal plane with velocity $\begin{pmatrix} 3 \\ -9 \end{pmatrix}$ ms^{-1}, when it collides with stationary particle Q. P has mass 5 kg, while Q has mass 10 kg.

a) If the particles coalesce, find the velocity of the new particle after the collision. *(3 marks)*

b) If the particles do not coalesce, and P has velocity $\begin{pmatrix} 2 \\ 0 \end{pmatrix}$ ms^{-1} immediately following the collision, find the velocity of Q immediately following the collision. *(3 marks)*

7 A pirate ship is travelling at a constant velocity of $(9\mathbf{i} + 8\mathbf{j})$ kmh^{-1}, where the unit vectors **i** and **j** are due east and north respectively. At 1350 hours the pirate ship is at a point O. It heads for a port with position vector $(63\mathbf{i} + 56\mathbf{j})$ km relative to O.

a) At what time should the pirate ship arrive at the port?

(3 marks)

Also at 1350 hours, a naval ship is at rest with position vector $(-18\mathbf{i} + 6\mathbf{j})$ km relative to O. The naval ship aims to intercept the pirate ship as it arrives at the port.

b) Find the naval ship's acceleration if it arrives at the port at the same time as the pirate ship. *(4 marks)*

8

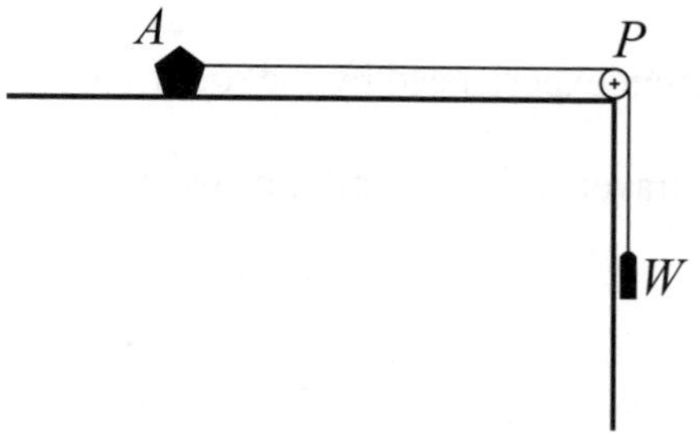

A particle, A, is attached to a weight, W, by a light inextensible string which passes over a smooth pulley, P, as shown. When the system is released from rest, with the string taut, A and W experience an acceleration of 4 ms^{-2}. A moves across a rough horizontal plane and W falls vertically. The mass of W is 1.5 times the mass of A.

a) Given that the mass of A is 0.2 kg, find the coefficient of friction, μ, between A and the horizontal plane.

(5 marks)

W falls for h m until it hits the ground and does not rebound. A continues to move until it reaches P with speed 3 ms^{-1}. The initial distance between A and P is $\frac{7}{4}h$ m.

b) Find the distance h.

(7 marks)

c) How did you use the information that the string is inextensible? *(1 mark)*

9 A boat is travelling in a river. The water is moving due east at a speed of 8 ms^{-1}. The velocity of the boat relative to the water is 5 ms^{-1}, on a bearing of 310°, as shown. The banks of the river are parallel.

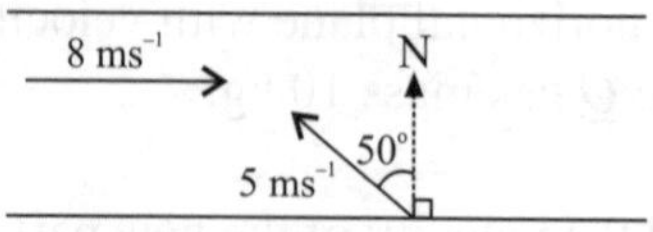

a) Find v, the magnitude of the resultant velocity of the boat.

(4 marks)

b) State one modelling assumption you have made about the boat.

(1 mark)

Algorithms

Welcome to the wonderful world of Decision Maths. And what a good decision it was too. This page is on algorithms, which aren't as scary as they sound. You've probably come across algorithms before, though you might not know it.

Algorithms are sets of Instructions

An algorithm is just a fancy mathematical name for a set of instructions for solving a problem. You come across lots of algorithms in everyday life — recipes, directions and assembly instructions are all examples of algorithms.

1) Algorithms start with an input (e.g. in a recipe, the input is the raw ingredients). You carry out the algorithm on the input, following the instructions in order.
2) Algorithms have an end result — something that you achieve by carrying out the algorithm (e.g. a cake). An algorithm is said to be correct if it achieves its end result.
3) Algorithms will stop when you've reached a solution, or produced your finished product — they're finite. They must have a stopping condition — an instruction that tells you to stop when you've reached a certain point.
4) Algorithms are often written so that computers could follow the instructions. Computer programming is an important application of Decision Maths.

Anyone should be able to follow an algorithm — not just the person who wrote it.

In Maths, the End Result is the Solution

Most mathematical algorithms are general — they work for a range of different inputs.

Algorithms can be used to solve mathematical problems too.

1) The input in a mathematical algorithm is the number (or numbers) you start with. Your end result (output) is the final number you end up with — this'll be the solution to the original problem.
2) Any number you put in will have a unique output — you won't get different sets of solutions for the same input.
3) It's a good idea to write down the numbers each instruction produces in a table — sometimes the algorithm will tell you when to do this. This table is called a trace table.

The Russian Peasant algorithm Multiplies two numbers

The Russian Peasant algorithm is a well-known algorithm that multiplies two numbers together. And I've no idea what Russian peasants have to do with it.

1) Write down the two numbers that you are multiplying in a table. Call them x and y.
2) Divide x by 2 and write down the result underneath x, ignoring any halves.
3) Multiply y by 2 and write down the result underneath y.
4) Repeat steps 2) - 3) for the numbers in the new row. Keep going until the number in the x-column is 1.
5) Work down your table and cross out every row that has an even value for x.
6) Add up the remaining numbers in the y-column (i.e. the ones that haven't been crossed out). This is the solution xy.

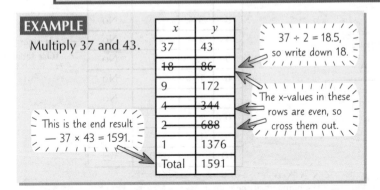

E.g. if $x = 11$, when you divide it by 2 you write down 5, not 5.5.

This is an example of a stopping condition.

EXAMPLE Multiply 37 and 43.

x	y
37	43
~~18~~	~~86~~
9	172
~~4~~	~~344~~
~~2~~	~~688~~
1	1376
Total	1591

$37 \div 2 = 18.5$, so write down 18.

The x-values in these rows are even, so cross them out.

This is the end result — $37 \times 43 = 1591$.

EXAMPLE Multiply 21 and 52.

x	y
21	52
~~10~~	~~104~~
5	208
~~2~~	~~416~~
1	832
Total	1092

This is the end result — $21 \times 52 = 1092$.

Feel the beat of the rithm of the night...

For your homework tonight, I would like you to make me a cake please. My favourite's carrot cake with cream cheese icing.

Pseudo-Code and Flow Charts

Right, now you know what an algorithm is, it's time to have some fun. Well, I say fun...

Algorithms can be written in **Pseudo-Code**...

1) Algorithms can be written in lots of different ways — e.g. a set of written instructions, a flow chart (see below), computer programming language or pseudo-code.
2) Pseudo-code is a bit like computer programming language, but less formal — it's written for a person, not a computer. It won't be written in full sentences though — just a set of brief instructions.
3) In the exam, you need to be able to follow an algorithm written in pseudo-code, but you won't have to write your own. You might have conditions like IF and THEN — you only do the THEN instruction if the IF bit's true.

There's an example of this in the questions on p.206 — make sure you can follow it.

EXAMPLE This is the pseudo-code for an algorithm that outputs the first 10 triangle numbers:

STEP 1: Input X = 1 and Y = 2

STEP 2: Print X

STEP 3: Let X = X + Y

STEP 4: Let Y = Y + 1

STEP 5: If Y < 12, then return to STEP 2. If Y = 12, then stop.

You start off with X = 1 and Y = 2, so you print 1, then X becomes 1 + 2 = 3 and Y becomes 3. 3 < 12, so you go back to step 2. This time, print 3, then X becomes 3 + 3 = 6 and Y becomes 4. You carry on like this until Y = 12.

Following this algorithm gives 1, 3, 6, 10, 15, 21, 28, 36, 45, 55.

...or in **Flow Charts**

Algorithms can be written as flow charts. There are three different types of boxes which are used for different things:

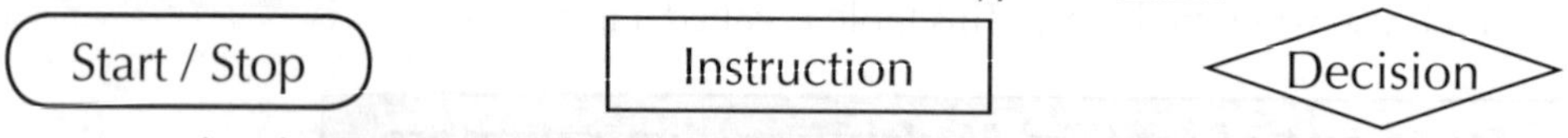

The boxes are connected with arrows to guide you through the flow chart. 'Decision' boxes will ask a question, and for each one you have a choice of arrows, one arrow for 'yes' and one for 'no', which will take you to another box. Sometimes flow charts will include a loop which lets you repeat steps until the algorithm is finished.

Put the **Results** from the flow chart into a **Trace Table**

It can sometimes be a bit tricky keeping track of the results of a flow chart, especially if you have to go round a loop lots of times. It's a good idea to put your results into a trace table — it's much easier to see the solutions that way.

EXAMPLE This flow chart works out the factors of a number, a.

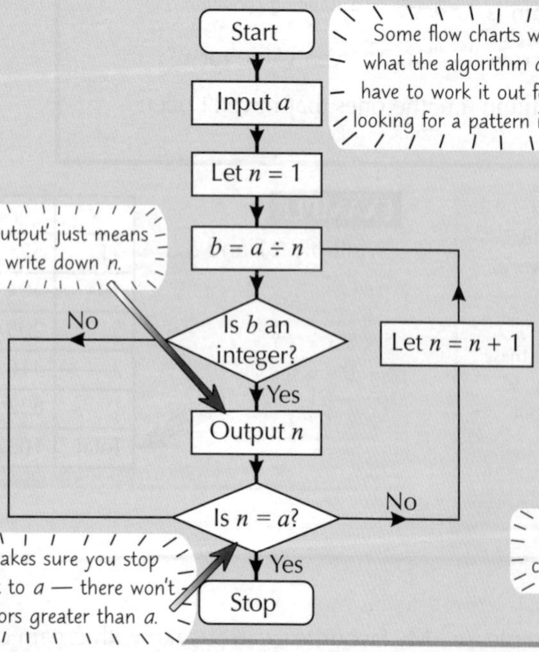

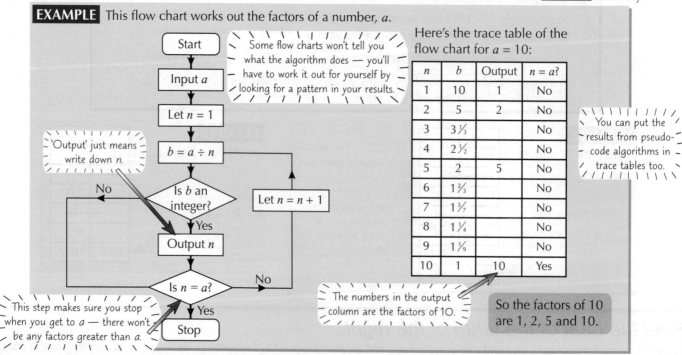

Some flow charts won't tell you what the algorithm does — you'll have to work it out for yourself by looking for a pattern in your results.

'Output' just means write down n.

This step makes sure you stop when you get to a — there won't be any factors greater than a.

Here's the trace table of the flow chart for $a = 10$:

n	b	Output	$n = a$?
1	10	1	No
2	5	2	No
3	$3\frac{1}{3}$		No
4	$2\frac{1}{2}$		No
5	2	5	No
6	$1\frac{2}{3}$		No
7	$1\frac{3}{7}$		No
8	$1\frac{1}{4}$		No
9	$1\frac{1}{9}$		No
10	1	10	Yes

You can put the results from pseudo-code algorithms in trace tables too.

The numbers in the output column are the factors of 10.

So the factors of 10 are 1, 2, 5 and 10.

Sorting

You've probably been able to sort things into alphabetical or numerical order since you were knee-high to a hamster, but in D1 you need to know how to sort things using an algorithm. At least it'll be easy to check your answer.

A Bubble Sort compares Pairs of numbers

The bubble sort is an algorithm that sorts numbers (or letters). It's pretty easy to do, but it can be a bit fiddly, so take care.

The Bubble Sort

1) **Look at the first two numbers in your list. If they're in the right order, you don't have to do anything with them. If they're the wrong way round, swap them. It might help you to make note of which numbers you swap each time.**
2) **Move on to the next pair of numbers (the first will be one of the two you've just compared) and repeat step 1. Keep going through the list until you get to the last two numbers. This set of comparisons is called a pass.**
3) **When you've finished the first pass, go back to the beginning of the list and start again. You won't have to compare the last pair of numbers, as the last number is now in place. Each pass has one less comparison than the one before it. When there are no swaps in a pass, the list is in order.**

If there are n numbers in your list, there will be $n - 1$ comparisons in the first pass.

Stop when there are No More Swaps

It's called the bubble sort because the highest numbers rise to the end of the list like bubbles (apparently). It'll all be a lot easier once you've been through an example...

EXAMPLE Use a bubble sort to write the numbers 14, 10, 6, 15, 9, 21, 17 in ascending order.

First pass:

14, 10, 6, 15, 9, 21, 17	14 and 10 compared and swapped
10, 14, 6, 15, 9, 21, 17	14 and 6 compared and swapped
10, 6, 14, 15, 9, 21, 17	14 and 15 compared — no swap
10, 6, 14, 15, 9, 21, 17	15 and 9 compared and swapped
10, 6, 14, 9, 15, 21, 17	15 and 21 compared — no swap
10, 6, 14, 9, 15, 21, 17	21 and 17 compared and swapped
10, 6, 14, 9, 15, 17, 21	End of first pass.

At the end of the first pass, the highest number has moved to the end of the list.

At the end of the second pass the list is: 6, 10, 9, 14, 15, 17, 21.
At the end of the third pass the list is: 6, 9, 10, 14, 15, 17, 21.
On the fourth pass there are no swaps, so the numbers are in ascending order.

You might have to make Lots of Passes and Comparisons

1) If there are n numbers in the list, the maximum number of passes you might need is n (the numbers will be in order after $n - 1$ passes, but you have to do the nth pass to check). On each pass, you'd only get one number in the right place.
2) On the first pass, you have to make $n - 1$ comparisons, with a maximum of $n - 1$ swaps. On the second pass, you'll have to make $n - 2$ comparisons, as one number is in place from the first pass. On the third pass, there'll be $n - 3$ comparisons etc.

So for a bubble sort with 7 numbers, max. number of comparisons (or swaps) is $6 + 5 + 4 + 3 + 2 + 1 = 21$
Or for a bubble sort with 50 numbers, max. number of comparisons (or swaps) is $\frac{1}{2} \times 49 \times 50 = 1225$

For big lists, use the formula $S_k = \frac{1}{2}k(k + 1)$ for the sum of the first k whole numbers (though if $n = 50$, you want the sum of the first 49 numbers).

3) If you have to make the maximum number of swaps, it means that the original list was in reverse order.
4) You can also use the algorithm to put numbers in descending order — on each comparison, just put the higher number first instead.

Double, double, toil and trouble, fire burn and cauldron bubble...

Bubble sorts are fairly easy, but they aren't the most efficient way of sorting. For a list of 100 numbers, you might have to do 100 passes with 4950 comparisons (or swaps). Fortunately, there are a few more sorting methods that are a bit quicker.

Sorting

The bubble sort can get a bit boring after a while — sometimes when you swap one pair of numbers, you know that you're going to have to make another swap on the next pass. Fortunately the shuttle sort finds a way around that.

*The **Shuttle Sort** is an **Improved** version of the **Bubble Sort***

The main problem with the bubble sort is that you have to make a lot of comparisons. The shuttle sort is a bit more efficient, as it reduces the overall number of comparisons needed.

The Shuttle Sort

1) **Look at the first two numbers in your list. If they're in the right order, leave them as they are. If they're the wrong way round, swap them. This is your first pass.**
2) **Move on to the next pair of numbers (the first will be one of the two you've just compared) and repeat step 1.**
3) **If you made a swap on the second comparison, compare the number you've just swapped to the first number in the list. If they're the wrong way round, swap them. This is your second pass.**
4) **Now move on to the next pair of numbers (the third and fourth in the list) and compare them. If you make a swap, compare the number you've just swapped to the number before it, and swap if you need to. Keep working backwards until either you get to a number it can't be swapped with or you reach the beginning of the list. This is another pass completed.**
5) **Continue through the list, repeating step 4 until you get to the last number in the list. Each time, compare backwards until you can't swap the number anymore.**

A pass in a shuttle sort is different from a pass in a bubble sort.

The shuttle sort usually needs fewer comparisons than the bubble sort — though you'll have to make the same number of swaps. After the first pass, the first 2 numbers will be in the right places. After the second pass, the first three numbers will be in the right places and so on. For a list of *n* numbers, you'll need to make *n* – 1 passes to get the list in order.

*The **Shuttle Sort** will have the **Same Number** of **Swaps** as the **Bubble Sort***

Even though you'll have to make the same number of swaps as you would using the bubble sort, you'll make fewer comparisons — this means the shuttle sort's more efficient.

EXAMPLE Use a shuttle sort to write the numbers 14, 10, 6, 15, 9, 21, 17 in ascending order.

This is the same set of numbers used in the bubble sort example on the previous page.

First pass:	14, 10, 6, 15, 9, 21, 17	14 and 10 compared and swapped
Second pass:	10, 14, 6, 15, 9, 21, 17	14 and 6 compared and swapped
	10, 6, 14, 15, 9, 21, 17	10 and 6 compared and swapped
Third pass:	6, 10, 14, 15, 9, 21, 17	14 and 15 compared — no swap
Fourth pass:	6, 10, 14, 15, 9, 21, 17	15 and 9 compared and swapped
	6, 10, 14, 9, 15, 21, 17	14 and 9 compared and swapped
	6, 10, 9, 14, 15, 21, 17	10 and 9 compared and swapped
	6, 9, 10, 14, 15, 21, 17	6 and 9 compared — no swap
Fifth pass:	6, 9, 10, 14, 15, 21, 17	15 and 21 compared — no swap
Sixth pass:	6, 9, 10, 14, 15, 21, 17	21 and 17 compared and swapped
	6, 9, 10, 14, 15, 17, 21	15 and 17 compared — no swap
	6, 9, 10, 14, 15, 17, 21	End of shuttle sort.

There are 7 numbers, so you have to make 7 – 1 = 6 passes. It was much quicker than the bubble sort — there were only 11 comparisons (instead of 18). You still needed to make 7 swaps though.

Not quite as exciting as a space shuttle — more exciting than a shuttlecock...

The shuttle sort is trickier than a bubble sort, but it's more efficient. Make sure you don't get them mixed up — or even worse, combine the two. Attempting to do a buttle sort in the exam might make the examiners laugh, but it won't get you any marks.

Sorting

Next up: the Shell sort. This algorithm was invented by Donald Shell in 1959 and regularly appears in 'Top Ten Algorithms of All Time' lists. OK, so I made that last bit up, but it is a pretty good algorithm.

*The **Shell Sort** shares the numbers between **Subsets***

To do a Shell sort, you have to divide the list of numbers into smaller lists (called subsets) that are easy to sort.

The Shell Sort

1) **Share the n numbers in your list between $n/2$ subsets. Put the first number in the first subset, the second number in the second subset etc. Write the first subset on one line, the next one underneath it and so on.**
2) **Sort each subset into order.**
3) **Put the subsets back together, working down your sets (so take the first number from the first subset, then the first number from the second subset etc.). This is one pass.**
4) **Share your new list between subsets — this time, you want half the number of subsets you had before, and sort each subset. Put your list back together like you did in step 3.**
5) **Repeat step 4) until you only have one subset — sort this subset using a shuttle sort (see p.202).**

If n is odd, ignore the remainder — so if n = 9, there would be 9/2 = 4.5 = 4 subsets.

If you had 5 subsets at step 1, you now want 5/2 = 2.5 = 2 subsets.

*Be **Careful** with the **Layout***

Make sure you get the layout right — if you have your subsets lined up in the wrong way, it's easy to make a mistake when you're putting the list back together. You don't have to write them out diagonally, it just makes them easier to read.

EXAMPLE Sort the numbers 12, 32, 24, 11, 17, 5, 19, 13, 22 into ascending order using the Shell sort.

1) n = 9, so you want 9/2 = 4.5 = 4 subsets:

First subset:	12		17		22	
Second subset:		32		5		
Third subset:			24		19	
Fourth subset:				11		13

Put the first item into the first subset, the second item into the second subset, the third item into the third subset and the fourth item into the fourth subset. Now put the fifth item into the first subset, the sixth item into the second subset, the seventh item into the third subset and the eighth item into the fourth subset. Finally, put the ninth item into the first subset.

2) Sorting these subsets into order gives:

First subset:	12		17		22	
Second subset:		5		32		
Third subset:			19		24	
Fourth subset:				11		13

3) Put this list back together:

12, 5, 19, 11, 17, 32, 24, 13, 22

This order comes from reading down the diagonal columns.

4) Now divide the list from step 3) into 4/2 = 2 subsets:

First subset:	12		19		17		24		22
Second subset:		5		11		32		13	

Putting these subsets in order gives:

First subset:	12		17		19		22		24
Second subset:		5		11		13		32	

This becomes: 12, 5, 17, 11, 19, 13, 22, 32, 24

5) Finally, arrange the list into 2/2 = 1 subset.
To order this, use a shuttle sort (see p.202).

After the shuttle sort, the list is:

5, 11, 12, 13, 17, 19, 22, 24, 32

The shuttle sort will be much quicker now the list is almost in order.

The Shell sort is pretty good — by breaking down the numbers into subsets, it makes them much quicker to sort. It works particularly well on big sets of numbers.

But I'm allergic to shellfish...

The shell suit was popular in the late 80s and early 90s. It was a brightly-coloured tracksuit made of a shiny (and flammable) material — wearing them on bonfire night was a bit of a risk. This probably won't be on the exam but it's useful information.

Sorting

Here's the last sorting method — it's called the quick sort, and with a name like that, you'd hope it was pretty damn quick. I'd rank its speed as midway between a sloth and a fighter pigeon (and those pigeons can be rather fast).

A **Quick Sort** breaks the list into **Smaller Lists**

The quick sort algorithm works by choosing a pivot (see below) which breaks down the list into two smaller lists, which are then broken down in the same way until the numbers are in order.

The Quick Sort

1) **Choose a pivot. Move any numbers that are less than the pivot to a new list on the left of it and the numbers that are greater to a new list on the right. Don't change the order of the numbers though.**
2) **Repeat step 1) for each of the smaller lists you've just made. You'll need to choose new pivots for the new lists.**
3) **When every number has been chosen as a pivot, you can stop, as the list is in order.**

Sometimes the smaller lists will only have one number in them — you don't need to do anything with these as they're already in order.

The pivot can be any number in the list — it's usually easiest to use the first number though.

Keep track of the **Pivots** you use

It's a good idea to circle or underline the pivots you're using at each step of the quick sort — it helps you keep track of where you're up to. In the example below, the numbers are written in grey when they're in the correct place.

EXAMPLE Sort the numbers 54, 36, 29, 56, 45, 39, 32, 27 into ascending order using a quick sort.

Use the first item in the list as the pivot. So the pivot is 54.

Now make new lists by moving numbers < 54 to the left, and numbers > 54 to the right. Don't reorder the numbers, just write them down in the order they appear in the original list on the correct side of the pivot.

These are all the numbers < 54 (in the same order as in the original list). → 36, 29, 45, 39, 32, 27 (l_1), 54, 56 (l_2) ← This is the only number > 54.

The list has been divided into smaller lists, l_1 and l_2. l_2 only has one item in it, so 56 is in the correct place (it's a pivot in a list of its own). Use the first number in l_1 (36) as the new pivot.
Rearranging around the new pivot gives:

29, 32, 27 (l_3), 36, 45, 39 (l_4), 54, 56 ← 54 and 56 are in the right place now.

This time, both lists have more than one item so there are two new pivots: 29 from list l_3 and 45 from list l_4.
Rearranging again using the new pivots gives:

27 (l_5), 29, 32 (l_6), 36, 39 (l_7), 45, 54, 56

The lists (l_5, l_6 and l_7) only have one item in them, so all the numbers are in the correct place — the list is now in order.

Like the Shell sort, the quick sort breaks the numbers down into subsets which are easier to sort. It can be a bit fiddly, but usually involves fewer steps than the other sorting algorithms.

Quick — let's get out of here...

Phew, that's the last sorting method you need for the exam. Any of them could come up, so make sure you know them all. Of course, you may be thinking that it would be much easier to sort them just by looking at them, but the point of these algorithms is that you could program them into a computer and sort huge lists of numbers really quickly.

D1 Section 1 — Practice Questions

That's the first section of D1 done and dusted — and I don't think it was that bad. Once you get your head round the fact that 'algorithm' is just a fancy word for 'set of instructions', you're over the first hurdle. Time now for a few questions to check you know your stuff. Here are some straightforward ones to start off with.

Warm-up Questions

1) For each of the following sets of instructions, identify the input and output.
   a) a recipe for vegetable soup,
   b) directions from Leicester Square to the Albert Hall,
   c) flat-pack instructions for building a TV cabinet.
2) Use the Russian Peasant algorithm to multiply 17 and 56.
3) What are diamond-shaped boxes in flow diagrams used for?
4) Use the flow chart on p.200 to work out the factors of 16.
5) Use a bubble sort to write the numbers 72, 57, 64, 54, 68, 71 in ascending order. How many passes do you need to make?
6) If you had to put a list of 10 numbers in order using a bubble sort, what is the maximum number of comparisons you'd need to make?
7) Use a shuttle sort to write the numbers 21, 11, 23, 19, 28, 26 in ascending order.
8) Sort the numbers 101, 98, 79, 113, 87, 108, 84 into ascending order using a Shell sort.
9) Sort the numbers 0.8, 1.2, 0.7, 0.5, 0.4, 1.0, 0.1 into ascending order using a quick sort. Write down the pivots you use at each step.

If those warm-up questions have left you hungry for more, here are some tasty exam-style questions for you to sink your teeth into. They'll need a bit more work than the warm-up questions, but it'll be worth it, I promise.

Exam Questions

1   77   83   96   105   78   89   112   80   98   94

a) Use a quick sort to arrange the list of numbers above into ascending order. You must clearly show the pivots you use at each stage.

(5 marks)

b) A list of six numbers is to be sorted into ascending order using a bubble sort.

(i) Which number(s) will definitely be in the correct position after the first pass?

(1 mark)

(ii) Write down the maximum number of passes and the maximum number of swaps needed to sort a list of six numbers into ascending order using a bubble sort.

(2 marks)

2 a) Rearrange the following set of numbers into ascending order using a shuttle sort.

1.3   0.8   1.8   0.5   1.2   0.2   0.9

State how many passes you made.

(5 marks)

b) How many comparisons and swaps were made on the first pass?

(1 mark)

D1 Section 1 — Practice Questions

If you thought I'd leave you hanging with just one page of questions, you're very much mistaken. I'd never be that mean to you — and to prove it, here's another page full of beautiful exam questions.

3 Consider the following algorithm:

Step 1: Input A, B with $A < B$
Step 2: Input $N = 1$
Step 3: Calculate $C = A \div N$
Step 4: Calculate $D = B \div N$
Step 5: If both C and D are integers, output N
Step 6: If $N = A$, then stop. Otherwise let $N = N + 1$ and go back to Step 3.

a) Carry out the algorithm with $A = 8$ and $B = 12$. Record your results.

(3 marks)

b) (i) What does this algorithm produce?

(1 mark)

(ii) Using your answer to part (i) or otherwise, write down the output that would be produced if you applied the algorithm to $A = 19$ and $B = 25$, and explain your answer. You do not need to carry out the algorithm again.

(2 marks)

4 Mark Dan Adam James Stella Helen Robert

Use a quick sort to list the above names in alphabetical order. Show clearly the pivots you use.

(4 marks)

5 The numbers 54, 71, 63, 72, 68, 59, 60, 55 are to be sorted into ascending order using a Shell sort. After the first pass, the order of the numbers is 54, 59, 60, 55, 68, 71, 63, 72

a) State the number of comparisons and the number of swaps made in the first pass.

(2 marks)

b) Complete the Shell sort, stating the new order after each pass.

(5 marks)

Graphs

You probably reckon you're an old pro at graphs. But the graphs coming up are rather different. For a start, there's not a scrap of squared paper in sight. Fret not — soon they'll be as innocuous to you as a bar chart.

Graphs have **Points** Connected by **Lines**

Here's the definition of a graph:

A **graph** is made up of **points** (called vertices or nodes) joined by **lines** (called edges or arcs).

Graphs can be used to model or solve real-life problems. Here are a few examples:

Use Graphs to Show **Connections** in **Geographical Problems**

In this graph, the vertices represent towns and the edges represent roads.

The graph doesn't show where the towns are in relation to each other — just how they are linked by roads.

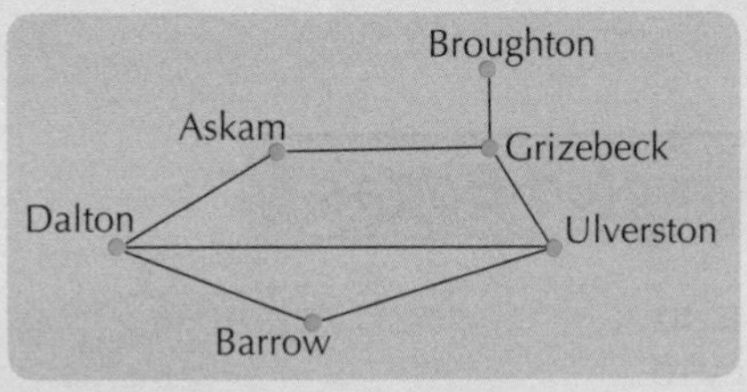

Use **Bipartite Graphs to** Model **Matching Problems**

Bipartite graphs have two sets of vertices. An edge only ever joins a vertex in one set to a vertex in the other set. You can never connect two vertices in the same set together. There's more on bipartite graphs in Section 6.

This graph shows the jobs a group of students would prefer to do at the end-of-term barbecue.

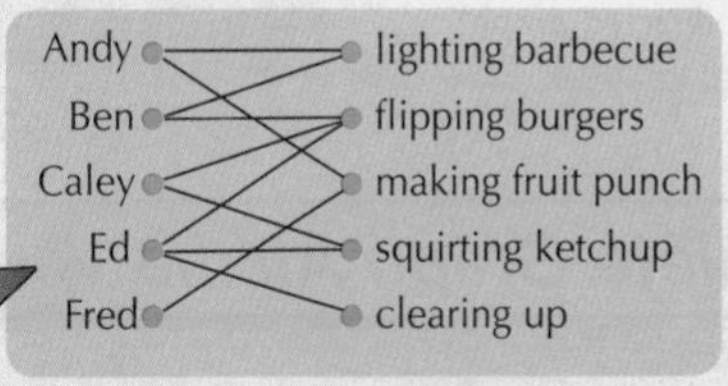

A **Path** is a Route in the Graph Which **Doesn't Repeat Any Vertices**

A path is a sequence of edges that flow on, end to end. The only thing is, you can't go through a vertex more than once.

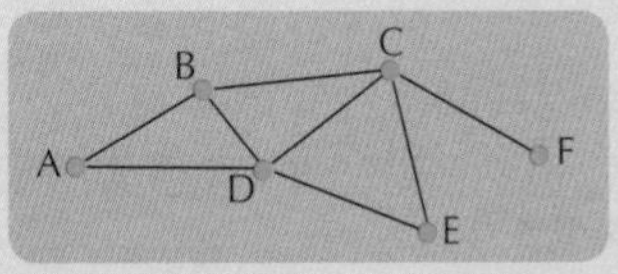

One possible path here is ABDECF. Another is CBAD.

DCECF isn't a path because you go through vertex C more than once.

A **Cycle** is a Path that **Brings you Back** to Your Starting Point

1) A cycle (or circuit) is a closed path. The end vertex is the same as the start vertex.
2) So on the graph above, ABDA is a cycle. Other cycles are ABCDA and CEDBC.

You still have to follow the rules for paths — so you can't go through a vertex twice. So ABDCEDA isn't a cycle because it goes through vertex D twice.

A **Hamiltonian Cycle** Goes Through **Each Vertex Exactly Once**

1) A Hamiltonian cycle is a cycle which goes through each vertex exactly once.
2) Like all cycles, it brings you back to the start vertex.
3) Not all graphs contain Hamiltonian cycles — those that do are called Hamiltonian graphs.
4) The one above isn't a Hamiltonian graph (but it would be if you removed vertex F and the edge attached to it — then it would contain the Hamiltonian cycle ABCEDA).

There are heaps of other examples on page 224.

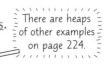

I've got geographical problems — I have no sense of direction...

Make sure you learn all the definitions — it's easy to get them muddled up, and you'll need them later on. These examples might look pretty simple, but you can use them to solve much harder problems.

Graphs

There are loads of different types of graph. And you need to know all their names. It's not hard — it just takes learning.

***Weighted Graphs** have a **Number** on Each Edge*

1) Weighted graphs, or networks, have a number associated with each edge (called the weight of the edge).
2) Weights often give you lengths — like in this network showing points in a nature reserve and the footpaths joining them. They sometimes give you costs or times too.

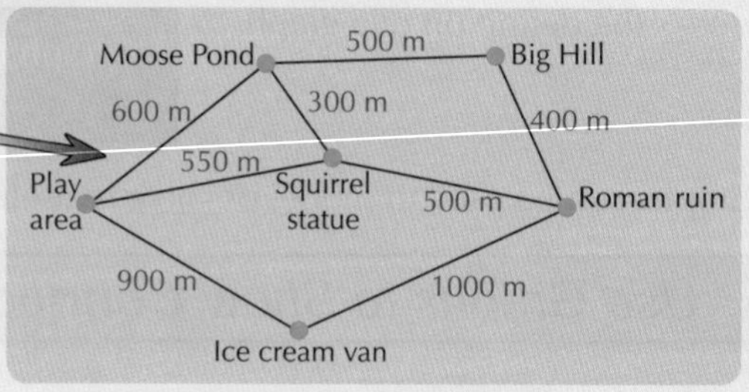

Directed Graphs** have **Directed Edges

Sometimes edges have directions, e.g. to show one-way streets. If they do, they're called directed edges and the graph is a directed graph or a digraph.

A graph without any directed edges is sometimes referred to as an undirected graph.

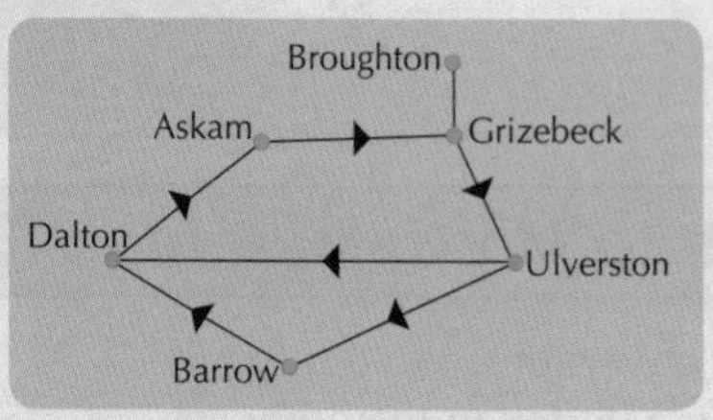

The edges on this digraph show the bus routes between the towns. A bus goes from Dalton to Askam, but not from Askam to Dalton.

There's no direction on the edge connecting Broughton and Grizebeck, so the buses run in both directions.

*In a **Complete Graph All** the Vertices are **Directly Connected***

1) Each vertex in a complete graph is joined directly to every other vertex by exactly one edge.

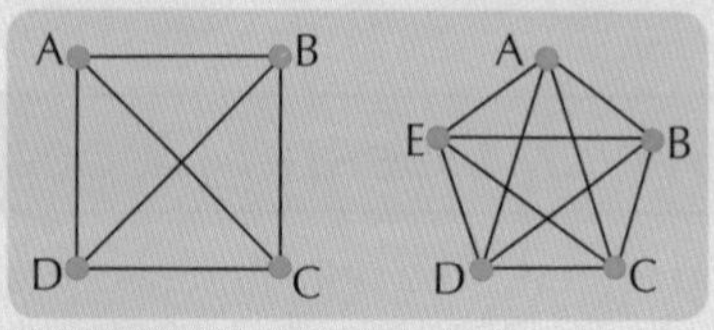

The notation K_n is used for a complete graph with n vertices. So the graphs here are K_4 (left) and K_5 (right).

2) In a complete bipartite graph, each vertex is joined directly to each vertex in the opposite set.

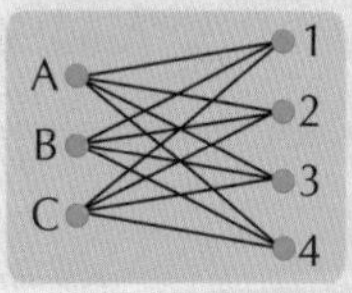

For complete bipartite graphs, the notation $K_{m,n}$ is used. m is the number of vertices in one set, n is the number in the other. The number of edges altogether is $m \times n$. The graph shown here is $K_{3,4}$.

*Graphs **Without Loops** or **Multiple Edges** are **Simple***

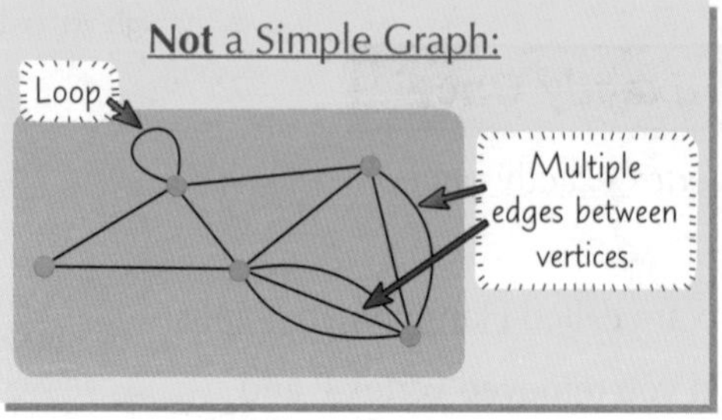

1) Graphs can have more than one edge between a pair of vertices. There can also be loops connecting vertices to themselves.
2) Graphs without any loops or multiple edges between vertices are called simple graphs (like the nature reserve and bus route examples above).

And your leg bone's connected to your foot bone...

A page of lovely definitions. You'll need them very, very soon, so learn them and then scribble them out from memory. Then check back to the page to see what you missed. Graphs are all over the place — the London Underground map is one.

Graphs

If these graph definitions were a tunnel, you'd now be seeing a dot of light in the distance.

Graphs Can be **Connected** or Not Connected

1) Two vertices are connected if there's a path between them — the path can pass through other vertices.
2) A graph is connected if all its vertices are connected.

This graph is connected.

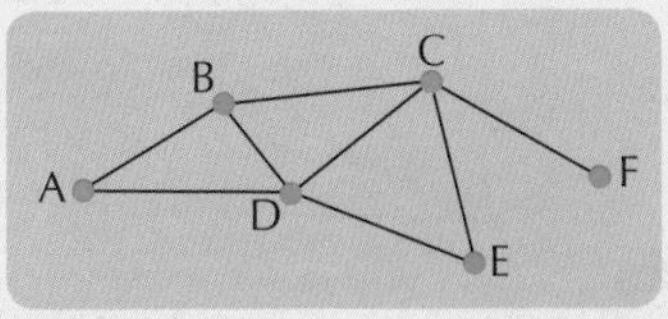

This graph is NOT connected — you can't get from some vertices to others. E.g. there's no path between B and C, or between E and A.

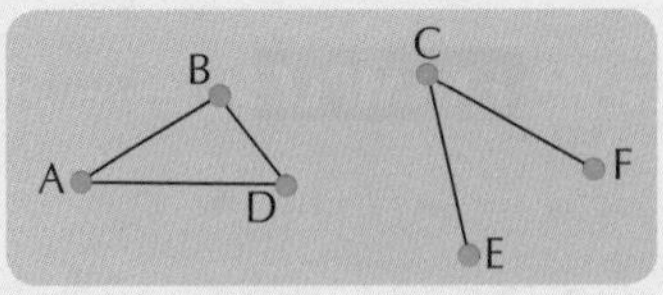

Trees are **Graphs** that have **No Cycles**

See p207 for more on cycles.

They also have to be connected graphs (see above).
Both graphs here are connected — but only the first is a tree (the graphical type of tree that is).

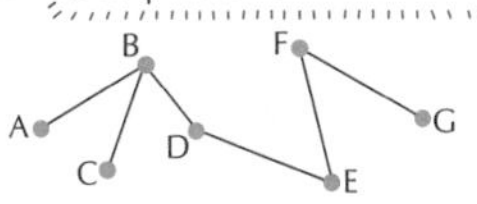

This is a tree — there are no cycles.

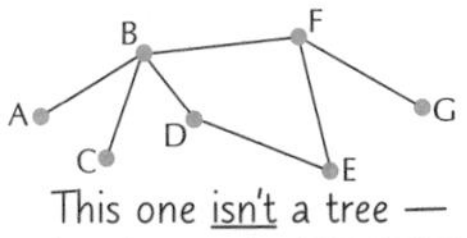

This one isn't a tree — there's a cycle (BDEFB.)

Spanning Trees are **Subgraphs** that are Also Trees

Subgraphs are just bits of another graph.

1) Spanning trees can't be just any old subgraph though — they have to include all the vertices.
2) So if you're asked to draw a spanning tree of a graph, you can only delete edges from the original graph.
3) The number of edges in a spanning tree is always one less than the number of vertices.
4) Like most things in this section, a few diagrams speak a thousand words...

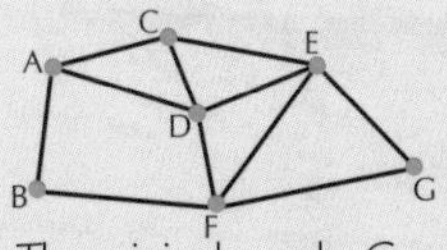

The original graph, G.

A spanning tree of G. There's no cycle, and it contains all the vertices from G.

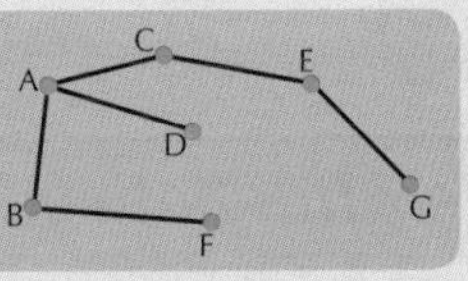

Another spanning tree of G. There are plenty more that could be drawn too.

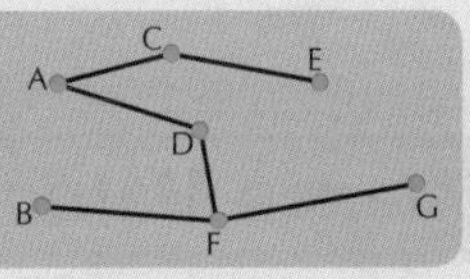

The **Degree** of a **Vertex** is the **Number of Lines** coming off it

1) Here's the formal definition:

The **degree** or **valency** of a vertex is the number of edges connected to it.

EXAMPLE Calculate the degree of each vertex in the graph below.

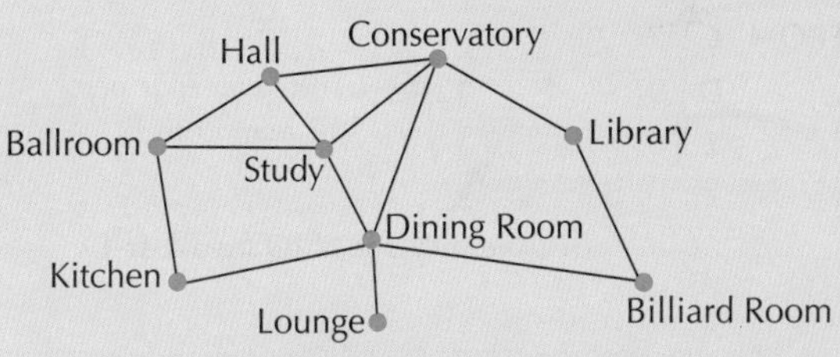

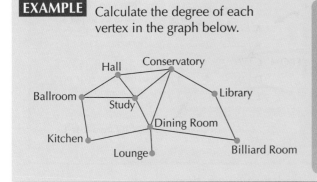

Vertex	Degree
Ballroom	3
Billiard Room	2
Conservatory	4
Dining Room	5
Hall	3
Kitchen	2
Library	2
Lounge	1
Study	4

Rules of Degrees

The sum of the degrees is always double the number of edges.
(It's a count of how many edge ends there are.)

So, the sum of degrees is always even.

Here, there are 13 edges and the sum of the degrees is 26 (2 × 13).

2) A vertex with an odd degree is odd, and one with an even degree is — wait for it — even.
So the Billiard Room, Conservatory, Kitchen, Library and Study are all even, and the rest are odd.

Now what would a tree do with a cycle anyway...

I didn't put the bit about the sum of degrees always being even in *just* because it's fascinating. (Although it certainly is.) It actually comes up quite often in exam questions. They don't ask you straight out though. They'll phrase it in a cunning way. E.g. Why can't you have a graph consisting of three vertices with odd degrees? Just stay calm and it'll be OK.

Graphs

We're now cruising at the correct altitude and you may unfasten your seat belts and move around the cabin. But don't leave the aircraft, it's cold out there and there's no floor.

Adjacency Matrices show the *Number of Links* between *Vertices*

1) To draw an adjacency (or incidence) matrix from a graph, go through each space in the matrix and count the number of direct connections from the vertex at the left of the row to the vertex at the top of the column.

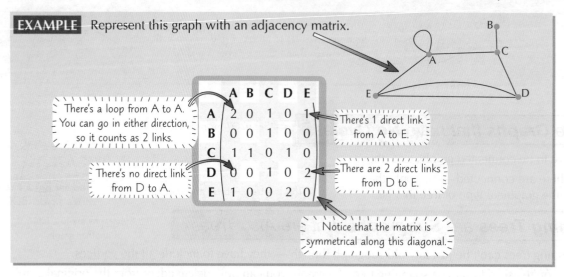

EXAMPLE Represent this graph with an adjacency matrix.

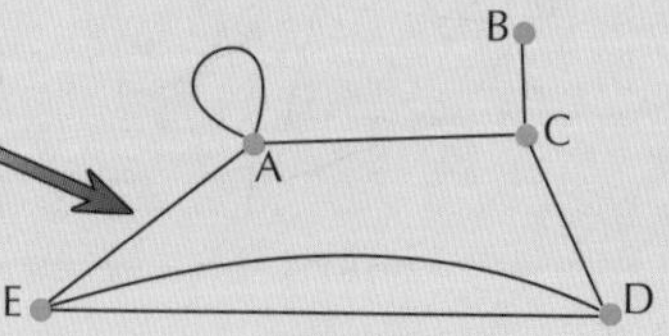

	A	B	C	D	E
A	2	0	1	0	1
B	0	0	1	0	0
C	1	1	0	1	0
D	0	0	1	0	2
E	1	0	0	2	0

There's a loop from A to A. You can go in either direction, so it counts as 2 links.

There's no direct link from D to A.

There's 1 direct link from A to E.

There are 2 direct links from D to E.

Notice that the matrix is symmetrical along this diagonal.

2) You might have to draw a graph from an adjacency matrix too. Mark the vertices first. Then go through each space in the matrix and draw edges in. But remember, one edge between A and B is the same as one between B and A — don't draw two.

Distance Matrices show the *Weights* between Vertices

1) To draw a distance matrix from a weighted digraph (p. 208), go through each space in the matrix and write down the weight between the two vertices. You only include direct links — don't start adding weights together.
2) Be really careful with directed edges. A weight on a directed edge only goes in one space of the matrix, as below.

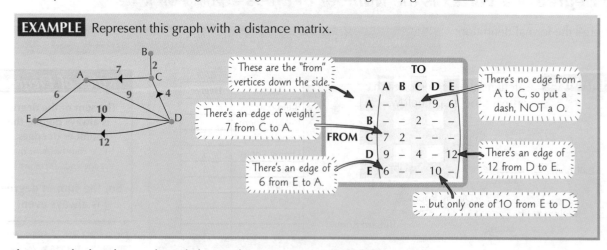

EXAMPLE Represent this graph with a distance matrix.

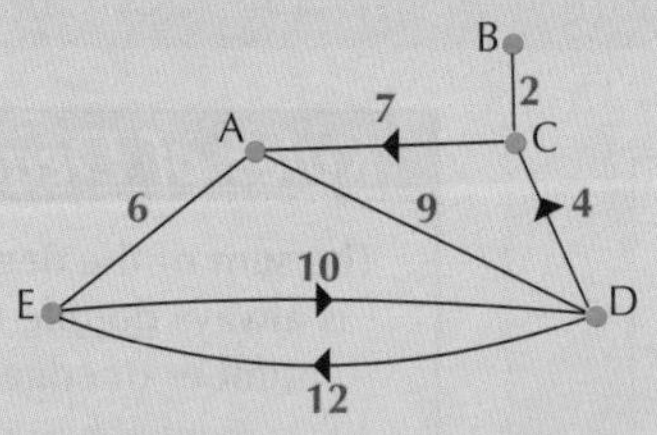

These are the "from" vertices down the side.

FROM \ TO	A	B	C	D	E
A	–	–	–	9	6
B	–	–	2	–	–
C	7	2	–	–	–
D	9	–	4	–	12
E	6	–	–	10	–

There's no edge from A to C, so put a dash, NOT a 0.

There's an edge of weight 7 from C to A.

There's an edge of 12 from D to E...

There's an edge of 6 from E to A.

... but only one of 10 from E to D.

3) If you're asked to draw a digraph from a distance matrix, mark the vertices, then go through the matrix, adding edges and weights in. If a weight only appears once in the matrix, the edge must be directed, so add an arrow.

Adjacency matrices — good for snuggling up...

The best thing you can do now is cover up the graphs above and practise drawing them from the matrices. Then draw the matrices from the graphs. And watch out for those blasted directed edges — they ruin the aesthetically pleasing symmetry.

Minimum Spanning Trees

The stuff you've seen so far in this section might seem like it's been dreamed up by bored mathematicians to provide you with useless facts to learn. But this minimum spanning tree stuff is actually rather handy in the real world.

A Minimum Spanning Tree is the Shortest way to Connect All the Points

See page 209 if you've forgotten what spanning trees are.

A **minimum spanning tree** (MST) is a spanning tree where the total length of the edges is as **small as possible**.

An MST is also known as a minimum connector.

Minimum spanning trees come in handy for cable or pipe-laying companies. If they need to connect several buildings in a town, say, they'd want to find the cheapest path — this may be the shortest route, or have the easiest ground to dig up.

Kruskal's Algorithm Finds Minimum Spanning Trees

Being absolutely certain that you've got the minimum spanning tree is tricky, so using an algorithm helps.

KRUSKAL'S ALGORITHM

1) List the edges in ascending order of weight.
2) Pick the edge of least weight — this starts the tree.
3) Look at the next edge in your list.
   - if it forms a cycle, DON'T use it and go on to the next edge.
   - if it doesn't form a cycle, add it to the tree.
4) Repeat step 3 until you've joined all the vertices.

This is a 'greedy algorithm'. You make the choice that seems best at each stage, without worrying about later choices.

If there are n vertices in a network, there will always be (n – 1) edges in an MST.

EXAMPLE Use Kruskal's algorithm to find a minimum spanning tree for this network.

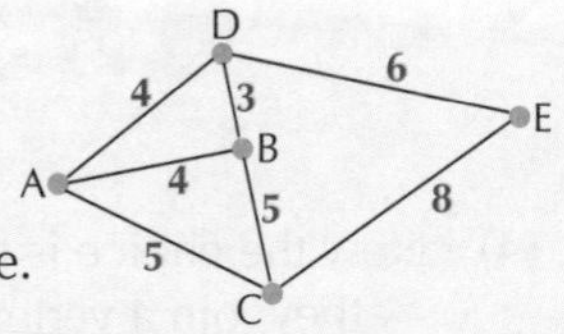

Edge	Weight	Used?
BD	3	✓
AD	4	✓
AB	4	×
BC	5	✓
AC	5	×
DE	6	✓
CE	8	×

You could put AD and AB in either order, because they're the same weight.

1) The shortest edge is BD, so that starts the tree.
2) The next edge, AD, doesn't form a cycle, so adds on. Here's the tree so far.

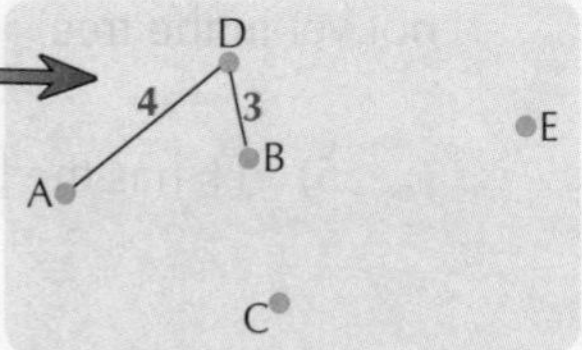

3) AB would form a cycle, so it's rejected.
4) BC doesn't form a cycle, so it's added on.
5) You continue down the list like this until all vertices are connected, meaning the MST is complete.

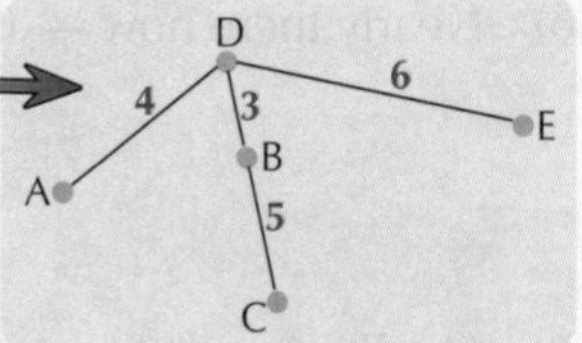

You'll often be asked to find the weight of your MST. This is easy — just add all the weights up. So the weight of this MST is 3 + 4 + 5 + 6 = 18.

There are often a few different MSTs that can be found for a network — and you might be asked to find them.

In the list above, AB could have been used instead of AD, or AC instead of BC. All the different combinations of these edges give three more MSTs:

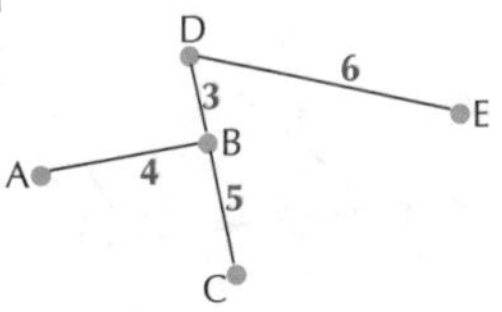

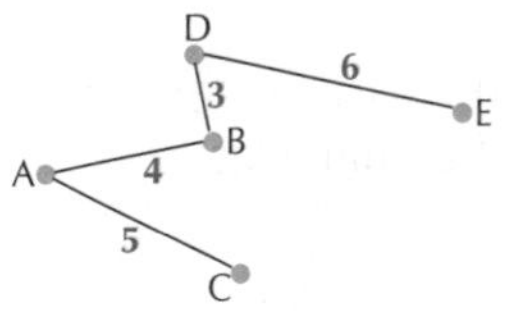

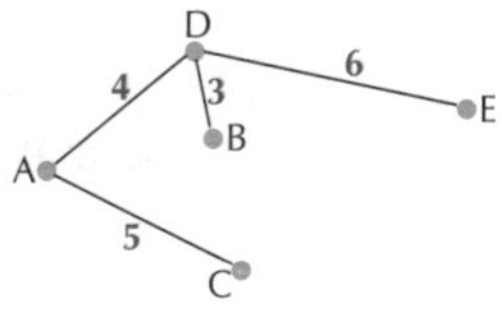

MST = Mathematical Stress and Tension...

All the minimum spanning trees for a network have the same total weight — if they didn't, they wouldn't all be minimums. Kruskal's algorithm works fine, but you can't do it straight from matrix forms of graphs, which is what computers often use. And if you've got a whopping network, this might be important. This is where Prim's algorithm is tops — see the next page...

Minimum Spanning Trees

Prim's algorithm does exactly the same job as Kruskal's algorithm. I'd love to say you only need to learn the one you like best, but that'd be a fib. You've got to learn them both, of course.

Prim's Algorithm *Finds Minimum Spanning Trees Too*

PRIM'S ALGORITHM

1) Pick a vertex, any vertex — this starts the tree.
2) Choose the edge of least weight that'll join a vertex in the tree to one not yet in the tree.
3) Repeat step 2 until you've joined all the vertices.

If there's more than one to pick from, just choose randomly.

EXAMPLE Use Prim's algorithm to find a minimum spanning tree for the network on the right.

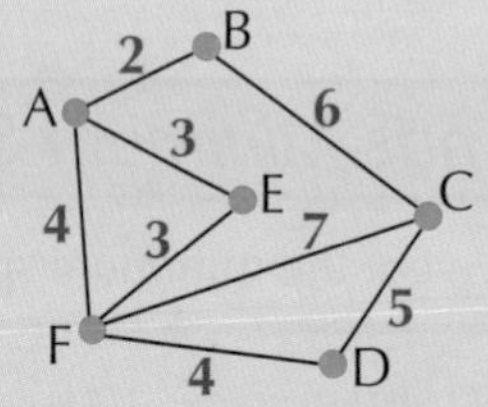

1) I'm going to randomly pick E to start the tree.

There are two edges that join E to vertices not yet in the tree. EA and EF. They're both the same weight (3), so I'll randomly pick EA.

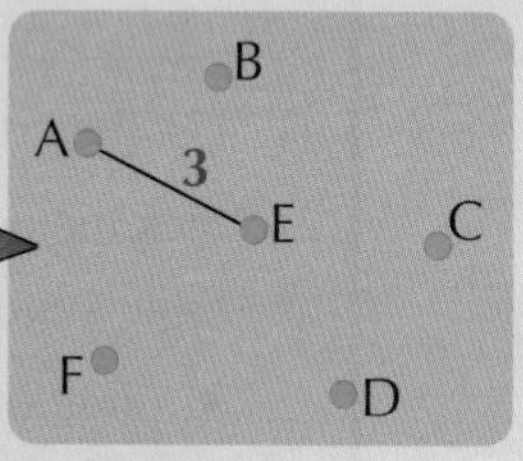

2) There are three edges that join a vertex in the tree (A or E) to a vertex outside the tree: AB (2), AF (4), EF (3).

3) AB has the least weight, so that's the one to add.

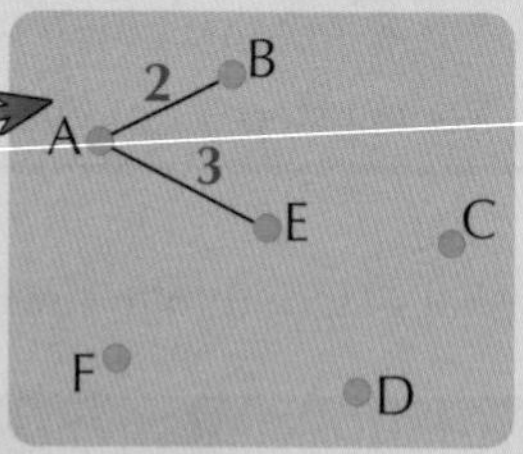

4) Now the choice is from edges AF (4), EF (3) or BC (6) — they join a vertex in the tree (A, B or E) to one not yet in the tree.

5) EF has the least weight, so it's added on next.

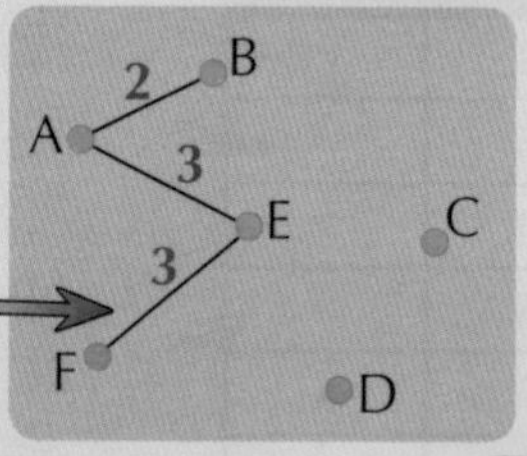

You don't have to check for cycles like you did with Kruskal's algorithm. Connecting to a vertex outside the tree will never make a cycle (one less thing to worry about).

6) Nearly there now — the next choice is from BC (6), FC (7) or FD (4).

7) FD has least weight, so that's the one to add.

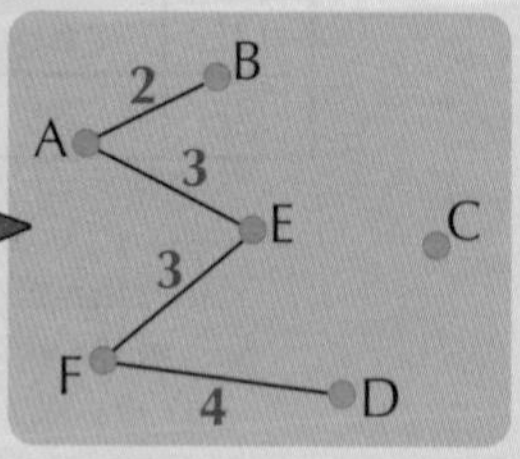

8) Finally, C could be joined by BC (6), FC (7) or DC (5).

9) DC has the least weight, so that's the final edge.

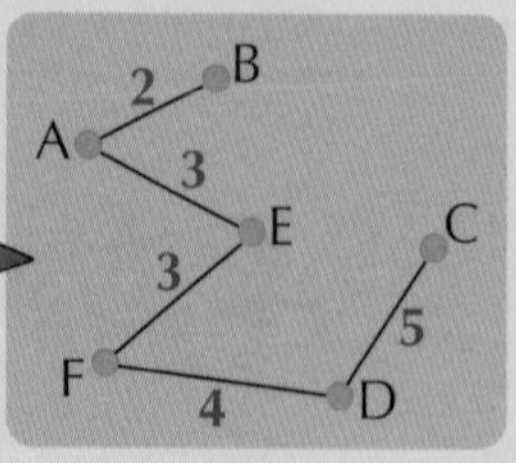

You might be asked to comment on how appropriate your solution is. E.g. imagine the edges are bus routes of different lengths, and you're considering cancelling some. Even though the towns A-F are still all connected in the minimum spanning tree, people who live at B and work at C are likely to be very cross.

Wasn't Mr Prim one of the Mr Men?...

Practice, practice, practice is definitely the key for this algorithm too. There are only three steps to memorise, so that's not too tricky. You've just got to apply them accurately — so check you've considered all the possible edges. Exam questions usually ask you to show your working, or to state the order that you add the edges in, so winging it with a different method won't do.

Minimum Spanning Trees

The easiest way of putting a graph into a computer is to use a matrix. And the reason why Prim's algorithm is so useful is that it can be used on a distance matrix. It looks like a horrible mess of crossing out and circling randomly, but stay with me, and it'll all come clear.

Prim's Algorithm *Can be Used on* ***Distance Matrices***

PRIM'S ALGORITHM

1) Pick any vertex to start the tree.
2) Cross out the row for the new vertex and circle the column for it.
3) Look for the smallest weight that's in ANY circled column AND isn't yet crossed out. Circle it. This is the next edge to add to the tree. The row it's in gives you the new vertex.
4) Repeat steps 2 and 3 until all the rows are crossed out.

EXAMPLE Use Prim's algorithm to find a minimum spanning tree for the graph represented by this distance matrix.

	A	B	C	D	E
A	–	4	–	3	3
B	4	–	5	–	6
C	–	5	–	8	7
D	3	–	8	–	2
E	3	6	7	2	–

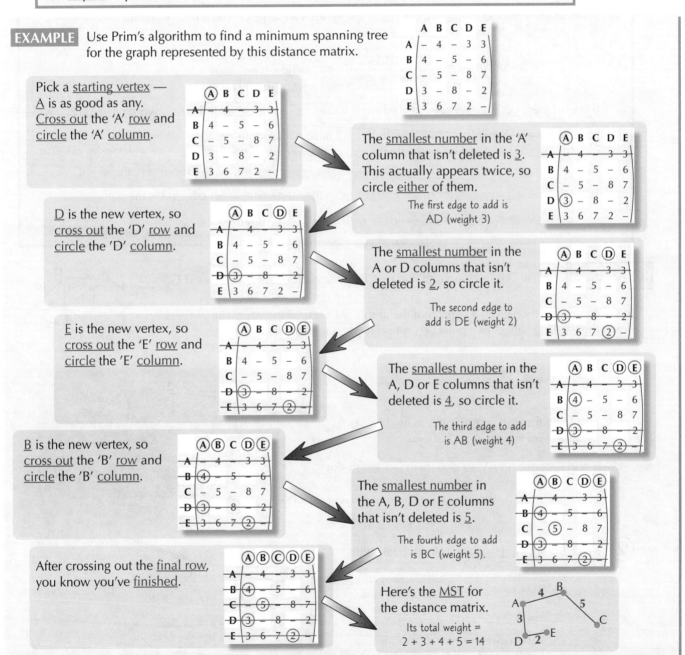

Pick a starting vertex — A is as good as any. Cross out the 'A' row and circle the 'A' column.

	Ⓐ	B	C	D	E
~~A~~	~~–~~	~~4~~	~~–~~	~~3~~	~~3~~
B	4	–	5	–	6
C	–	5	–	8	7
D	3	–	8	–	2
E	3	6	7	2	–

The smallest number in the 'A' column that isn't deleted is 3. This actually appears twice, so circle either of them.

The first edge to add is AD (weight 3)

	Ⓐ	B	C	D	E
~~A~~	~~–~~	~~4~~	~~–~~	~~3~~	~~3~~
B	4	–	5	–	6
C	–	5	–	8	7
D	③	–	8	–	2
E	3	6	7	2	–

D is the new vertex, so cross out the 'D' row and circle the 'D' column.

	Ⓐ	B	C	Ⓓ	E
~~A~~	~~–~~	~~4~~	~~–~~	~~3~~	~~3~~
B	4	–	5	–	6
C	–	5	–	8	7
~~D~~	~~③~~	~~–~~	~~8~~	~~–~~	~~2~~
E	3	6	7	2	–

The smallest number in the A or D columns that isn't deleted is 2, so circle it.

The second edge to add is DE (weight 2)

	Ⓐ	B	C	Ⓓ	E
~~A~~	~~–~~	~~4~~	~~–~~	~~3~~	~~3~~
B	4	–	5	–	6
C	–	5	–	8	7
~~D~~	~~③~~	~~–~~	~~8~~	~~–~~	~~2~~
E	3	6	7	②	–

E is the new vertex, so cross out the 'E' row and circle the 'E' column.

	Ⓐ	B	C	Ⓓ	Ⓔ
~~A~~	~~–~~	~~4~~	~~–~~	~~3~~	~~3~~
B	4	–	5	–	6
C	–	5	–	8	7
~~D~~	~~③~~	~~–~~	~~8~~	~~–~~	~~2~~
~~E~~	~~3~~	~~6~~	~~7~~	~~②~~	~~–~~

The smallest number in the A, D or E columns that isn't deleted is 4, so circle it.

The third edge to add is AB (weight 4)

	Ⓐ	B	C	Ⓓ	Ⓔ
~~A~~	~~–~~	~~4~~	~~–~~	~~3~~	~~3~~
B	④	–	5	–	6
C	–	5	–	8	7
~~D~~	~~③~~	~~–~~	~~8~~	~~–~~	~~2~~
~~E~~	~~3~~	~~6~~	~~7~~	~~②~~	~~–~~

B is the new vertex, so cross out the 'B' row and circle the 'B' column.

	Ⓐ	Ⓑ	C	Ⓓ	Ⓔ
~~A~~	~~–~~	~~4~~	~~–~~	~~3~~	~~3~~
~~B~~	~~④~~	~~–~~	~~5~~	~~–~~	~~6~~
C	–	5	–	8	7
~~D~~	~~③~~	~~–~~	~~8~~	~~–~~	~~2~~
~~E~~	~~3~~	~~6~~	~~7~~	~~②~~	~~–~~

The smallest number in the A, B, D or E columns that isn't deleted is 5.

The fourth edge to add is BC (weight 5).

	Ⓐ	Ⓑ	C	Ⓓ	Ⓔ
~~A~~	~~–~~	~~4~~	~~–~~	~~3~~	~~3~~
~~B~~	~~④~~	~~–~~	~~5~~	~~–~~	~~6~~
C	–	⑤	–	8	7
~~D~~	~~③~~	~~–~~	~~8~~	~~–~~	~~2~~
~~E~~	~~3~~	~~6~~	~~7~~	~~②~~	~~–~~

After crossing out the final row, you know you've finished.

	Ⓐ	Ⓑ	Ⓒ	Ⓓ	Ⓔ
~~A~~	~~–~~	~~4~~	~~–~~	~~3~~	~~3~~
~~B~~	~~④~~	~~–~~	~~5~~	~~–~~	~~6~~
~~C~~	~~–~~	~~⑤~~	~~–~~	~~8~~	~~7~~
~~D~~	~~③~~	~~–~~	~~8~~	~~–~~	~~2~~
~~E~~	~~3~~	~~6~~	~~7~~	~~②~~	~~–~~

Here's the MST for the distance matrix.

Its total weight = 2 + 3 + 4 + 5 = 14

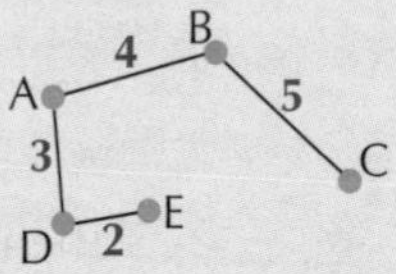

This method also calculates the birthdate of your one true love...

When you're doing this yourself, you don't have to draw out the matrix a zillion times like I've done. Phew, I hear you say. I just wanted you to see all my steps. It took me ages, so do admire them all, and then have a practice for yourself.

Dijkstra's Algorithm

This is another of those algorithms that look really, really complicated. But when you've learnt the steps, you can string them together pretty rapidly. This one does a different job from the last two, so don't just skim the first bit.

Dijkstra's Algorithm *Finds the* ***Shortest Path*** *between Two Vertices*

1) If you're driving between two cities with a complicated road network between them, it's good to be able to work out which route is quickest (just like satnavs do).
2) Dijkstra's algorithm is a foolproof way to do this. Basically, you label each vertex with the length of the shortest path found so far from the starting point. If you find a shorter path, then you change the label. You keep doing this until you're sure that you've got the shortest distance to it.

Weights can also represent costs — you might want to find the cheapest route.

DIJKSTRA'S ALGORITHM

1) Give the Start vertex the final value '0'.

   Once you've given a vertex a final label, you can't change it.

2) Find all the vertices directly connected to the vertex you've just given a final value to. Give each of these vertices a working value.

   Working value = Final value at previous vertex + weight of edge between previous vertex and this one.

   If one of these vertices already has a working value, replace it ONLY if the new working value is lower.

3) Look at the working values of vertices that don't have a final value yet. Pick the smallest and make this the final value of that vertex.

   If two vertices have the same smallest working value, pick either.

4) Now repeat steps 2 and 3 until the End vertex has a final value (this is the shortest path length).
5) Trace the route backwards (from the End vertex to the Start vertex). An edge is on the path if:

   Weight of edge = Difference in final values of the edge's vertices

EXAMPLE: Use Dijkstra's algorithm to find the shortest route between A and G.

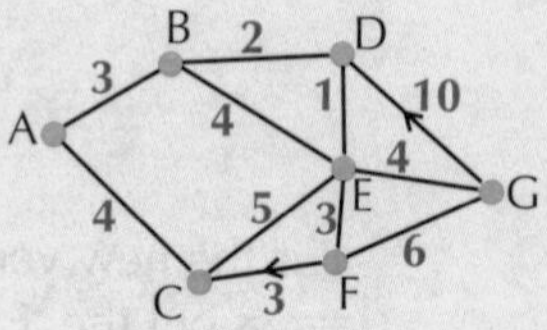

In the exam, you'll be given a version of the graph. At each vertex you write down the working value, then put a box round it when it becomes a final value.

① First label the start vertex with the final value '0'.
Put a box round the O as it's the final value.

B and C connect to A, so give them working values.

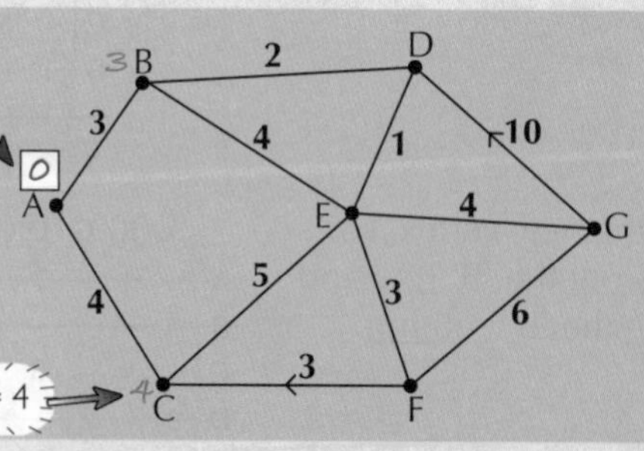

previous final value + connecting edge weight = O + 4 = 4

② Make the smallest working value a final value.

3 is the lowest working value, so make that a final value by putting it in a box.

Then give the vertices connecting to B (i.e. D and E) working values.

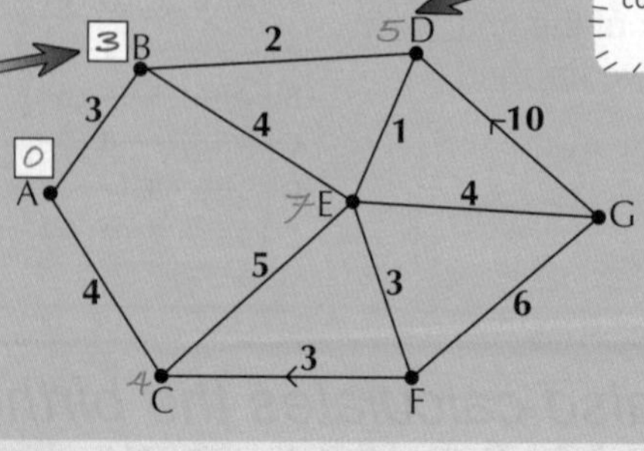

previous final value + connecting edge weight = 3 + 2 = 5

Dijkstra's Algorithm

③ Again, make the smallest working value a final value.

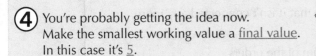

4 is the lowest working value, so make that a final value.

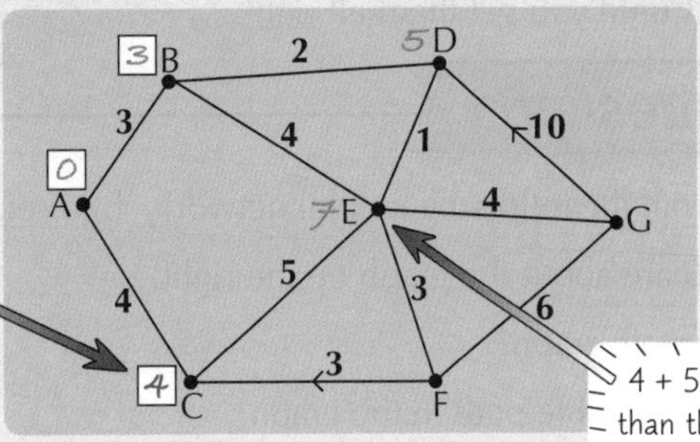

4 + 5 = 9, but this is greater than the current working value for E, so you DON'T replace it.

Then give all vertices without final values connecting to C a working value. It's only E, because F is connected by a directed edge that only goes from F to C, not from C to F.

④ You're probably getting the idea now. Make the smallest working value a final value. In this case it's 5.

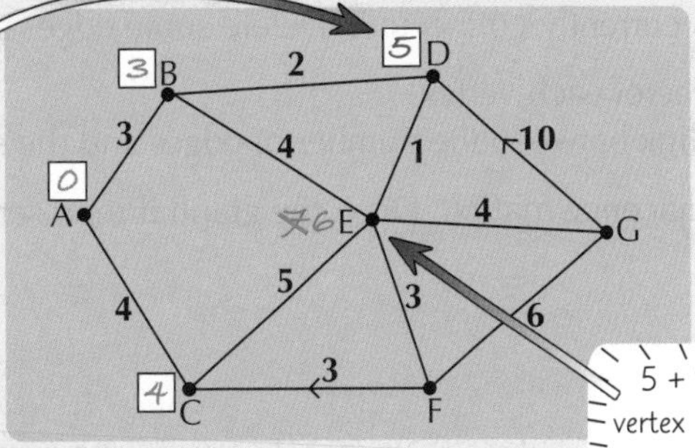

Give all vertices without final values connecting to D a working value. It's only E in this case. (G is connected by a directed edge running in the opposite direction. And of course B already has a final value.)

5 + 1 = 6 (you're coming from vertex D this time). This is smaller than the current working value for E, so you DO replace it.

⑤ E has the smallest working value (6), (in fact the only working value) so make that its final value.

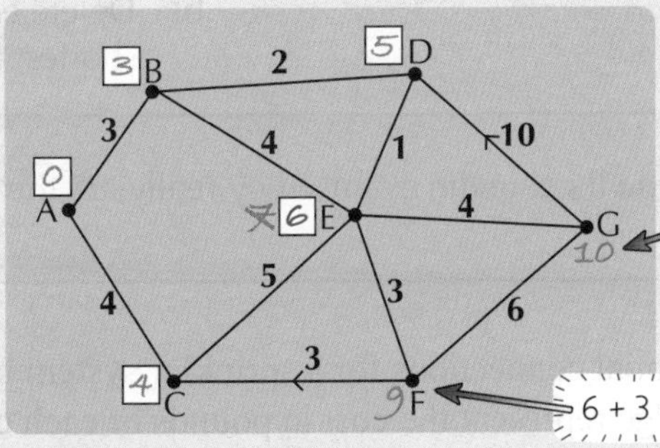

6 + 4 = 10

6 + 3 = 9

Give the vertices connecting to E (F and G) working values.

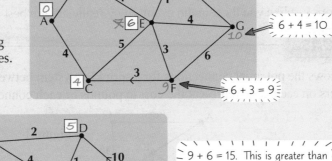

⑥ F has the smallest working value of 9, so make that its final value.

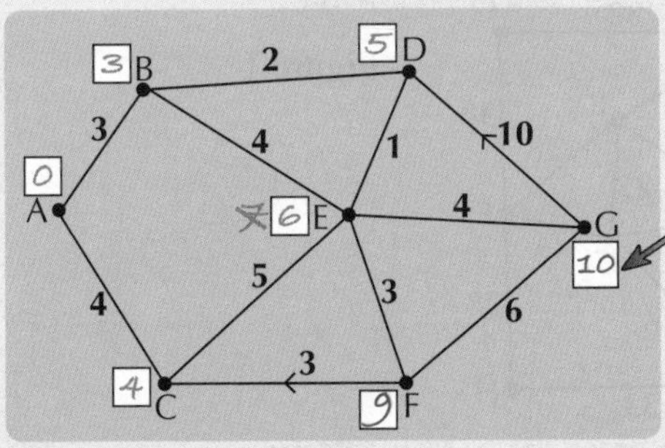

9 + 6 = 15. This is greater than the current working value for G, so you leave it as 10.

G (the End vertex) is the only vertex left. Make its working value the final value — this is the length of the shortest route.

⑦ Now it's time to figure out the route. An edge is on the path if:

Weight of edge = Difference in final values of edge's vertices

Working backwards from G (the End vertex):

- The edge EG is on the path, because the difference in the final values of E and G is 4, which is the weight of the edge EG.
- The edge DE is on the path, because the difference in the final values of D and E is 1, which is the weight of the edge DE.
- And so on, all the way back to A.

So the shortest route from A to G is ABDEG.

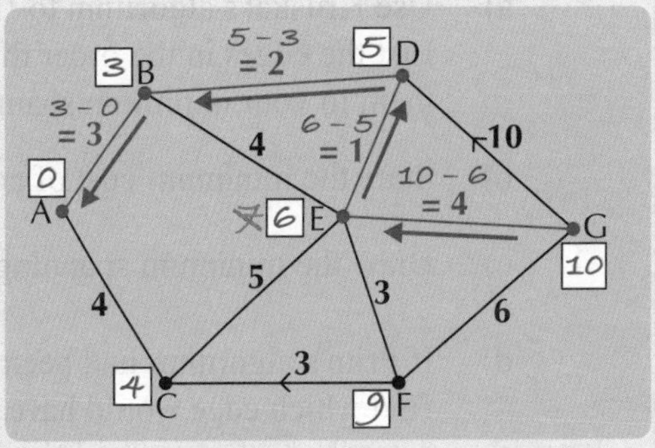

It's a bit like the Time Warp — now get it right...

Sometimes there's more than one shortest route. If there is, you'll find two possible edges leading off from a vertex when you're tracing the route back. Remember to read the question carefully — they might want both of the shortest routes.

D1 Section 2 — Practice Questions

Right, now for this algorithm. 1) Try the questions. 2) Check your answers. 3) Reread the page on any you got wrong. 4) Repeat steps 1-3 until you get them all right.

Warm-up Questions

1) Explain what the following are; a) network, b) directed graph, c) tree, d) spanning tree

Questions 2–6 are about the graph on the right.

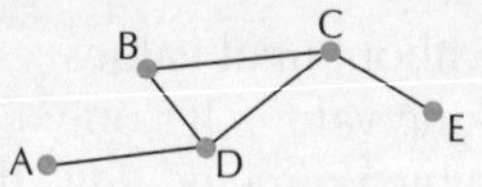

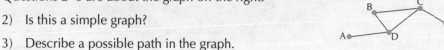

2) Is this a simple graph?
3) Describe a possible path in the graph.
4) Describe a possible cycle in the graph.
5) The graph is currently connected. Delete some edges so that it isn't connected any more.
6) List the degree of each vertex.
Explain the link between the number of edges and the sum of the orders.
7) Here's an adjacency matrix. Draw the graph it represents.

	A	B	C	D
A	0	0	1	1
B	0	2	0	1
C	1	0	0	1
D	1	1	1	0

8) Using Dijkstra's algorithm on the graph on page 214: a) Find the shortest route from A to F.
b) Delete edge BD. Now find the shortest route from A to G.

Now for some questions just like you'll get in the exam. They really are the best sort of practice you can do.

Exam Questions

1 Figure 1 shows the potential connections for a sprinkler system between greenhouses at a plant nursery. The numbers on each edge represent the cost in pounds of each connection.

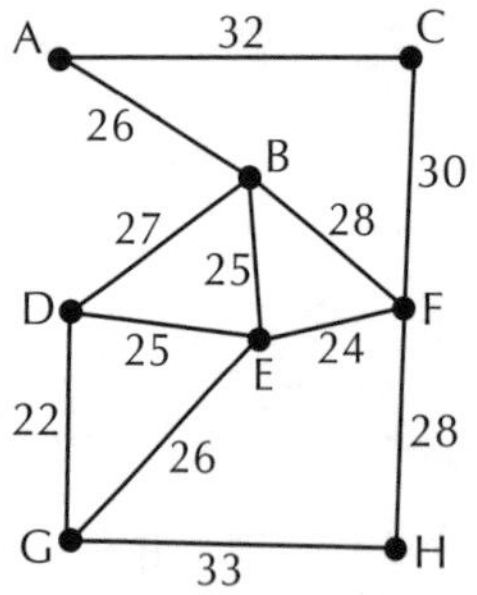

Figure 1

a) Use Kruskal's algorithm to find a minimum spanning tree for the network in Figure 1. List the edges in the order that you consider them and state whether you are adding them to your minimum spanning tree.
(3 marks)

b) State the minimum cost of connecting the sprinkler system.
(1 mark)

c) Draw the minimum spanning tree obtained in a).
(2 marks)

d) If Prim's algorithm had been used to find the minimum spanning tree, starting from E, find which edge would have been the final edge added. Show your working.
(2 marks)

e) State two advantages of Prim's algorithm over Kruskal's algorithm for finding a minimum spanning tree.
(2 marks)

D1 Section 2 — Practice Questions

Keep going. You can do it. No need for gas and air yet.

2

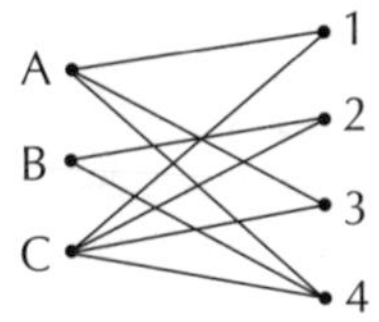

Figure 2

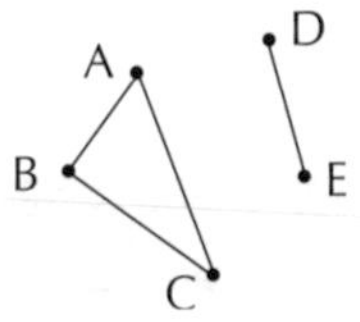

Figure 3

a) Name the type of graph drawn in Figure 2. *(1 mark)*

b) State the number of edges that would need to be added to Figure 3 to make the graph connected. *(1 mark)*

c) What is the sum of the orders of the vertices in Figure 3? *(1 mark)*

d) Explain why it is impossible to add edges to Figure 3 so that all vertices have an odd order. *(2 marks)*

3 The table shows the lengths, in miles, of the roads between five towns.

	A	B	C	D	E
A	–	14	22	21	18
B	14	–	19	21	20
C	22	19	–	21	15
D	21	21	21	–	24
E	18	20	15	24	–

a) Use Prim's algorithm, starting from A, to find a minimum spanning tree for this table. Write down the edges in the order that they are selected. *(3 marks)*

b) Draw your tree and state its total weight. *(2 marks)*

c) State the number of other spanning trees that are the same length as your answer in part (b). *(1 mark)*

4 The diagram below shows a network of forest paths. The number on each edge represents the time, in minutes, required to walk along the path.

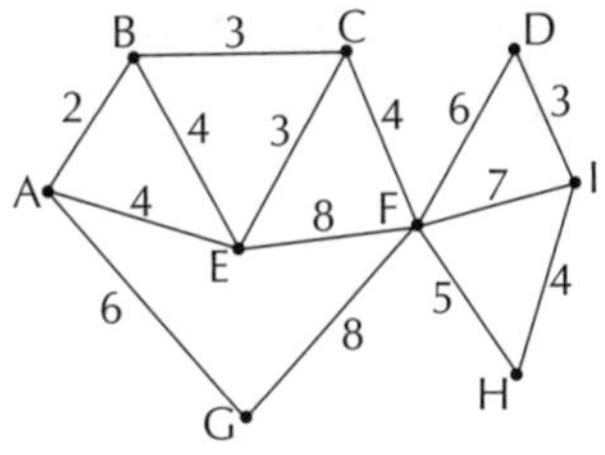

a) Write down the number of edges in a minimum spanning tree of the network shown. *(1 mark)*

b) Use Dijkstra's algorithm to find the fastest route from A to I. State how long the route will take. *(6 marks)*

c) A new path, requiring x minutes to walk along, is to be made between G and H. The new path reduces the time required to walk between A and I. Find and solve an inequality for x. *(2 marks)*

Traversable Graphs

You know that puzzle where you have to draw the shape on the right without taking your pencil off the paper? Well, you're about to find out the logic behind it and exactly which points you can start drawing from.

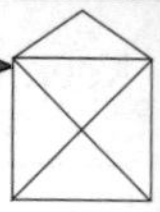

Graphs can be Eulerian...

If all the vertices in a graph have an even degree, the graph is Eulerian.

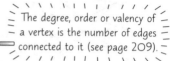

The degree, order or valency of a vertex is the number of edges connected to it (see page 209).

1) These three graphs are all Eulerian. Every vertex is even.

The numbers show the degrees.

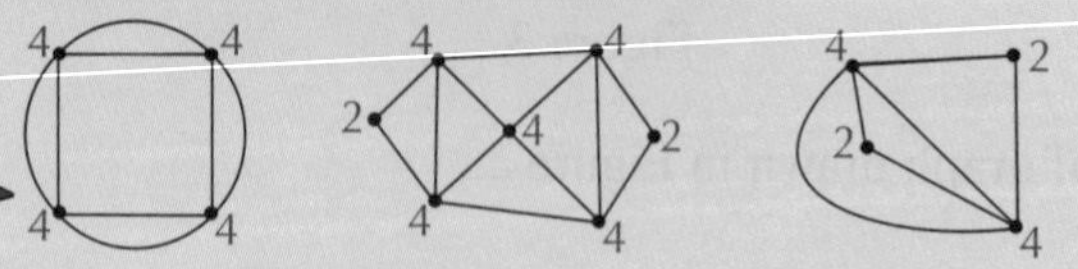

2) Eulerian graphs are traversable. This means it's always possible for you to start at any point, draw along each edge exactly once without taking your pen off the paper, and end up back at your starting position (this is an Eulerian cycle). Not every route works, but there'll definitely be some that do.
3) Or to look at it another way, if the graph represents roads, it's possible to walk down each of them exactly once before getting back to your starting point.

EXAMPLE Find a route that traverses the graph on the right.

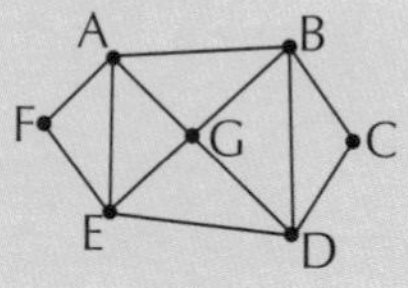

The graph is Eulerian. So you can start at any point.
A possible route is: AGDBCDEGBAFEA.
Another route is: EDCBDGAFEABGE.

No matter which route you take, it'll always involve passing through the same number of vertices (13 in this case).

...Semi-Eulerian...

If exactly two vertices have an odd degree, and the rest are even, the graph is semi-Eulerian.

1) These three graphs are all semi-Eulerian. There are exactly two odd vertices.

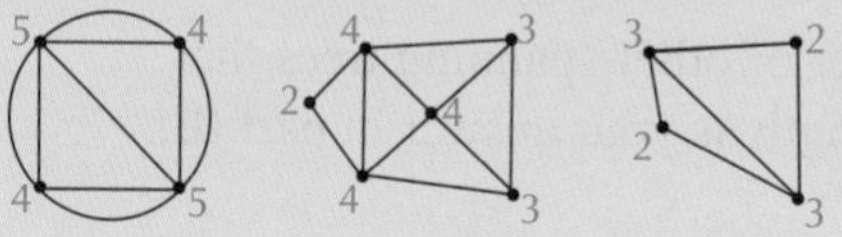

A route that goes along each edge once, but doesn't end up back at its starting point is an Eulerian trail.

2) Semi-Eulerian graphs are semi-traversable. This means it's possible to go along every edge on the graph exactly once, but ONLY if you start at one odd vertex and end up at the other odd vertex.

EXAMPLE Find a route that traverses the graph on the right.

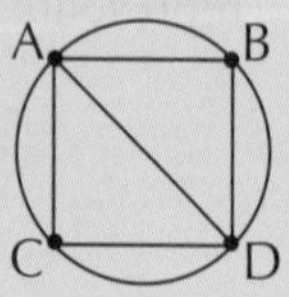

This graph is semi-Eulerian, so you have to start and end at the odd vertices (A and D).
A possible route is: ABDCA around the square, then ABDCA around the circle, then across the diagonal to D.

Remember — the sum of degrees is always even (see page 209). This means there'll always be an even number of odd vertices in a graph. So you'll never have 3 or 5 odd vertices, but you might have 4 or 6.

...or Neither

1) If a graph has more than two odd vertices, you can't traverse it.
2) There's no route that travels along each edge exactly once. You have to go along some of them twice.

Some lovely non-Eulerian graphs. Try to find a route through them that goes along each edge exactly once. No taking your pen off the paper. See — you can't, can you?

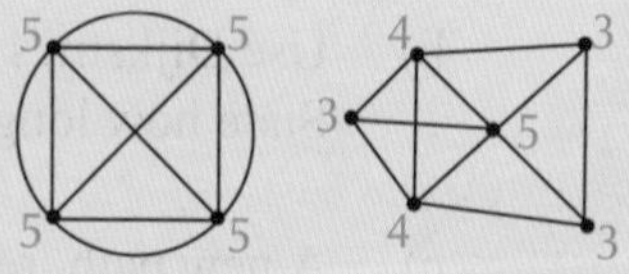

Eul 'er now and it might stop 'er squeaking...

Right. So, Eulerian = traverse from any point and get back to starting point. Semi-Eulerian = traverse from one odd vertex to the other odd vertex. Non-Eulerian = you can't traverse. You have to go over some edges twice. And that's about it.

Route Inspection Problems

The route inspection problem is also called the Chinese postman problem. (It's named after a Chinese guy called Kwan Mei-Ko who discovered it in the 1960s. He wasn't a postman though — he was a mathematician.)

The **Route Inspection Algorithm** Finds the **Shortest Route** Covering **All Edges**

1) Route inspection problems ask you to find the shortest route through a network that goes along each edge before returning to the starting point.
2) It's the route that, say, a railway engineer would take if he had to inspect all the tracks.
3) In postman terms, the postman wants to find the shortest route that allows him to deliver letters to every street in a city, and brings him back to his starting point for a cup of tea.
4) And you'll never guess what. There are algorithms for solving these types of problems.
5) The first step is always to consider whether the graph is Eulerian, semi-Eulerian, or neither.

See the previous page.

A pre-1960s postman still manages to raise a smile, despite walking further than he has to each day.

Eulerian Graphs are Most **Straightforward**

Remember — in an Eulerian graph, all vertices have an even degree.

Length of inspection route in an Eulerian graph = Weight of network

1) Eulerian graphs are traversable. So whatever point you start from, you can travel along each edge exactly once, and end up back at your start point.
2) Because you've gone down each edge once, you find the length of the route by just adding up all the edge weights.

EXAMPLE Find an inspection route for the network on the right. Your route must start and finish at A. State the length of the route.

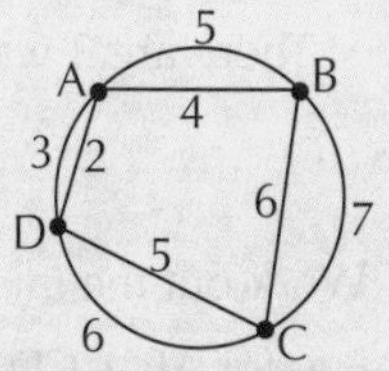

The graph is Eulerian — the vertices all have even degrees (they're all 4). So the graph is traversable, and a possible route is: ABCDABCDA (once round the quadrilateral, then once around the circle).

Length of the route = sum of weights = 4 + 6 + 5 + 2 + 5 + 7 + 6 + 3 = 38

Semi-Eulerian Graphs are a **Bit Trickier**

In a semi-Eulerian graph, exactly two vertices have an odd degree, remember.

In semi-Eulerian networks, you have to repeat the shortest path between the two odd vertices in an inspection route. You can think of it as adding edges to make the network Eulerian so that it can be traversed.

This formula lets you find the length of the inspection route:

Length of inspection route in a semi-Eulerian graph = Weight of network + weight of the shortest path between the two odd vertices

EXAMPLE Find an inspection route for the network on the right. Your route must start and finish at B. State the length of the route.

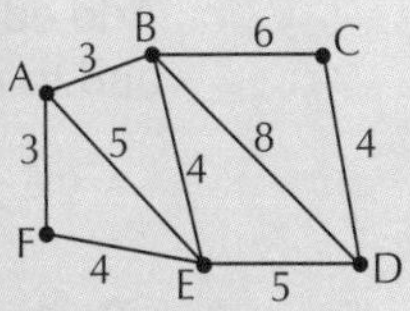

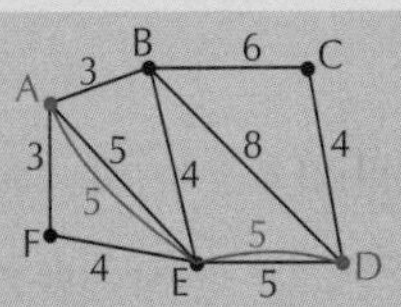

The graph is semi-Eulerian — the odd vertices are A and D. The shortest path between them is AED, of length 10 (5 + 5). So extra edges AE and ED are added.

This is worked out just by looking at the different possibilities, i.e. 'by inspection'.

A possible path is BCDEFABDEAEB.

The weight of the network = 3 + 6 + 4 + 5 + 4 + 3 + 5 + 4 + 8 = 42

Length of inspection route = Weight of network + weight of shortest path between odd vertices
= 42 + 10 = 52

Rather than use the formula, you could just find the length of your path by adding all the weights.

You only have to do this if you want to start and end at the same vertex. If you can pick and choose, just start at one odd vertex and end at the other. Then the length of the route will equal the network's weight (see the previous page).

Of course, during postal strikes, the postman stays put...

In the exam, you'll either be able to figure out the distance between the odd vertices by just looking, or you'll have worked it out earlier in the question using Dijkstra's algorithm (p.214). Right, now to find out what to do with non-Eulerian graphs...

Route Inspection Problems

The previous page told you how to find the shortest inspection route for Eulerian and semi-Eulerian networks. Now it's time to find out how to do it with non-Eulerian networks. It's a teensy bit more complicated — but I guess that's why it comes up more often in exam questions.

Learn the **Route Inspection Algorithm** for **Non-Eulerian** Networks

1) Non-Eulerian networks have more than two odd vertices.
2) But the exam board promises not to give you a network with more than four odd vertices, and networks never have odd numbers of odd vertices (see p.218). So you only have to worry about doing this stuff with four odd vertices.

FINDING THE SHORTEST INSPECTION ROUTE FOR A NON-EULERIAN NETWORK:
With this method you can start at any vertex, and you'll end up at the same one.
Pair the four odd vertices in all the possible ways, find the pairing that gives you the smallest total, then repeat the paths between these pairs of vertices.

This'll become clearer with an example:

EXAMPLE Find an inspection route for the network below. Your route must start and finish at E. State the length of the route.

1) Pick out the vertices with odd degrees.

They're marked in pink on this network — A, B, C, D

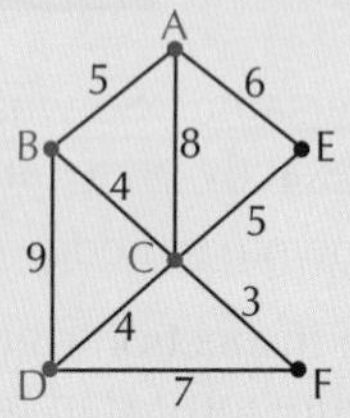

2) Pair the vertices in all the ways possible.

There are 3 ways to pair 4 vertices: AB + CD
AC + BD
AD + BC

3) Work out the minimum total distance for each pairing.

E.g. for AB + CD, you find the smallest distance between AB, and the smallest distance between CD, and add them together:

AB + CD = 5 + 4 = 9
AC + BD = 8 + 8 = 16
AD + BC = 12 + 4 = 16

There are loads of paths from A to D. By inspection, ACD is shortest (8 + 4 = 12), so use that one.

4) Pick the pairing with the smallest total distance.

With a length of 9, it's AB + CD that has the smallest total. AB and CD will be the paths you repeat in the inspection route.

5) Add extra edges along each path in your pair.

Add edges to repeat the path between A and B and to repeat the path between C and D.

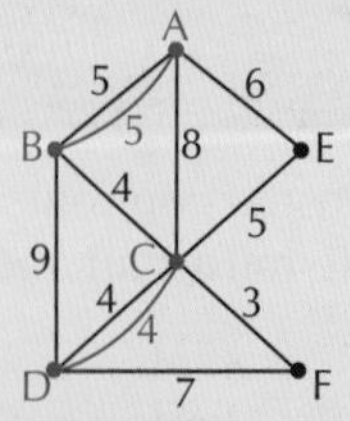

6) The graph is now Eulerian — so you can find an inspection route through it.

A possible route starting and finishing at E = EABDCABCFDCE. ⟸ I've marked the pink paths for you.

Alternatively, you could just add up all the edges on your path.

7) Add the lengths of the new edges to the weight of the network to get the length of the inspection route.

The length of the route = weight of network + length of extra edges
= 51 + 9 = 60

Weight of network = 5 + 6 + 8 + 4 + 5 + 9 + 4 + 7 + 3 = 51

He ain't heavy — so he's the best pairing possible...

Even if it's screamingly obvious which pairing is going to have the smallest value, always show the examiner that you've considered all the possible pairings and didn't just get lucky this once. Oh, and the most direct path mightn't be the shortest, so do check carefully, or it'll cock the whole thing up, and you'll shed marks like a German Shepherd sheds fur.

Route Inspection Problems

Those examiners don't want you thinking they're the least bit cuddly. After asking you to find an inspection route for a network starting and finishing at a certain point, they'll often tell you to find the minimum inspection route if you can start and finish wherever you like.

Starting and Finishing at **Different Odd Vertices** *Always* **Shortens** *the Route*

In a network with four odd vertices, the shortest route which goes down each path at least once always involves starting at one odd vertex and ending at a different odd vertex.

FINDING THE SHORTEST INSPECTION ROUTE STARTING AND ENDING AT DIFFERENT POINTS OF YOUR CHOICE:

You start at one odd vertex and end at another odd vertex, so you only have to repeat the path between one pair of vertices. You want this path to be as short as possible.

EXAMPLE A feather duster salesman wants to travel along each street in a housing estate. He can start his journey at any point, and end it at any point. The graph represents the streets in the estate, and the numbers represent the lengths of each street in hundreds of metres.

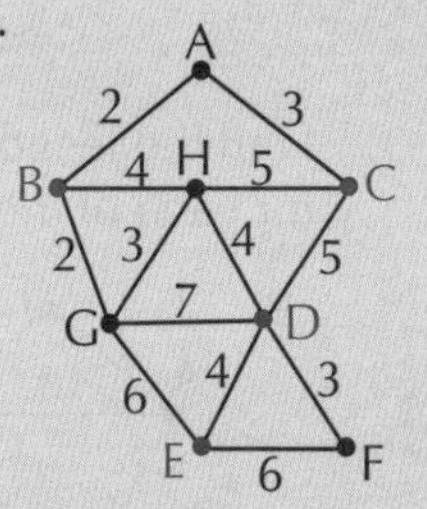

(a) State the vertices that the salesman could start at to minimise his journey.

There are four odd vertices, and you're going to start and end at two of them, so this leaves one pair of odd vertices. You have to repeat the path between these two vertices, so make sure it's the shortest possible.

The odd vertices are B, C, D, E.
The distance between each possible pair is:
BC = 5, DE = 4, BD = 8, CE = 9, BE = 8, CD = 5

The distance between D and E is shortest, at only 4. So that's the path you need to repeat.
So you start at either of vertices B or C, and end at the other.

(b) Find the length of his journey.

You just have to repeat the path between D and E.

So, total length of journey = weight of network + path between D and E
= 54 + 4 = 58

Always check the units — the weights represent hundreds of metres, so the journey is actually 5800 metres long, or 5.8 km.

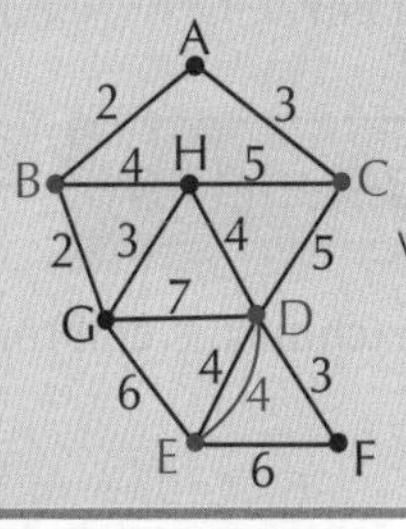

Weight of network
= 54

Walking on **Both Sides** *of the Street Makes it an* **Eulerian Network**

1) Sometimes, the usual 'inspection route' or 'Chinese postman' problem is changed so that the person has to go down each path twice. Perhaps they'll be inspecting each pavement, or delivering leaflets to both sides of the streets.
2) This actually makes it easier to solve. It effectively doubles the edges at each vertex, making all the vertices effectively even. The network is now Eulerian, so you can traverse it, and the length of the inspection route will just be double the weight of the network.

EXAMPLE The feather duster salesman decides to go down each street twice, once on each side. What is the length of his new route?

Weight of original network = 54.
Weight of doubled network = 54 × 2 = 108
Length of new route = 10 800 metres, or 10.8 km.

B was an odd vertex with 3 edges connected to it. It's now effectively even with 6 edges connected to it.

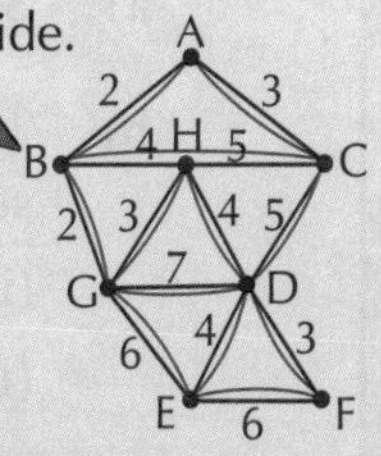

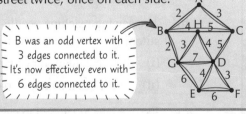

How's your network? — dunno, the fish just sort of get caught in it...

Remember — always read the question carefully to see what it's asking. Don't just scan it, say "Ah, Chinese postman" and jump in there. The examiners might have added a subtle twist. But to be kind, they'll often just tell you the network's weight.

D1 Section 3 — Practice Questions

Right. Time to see if you have the information from this section stuck in your head...

Warm-up Questions

1) Say whether each of these graphs is Eulerian, semi-Eulerian or neither.

a)

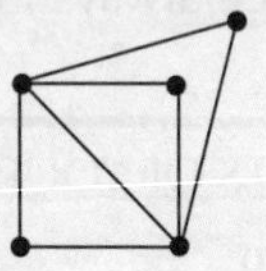

b)

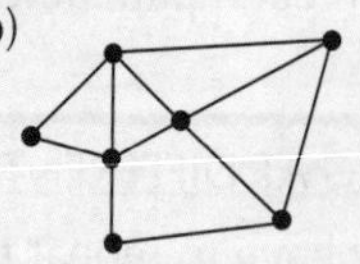

c)

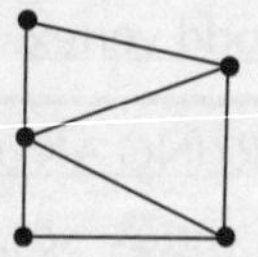

d)

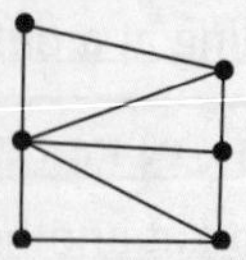

2) Identify the odd vertices in the graph on the right.
Write down all the possible ways of pairing the odd vertices.

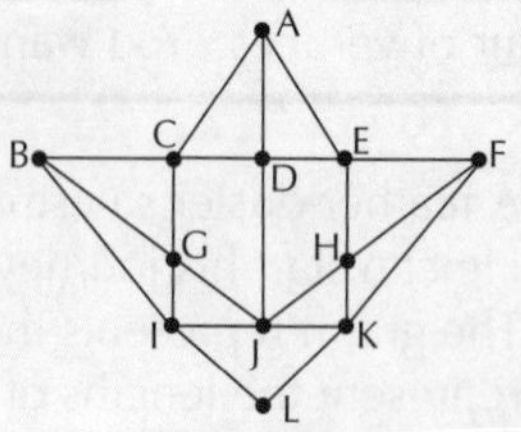

3) Find the length of the shortest "Chinese postman" route for each of these networks. Start and end at vertex A.

a)

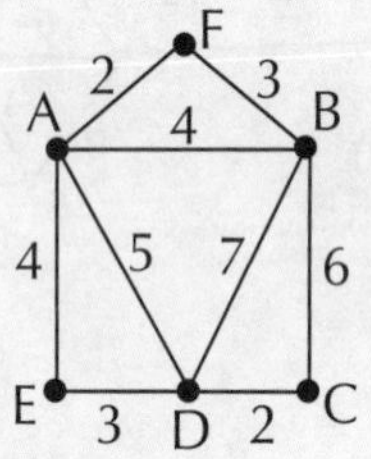

b)

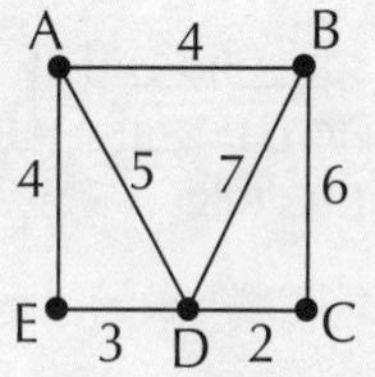

c)

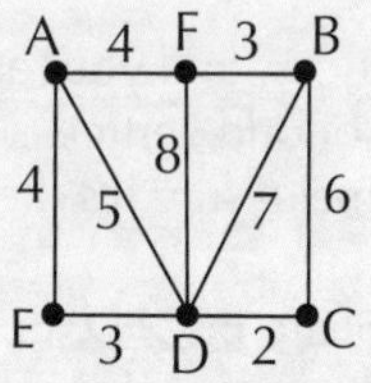

4) Repeat question 3. But this time you can start and finish at any vertices.
State which vertices you're starting and finishing at.

That's all the main skills brushed up on. Now it's time to see if you can apply them to some exam-style questions.

Exam Questions

1 A machinist is embroidering logos on sportsbags. The two logos are shown below.

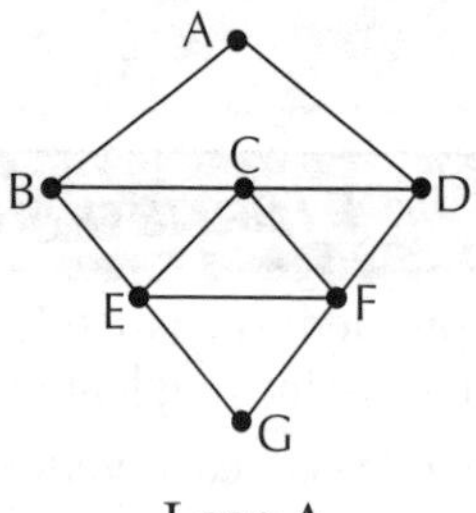

Logo A

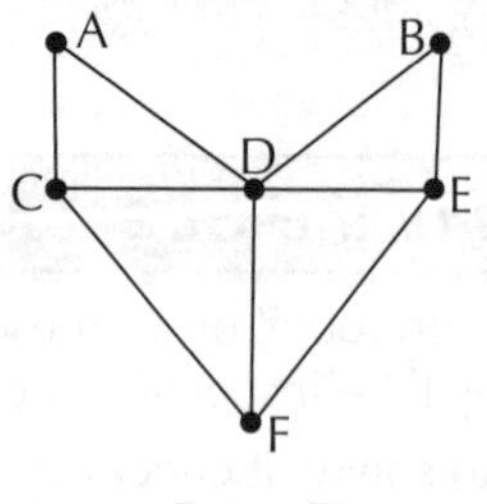

Logo B

a) Say whether each logo consists of an Eulerian graph, a semi-Eulerian graph or neither. *(2 marks)*

b) The sewing machine needle is positioned at any starting point, and sews a route without stitching any line more than once. It can be lifted and moved to a new starting point.

(i) For each logo, how many times must the needle be lifted? *(2 marks)*

(ii) For logo A, state an efficient starting vertex. *(1 mark)*

c) Extra edges are added to each logo to make them Eulerian.
State the minimum number of edges that must be added to each logo. *(2 marks)*

D1 Section 3 — Practice Questions

These are your typical "Chinese postman" problems — of course they contain subtle variations. Just like in real exams.

2 The diagram on the right shows all the streets in a town, and their lengths in metres.

Angus is considering moving to the town and wants to walk down each street at least once. He parks his car at K.

a) Explain why it is not possible to walk down each street only once and return to the starting point. *(1 mark)*

b) Find the length of the shortest route Angus could follow, starting and finishing at K. *(6 marks)*

c) Angus's friend offers to drop him off at any point, and after he's walked down each street at least once, to pick him up from any point.

(i) Find the length of the optimal route for Angus. *(2 marks)*

(ii) State the vertices from which Angus could start in order to achieve this optimal route. *(1 mark)*

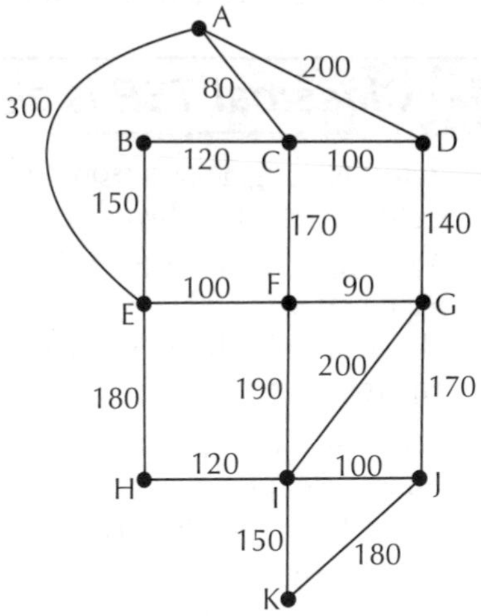

Total length of all the roads = 2740 m

3 The diagram below shows the paths in a park, and the time taken to walk them in minutes.

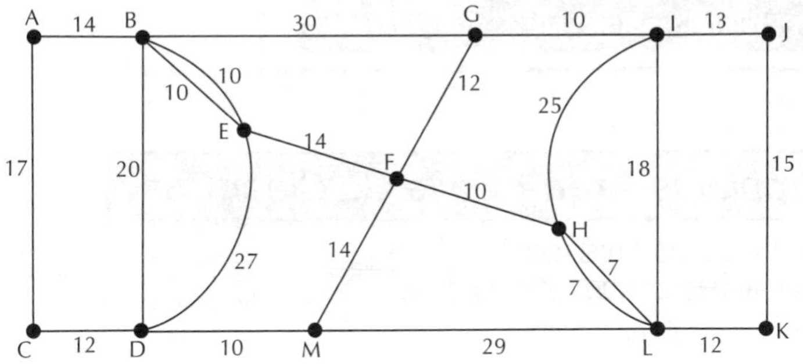

Alice the park keeper needs to walk down each path to check for storm-damaged trees. She parks her car at F.

Time required to walk along all paths = 336 minutes

a) Find the time for the shortest route Alice could follow, starting and finishing at F. *(6 marks)*

b) If Alice starts at point B, and can finish at any point:

(i) What point should she end at for the optimal route? Show your working. *(2 marks)*

(ii) How long will it take her to walk along all the paths now? *(1 mark)*

4 The diagram on the right shows the distances between towns in miles. The total road distance is 106 miles.

Jamie is inspecting the hedgerows along the roads. He needs to go along each road, starting and finishing at A.

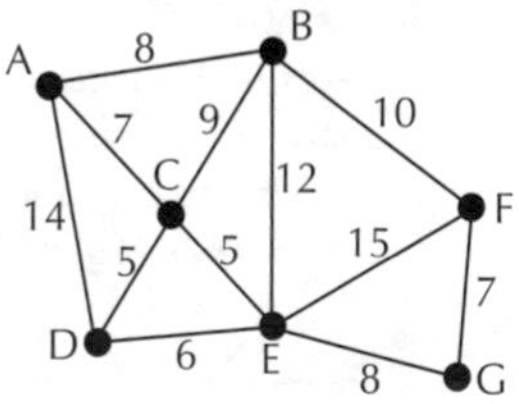

a) Find the length of the optimal 'Chinese postman' route for Jamie. *(6 marks)*

b) There are ice-cream shops at points C and E. If Jamie follows his optimal route, how many times will he pass an ice-cream shop? *(2 marks)*

c) Jamie decides it would be better if he went along each road twice. What is the length of his new optimal route? *(2 marks)*

Travelling Salesperson Problem

In the last section, you met the Chinese Postman problem. Well, this section's all about the Travelling Salesperson problem (TSP). I'm not sure where the salesperson is from exactly, but he or she sure does get around. Nobody knows what they sell either — I reckon it's black-market hamsters.

The **Classical TSP** is all about Finding the **Shortest Hamiltonian Cycle**

1) In a Travelling Salesperson problem, you need to visit every vertex in a network.
2) A classical TSP aims to find the shortest route that allows you to visit each vertex exactly once and end up back at your starting point — classical TSPs always involve complete networks (see page 208). The route you find will be a Hamiltonian cycle (see page 207).
3) There isn't an algorithm guaranteed to find the shortest Hamiltonian cycle. And it's not practical to test each one as the number of possible Hamiltonian cycles for complete graphs rises dramatically each time you add a vertex.
4) Instead, you find a lower and upper bound, then a reasonably good solution between them. Luckily, there are algorithms for finding the lower and upper bound.

EXAMPLE

Two possible Hamiltonian cycles starting from A are: AEDCBA (length 13) and ACBDEA (length 15).

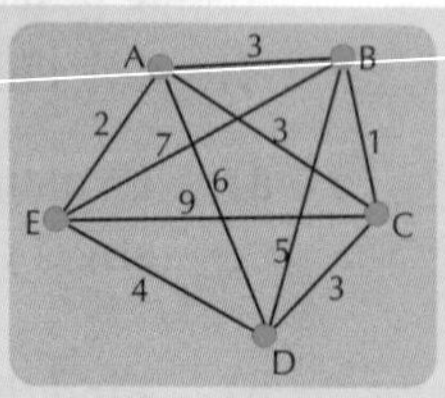

10 vertices means well over 100, 000 Hamiltonian cycles — a lot for even a computer to consider

TACKLING CLASSICAL TRAVELLING SALESPERSON PROBLEMS

1) Find a Hamiltonian cycle using the Nearest Neighbour Algorithm (see the next page) — this is the upper bound.
2) Find the lower bound by using the Lower Bound Algorithm (unfortunately, it doesn't have a more interesting name, but see page 227).
3) The weight of the minimum Hamiltonian cycle is:

lower bound $\leq$ minimum weight of Hamiltonian cycle $\leq$ upper bound

Practical Travelling Salesperson Problems **Aren't** Quite So Clear Cut

1) Practical TSPs involve someone wanting to find the shortest distance by which they can visit every vertex of a network and end up back at their starting point.
2) This sounds like it's exactly the same as the classical problem, but it can have big differences.

E.g. a tourist wanting to visit all the landmarks in a city, starting and finishing at his hotel.

There **Isn't** Always a **Hamiltonian Cycle**

1) If a graph isn't complete, it might not have any Hamiltonian cycles. So to visit all the vertices, you have to repeat some of them.
2) You have to add edges to turn the graph into a complete graph which you can tackle using the classical method.

EXAMPLE

There's no Hamiltonian cycle in this graph, so a salesperson visiting cities A-D would need to go through one of them twice (e.g. ACBDCA)....

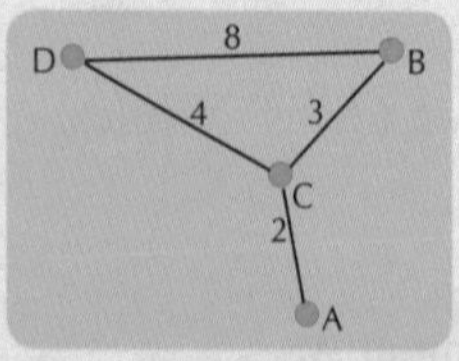

...so add edges to make a complete graph.

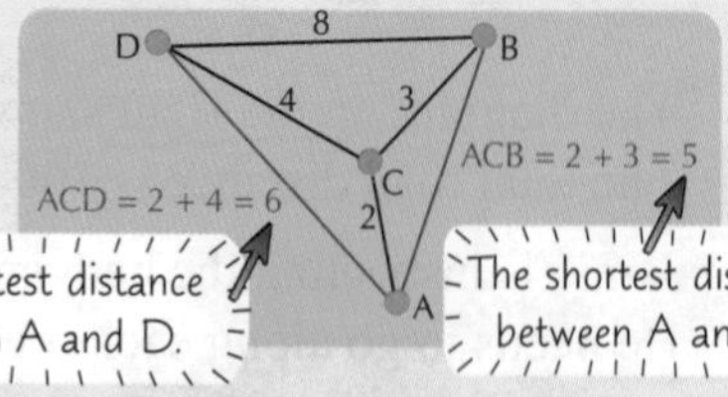

The shortest distance between A and D.

The shortest distance between A and B.

3) By inspection, the Hamiltonian cycle ADCBA is as short as you can get, with a weight of 18. You have to interpret this in the context of the practical problem though — the actual route would be ACDCBCA. This only involves the original edges, but it still has a weight of 18.

Just a TSP of sugar helps the medicine go down...

This is one of the bits of D1 that could come in really handy. Next time you're planning a holiday, make a list of all the things you want to see, find out the distances between them, then use the algorithms from this section to work out the minimum Hamiltonian cycle. Your friends might think you're a bit weird, but they'll thank you for it at the end of the day.

Travelling Salesperson Problem

The Nearest Neighbour algorithm finds a reasonable solution to the Travelling Salesperson problem — it might not be the best one though. In any case, it gives you an upper bound to try to beat. Think of it as a competition — whoever finds the best lower bound wins.

The Length of **Any Known Tour** is an **Upper Bound**

The SMALLER the upper bound the better.

This is because you want the interval between the upper and lower bounds to be as small as possible.

1) If the first tour you find has a length of 20, this is your initial upper bound. You know it's a possible solution, but you don't know if it's the optimum one.
2) If your second tour has a length of 22, then the upper bound stays at 20 — it's still the best so far.
3) But if you find a tour with a length of 17, then this is an improvement so it becomes the new upper bound.

The **Nearest Neighbour Algorithm** Finds a Hamiltonian Cycle

With the Nearest Neighbour algorithm, you basically choose the unvisited vertex closest to you each time. Once you've been to all the vertices, you go along the edge that takes you straight back to the starting point.

In the exam, you might have to make the graph complete (if it isn't already).

THE NEAREST NEIGHBOUR ALGORITHM
1) Choose a starting vertex (you'll usually be told which one to use in an exam question).
2) Choose the nearest unused vertex.
3) Repeat Step 2 until you've visited each vertex.
4) Finish the Hamiltonian cycle by returning to the starting vertex.

EXAMPLE The graph shows the durations in hours of train journeys between five Russian towns. Ben wants to visit all five towns, spending the minimum amount of time on trains. Apply the Nearest Neighbour algorithm, starting at vertex A, to find an upper bound for Ben's optimum route.

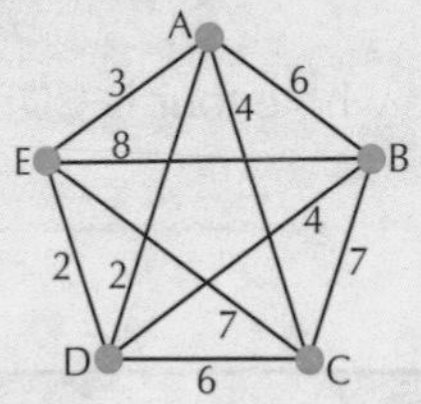

The graph is complete and shows the minimum durations between each town.
Applying the Nearest Neighbour algorithm from A:
The closest vertex to A is D (2 hours).
The closest unused vertex to D is E (2 hours).
The closest unused vertex to E is C (7 hours).
The closest unused vertex to C is B (7 hours).
That's all the vertices visited, so back to A (6 hours).
Upper bound = total duration of ADECBA = 24 hours.

The starting vertex for the Nearest Neighbour algorithm is just for applying the algorithm. The tour doesn't have to start there. E.g. Ben could start this tour at D and go DBEACD.

Starting from a **Different Vertex** Sometimes Gives a **Different Result**

1) Applying the Nearest Neighbour algorithm from vertex C on the above graph gives you the tour CADEBC, which has a duration of 23 hours.
2) This is shorter than the first one, so it becomes the new upper bound.

Ideally, you'd try the algorithm from each vertex, then pick the smallest upper bound. But you won't be expected to do all that in an exam.

Travelling salesperson problem — it's political correctness gone mad...

It's a bit disappointing that there's not a magical algorithm to find the optimum solution, like there was with the Chinese Postman problem and the finding the shortest route with Dijkstra's algorithm thingymajig. But you can't have everything. Although if I was wishing for things, I don't think an optimum solution to a TSP would be top of my list.

Travelling Salesperson Problem

Complete graphs with more than about 5 vertices start to get ridiculously complicated when they're drawn out, so you'll often be given the information in the form of a matrix.

You can use the *Nearest Neighbour Algorithm* Directly on a *Matrix*

1) Exam questions often give you the weights between vertices in distance matrix form (see p210).
2) You might have to complete the matrix — fill in missing distances or work out the shortest route between vertices if there isn't a direct path between them.
3) Here's an example of how to apply the Nearest Neighbour algorithm to the matrix:

EXAMPLE Poppy is on holiday in New York and wants to see landmarks A-F. The time that it takes in minutes to get between each place on public transport are given in the matrix.
Use the Nearest Neighbour algorithm starting from C to find an upper bound for Poppy's tour.

From \ To	A	B	C	D	E	F
A	—	5	35	5	10	15
B	10	—	10	20	20	35
C	45	15	—	10	30	25
D	20	10	5	—	5	10
E	40	25	15	35	—	40
F	35	20	25	30	20	—

① Find vertex C in the 'From' column, and look across to find the lowest number (the time taken to get to the closest vertex).
It's 10 mins to D, so D is the next vertex in the tour.
You don't want to go to D again, so cross out this column. You don't want to go to C again until the end, so cross out that column too so you don't get muddled.

From \ To	A	B	C	D	E	F
A	—	5	35	5	10	15
B	10	—	10	20	20	35
(C)	45	15	—	(10)	30	25
D	20	10	5	—	5	10
E	40	25	15	35	—	40
F	35	20	25	30	20	—

The times differ depending on which direction you're going in this matrix.

② Now you're at D, so find D in the 'From' column, and look across to find the shortest time to an unvisited vertex (i.e. not C). This is 5 mins to E, so E is the next vertex in the tour. So far the tour is CDE. Now cross out column E so you don't accidentally revisit it.

From \ To	A	B	C	D	E	F
A	—	5	35	5	10	15
B	10	—	10	20	20	35
(C)	45	15	—	(10)	30	25
(D)	20	10	5	—	(5)	10
E	40	25	15	35	—	40
F	35	20	25	30	20	—

③ Keep going with this until all the vertices are visited, then return to the start vertex (C). F is the last vertex visited — it's then 25 mins back to C.

The whole tour is CDEBAFC, which takes 90 minutes. So 90 minutes is an upper bound.

Sometimes the Nearest Neighbour Algorithm is a *Bit Rubbish*

1) The Nearest Neighbour algorithm is a greedy algorithm, so you choose the best option at any one moment, rather than thinking about what's best in the long run.
2) This can mean you end up having to go along really long edges at the end. You'll often be able to see a better route just by looking.

EXAMPLE Use the Nearest Neighbour algorithm, starting from vertex A, to find an upper bound for the classical Travelling Salesperson problem for the network below.

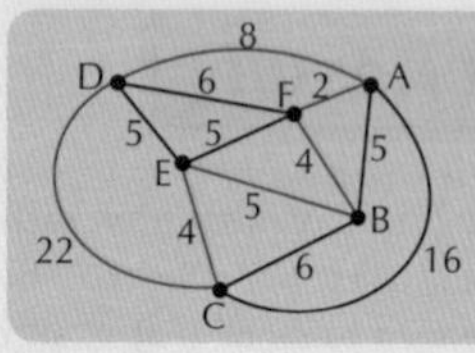

Applying the algorithm from A gives AFBECDA, which has a weight of 45.

But that route zigzags, then goes along the really long route from C to D. Using common sense, you'd go in the circular route ABCEDFA (or AFDECBA). This has a weight of only 28.

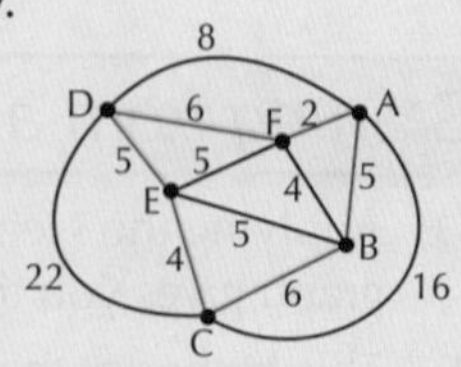

Everybody needs good Nearest Neighbour algorithms...

If you're not happy with your nearest neighbours, a good way to annoy them is to stand outside their window and change their TV channel with your remote. It'll drive them mad. Deny all knowledge about it if they mention it though.

Travelling Salesperson Problem

Now to find the lower bound. It's a completely different method from finding the upper bound, and it even involves your old friends from D1 Section 2 — Prim and Kruskal.

The Lower Bound Algorithm Involves Finding a Minimum Spanning Tree

1) The Lower Bound algorithm calculates the minimum possible weight for a Hamiltonian cycle in the network.
2) There might not be a Hamiltonian cycle of this weight, but there definitely won't be a shorter one.

Edges joined to a vertex, are said to be incident to it.

THE LOWER BOUND ALGORITHM

1) Choose a vertex, say A (you'll be told which one in an exam question). Find the two lowest weight edges joined to vertex A. Call their weights x and y.
2) Delete vertex A and all the edges joined to it. This is your 'reduced network'. Now find a minimum spanning tree (minimum connector) for the rest of the network and work out its weight. Call this W.
3) The Lower Bound = $W + x + y$

You do this with either Kruskal's or Prim's algorithm (see p211-213).

EXAMPLE By deleting vertex E, find a lower bound for the Travelling Salesperson problem on this network.

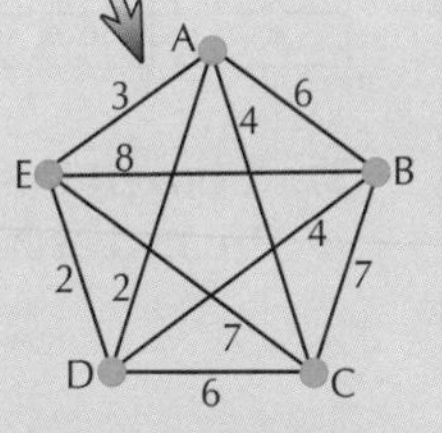

1) The two lowest weights joined to E are AE (3) and DE (2). So $x = 3$ and $y = 2$.

2) Delete vertex E and the incident edges. Now find a minimum spanning tree for the reduced network. I'm going to use Prim's algorithm, because I like it best.

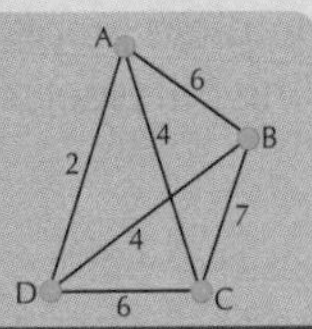

PRIM'S ALGORITHM

1) Pick a vertex, any vertex — this starts the tree. I'll start with A, ta.
2) Choose the edge of least weight that'll join a vertex in the tree to one not yet in the tree. Repeat this until you've joined all the vertices. That'll be AD (2), then AC (4), then BD (4).

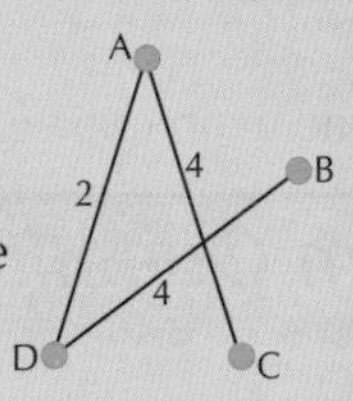

The weight, W, of the minimum spanning tree is 2 + 4 + 4 = 10.

3) The Lower Bound = $W + x + y$ = 10 + 3 + 2 = 15

3) If you're asked to find a lower bound from a distance matrix, start by crossing out the row and column of the deleted vertex. Then just use Prim's algorithm to find the minimum spanning tree (as on page 213).

The Largest Lower Bound is the Best

lower bound ≤ minimum weight of Travelling Salesperson tour ≤ upper bound

1) A small range of possible weights for the optimum tour is more useful than a large range — so you want your lower bound to be as large as possible.
2) To get the best lower bound, you should repeat the algorithm deleting a different vertex each time. Then you can pick the largest of the lower bounds.
3) If you find a tour which has the same weight as your lower bound, you know you've stumbled upon an optimum tour.
4) And if your lower bound and your upper bound are the same, you know that value is the weight of the optimum tour.

On the network above, you get the lower bounds below. (You check them and I'll send you a chocolate bunny if you find an error.)
Deleting A = lower bound 17
Deleting B = lower bound 18
Deleting C = lower bound 18
Deleting D = lower bound 17
So 18 is the best lower bound.

The end of the section — bet you're bounding about with joy...

Solving a Travelling Salesperson problem has lots of bits. But exam questions tend to break it down into different parts, and often only ask you to do part of the process, say find an upper bound. If you're asked to find a lower bound, you might have already found a minimum spanning tree for the network earlier in the question, so you can use that as a starting point.

D1 Section 4 — Practice Questions

First the hors d'oeuvres...

Warm-up Questions

1) a) Give three different Hamiltonian cycles for this network that start at A.
   b) Apply the Nearest Neighbour algorithm from each vertex in turn. What is your best upper bound?
   c) Apply the lower bound algorithm five times, deleting each vertex in turn. What is your best lower bound?
   d) What do your answers to b) and c) tell you about the optimum solution to the Travelling Salesperson problem for this network?

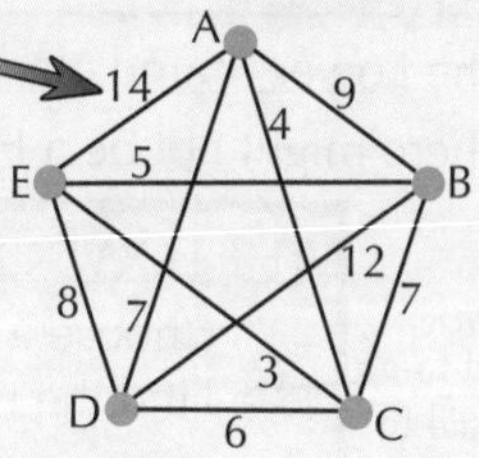

	A	B	C	D	E	F
A	—	4	6	5	8	12
B	4	—	14	22	6	11
C	6	14	—	18	3	5
D	5	22	18	—	13	15
E	8	6	3	13	—	20
F	12	11	5	15	20	—

2) a) Apply the Nearest Neighbour algorithm starting from F to find an upper bound for this network.
   b) Find a lower bound for this network by deleting vertex B.

3) Convert this graph into a complete graph so that it can be treated as a classical Travelling Salesperson problem.

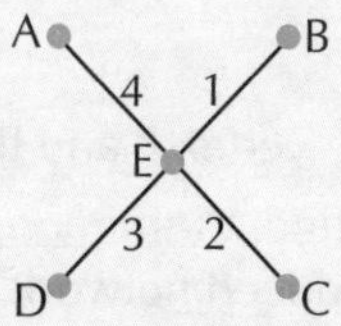

Now the main course. (Don't save room for pudding. There isn't any.)

Exam Questions

1 A network is shown below.

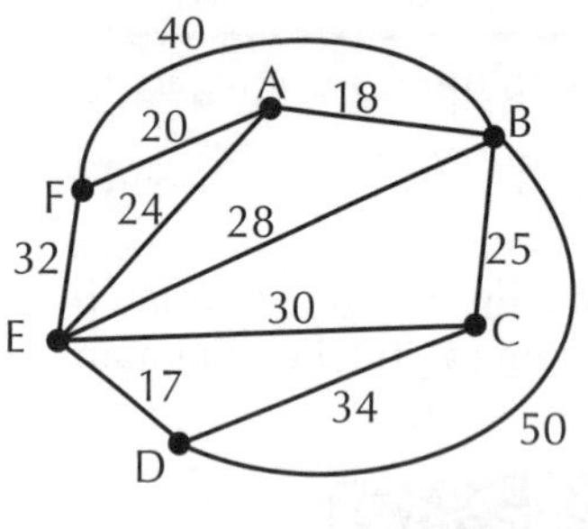

a) (i) State how many edges you would need to add to make this network complete. *(1 mark)*

(ii) Complete the network. *(3 marks)*

b) Find a minimum spanning tree for this network using Kruskal's algorithm. Draw your tree and state its weight. *(5 marks)*

c) Use your answer to part b) to find a lower bound for the travelling salesperson problem. Delete vertex C and all edges joined to it when forming your reduced network. *(3 marks)*

d) Find an upper bound for the travelling salesperson problem on this network by applying the Nearest Neighbour algorithm. Start from vertex A. *(3 marks)*

D1 Section 4 — Practice Questions

Travelling Salesperson exam questions aren't usually too tricky. But you still need to practise, practise, practise.

2 The diagram on the right shows all the streets in a town, and their lengths in metres. Max wants to display a poster at each intersection.

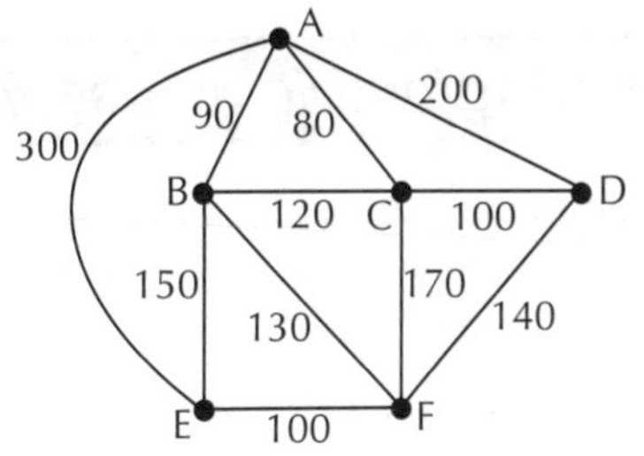

a) Complete the matrix on the right to show the distance between each intersection. *(2 marks)*

	A	B	C	D	E	F
A	—	90	80	170	210	
B	90	—	120		150	130
C	80	120	—	100		170
D	170		100	—		140
E	210	150			—	100
F		130	170	140	100	—

b) Find the length of tour ABCDEFA. *(2 marks)*

c) Find the length of the tour produced when the Nearest Neighbour algorithm is applied from vertex A. *(3 marks)*

d) Which of your answers to b) and c) is the best upper bound for Max's tour around the town? *(1 mark)*

e) Find a lower bound for the distance by deleting vertex D. *(6 marks)*

3 The diagram below shows the time taken to walk between park benches A-G in minutes. Alice the park keeper needs to inspect each bench for rot.

	A	B	C	D	E	F	G
A	—	15	10	7	6	7	4
B	15	—	8	5	12	13	14
C	10	8	—	6	15	18	2
D	7	5	6	—	7	5	13
E	6	12	15	7	—	3	10
F	7	13	18	5	3	—	6
G	4	14	2	13	10	6	—

a) Give an example of a Hamiltonian cycle for this park. *(2 marks)*

b) Starting from A, use Prim's algorithm to find a minimum spanning tree. Show the order in which edges are added and draw your minimum spanning tree, giving its total weight. *(6 marks)*

c) Using your answer to part b), calculate a lower bound for the duration of Alice's tour. You should delete vertex B and the edges incident to it. *(3 marks)*

d) Use the Nearest Neighbour method starting from F to find an upper bound for Alice's tour. *(3 marks)*

e) Using your previous answers, draw a conclusion about the duration of the optimum tour. *(2 marks)*

Linear Programming

Linear programming is a way of solving problems that have lots of inequalities, often to do with money or business. So if you're a budding entrepreneur, pay attention — this section could help you make your first million. Or just pass D1.

Linear Programming problems use Inequalities

The aim of linear programming is to produce an optimal solution to a problem, e.g. to find the solution that gives the maximum profit to a manufacturer, based on conditions that would affect it, such as limited time or materials. Before you start having a go at linear programming problems, there are a few terms you need to know.

1) In any problem, you'll have things that are being produced (or bought or sold etc.) — e.g. jars of jam or different types of books. The amount of each thing is represented by x, y, z etc — these are called the decision variables.
2) The constraints are the factors that limit the problem, e.g. a limited amount of workers available. The constraints are written as inequalities in terms of the decision variables. Most problems will have non-negativity constraints. This just means that the decision variables can't be negative. It makes sense really — you can't have –1 books.
3) The objective function is what you're trying to maximise or minimise (e.g. maximise profit or minimise cost). It's usually in the form of an function written in terms of the decision variables.
4) A feasible solution is a solution that satisfies all the constraints. It'll give you a value for each of the decision variables. On a graph, the set of feasible solutions lie in the feasible region (see p.231).
5) You're aiming to optimise the objective function — that's finding a solution within the feasible region that maximises (or minimises) the objective function. This is the optimal solution, and there can be more than one.

Put the Information you're given into a Table

Linear programming questions can look a bit confusing because you're given a lot of information in one go. But if you put all the information in a table, it's much easier to work out the inequalities you need.

EXAMPLE

A company makes garden furniture, and produces both picnic tables and benches. It takes 5 hours to make a picnic table and 2 hours to paint it. It takes 3 hours to make a bench and 1 hour to paint it. In a week, there are 100 hours allocated to construction and 50 hours allocated to painting. Picnic tables are sold for a profit of £30 and benches are sold for a profit of £10. The company wants to maximise their weekly profit.

Putting this information into a table gives:

Item of furniture	Construction time (hours)	Painting time (hours)	Profit (£)
Picnic table	5	2	30
Bench	3	1	10
Total time available:	100	50	

Now use the table to identify all the different parts of the problem and come up with the inequalities:

- The decision variables are the number of picnic tables and the number of benches, so let x = number of picnic tables and y = number of benches.
- The constraints are the number of hours available for each stage of manufacture. Making a picnic table takes 5 hours, so x tables will take $5x$ hours. Making a bench takes 3 hours, so y benches will take $3y$ hours. There are a total of 100 hours available. From this, you get the inequality $5x + 3y \leq 100$. Using the same method for the painting hours produces the inequality $2x + y \leq 50$. You also need $x, y \geq 0$.

  Don't forget the non-negativity constraints.
- The objective function is to maximise the profit. Each picnic table makes a profit of £30, so x tables make a profit of £$30x$. Each bench makes a profit of £10, so y benches make a profit of £$10y$. Let P be the profit, then the aim is to maximise $P = 30x + 10y$.

I'd like −3 picnic tables please...

In fancy examiner speak, the example above could be written like this: 'maximise $P = 30x + 10y$ subject to the constraints $5x + 3y \leq 100$, $2x + y \leq 50$ and $x, y \geq 0$'. Watch out for constraints such as 'there have to be at least twice as many benches as picnic tables — this would be written as $2x \leq y$. You might have to think about this one to get your head round it.

Feasible Regions

If you're a fan of drawing graphs, you'll love this page. It's a bit like the graphical inequality problems you came across at GCSE. Even if you're not that keen on graphs, or if the mere thought of them brings you out in a rash, don't worry — they're only straight line graphs.

Drawing Graphs can help solve Linear Programming Problems

Plotting the constraints on a graph is probably the easiest way to solve a linear programming problem — it helps you see the feasible solutions clearly. Get your ruler and graph paper ready.

1) Draw each of the constraints as a line on the graph. All you have to do is change the inequality sign to an equals sign and plot the line. If you find it easier, rearrange the equation into the form $y = mx + c$.
2) Then you have to decide which bit of the graph you want — whether the solution will be above or below the line. This will depend on the inequality sign — rearrange the inequality into the form $y = mx + c$, then think about which sign you'd use. For $y \leq mx + c$ (or <), you want the bit underneath the line, and if it's $y \geq mx + c$ (or >) then you want the bit above the line. If you're not sure, put the coordinates of a point in one region (e.g. the origin) into the equation and see if it satisfies the inequality.
3) Once you've decided which bit you want, shade the region you don't want. This way, when you put all the constraints on the graph, the unshaded region (the bit you want) is easy to see.
4) If the inequality sign is < or >, use a dotted line — this means you don't include the line in the region. If the inequality sign is ≤ or ≥ then use a normal line, so the line is included in the range of solutions.
5) Once you've drawn all the constraints on the graph, you'll be able to solve the problem. Don't forget the non-negativity constraints — they'll limit the graph to the first quadrant.

The Unshaded Area is the Feasible Region

Your finished graph should have an area, bounded by the lines of the constraints, that hasn't been shaded. This is the feasible region — the coordinates of any point inside the unshaded area will satisfy all the constraints.

EXAMPLE

On a graph, show the constraints $x + y \leq 5$, $3x - y \geq 2$, $y > 1$ and $x, y \geq 0$. Label the feasible region R.

Rearranging the inequalities into '$y = mx + c$' form and choosing the appropriate inequality sign gives: $y \leq 5 - x$, $y \leq 3x - 2$ and $y > 1$. The non-negativity constraints $x, y \geq 0$ are represented by the x- and y- axes.

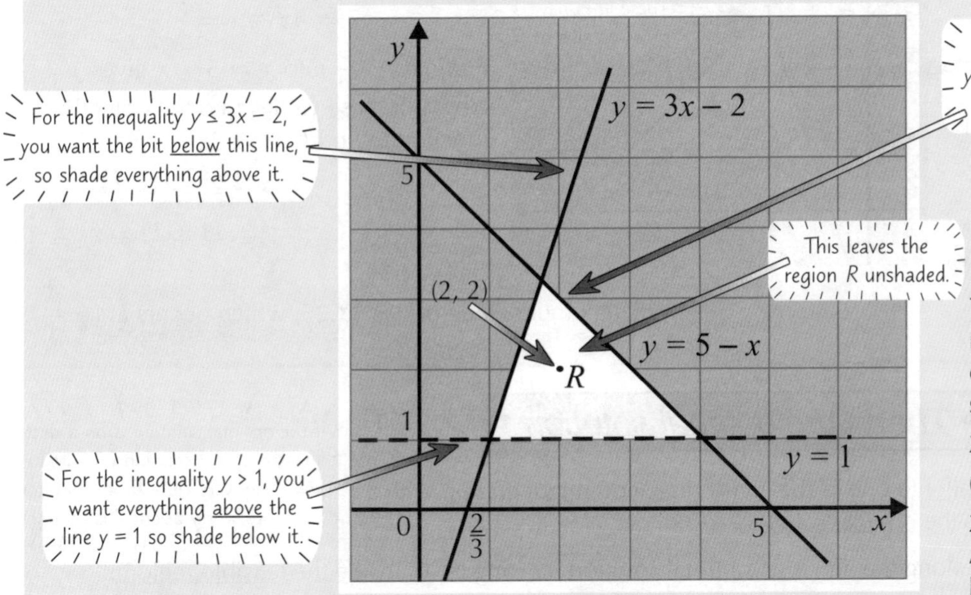

For the inequality y ≤ 3x – 2, you want the bit below this line, so shade everything above it.
For the inequality y ≤ 5 – x, you want the bit below this line so shade everything above it.
This leaves the region R unshaded.
For the inequality y > 1, you want everything above the line y = 1 so shade below it.
y
x
y = 3x – 2
y = 5 – x
y = 1
(2, 2)
R
5
1
0
2/3
5

To check that R is the feasible region, take a point inside R, e.g. (2, 2) and check that it satisfies all the inequalities:

$2 + 2 = 4 \leq 5$

$(3 \times 2) - 2 = 4 \geq 2$

$2 > 1$, and $2 \geq 0$.

All inequalities are satisfied so R is the feasible region.

It's feasible that I might become a film star...

...but not very likely — I'm a terrible actor. Anyway, although it's possible that you might get a linear programming problem with more than two variables, you won't have to draw graphs for these ones. You'll only have to graph 2-variable problems.

Optimal Solutions

Don't throw away your graphs just yet — you still need them for the next few pages. You're now getting on to the really useful bit — actually solving the linear programming problem.

Draw a line for the **Objective Function**

All the points in the feasible region (see previous page) satisfy all the constraints in the problem. You need to be able to work out which point (or points) also optimises the objective function. The objective function is usually of the form $Z = ax + by$, where Z either needs to be maximised (e.g. profit) or minimised (e.g. cost) to give the optimal solution.

The Objective Line Method

1. **Draw the straight line $Z = ax + by$, choosing a fixed value of Z (a and b will be given in the question). This is called the objective line.**
2. **Move the line to the right, keeping it parallel to the original line. As you do this, the value of Z increases (if you move the line to the left, the value of Z decreases).**
3. **If you're trying to maximise Z, the optimal solution will be the last point within the feasible region that the objective line touches as you slide it to the right.**
4. **If you're trying to minimise Z, the optimal solution will be the last point within the feasible region that the objective line touches as you slide it to the left.**

This is sometimes called the ruler method, as a good way to do it is to slide a ruler over the graph parallel to the objective line.

The objective lines have the **Same Gradient**

When you draw your first objective line, you can use any value for Z. Pick one that makes the line easier to draw — e.g. let Z be a multiple of both a and b so that the intercepts with the axes are easy to find.

EXAMPLE Using the example from the previous page, maximise the profit $P = 2x + 3y$.

First, choose a value for P, say $P = 6$. This means that the objective line goes through (3, 0) and (0, 2). Draw this line on the graph.

Slide the objective line to the right until it reaches the last point within R. At this point, P is maximised.

This is the first objective line, where $P = 6$.

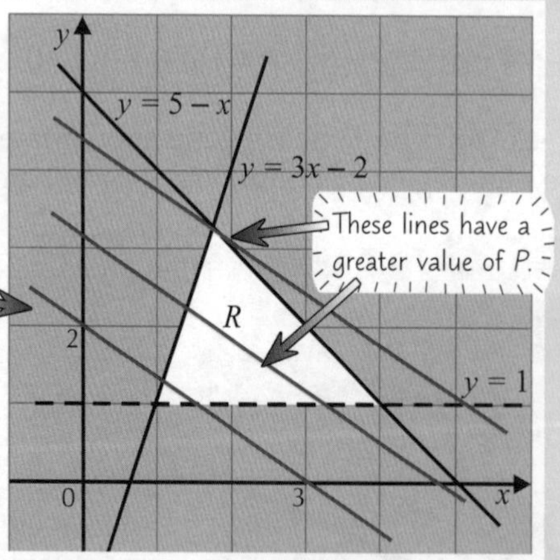

From the diagram, you can see that the last point the objective line touches is the intersection of the lines $y = 5 - x$ and $y = 3x - 2$.

To find the point of intersection, solve these simultaneous equations. This will give you the optimal solution.

Substituting $y = 5 - x$ into $y = 3x - 2$ gives: $5 - x = 3x - 2$

$$7 = 4x \Rightarrow x = \tfrac{7}{4}$$

Putting $x = \frac{7}{4}$ into $y = 5 - x$ gives $y = \frac{13}{4}$.

Now put these values into the objective function to find P: $P = 2x + 3y$

$$= 2\left(\tfrac{7}{4}\right) + 3\left(\tfrac{13}{4}\right)$$

$$= \tfrac{14}{4} + \tfrac{39}{4} = \tfrac{53}{4} = 13.25$$

So the maximum value of P is 13.25, which occurs at $\left(\frac{7}{4}, \frac{13}{4}\right)$.

There might be **More Than One** optimal solution

1) If the objective line is parallel to one of the constraints, you might end up with a section of a line that gives the optimal solution.
2) If this happens, any point along the line is an optimal solution (as long as it's inside the feasible region).
3) This shows that there can be more than one optimal solution to a problem.

If the optimal solution is on a dotted line, the actual solution will just be a point very very close to it. You don't need to worry about this though.

Maximise your chance of passing D1...

Make sure you're happy with solving simultaneous equations — if not, have a look back at page 19. You won't have to solve any tricky quadratic equations for this section, but you need to be able to solve linear simultaneous equations quick-smart.

Optimal Solutions

If you object to using the objective line, there is another method. This one uses a lot more simultaneous equations, but you don't have to worry about keeping the ruler parallel or stopping global warming or anything like that.

Optimal Solutions are found at Vertices

The optimal solution for the example on the previous page was found at a vertex of the feasible region. This isn't a coincidence — if you have a go at some more linear programming problems, you'll find that the optimal solutions always occur at a vertex (or an edge) of the feasible region. This gives you another way to solve the problem.

The Vertex Method

1. **Find the x- and y-values of the vertices of the feasible region. You do this by solving the simultaneous equations of the lines that intersect at each vertex.**
2. **Put these values into the objective function $Z = ax + by$ to find the value of Z.**
3. **Look at the Z values and work out which is the optimal value. Depending on your objective function, this might be either the smallest (if you're trying to minimise Z) or the largest (if you're trying to maximise Z).**

If two vertices A and B produce the same Z value, this means that all points along the edge AB are also optimal solutions.

Test Every vertex

Even if it looks obvious from the graph, you still have to test each vertex of the feasible region. Sometimes the origin will be one of the vertices — it's really easy to test, as the objective function will just be equal to 0 there. Don't forget vertices on the x- and y-axes too.

You can sometimes just read off the coordinates from your graph (as long as it's accurate).

EXAMPLE

Minimise $Z = 8x + 9y$, subject to the constraints $2x + y \geq 6$, $x - 2y \leq 2$, $x \leq 4$, $y \leq 4$ and $x, y \geq 0$.

Drawing these constraints on a graph produces the diagram below, where A, B, C and D are the vertices of the feasible region R:

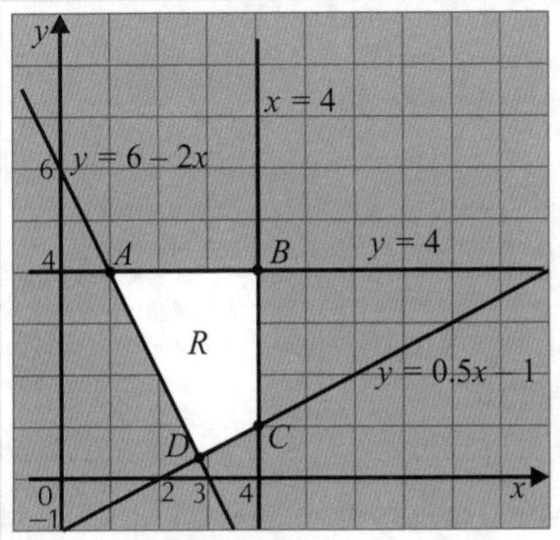

Point A is the intersection of the lines $y = 6 - 2x$ and $y = 4$, so A has coordinates (1, 4).

Point B is the intersection of the lines $x = 4$ and $y = 4$, so B has coordinates (4, 4).

Point C is the intersection of the lines $x = 4$ and $y = 0.5x - 1$, so C has coordinates (4, 1).

Point D is the intersection of the lines $y = 6 - 2x$ and $y = 0.5x - 1$, which has coordinates $\left(\frac{14}{5}, \frac{2}{5}\right)$.

You need to use simultaneous equations to find the coordinates of D.

Putting these values into the objective function $Z = 8x + 9y$:

At A, $Z = (8 \times 1) + (9 \times 4) = 44$.

At B, $Z = (8 \times 4) + (9 \times 4) = 68$.

At C, $Z = (8 \times 4) + (9 \times 1) = 41$.

At D, $Z = (8 \times 2.8) + (9 \times 0.4) = 26$.

So the minimum value of Z is 26, which occurs when $x = \frac{14}{5}$ and $y = \frac{2}{5}$.

In this example, it was really easy to find the coordinates of A, B and C, as at least one of the values was given by the equation of the line. D was a bit harder, as it involved simultaneous equations, but it wasn't too bad.

I'm getting vertigo...

Some questions might give you two different objective functions and ask you to minimise cost and maximise profit for the same set of constraints. The vertex method is really useful here, as once you've worked out the coordinates of the vertices, you can easily put the values into both objective functions without having to do any more work (or draw on confusing lines).

Optimal Integer Solutions

The methods on the previous two pages are all very well and good, but I can't exactly make 2.5 teddy bears, even if it does maximise my profit. There must be a better way, one that doesn't involve mutilating soft toys...

Some problems need **Integer Solutions**

1) Sometimes it's fine to have non-integer solutions to linear programming problems — for example, if you were making different fruit juices, you could realistically have 3.5 litres of one type of juice and 4.5 litres of another.
2) However, if you were making garden furniture, you couldn't make 3.5 tables and 4.5 benches — so you need integer solutions. Sometimes you'll be asked to interpret your solutions and say how realistic they are in context.
3) If you're not asked to interpret your solution, you might have to decide whether a problem needs integer solutions. It's common sense really — just think about whether you can have fractions of the decision variables.

You can use the **Objective Line Method** or the **Vertex Method**

Some problems have optimal integer solutions that are far away from the vertices — but you don't need to worry about these for D1.

Both of the methods covered on pages 232-233 can be used to find an optimal integer solution — it just depends on how clear your graph is.

1) You use the objective line method in exactly the same way as before, but instead of looking for the last vertex the line touches, you need to look for the last point with integer coordinates in the feasible region. This might be hard to do if your graph isn't very accurate, or if the scale isn't clear.
2) The other way to find the optimal integer solution is to use the vertex method to find which vertex to use. Then, consider all the points with integer coordinates that are close by. Make sure you check whether these points still satisfy the constraints though — test this before you put the values into the objective function.

The **Optimal Integer Solution** must be **Inside** the **Feasible Region**

It's easy to forget that not all the solutions near the optimal vertex will be inside the feasible region — you can check either by eye on an accurate graph, or put the coordinates into each of the constraints.

EXAMPLE The optimal solution to the problem on the previous page occurred at $\left(\frac{14}{5}, \frac{2}{5}\right)$ (= (2.8, 0.4)). Find the optimal integer solution.

Looking at the integers nearby gives you the points (3, 0), (3, 1), (2, 0) and (2, 1) to test. However, the point (3, 0) doesn't satisfy the constraint $x - 2y \leq 2$, and (2, 0) and (2, 1) don't satisfy $2x + y \geq 6$, so the optimal integer solution is at (3, 1), where $Z = (8 \times 3) + (9 \times 1) = 33$.

EXAMPLE

A company makes baby clothes. It makes x sets of girls' clothes and y sets of boys' clothes, for a profit of £6 and £5 respectively, subject to the constraints $x + y \leq 9$, $3x - y \leq 9$, $y \leq 7$ and $x, y \geq 0$. Maximise the profit, $P = 6x + 5y$

This example uses the vertex method, but you could also use the objective line method.

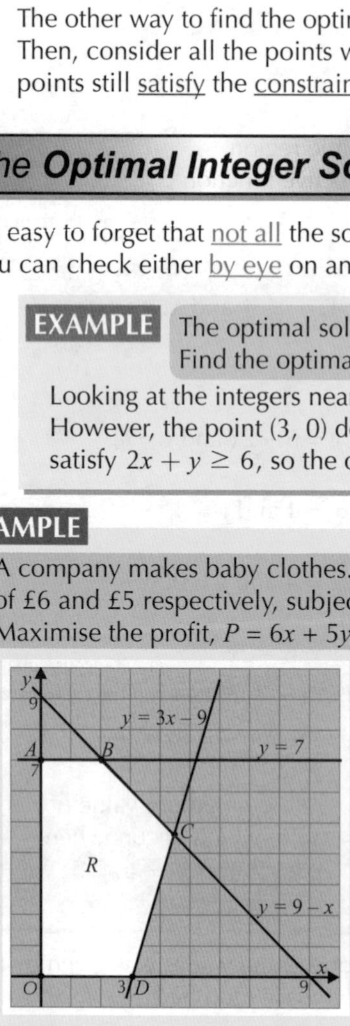

The feasible region is the area $OABCD$, with coordinates $O(0, 0)$, $A(0, 7)$, $B(2, 7)$, $C\left(\frac{9}{2}, \frac{9}{2}\right)$ and $D(3, 0)$.

The value of P at each vertex is O: £0, A: £35, B: £47, C: £49.50 and D: £18.

The maximum value of P is £49.50, which occurs at $\left(\frac{9}{2}, \frac{9}{2}\right)$ (= (4.5, 4.5)). However, making 4.5 sets of clothes is not possible, so an integer solution is needed.

The integer coordinates near C are (4, 5), (5, 5), (5, 4) and (4, 4). (5, 5) and (5, 4) don't satisfy the constraint $3x - y \leq 9$ so are outside the feasible region. At (4, 5), P = £49, and at (4,4), P = £44, so £49 is the maximum profit.

So the company needs to make 4 sets of girls' clothes and 5 sets of boys' clothes to make the maximum profit of £49.

My solution to a problem is to close my eyes and hope it'll go away...

Oo, I do like a nice integer now and then. Like a good cup of tea, integers can really brighten your morning. They're harder to dunk biscuits in, but they do provide sensible solutions to linear programming problems. Watch out for them in the exam.

D1 Section 5 — Practice Questions

This section isn't too bad really — once you've got your head round turning the constraints into inequalities, all you have to do is draw a graph and Bob's your uncle. I wish I had an Uncle Bob. Anyway, here are some warm-up questions for you to have a go at.

Warm-up Questions

1) Give a brief definition of:
   a) decision variables
   b) objective function
   c) optimal solution
2) What does a dotted line on a graph show?
3) What is an objective line?
4) Give one example of a problem that doesn't need integer solutions, and one example that does.
5) A company makes posters in two sizes, large and small. It takes 10 minutes to print each large poster, and 5 minutes to print each small poster. There is a total of 250 minutes per day allocated to printing. It takes 6 minutes to laminate a large poster and 4 minutes for a small poster, with a total of 200 minutes laminating time. The company want to sell at least as many large posters as small, and they want to sell at least 10 small posters. Large posters are sold for a profit of £6 and small posters are sold for a profit of £3.50.
   a) Write this out as a linear programming problem. Identify the decision variables, constraints and objective function.
   b) Show the constraints for this problem graphically. Label the feasible region *R*.
   c) Maximise the profit, using either the objective line method or the vertex method. Don't worry about integer solutions for now.
   d) Use your answer to part c) to find the optimal integer solution.

Exam questions try and scare you by throwing a lot of information at you all at once. Don't let them bully you, just take them one bit at a time and you'll soon get the better of them. Here are a few for you to do.

Exam Questions

1 Anna is selling red and white roses at a flower stall. She buys the flowers from a wholesaler, where red roses cost 75p each and white roses cost 60p each. Based on previous sales, she has come up with the following constraints:

- She will sell both red roses and white roses.
- She will sell more red roses than white roses.
- She will sell a total of at least 100 flowers.
- The wholesaler has 300 red roses and 200 white roses available.

Let x be the number of red roses she buys and y be the number of white roses she buys. Formulate this information as a linear programming problem. Write out the constraints as inequalities and identify a suitable objective function, stating how it should be optimised. You do not need to solve this problem.

(7 marks)

D1 Section 5 — Practice Questions

In the words of some little Dickensian orphan, 'please sir, I want some more'. Well, I'd be more than happy to oblige — here's another page full of exciting exam questions.

2 A company sells three packs of craft paper, bronze, silver and gold. Each pack is made up of three different types of paper, tissue paper, sugar paper and foil.

- The gold pack is made up of 6 sheets of foil, 15 sheets of sugar paper and 15 sheets of tissue paper.
- The silver pack is made up of 2 sheets of foil, 9 sheets of sugar paper and 4 sheets of tissue paper.
- The bronze pack is made up of 1 sheet of foil, 6 sheets of sugar paper and 1 sheet of tissue paper.
- Each day, there are 30 sheets of foil available, 120 sheets of sugar paper available and 60 sheets of tissue paper available.
- The company is trying to reduce the amount of foil used, so it uses at least three times as many sheets of sugar paper as of foil.

The company makes x gold packs, y silver packs and z bronze packs in a day.

(a) Apart from the non-negativity constraints, write out the other four constraints as inequalities in terms of x, y and z. Simplify each inequality where possible.

(8 marks)

(b) On Monday, the company decides to make the same number of silver packs as bronze packs.

(i) Show that your inequalities from part (a) become

$$2x + y \leq 10$$
$$x + y \leq 8$$
$$3x + y \leq 12$$
$$2y \geq x$$

(3 marks)

(ii) On graph paper, draw a graph showing the constraints from part (i) above, as well as the non-negativity constraints. Label the feasible region R.

(5 marks)

(iii) Use your graph to work out the maximum number of packs the company can make on Monday.

(2 marks)

(iv) Gold packs are sold for a profit of £3.50, silver packs are sold for a profit of £2 and bronze packs are sold for a profit of £1. Use your answers to parts (ii) and (iii) to maximise the profit they make, and state how many of each type of pack they need to sell.

(3 marks)

3 The graph below shows the constraints of a linear programming problem. The feasible region is labelled R.

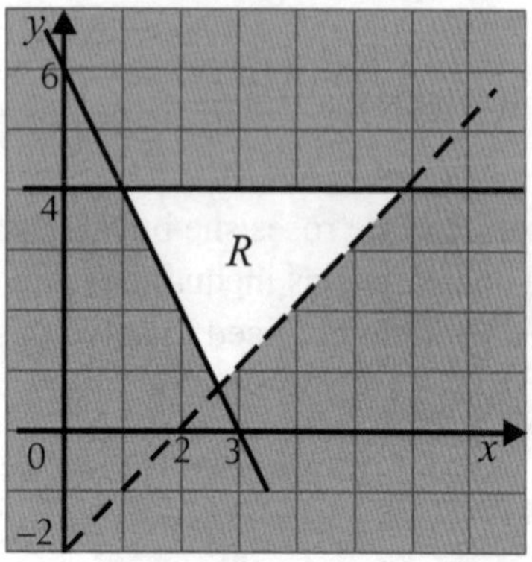

(a) Including the non-negativity constraints, find the inequalities that produce R.

(5 marks)

(b) Find the coordinates of each vertex of R.

(4 marks)

The aim is to minimise $C = 4x + y$.

(c) Find the optimal solution and state where this value occurs.

(3 marks)

Matchings

Bipartite graphs are a really useful way of allocating people to tasks, taking into account what they want to do, what they can do, and what they won't do. For example, on the frequent occasions that I have to use my superpowers, I'd prefer to fly, am prepared to turn invisible but will not dissolve into a puddle of water.

Bipartite Graphs have Two Sets of Vertices

The points in a graph are called vertices (or nodes) and the lines are called edges (or arcs) — this was covered in Section 2.

1) A bipartite graph is made up of two sets of vertices that are linked by edges. The edges go from one set of vertices to the other — vertices within the same set can't be joined to each other.
2) In lots of the examples you'll come across, one set of vertices will be the people and the other will be the jobs or tasks they have to do. You'll be told who can do which job.
3) There doesn't have to be the same number of vertices in each set.

Sometimes the information will be in a table or an adjacency matrix (see below).

EXAMPLE

Jenny, Katie, Lizzie, Martyn and Nikki are planning a picnic. Jenny can bring sandwiches and crisps, Katie can bring sandwiches and pork pies, Lizzie can bring drinks, Martyn can bring pork pies and biscuits and Nikki can bring biscuits, quiche and crisps. Draw a bipartite graph to show this information.

First, list all the people on one side of the graph, and all the food (and drink) on the other side. Then draw lines connecting each person to all the items they can bring.

The lines show what each person can bring.

In this example, there are more vertices on the right than on the left.

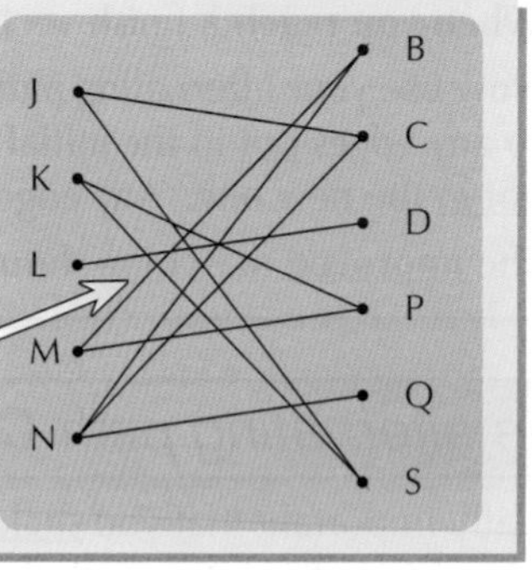

A Matching links One Person to One Task

Once you've drawn your bipartite graph showing who can do what, you need to work out a solution that assigns one person to one job. This is called a matching.

1) In a matching, you can only have one edge for each vertex — so each person only does one job (and each job only has one person doing it). Matchings are one-to-one.
2) You won't always be able to match all the vertices in one set with all the vertices in the other — it depends on who can do which task.
3) If there are the same number of people as jobs, and each job is assigned to a person, the matching is said to be complete. So if there are x vertices in each set, a complete matching has x edges.
4) It's not always possible to have a complete matching — e.g. if there are two jobs that can only be done by the same person, one of the jobs won't be included in the matching.
5) If a complete matching can't be done, you might have to find a maximum (or 'maximal') matching — a matching that has the greatest number of edges possible (so as many jobs as possible are being done). There can be more than one possible maximum matching.

You don't need to worry about the difference between 'maximal' and 'maximum' for D1

EXAMPLE Andy, Ben, Carys and Daniel are going to a theme park. They can only afford to go on one ride each, and they each want to try out a different ride. Their choices are shown in the adjacency matrix. Draw a bipartite graph to show this information, then use it to find a complete matching.

	Big Wheel	Log Flume	Roller-coaster	Teacups
Andy	1	0	0	1
Ben	0	0	1	1
Carys	0	1	1	0
Daniel	1	0	0	0

1 means there is an edge from Andy to the big wheel.

Bipartite graph

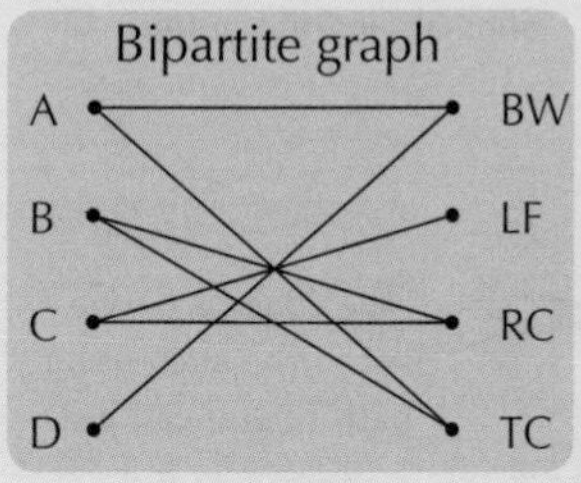

Complete matching

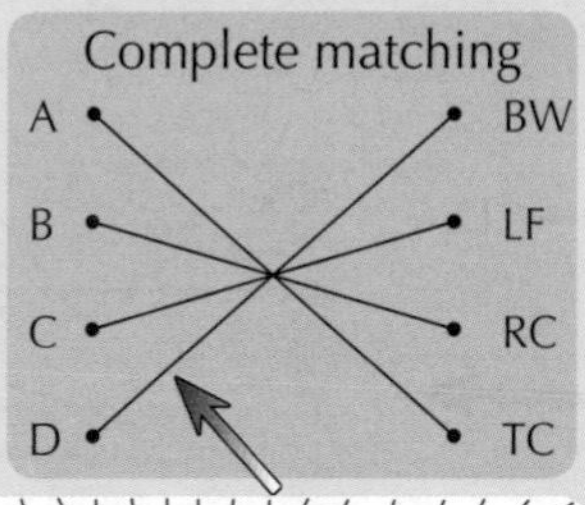

Start with Daniel, who only wants to go on the big wheel, and work from there.

After a bad experience, Andy flatly refuses to go on the teacups again, so a complete matching is no longer possible. Find a maximum matching for their next visit to the theme park.

Maximum matching

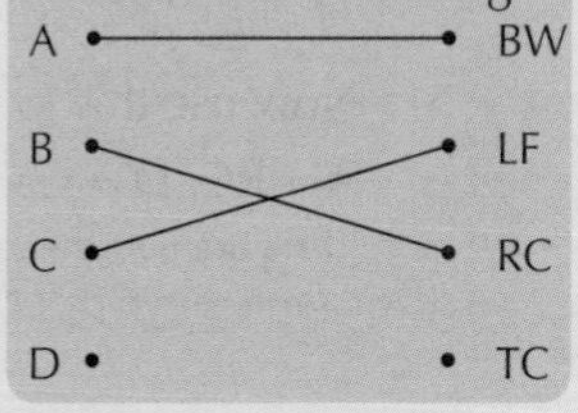

Add one teaspoon of bipartite of soda...

If you're thinking the graphs on this page look familiar, you're dead right — you came across bipartite graphs on page 207.

Alternating Paths

Sometimes your original matching can be improved. To do this, you need to be able to find an alternating path.

Start with an ***Initial Matching*** and find an ***Alternating Path***

Draw any matching from a bipartite graph. This is the initial matching — you're trying to improve it.

The Alternating Path Method

1) **An alternating path starts at a vertex on one side of the graph that isn't included in the initial matching and finishes at a vertex on the other side of the graph that also isn't in the initial matching.**
2) **To get from the start vertex to the finishing vertex, you have to alternate between edges that are not in and in the initial matching. So the first edge you use (from the unmatched starting vertex) is not in the initial matching, the second edge is, the third one isn't and so on until you get to a finish vertex.**
3) **When you reach a finish vertex (one not in the initial matching), you can stop — you've made a 'breakthrough'.**
4) **Now use your alternating path to construct an improved matching. Take the path and change the status of the edges, so any edges not in the initial matching are in the new matching, and the edges that were in the initial matching are not in the new one. Any edges in the initial matching that aren't in the alternating path just stay as they are.**
5) **The improved matching should include two vertices that weren't in the initial matching, and have one extra edge.**

The alternating path ***Can't Change*** the ***Original Information***

It's really important that you stick to the information you were given in the first place — you can't make people do jobs in the alternating path that they weren't doing in the bipartite graph. That wouldn't be fair.

EXAMPLE

Anne, Dick, George, Julian and Timmy are on an adventure holiday.
They have a choice of five activities: abseiling, canoeing, diving, mountain biking and rock-climbing
Anne wants to go mountain biking or rock-climbing, Dick wants to go canoeing or diving, George wants to go abseiling or mountain biking, Julian wants to go diving or rock-climbing and Timmy just wants to go abseiling.
Find an alternating path that improves on this initial matching:
Anne – mountain biking, Dick – canoeing, George – abseiling, Julian – rock climbing

1) Start by drawing a bipartite graph so you see all the preferences:

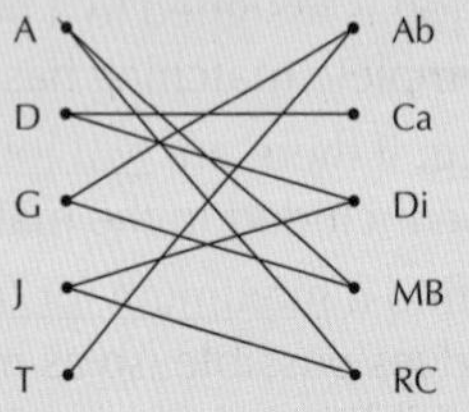

2) And draw the initial matching:

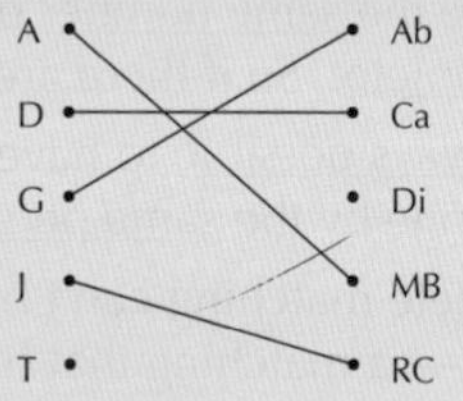

3) For the alternating path, start at T and find a path that connects it to Di, the other unmatched vertex (it won't be a direct path, as Timmy doesn't want to go diving).

One alternating path goes like this:

T - - - Ab —— G - - - MB —— A - - - RC —— J - - - Di

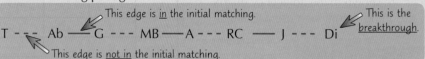

This edge is in the initial matching. (Ab — G)
This edge is not in the initial matching. (T - - - Ab)
This is the breakthrough. (Di)

Don't forget that every edge in your alternating path has to be valid, i.e. in the original bipartite graph.

4) Changing the status of edges gives:

T —— Ab - - - G —— MB - - - A —— RC - - - J —— Di

5) Now use this to construct the improved matching:

A = RC, D = Ca, G = MB, J = Di, T = Ab

There are now the same number of edges as there are vertices in each set, so this is a complete matching.

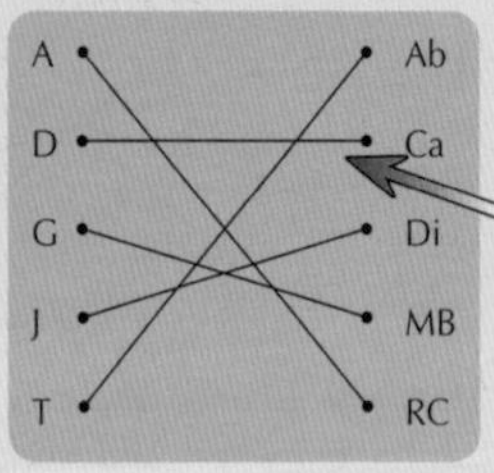

The edge D — Ca isn't in the alternating path so you can just leave it as it is.

You put your left hand in...

... your right hand not in. In, not in, in, not in and shake it all about.

Maximum Matchings

Once you're happy with finding alternating paths and improved matchings (see previous page), you can use that method to find a 'maximum matching' — one that uses the greatest number of edges possible.

Repeat the *Alternating Path* method to get a *Maximum Matching*

To find the maximum matching, you need to keep finding alternating paths and improved matchings — this is the maximum matching algorithm.

The Maximum Matching Algorithm

1) **Start with any initial matching.**
2) **Try to find an alternating path (using the method on p.238). If you find one, use this to form an improved matching. If there isn't one, then this is a maximum matching — so stop.**
3) **If there are no unmatched vertices, stop — the matching is complete. If there are still unmatched vertices, repeat step 2) using the improved matching as the new initial matching.**

There might be more than one maximum matching.

Use Trees to show different Paths

In some alternating paths, you have a choice between different vertices. You should draw a tree diagram to show the possible paths, then pick the route that gets you to a breakthrough the fastest.

EXAMPLE

A music school offers tuition in the following instruments: clarinet, flute, piano, saxophone, trumpet and violin. Six children want to start lessons, but the school only takes on one pupil per instrument. Their preferences are:

Child	First choice	Second choice	Third choice
Chad	Saxophone	Trumpet	—
Elly	Trumpet	Piano	Clarinet
Karen	Flute	Piano	—
Mike	Trumpet	Saxophone	—
Peter	Clarinet	Violin	Piano
Stuart	Trumpet	Clarinet	—

Draw a bipartite graph to show this information.

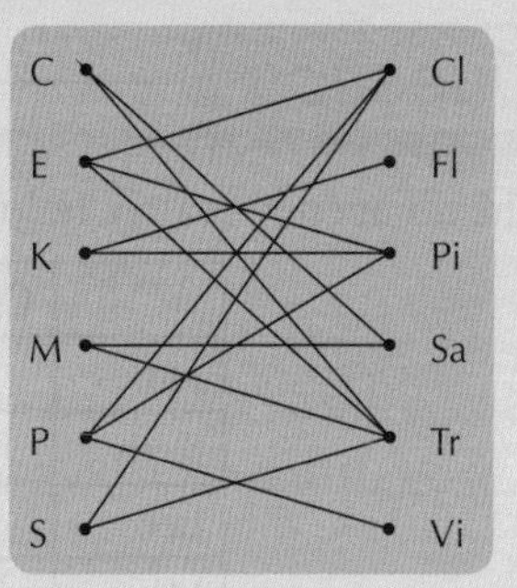

Initially, Chad, Elly, Karen and Peter are matched to their first choice. Draw this initial matching, then use the maximum matching algorithm to improve the matching as much as possible.

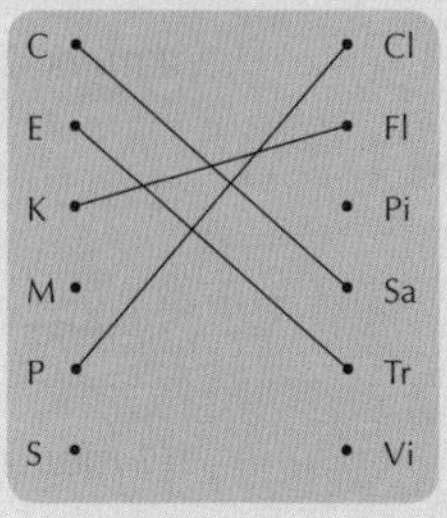

Initial matching

1) From the initial matching, find an alternating path from Mike to an unmatched instrument:

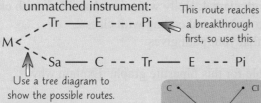

M < Tr — E - - - Pi
M < Sa — C - - - Tr — E - - - Pi

This route reaches a breakthrough first, so use this.

Use a tree diagram to show the possible routes.

This produces the matching:
C = Sa, E = Pi, K = Fl, M = Tr
(P = Cl doesn't change)

There are still two unmatched vertices, we need another alternating path.

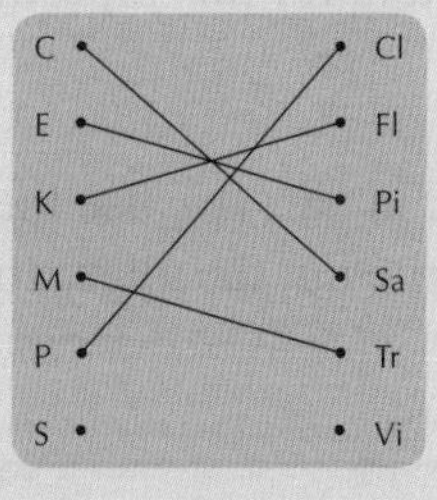

2) Using the new matching, find an alternating path from Stuart to an unmatched instrument:

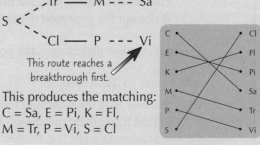

S < Tr — M - - - Sa
S < Cl — P - - - Vi

This route reaches a breakthrough first.

This produces the matching:
C = Sa, E = Pi, K = Fl,
M = Tr, P = Vi, S = Cl

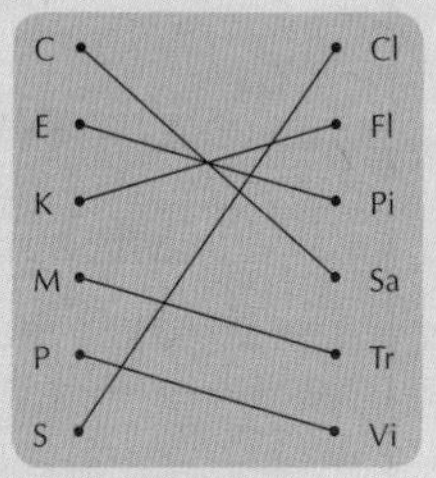

There are no more unmatched vertices, so this is a complete matching.

If only marks grew on trees...

I bet you never thought AS-level Maths would be about joining the dots. Well, technically it's drawing edges between vertices, but I'd rather think of it as joining the dots — much more fun that way. Right, I'm off to do some potato printing now.

D1 Section 6 — Practice Questions

That's the end of Section 6 — and it's also the end of D1. Just one page of practice questions, then you're ready to have a go at some practice exams to test your powers of decisiveness.

Warm-up Questions

1) Elizabeth, Jane, Kitty, Lydia and Mary are going to an art gallery. Elizabeth likes Renaissance art, portraits and sculptures, Jane likes portraits, sculptures and modern art, Kitty likes the cafe, Lydia likes modern art and the cafe and Mary likes Renaissance art.
   a) Draw this information on a bipartite graph.
   b) Use your bipartite graph to find a complete matching.

2) At a school dance, Alice, Bella, Charlotte, Daisy, Evie and Felicity have to be paired with Gerwyn, Hector, Iago, Jason, Kyle and Liam. The bipartite graph below shows who the girls want to dance with.

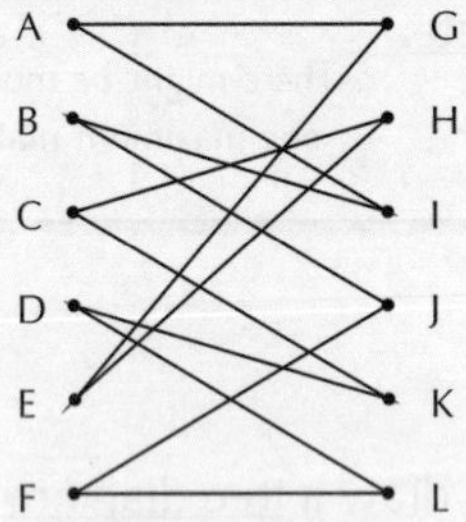

The initial matching pairs Alice with Gerwyn, Bella with Jason, Charlotte with Hector and Daisy with Liam.

   a) Draw the initial matching.
   b) Using the maximum matching algorithm, find alternating paths so that everyone has someone to dance with and they aren't left sitting by themselves and looking sad.

Exam questions on matchings are all pretty similar, so it's a good idea to get some practice in now.

Exam Question

1 The adjacency matrix below shows five tutors and the classes they want to teach:

	Class 1	Class 2	Class 3	Class 4	Class 5
James	0	1	0	1	0
Kelly	1	1	0	0	0
Lee	0	0	1	0	1
Mia	0	0	0	1	0
Nick	1	0	0	1	0

a) Draw a bipartite graph to represent this information. *(2 marks)*

b) An initial matching pairs James with class 4, Kelly with class 1 and Lee with class 3.
Find an alternating path starting from Mia and ending with class 2.
Write out the improved matching that your path gives. *(3 marks)*

c) A complete matching is not possible for this bipartite graph. Explain why. *(1 mark)*

Mia agrees to teach class 3.

d) Starting with the matching found in part (b), use the maximum matching algorithm to find a complete matching. Write out the alternating path and the final matching. *(3 marks)*

General Certificate of Education
Advanced Subsidiary (AS) and Advanced Level

Decision Mathematics D1 — Practice Exam One

Time Allowed: 1 hour 30 min

Graphical calculators may be used for this exam.

Give any non-exact numerical answers to an appropriate degree of accuracy.

There are 75 marks available for this paper.

1 Frederick, Vicky, Dexter, Rhiannon, Janet, Nahla, Sophia, Anthony, Luke, Paul, Samira

a) Sort these names into alphabetical order using a quick sort. Make all your pivots clear. *(5 marks)*

b) State how many passes you would need to make to sort these names using a shuttle sort. *(1 mark)*

2 The algorithm below calculates square roots correct to 2 decimal places.

Line 10:	Input A
Line 20:	Let N = 1
Line 30:	Let P = A ÷ N
Line 40:	Let Q = P + N
Line 50:	Let R = Q ÷ 2
Line 60:	If $\lvert R - N \rvert < 0.01$, then go to Line 100
Line 70:	If $\lvert R - N \rvert \geq 0.01$, then go to Line 80
Line 80:	Let N = R
Line 90:	Go to Line 30
Line 100:	Write R to 2 decimal places
Line 110:	Stop.

a) Trace the algorithm for A = 2. *(6 marks)*

b) State the result you would obtain if you traced the algorithm for A = 9. *(1 mark)*

3 Kalim, Louisa, Melissa, Nathan, Olivia and Priya need to be matched to six jobs numbered 1 - 6.
The table on the right shows the jobs that each person is willing to do.

Name	Jobs
Kalim	2, 5
Louisa	1, 4
Melissa	2, 6
Nathan	1, 3
Olivia	3, 5
Priya	2, 4

a) Show this information on a bipartite graph. *(2 marks)*

Initially, Kalim is matched to job 5, Louisa is matched to job 1, Melissa is matched to job 2, Nathan is matched to job 3 and Priya is matched to job 4.

b) Starting with the initial matching, use an alternating path to find a complete matching for the information above. *(4 marks)*

4 Natalie makes greetings cards to sell at craft fairs. She makes x birthday cards and y blank cards. Each birthday card costs 40p to make and each blank card costs 50p to make. She has a total of £8 she can spend on materials for the cards.

a) Write out this constraint as an inequality in terms of x and y. *(1 mark)*

b) Another two constraints are represented by the inequalities $y \leq 8$ and $y \leq x$.
Write out in words what these constraints mean in terms of the number of birthday cards and blank cards. *(2 marks)*

c) On graph paper, draw these three constraints, together with the non-negativity constraints $x, y \geq 0$ and label the feasible region R. *(4 marks)*

Natalie makes a profit of £0.75 on each birthday card and £1.50 on each blank card. Assume she sells all the cards she makes. She wants to maximise the profit, P.

d) Write out the objective function, P, in terms of x and y. *(1 mark)*

e) Use the graph from part (c) to optimise the objective function. State how many birthday cards and how many blank cards Natalie should make, and write down the maximum profit. *(5 marks)*

5 The network below represents the main roads between towns in an area.
The weights on the edges represent distances in miles, and the vertices represent road junctions.

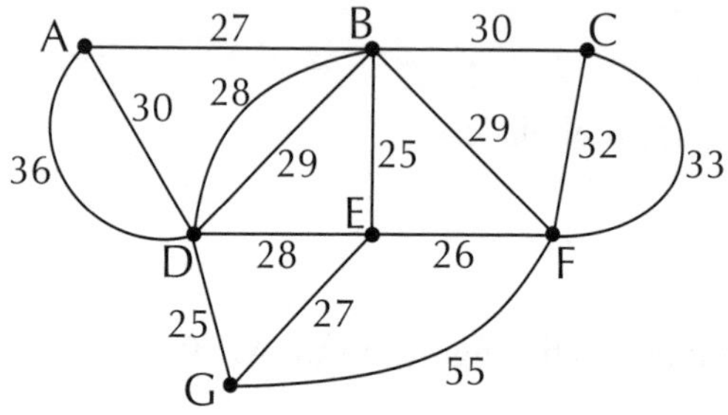

Total length of roads = 460 miles

a) Duncan works in town C and lives in town G. On a copy of the diagram, use Dijkstra's algorithm to find Duncan's shortest route home. Write down the route and give its length. *(7 marks)*

b) Duncan drives a truck which paints lines on roads. He is going to paint white lines on all the roads. He uses the shortest route that travels down each road, starting from C and ending at G.
(i) Which roads does Duncan need to drive along more than once?
(ii) How many miles will Duncan drive? *(2 marks)*

c) The next day Duncan paints another set of lines on all the roads.
He has to start and finish at work (C).
Find the length of the shortest route that Duncan can use. Show all your working. *(5 marks)*

6 The diagram shows the various cycle paths in a park. The bike hire shop is at G.
A floodlighting system is to be installed so that the cycle paths can be used after dark. There will be a light at each intersection. The number on each edge represents the distance, in metres, between two intersections.

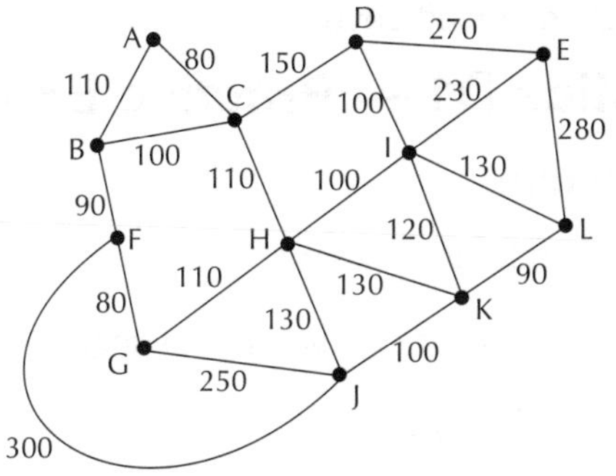

a) Cabling must be laid along the paths so that each intersection is connected.
Starting from the bike hire shop (G), use Prim's algorithm to find a cabling layout that will use a minimum amount of cable. List the paths in the order they are added. *(5 marks)*

b) Draw your minimum spanning tree and state its length. *(4 marks)*

c) The warden used Kruskal's algorithm to find the same minimum spanning tree.
Find the tenth and the eleventh edges that the warden added to his spanning tree. *(3 marks)*

7 a) Nicola has worked out the following lower bounds for a travelling salesperson problem:
187, 213, 195, 199, 207, 182.

(i) State the best lower bound. *(1 mark)*

She has found the following upper bounds for the same problem: 234, 218, 226, 217, 229, 235.

(ii) State the best upper bound. *(1 mark)*

(iii) From your answers to parts (i) and (ii), what can you conclude about the weight of the optimum tour of this problem? *(1 mark)*

b) Stuart is a security guard for an office block. At the end of each day, he has to go round every office and lock up. The time (in minutes) it takes him to walk between offices are shown in the matrix below.

	A	**B**	**C**	**D**	**E**	**F**
A	–	6	7	5	11	2
B	6	–	4	12	8	6
C	7	4	–	9	10	13
D	5	12	9	–	3	7
E	11	8	10	3	–	1
F	2	6	13	7	1	–

(i) Give an example of a Hamiltonian cycle for the office block, starting from office A.
State its total weight. *(2 marks)*

(ii) Find an upper bound for Stuart's quickest route using the Nearest Neighbour algorithm.
Start at office B. Compare this value with your answer to part (i), and state which upper bound is better. *(4 marks)*

(iii) Use Prim's algorithm to find a minimum spanning tree, starting from office C.
Show the order in which the edges are added and draw your tree, stating its weight. *(5 marks)*

(iv) Using your answer to part (iii), find a lower bound for Stuart's route.
Delete vertex D and all edges incident to it. *(3 marks)*

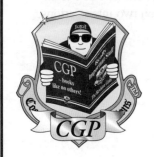

General Certificate of Education
Advanced Subsidiary (AS) and Advanced Level

Decision Mathematics D1 — Practice Exam Two

Time Allowed: 1 hour 30 min

Graphical calculators may be used for this exam.

Give any non-exact numerical answers to an appropriate degree of accuracy.

There are 75 marks available for this paper.

1 281 276 255 290 263 287 296 259 271

a) Sort these numbers into ascending order using a Shell sort. Show the order after each pass. *(5 marks)*

b) State the number of comparisons and swaps made on the first pass. *(2 marks)*

c) The original list is sorted into descending order using a bubble sort. Write down the maximum number of comparisons and the maximum number of passes needed. *(2 marks)*

2 Carlos, Diego, Edgar, Fatima and Gino are being assigned to tasks 1 - 5.
The graph below left shows the tasks they are able to do, and the graph below right shows the initial matching.

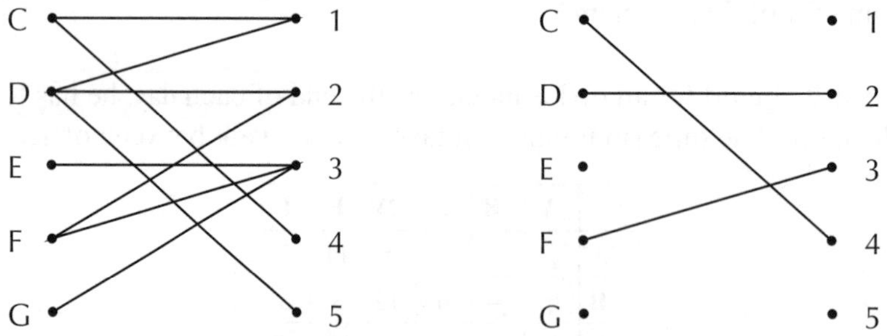

a) Find an alternating path from Edgar to task 1. Draw on a graph the improved matching given by the alternating path. *(3 marks)*

b) It is not possible to obtain a complete matching. Explain why. *(1 mark)*

Edgar is trained to carry out task 5.

c) Using the matching found in part a), find a complete matching for this information. *(2 marks)*

3 An algorithm for finding the next multiple of 9 from any integer A is represented as the flow chart below:

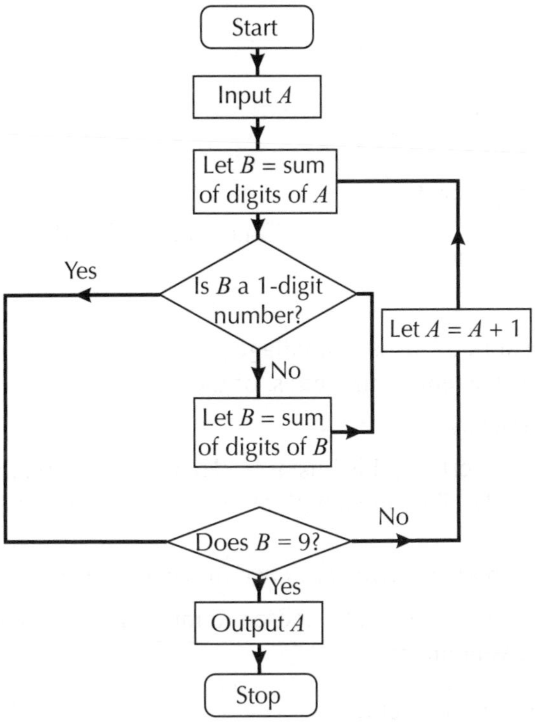

a) Use the algorithm to show that the output for $A = 977$ is 981. State the values of A and B at each pass through the flow chart. *(4 marks)*

b) Write down the output when $A = 978$. *(1 mark)*

c) Explain why no more than 9 passes through the algorithm will ever be needed. *(1 mark)*

4 Scott draws a graph that has 8 vertices and 16 edges. All the vertices are even, and the graph is connected.

a) Is this graph Eulerian? Explain your answer. *(2 marks)*

b) Explain what he would have to do to make the graph semi-Eulerian. *(1 mark)*

c) Scott draws a Hamiltonian cycle for his graph. How many edges does he use? *(1 mark)*

d) How many edges will he need to draw a minimum spanning tree for the graph? *(1 mark)*

5 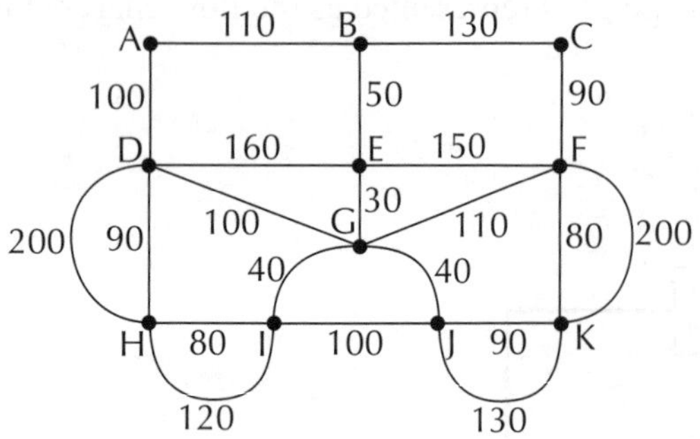

Total length of roads = 2200 m

The diagram shows the network of roads in a village.
The number on each edge is the length, in metres, of the road.

Rufus runs a roof repair company.

a) Rufus delivers leaflets advertising his business. He needs to walk along each road at least once, starting and finishing at H. Find the length of an optimal route for Rufus. *(6 marks)*

b) If Rufus can start and finish the delivery at two distinct vertices,

(i) state which two vertices should be chosen to minimise the length of his new route. Give a reason for your answer. *(2 marks)*

(ii) Find the length of his new route. *(1 mark)*

c) Rufus lives at H and needs to repair a roof at C. Use Dijkstra's algorithm to find the minimum driving distance between H and C. State the corresponding route. *(6 marks)*

6 The table shows the distances, in metres, between bird boxes in a reserve.

	A	B	C	D	E	F
A	–	140	167	205	150	173
B	140	–	145	148	159	170
C	167	145	–	210	195	180
D	205	148	210	–	155	178
E	150	159	195	155	–	185
F	173	170	180	178	185	–

a) Use Prim's algorithm, starting from A, to find a minimum spanning tree for this table of distances. You must list the edges that form your tree in the order that they are selected. *(3 marks)*

b) Draw your tree and calculate its total weight. *(2 marks)*

c) One advantage of Prim's algorithm over Kruskal's is that you don't have to check for cycles. Explain why you don't have to check for cycles in Prim's algorithm. *(2 marks)*

7 The diagram below shows a network. The distances between the vertices are given on the edges.

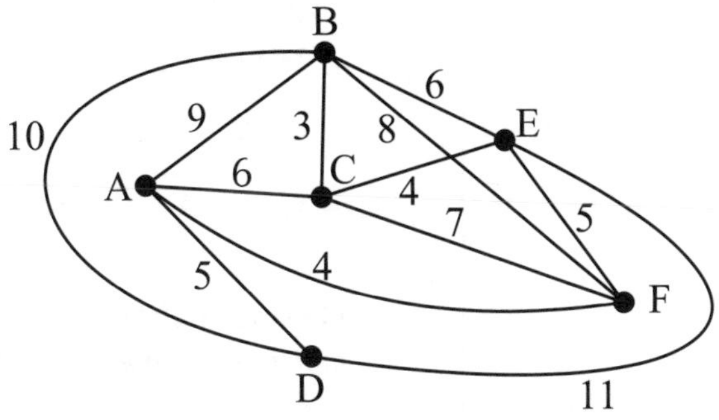

a) Add edges to this network to make the diagram complete. *(2 marks)*

b) Use the nearest neighbour algorithm to find an upper bound for your completed network. Start from vertex A. *(3 marks)*

c) Find a minimum spanning tree for your network using Kruskal's algorithm. Draw your tree and state its weight. *(4 marks)*

d) Hence find a lower bound for the completed network by deleting vertex B and all edges incident to it. *(3 marks)*

e) Using your answers to parts b) and d), draw a conclusion about the duration of the optimum tour. *(1 mark)*

8 A charity is selling tickets to its annual charity ball. It has two corporate packages available: business class and premier.

Each business class package includes 5 tickets and each premier package includes 10 tickets.

Each business class package includes 2 bottles of wine and each premier package includes 8 bottles of wine.

There are 300 tickets and 160 bottles of wine available.

At least 5 of each type of package are sold, and at least 20 packages are sold in total.

The charity sells x business class packages and y premier packages.

a) 5 constraints are required when expressing this information as a linear programming problem.

(i) Show why the following constraints are required: $x + 2y \leq 60$, $x + 4y \leq 80$ *(2 marks)*

(ii) State three other constraints that must also be used. *(2 marks)*

b) Each business class package makes a profit of £150 and each premier package makes a profit of £200. The charity wants to calculate the minimum and maximum profit from the ball.

(i) On graph paper, draw a diagram to represent this linear programming problem. Clearly label the feasible region and draw on an objective line. *(6 marks)*

(ii) Find the maximum profit from the ball. Write down how many of each type of package need to be sold to make this profit. *(2 marks)*

(iii) Find the minimum profit from the ball. Write down how many of each type of package need to be sold to make this profit. *(2 marks)*

Answers

C1 Section 1 — Algebra Fundamentals

Warm-up Questions

1) a) a & b are constants, x is a variable.
   b) a & b are constants, x is a variable.
   c) a, b & c are constants, y is a variable.
   d) a is a constant, x & y are variables.
2) A, C and D are identities.
3) a) $x = \pm\sqrt{5}$ b) $x = -2 \pm \sqrt{3}$
4) a) $2\sqrt{7}$ b) $\frac{\sqrt{5}}{6}$ c) $3\sqrt{2}$ d) $\frac{3}{4}$
5) a) $\frac{8}{\sqrt{2}} = \frac{8}{\sqrt{2}} \times \frac{\sqrt{2}}{\sqrt{2}} = \frac{8\sqrt{2}}{2} = 4\sqrt{2}$
   b) $\frac{\sqrt{2}}{2} = \frac{\sqrt{2}}{(\sqrt{2})^2} = \frac{1}{\sqrt{2}}$
6) $136 + 24\sqrt{21}$
7) $3 - \sqrt{7}$
8) a) $a^2 - b^2$ b) $a^2 + 2ab + b^2$
   c) $25y^2 + 210xy$ d) $3x^2+10xy+3y^2+13x+23y+14$
9) a) $xy(2x + a + 2y)$ b) $a^2x(1 + b^2x)$
   c) $8(2y + xy + 7x)$ d) $(x - 2)(x - 3)$
10) a) $\frac{52x + 5y}{60}$ b) $\frac{5x - 2y}{x^2y^2}$ c) $\frac{x^3 + x^2 - y^2 + xy^2}{x(x^2 - y^2)}$
11) a) $\frac{3a}{2b}$ b) $\frac{2(p^2 + q^2)}{p^2 - q^2}$ c) 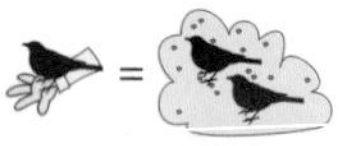

Exam Questions

1 a) $(5\sqrt{3})^2 = (5^2)(\sqrt{3})^2 = 25 \cdot 3$
$= 75$ ***[1 mark]***
b) $(5 + \sqrt{6})(2 - \sqrt{6}) = 10 - 5\sqrt{6} + 2\sqrt{6} - 6$ ***[1 mark]***
$= 4 - 3\sqrt{6}$ ***[1 mark]***

2 Multiply top and bottom by $3 + \sqrt{5}$ to 'rationalise the denominator':

$\frac{5 + \sqrt{5}}{3 - \sqrt{5}} = \frac{(5 + \sqrt{5})(3 + \sqrt{5})}{(3 - \sqrt{5})(3 + \sqrt{5})}$ ***[1 mark]***

$= \frac{15 + 5\sqrt{5} + 3\sqrt{5} + 5}{9 - 5}$ ***[1 mark]***

$= \frac{20 + 8\sqrt{5}}{4}$ ***[1 mark]***

$= 5 + 2\sqrt{5}$ ***[1 mark]***

3 $\sqrt{18} - 2\sqrt{8} + \frac{4}{\sqrt{2}}$

$= \sqrt{9 \times 2} - 2\sqrt{4 \times 2} + \frac{4\sqrt{2}}{\sqrt{2} \times \sqrt{2}}$

$= 3\sqrt{2} - 4\sqrt{2} + 2\sqrt{2}$

$= \sqrt{2}$

[4 marks available — 1 mark each for correctly simplifying each initial term, 1 mark for correct final answer]

4 $\frac{4\sqrt{32}}{\sqrt{2}} - 2\sqrt{3} + \frac{4\sqrt{5}}{\sqrt{20}} + \sqrt{12}$

$= \frac{4\sqrt{16 \times 2}}{\sqrt{2}} - 2\sqrt{3} + \frac{4\sqrt{5}}{\sqrt{4 \times 5}} + \sqrt{4 \times 3}$

$= \frac{4\sqrt{16}\sqrt{2}}{\sqrt{2}} - 2\sqrt{3} + \frac{4\sqrt{5}}{\sqrt{4}\sqrt{5}} + \sqrt{4}\sqrt{3}$

$= 16 - 2\sqrt{3} + 2 + 2\sqrt{3}$

$= 18$

[5 marks available — 1 mark each for simplifying each initial term, 1 mark for correct final answer]

C1 Section 2 — Quadratics

Warm-up Questions

1) a) $(x + 1)^2$ b) $(x - 10)(x - 3)$
   c) $(x + 2)(x - 2)$ d) $(3 - x)(x + 1)$
   e) $(2x + 1)(x - 4)$ f) $(5x - 3)(x + 2)$
2) a) $(x - 2)(x - 1) = 0$, so $x = 2$ or 1
   b) $(x + 4)(x - 3) = 0$, so $x = -4$ or 3
   c) $(2 - x)(x + 1) = 0$, so $x = 2$ or -1
   d) $(x + 4)(x - 4) = 0$, so $x = \pm 4$
   e) $(3x + 2)(x - 7) = 0$, so $x = -2/3$ or 7
   f) $(2x + 1)(2x - 1) = 0$, so $x = \pm 1/2$
   g) $(2x - 3)(x - 1) = 0$, so $x = 3/2$ or 1
3) a) $(x-2)^2 - 7$; minimum value $= -7$ at $x = 2$, and this crosses the x-axis at $x = 2 \pm \sqrt{7}$
   b) $\frac{21}{4} - \left(x + \frac{3}{2}\right)^2$; maximum value $= 21/4$ at $x = -3/2$, and this crosses the x-axis at $-\frac{3}{2} \pm \frac{\sqrt{21}}{2}$.
   c) $2(x-1)^2 + 9$; minimum value $= 9$ at $x = 1$, and this doesn't cross the x-axis.
   d) $4\left(x - \frac{7}{2}\right)^2 - 1$; minimum value $= -1$ at $x = 7/2$, crosses the x-axis at $x = \frac{7}{2} \pm \frac{1}{2}$ i.e. $x = 4$ or 3
4) a) $b^2 - 4ac = 16$, so 2 roots
   b) $b^2 - 4ac = 0$, so 1 root
   c) $b^2 - 4ac = -8$, so no roots

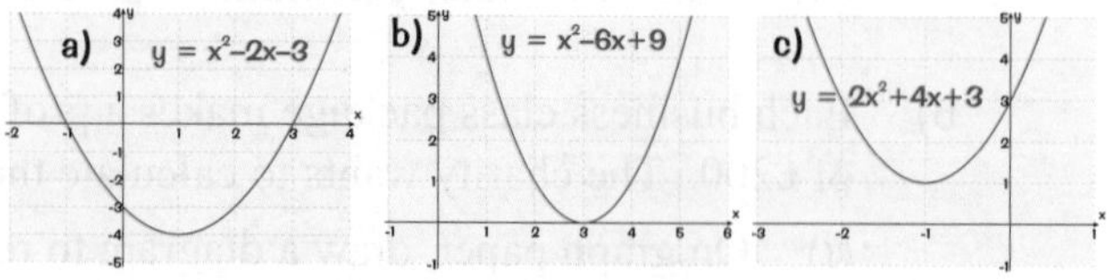

5) a) Using the quadratic formula with $a = 3$, $b = -7$ and $c = 3$:

$x = \frac{-(-7) \pm \sqrt{(-7)^2 - (4 \times 3 \times 3)}}{(2 \times 3)}$

$= \frac{7 \pm \sqrt{49 - 36}}{6}$

$= \frac{7 \pm \sqrt{13}}{6}$

so $x = \frac{7 + \sqrt{13}}{6}, x = \frac{7 - \sqrt{13}}{6}$.

Answers

b) Using the quadratic formula with $a = 2$, $b = -6$ and $c = -2$:

$$x = \frac{-(-6) \pm \sqrt{(-6)^2 - (4 \times 2 \times -2)}}{(2 \times 2)}$$
$$= \frac{6 \pm \sqrt{36 + 16}}{4}$$
$$= \frac{6 \pm \sqrt{52}}{4}$$
$$= \frac{6 \pm 2\sqrt{13}}{4}$$
$$= \frac{3 \pm \sqrt{13}}{2}$$
so $x = \frac{3 + \sqrt{13}}{2}, x = \frac{3 - \sqrt{13}}{2}$.

c) Using the quadratic formula with $a = 1$, $b = 4$ and $c = -6$:

$$x = \frac{-(4) \pm \sqrt{(4)^2 - (4 \times 1 \times -6)}}{(2 \times 1)}$$
$$= \frac{-4 \pm \sqrt{16 + 24}}{2}$$
$$= \frac{-4 \pm \sqrt{40}}{2}$$
$$= \frac{-4 \pm 2\sqrt{10}}{2}$$
$$= -2 \pm \sqrt{10}$$
so $x = -2 + \sqrt{10}, x = -2 - \sqrt{10}$.

6) $k^2 - (4 \times 1 \times 4) > 0$, so $k^2 > 16$ and so $k > 4$ or $k < -4$.

Exam Questions

1 For equal roots, $b^2 - 4ac = 0$ ***[1 mark]***

$a = 1$, $b = 2k$ and $c = 4k$

so $(2k)^2 - (4 \times 1 \times 4k) = 0$ ***[1 mark]***

$4k^2 - 16k = 0$

$4k(k - 4) = 0$ ***[1 mark]***

so $k = 4$ (as k is non-zero). ***[1 mark]***

2 a) For distinct real roots, $b^2 - 4ac > 0$ ***[1 mark]***

$a = p$, $b = p + 3$ and $c = 4$

so $(p + 3)^2 - (4 \times p \times 4) > 0$ ***[1 mark]***

$p^2 + 6p + 9 - 16p > 0$

$p^2 - 10p + 9 > 0$ ***[1 mark]***

b) $p^2 - 10p + 9$ is a u-shaped quadratic, which crosses the x-axis when $p^2 - 10p + 9 = 0$, that is when $(p - 9)(p - 1) = 0$ ***[1 mark]***, which occurs at $p = 9$ and $p = 1$ ***[1 mark]***. As it's u-shaped, $p^2 - 10p + 9 > 0$ when p is outside these values ***[1 mark]***, that is when $p < 1$ or $p > 9$ ***[1 mark]***.

3 Expanding the brackets on the RHS gives the quadratic $mx^2 + 4mx + 4m + p$. Equating the coefficients of x^2 gives $m = 5$ ***[1 mark]***. Equating the coefficients of x gives $n = 4m$, so $n = 20$ ***[1 mark]***. Equating the constant terms gives $14 = 4m + p \Rightarrow p = -6$ ***[1 mark]***.

4 a) $a = 12 \div 2 = 6$ ***[1 mark]***, so the completed square is $(x - 6)^2 + b = x^2 - 12x + 36 + b$. Equating coefficients gives $15 = 36 + b$, so $b = 15 - 36 = -21$ ***[1 mark]***. The final expression is $(x - 6)^2 - 21$.

b) (i) The minimum occurs when the expression in brackets is equal to 0, which means the minimum is the value of b, which from (a) above is -21 ***[1 mark]***.

(ii) From above, the minimum occurs when the expression in brackets is equal to 0, i.e. when $x = 6$ ***[1 mark]***.

Part (b) is dead easy once you've completed the square — you can just take your values straight from there.

5 a) Put $a = 1$, $b = -14$ and $c = 25$ into the quadratic formula:

$$x = \frac{-(-14) \pm \sqrt{(-14)^2 - (4 \times 1 \times 25)}}{(2 \times 1)}$$
$$= \frac{14 \pm \sqrt{196 - 100}}{2}$$
$$= \frac{14 \pm \sqrt{96}}{2}$$
$$= \frac{14 \pm 4\sqrt{6}}{2}$$
$$= 7 \pm 2\sqrt{6}$$
so $x = 7 + 2\sqrt{6}, x = 7 - 2\sqrt{6}$

[3 marks available — 1 mark for putting correct values of a, b and c into the quadratic formula, 1 mark each for final x-values.]

b)

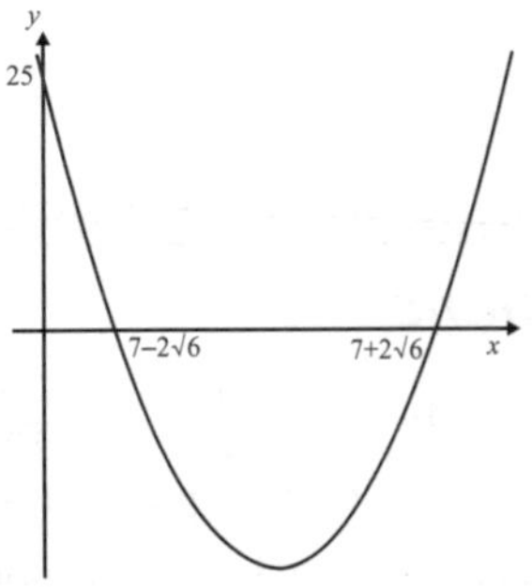

[3 marks available — 1 mark for drawing u-shaped curve, 1 mark for using answers from a) as x-axis intercepts and 1 mark for correct y-axis intercept (0, 25).]

c) As the graph is u-shaped, the inequality is < 0 when x is between the two intercepts, i.e. $7 - 2\sqrt{6} \leq x \leq 7 + 2\sqrt{6}$ ***[1 mark]***.

6 a) (i) $m = 10 \div 2 = 5$ ***[1 mark]***, so the expression becomes $-(5 - x)^2 + n = -25 + 10x - x^2 + n$. Equating coefficients gives $-27 = -25 + n$, so $n = -27 - (-25) = -2$ ***[1 mark]***. The final expression is $-(5 - x)^2 - 2$.

(ii) $(5 - x)^2 \geq 0$ for all values of x, so $-(5 - x)^2 \leq 0$. Therefore: $-(5 - x)^2 - 2 < 0$ for all x, i.e. the function is always negative. ***[1 mark]***

b) (i) The y-coordinate is the maximum value, which is -2 ***[1 mark]***, and this occurs when the expression in the brackets = 0.

The x-value that makes the expression in the brackets 0 is 5 ***[1 mark]***, so the coordinates of the maximum point are (5, –2).

Answers

(ii)

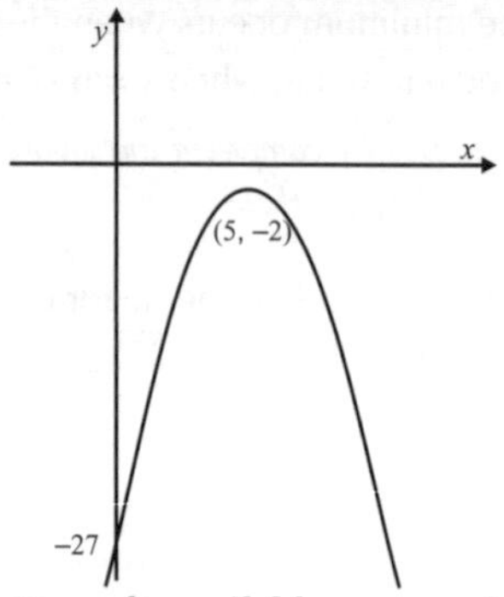

[2 marks available — 1 mark for drawing n-shaped curve that sits below the x-axis, 1 mark for correct y-axis intercept (0, –27)]

(iii) The line of symmetry of the curve is the line parallel to the y-axis which passes through (5, –2).
i.e. the line $x = 5$ ***[1 mark]***.

C1 Section 3 — Simultaneous Equations and Inequalities

Warm-up Questions

1) a) $x = -3, y = -4$ b) $x = -\frac{1}{6}, y = -\frac{5}{12}$
2) a) The line and the curve meet at the points (2, –6) and (7, 4).
   b) The line is a tangent to the parabola at the point (2, 26).
   c) The equations have no solution and so the line and the curve never meet.
3) a) $\left(\frac{1}{4}, -\frac{13}{4}\right)$ b) (4, 5) c) (–5, –2)
4) a) $x > -\frac{38}{5}$ b) $y \leq \frac{7}{8}$ c) $y \leq -\frac{3}{4}$
5) a) $x > \frac{5}{2}$ b) $x > -4$ c) $x \leq -3$
6) a) $-\frac{1}{3} \leq x \leq 2$ b) $x < 1 - \sqrt{3}$ or $x > 1 + \sqrt{3}$
   c) $x \leq -3$ or $x \geq -2$
7) a) $x \leq -3$ and $x \geq 1$ b) $x < -½$ and $x > 1$
   c) $-3 < x < 2$

Exam Questions

1 a) $3x + 2 \leq x + 6$
$2x \leq 4$ ***[1 mark]***
$x \leq 2$ ***[1 mark]***

b) $20 - x - x^2 > 0$
$(4 - x)(5 + x) > 0$ ***[1 mark]***
The graph crosses the x-axis at $x = 4$ and $x = -5$ ***[1 mark]***. The coefficient of x^2 is negative so the graph is n-shaped ***[1 mark]***.
So $20 - x - x^2 > 0$ when $-5 < x < 4$ ***[1 mark]***.

c) From above, x will satisfy both inequalities when $-5 < x \leq 2$ ***[1 mark]***.
For this bit, all you need to do is use your answers to parts a) and b) and work out which values of x fit in them both.

2 a) $3 \leq 2p + 5 \leq 15$
This inequality has 3 parts. Subtract 5 from each part to give: $-2 \leq 2p \leq 10$ ***[1 mark]***.
Now divide each part by 2 to give: $-1 \leq p \leq 5$
[1 mark for $-1 \leq p$ and 1 mark for $p \leq 5$].

b) $q^2 - 9 > 0$
$(q + 3)(q - 3) > 0$ ***[1 mark]***
The function is 0 at $q = -3$ and $q = 3$ ***[1 mark]***.
The coefficient of x^2 is positive so the graph is u-shaped.
So $q^2 - 9 > 0$ when $q < -3$ or $q > 3$
[1 mark for each correct inequality].
Use D.O.T.S. (Difference of Two Squares) to factorise the quadratic — remember that $a^2 - b^2 = (a + b)(a - b)$.

3 a) $(3x + 2)(x - 5)$ ***[1 mark]***
b) $(3x + 2)(x - 5) \leq 0$
The function is 0 at $x = -\frac{2}{3}$ and $x = 5$ ***[1 mark]***.
The coefficient of x^2 is positive so the graph is u-shaped, meaning the function is less than or equal to 0 between these x-values.
I.e. $3x^2 - 13x - 10 \leq 0$ when $-\frac{2}{3} \leq x \leq 5$
[2 marks, 1 for $-\frac{2}{3} \leq x$ and 1 for $x \leq 5$].

4 First, take the linear equation and rearrange it to get x on its own: $x = 6 - y$ ***[1 mark]***. Now substitute the equation for x into the quadratic equation:
$(6 - y)^2 + 2y^2 = 36$ ***[1 mark]***
$36 - 12y + y^2 + 2y^2 = 36$
$3y^2 - 12y = 0$
$y^2 - 4y = 0$
$y(y - 4) = 0$ ***[1 mark]***
so $y = 0$ and $y = 4$. ***[1 mark]***.
Now you've got the y-values, put them back into the equation for x ($x = 6 - y$) to find the x-values.
When $y = 0$, $x = 6 - y = 6 - 0 = 6$.
When $y = 4$, $x = 6 - y = 6 - 4 = 2$.
So solutions are $x = 6, y = 0$ ***[1 mark]***
and $x = 2, y = 4$ ***[1 mark]***.

5 a) At points of intersection, $-2x + 4 = -x^2 + 3$ ***[1 mark]***
$x^2 - 2x + 1 = 0$
$(x - 1)^2 = 0$ ***[1 mark]***
so $x = 1$ ***[1 mark]***. When $x = 1$, $y = -2x + 4 = 2$, so there is one point of intersection at (1, 2) ***[1 mark]***.

Answers

b)

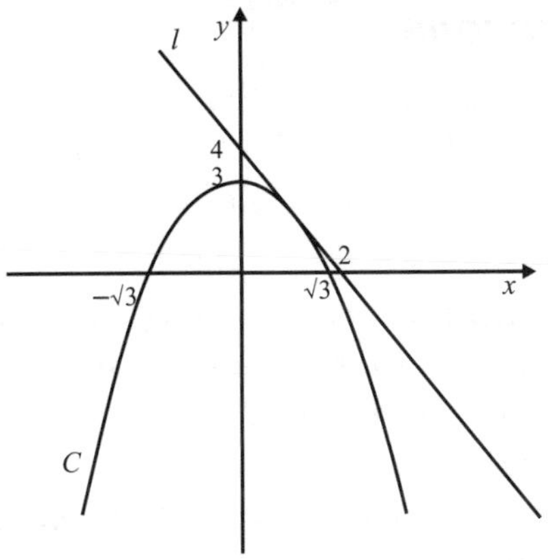

[5 marks available — 1 mark for drawing n-shaped curve, 1 mark for x-axis intercepts at ±√3, 1 mark for maximum point of curve and y-axis intercept at (0, 3). 1 mark for line crossing the y-axis at (0, 4) and the x-axis at (2, 0). 1 mark for line and curve touching in one place.]

6 a)

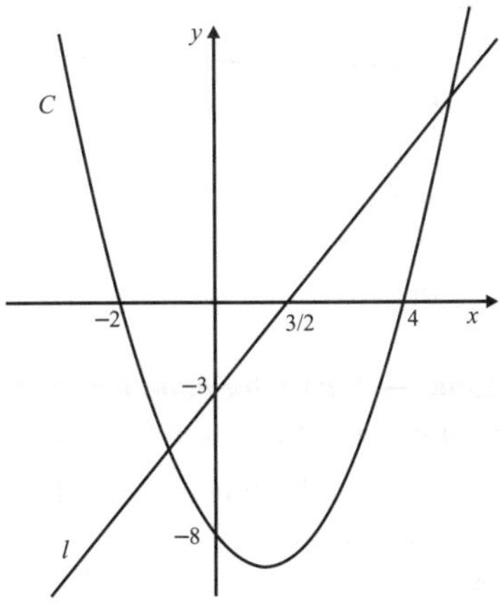

[5 marks available — 1 mark for drawing u-shaped curve, 1 mark for x-axis intercepts at –2 and 4 , 1 mark for y-axis intercept at (0, –8). 1 mark for line crossing the y-axis at (0, –3) and the x-axis at (3/2, 0). 1 mark for line and curve touching in two places.]

b) At points of intersection,

$$2x - 3 = (x + 2)(x - 4)$$

$$2x - 3 = x^2 - 2x - 8 \quad \textbf{\textit{[1 mark]}}$$

$$0 = x^2 - 4x - 5 \quad \textbf{\textit{[1 mark]}}$$

c) $x^2 - 4x - 5 = 0$

$(x - 5)(x + 1) = 0$ ***[1 mark]***

so $x = 5$, $x = -1$ ***[1 mark]***.

When $x = 5$, $y = (2 \times 5) - 3 = 7$ and when $x = -1$, $y = (2 \times -1) - 3 = -5$, so the points of intersection are (5, 7) ***[1 mark]*** and (–1, –5) ***[1 mark]***.

C1 Section 4 — More Polynomials

Warm-up Questions

1) a) $f(x) = (x + 2)(3x^2 - 10x + 15) - 36$

b) $f(x) = (x + 2)(x^2 - 3) + 10$

c) $f(x) = (x + 2)(2x^2 - 4x + 14) - 31$

2) a) (i) You just need to find f(–1). This is $-6 - 1 + 3 - 12 = -16$.

(ii) Now find f(1). This is $6 - 1 - 3 - 12 = -10$.

b) (i) $f(-1) = -1$

(ii) $f(1) = 9$

c) (i) $f(-1) = -2$

(ii) $f(1) = 0$

3) a) You need to find f(–2). This is $3(-2)^3 + 7(-2)^2 - 12(-2) + 14$ $= -24 + 28 + 24 + 14 = 42$.

b) You need to find f(–4/2) = f(–2). You found this in part a, so remainder = 42. You might also have noticed that $2x + 4$ is a multiple of $x + 2$ (from part a), so the remainder must be the same.

c) You need to find f(3). This is $3(3)^3 + 7(3)^2 - 12(3) + 14$ $= 81 + 63 - 36 + 14 = 122$.

d) You need to find f(6/2) = f(3). You found this in part c), so remainder = 122. You might also have noticed that $2x - 6$ is a multiple of $x - 3$, so the remainder must be the same.

4) a) You need to find f(1) — if f(1) = 0, then $(x - 1)$ is a factor: $f(1) = 2 - 6 + 4 = 0$, so $(x - 1)$ is a factor.

b) You need to find f(–1) — if f(–1) = 0, then $(x + 1)$ is a factor: $f(-1) = -2 + 6 + 4 = 8$, so $(x + 1)$ is not a factor.

c) You need to find f(2) — if f(2) = 0, then $(x - 2)$ is a factor: $f(2) = 16 - 12 + 4 = 8$, so $(x - 2)$ is not a factor.

d) The remainder when you divide by $(2x - 2)$ is the same as the remainder when you divide by $x - 1$. $(x - 1)$ is a factor (i.e. remainder = 0), so $(2x - 2)$ is also a factor.

5) If $f(x) = 2x^3 + 5x^2 + cx + d$, then to make sure f(x) is exactly divisible by $(x - 2)(x + 3)$, you have to make sure $f(2) = f(-3) = 0$.
$f(2) = 16 + 20 + 2c + d = 0$, i.e. $\underline{2c + d = -36}$.
$f(-3) = -54 + 45 - 3c + d = 0$, i.e. $\underline{3c - d = -9}$.
Add the two underlined equations to get: $5c = -45$, and so $c = -9$. Then $d = -18$.

Exam Questions

1 a) (i) Remainder $= f(1) = 2(1)^3 - 5(1)^2 - 4(1) + 3$ ***[1 mark]***
$= -4$ ***[1 mark]***.

(ii) Remainder $= f\left(-\frac{1}{2}\right) = 2\left(-\frac{1}{8}\right) - 5\left(\frac{1}{4}\right) - 4\left(-\frac{1}{2}\right) + 3$
[1 mark] $= \frac{7}{2}$ ***[1 mark]***.

b) If f(–1) = 0 then $(x + 1)$ is a factor.
$f(-1) = 2(-1)^3 - 5(-1)^2 - 4(-1) + 3$ ***[1 mark]***
$= -2 - 5 + 4 + 3 = 0$, so $(x + 1)$ is a factor of f(x).
[1 mark]

Answers

c) $(x + 1)$ is a factor, so divide $2x^3 - 5x^2 - 4x + 3$ by $x + 1$:
$2x^3 - 5x^2 - 4x + 3 - \underline{2x^2}(x + 1) = 2x^3 - 5x^2 - 4x + 3 - 2x^3 - 2x^2$
$= -7x^2 - 4x + 3$.
$-7x^2 - 4x + 3 - (\underline{-7x})(x + 1) = -7x^2 - 4x + 3 + 7x^2 + 7x$
$= 3x + 3$. Finally $3x + 3 - \underline{3}(x + 1) = 0$.
so $2x^3 - 5x^2 - 4x + 3 = (2x^2 - 7x + 3)(x + 1)$.

Alternatively, you could have factorised using the method on p.26.

Factorising the quadratic expression gives:
$f(x) = (2x - 1)(x - 3)(x + 1)$.

[4 marks available — 1 mark for correct working used to find quadratic factor, 1 mark for correct quadratic factor, 1 mark for attempt to factorise quadratic, 1 mark for correct factorisation of quadratic.]

2 a) $f(p) = (4p^2 + 3p + 1)(p - p) + 5$
$= (4p^2 + 3p + 1) \times 0 + 5$
$= 5$ ***[1 mark]***.

b) $f(-1) = -1$.
$f(-1) = (4(-1)^2 + 3(-1) + 1)((-1) - p) + 5$
$= (4 - 3 + 1)(-1 - p) + 5$
$= 2(-1 - p) + 5 = 3 - 2p$ ***[1 mark]***
So: $3 - 2p = -1$, $p = 2$ ***[1 mark]***.

c) $f(x) = (4x^2 + 3x + 1)(x - 2) + 5$
$f(1) = (4 + 3 + 1)(1 - 2) + 5 = -3$ ***[1 mark]***.

C1 Section 5 — Coordinate Geometry

Warm-up Questions

1) a) (i) $y + 1 = 3(x - 2)$ (ii) $y = 3x - 7$ (iii) $3x - y - 7 = 0$

b) (i) $y + \frac{1}{3} = \frac{1}{5}x$ (ii) $y = \frac{1}{5}x - \frac{1}{3}$
(iii) $3x - 15y - 5 = 0$

2) a) $M = (\frac{2 + 12}{2}, \frac{5 - 1}{2}) = (7, 2)$

b) $l = \sqrt{(7 - 2)^2 + (2 - 5)^2} = \sqrt{25 + 9} = \sqrt{34}$

3) a) $y = \frac{3}{2}x - 4$ b) $y = -\frac{1}{2}x + 4$

4) The equation of the required line is $y = \frac{3}{2}x + \frac{15}{2}$.

5)

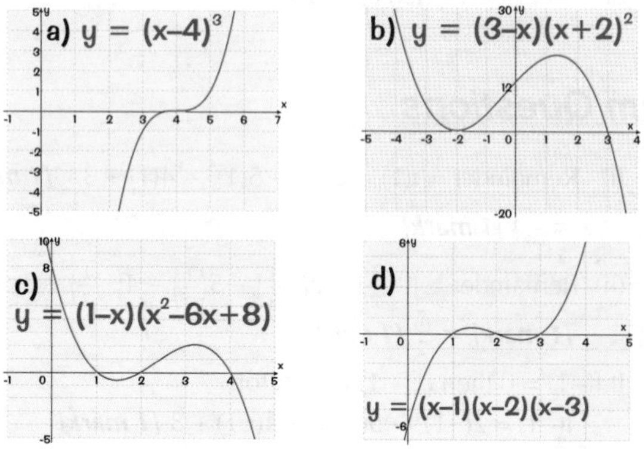

6) a) 3, (0, 0)

b) 2, (2, –4)

c) 5, (–3, 4)

Exam Questions

1 a) Substitute $(1, k)$ into $5y - x = 9$:
$5k - 1 = 9 \Rightarrow 5k = 10 \Rightarrow k = 2$. ***[1 mark]***

b) $l = \sqrt{(x_2 - x_1)^2 + (y_2 - y_1)^2} = \sqrt{(6 - 1)^2 + (3 - 2)^2} = \sqrt{26}$.
Mid-Point $= \left(\frac{1 + 6}{2}, \frac{2 + 3}{2}\right) = (3.5, 2.5)$.

[3 marks available — 1 mark for correct use of formula for length, 1 mark for correct answer for length, 1 mark for correct answer for mid-point.]

c) First find the equation of the line BC:
The gradient of BC is $m = \frac{3 - 0}{6 + 1} = \frac{3}{7}$ ***[1 mark]***, and so BC has equation:
$y - y_1 = m(x - x_1)$
$y - 0 = \frac{3}{7}(x + 1)$ ***[1 mark]***
$\Rightarrow 7y = 3x + 3$ ***[1 mark]***.
Now substitute $(2p + 3, p + 1)$ into your equation to find p:
$7(p + 1) = 3(2p + 3) + 3$ ***[1 mark]***
$7p + 7 = 6p + 12$
$\Rightarrow p = 5$ ***[1 mark]***.

2 a) Given the information in the question, you can write down the equation of Q as:
$(x - 3)^2 + (y - 4)^2 = 4$
Multiplying out the brackets:
$x^2 - 6x + 9 + y^2 - 8y + 16 = 4$
$\Rightarrow x^2 - 6x + y^2 - 8y = -21$
So $k = -21$.
[3 marks available — 1 mark for equation of circle, 1 mark for multiplying out brackets, 1 mark for finding k.]

b) $x + y = 9 \Rightarrow x = 9 - y$. Substitute into equation of circle:
$(9 - y - 3)^2 + (y - 4)^2 = 4$
$(6 - y)^2 + (y - 4)^2 = 4$
$36 - 12y + y^2 + y^2 - 8y + 16 = 4$
$2y^2 - 20y + 48 = 0$
$y^2 - 10y + 24 = 0$
$(y - 6)(y - 4) = 0$
$\Rightarrow y = 6, y = 4$,
$x = 9 - y \Rightarrow x = 3, x = 5$.
So $A = (3, 6)$, $B = (5, 4)$.
[4 marks available — 1 mark for substituting equation of line into equation of circle, 1 mark for forming quadratic in x or y, 1 mark for correct coordinates of A, 1 mark for correct coordinates of B.]

c) Mid-Point $= \left(\frac{3 + 5}{2}, \frac{6 + 4}{2}\right) = (4, 5)$ ***[1 mark]***.

d) The required distance is d on the diagram:

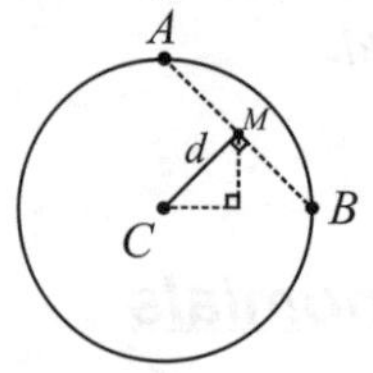

From part c), you know that M, the mid-point of AB, has coordinates (4, 5). You also know that C has coordinates (3, 4), so you can use the length formula to find d:
$d = \sqrt{(4 - 3)^2 + (5 - 4)^2}$ ***[1 mark]***
$= \sqrt{4 - 2} = \sqrt{2}$ ***[1 mark]***

Answers

3 a) Using the formula $y - y_1 = m(x - x_1)$, with the coordinates of point S for the x- and y- values and $m = -2$,

$y - (-3) = -2(x - 7)$ ***[1 mark]***

$y + 3 = -2x + 14$ ***[1 mark]***

$y = -2x + 11$ ***[1 mark]***

b) Putting $x = 5$ into $y = -2x + 11$ gives $y = 1$ ***[1 mark]***, so T does lie on the line.

4 a) Rearrange equation and complete the square:

$x^2 - 2x + y^2 - 10y + 21 = 0$ ***[1 mark]***

$(x - 1)^2 - 1 + (y - 5)^2 - 25 + 21 = 0$ ***[1 mark]***

$(x - 1)^2 + (y - 5)^2 = 5$ ***[1 mark]***

Compare with $(x - a)^2 + (y - b)^2 = r^2$:

centre = (1, 5) ***[1 mark]***, radius = $\sqrt{5}$ ***[1 mark]***.

b) The point (3, 6) and centre (1, 5) both lie on the diameter.

Gradient of the diameter = $\frac{6-5}{3-1} = 0.5$.

Q $(q, 4)$ also lies on the diameter, so $\frac{4-6}{q-3} = 0.5$.

$-2 = 0.5q - 1.5$

So $q = (-2 + 1.5) \div 0.5 = -1$.

[3 marks available — 1 mark for finding the gradient of the diameter, 1 mark for linking this with the point Q, and 1 mark for correct calculation of q.]

c) Tangent at Q is perpendicular to the diameter at Q, so

gradient $m = -\frac{1}{0.5} = -2$

$y - y_1 = m(x - x_1)$, and $(-1, 4)$ is a point on the line, so:

$y - 4 = -2(x + 1)$

$y - 4 = -2x - 2$

$2x + y - 2 = 0$ is the equation of the tangent.

[5 marks available — 1 mark for gradient = −1 ÷ gradient of diameter, 1 mark for correct value for gradient, 1 mark for substituting Q in straight-line equation, 2 marks for correct substitution of values in the correct form, or 1 mark if not in the form ax + by + c = 0.]

5 a) Just rearrange into the form $y = mx + c$ and read off m:

$3y = 15 - 4x$

$y = -\frac{4}{3}x + 5$ ***[1 mark]***

so the gradient of the line l is $-\frac{4}{3}$ ***[1 mark]***.

b) Gradient of the line $= -1 \div -\frac{4}{3} = \frac{3}{4}$ ***[1 mark]***

So $y = \frac{3}{4}x + c$.

Now use the x- and y- values of R to find c:

$1 = \frac{3}{4}(3) + c$

$1 = \frac{9}{4} + c$

$\Rightarrow c = -\frac{5}{4}$ ***[1 mark]***

so the equation of the line is $y = \frac{3}{4}x - \frac{5}{4}$ ***[1 mark]***

6 a) Radius is 4 ***[1 mark]***, as the circle touches the y-axis. Equation is therefore $(x - 4)^2 + (y - 10)^2 = 16$ ***[1 mark]***.

b) Substitute (8, 10) into equation of circle:

$(8 - 4)^2 + (10 - 10)^2 = 4^2 + 0^2 = 16$ ***[1 mark]***. So (8, 10) lies on the circle.

c) The gradient of the normal to the circle at this point is given by $m_n = \frac{y_2 - y_1}{x_2 - x_1} = \frac{10 + \sqrt{7} - 10}{1 - 4} = -\frac{\sqrt{7}}{3}$ ***[1 mark]***, and so gradient of tangent to circle at this point is $m_t = \frac{3}{\sqrt{7}}$ ***[1 mark]***. The equation of the tangent is then given by:

$y - y_2 = m_t(x - x_2) \Rightarrow y - (10 + \sqrt{7}) = \frac{3}{\sqrt{7}}(x - 1)$

[1 mark] $\Rightarrow \sqrt{7}y - 10\sqrt{7} - 7 = 3x - 3$

$\Rightarrow -3x + \sqrt{7}y = 10\sqrt{7} + 4$ ***[1 mark]***.

(or $y = \frac{3}{\sqrt{7}}x + \frac{10\sqrt{7} + 4}{\sqrt{7}}$ if you're a fan of $y = mx + c$)

C1 Section 6 — Differentiation

Warm-up Questions

1) $\frac{d}{dx}(x^n) = nx^{n-1}$

2) a) $\frac{dy}{dx} = 2x$ b) $\frac{dy}{dx} = 4x^3 + 1$

3) They're the same.

4) a) $\frac{dy}{dx} = 4x = 8$

b) $\frac{dy}{dx} = 8x - 1 = 15$

c) $\frac{dy}{dx} = 3x^2 - 14x = -16$

5) Differentiate $v = 17t^2 - 10t$ to give: $\frac{dv}{dt} = 34t - 10$ so, when $t = 4$, $\frac{dv}{dt} = 126$ ml/s.

6) The tangent and normal must go through (5, 0).

Differentiate to find $\frac{dy}{dx} = 2x - 3$, so gradient at (5, 0) is 7.

Therefore tangent can be written $y_T = 7x + c_T$; putting $x = 5$ and $y = 0$ gives $0 = 7 \times 5 + c_T$, so $c_T = -35$, and the equation of the tangent is $y_T = 7x - 35$.

The gradient of the normal must be $-\frac{1}{7}$, so the equation of the normal is $y_N = -\frac{1}{7}x + c_N$

Substituting in the coordinates of the point (5, 0) gives $0 = -\frac{5}{7} + c_N \Rightarrow c_N = \frac{5}{7}$; so the normal is

$y_N = -\frac{1}{7}x + \frac{5}{7} = \frac{1}{7}(5 - x)$.

7) For both lines, when $x = 4$, $y = 2$, so they meet at (4, 2). Differentiating the first curve gives $\frac{dy}{dx} = x^2 - 4x - 4$, which at $x = 4$ is equal to -4. Differentiating the other curve gives $\frac{dy}{dx} = \frac{1}{4}$. If you multiply these two gradients together you get -1, so the two curves are perpendicular at $x = 4$.

8) a) A point (on a graph) where $\frac{dy}{dx}$ (the gradient) = 0.

b) $y = x^3 - 6x^2 + 9x + 1 \Rightarrow \frac{dy}{dx} = 3x^2 - 12x + 9$, so set this equal to zero (and divide by 3) to get that the stationary points are where $x^2 - 4x + 3 = 0$, i.e. $(x - 1)(x - 3) = 0$. Sub $x = 1$ and $x = 3$ back into the original equation to get the corresponding y-values $y = 5$ and $y = 1$, and so the stationary points are (1, 5) and (3, 1).

9) $\frac{dy}{dx} = 12x - 6$, so function is increasing when $x > 0.5$, and decreasing when $x < 0.5$.

Answers

10) Maximum is when $\frac{dh}{dm} = 0$.

So, $\frac{m}{5} - \frac{m^2}{100} = m\left(\frac{1}{5} - \frac{m}{100}\right) = 0$

Since $m \neq 0$, $\frac{1}{5} - \frac{m}{100} = 0 \Rightarrow \frac{m}{100} = \frac{1}{5}$

$\Rightarrow m = \frac{100}{5} = 20\text{ g}$

Those questions covered the basics of differentiation, so if you got them all correct, bravely venture into the murky realm of Exam Questions. If you struggled, have a cuppa to fuel your noggin — then read the section again until it all makes sense.

Exam Questions

1 a) Differentiate each term:

$\frac{dy}{dx} = 7x^6 + 1$ ***[1 mark]***. At $x = 1$, $\frac{dy}{dx} = 8$ ***[1 mark]***.

b) This is a second-order derivative — just differentiate the answer for part a)

$\frac{d^2y}{dx^2} = \frac{d}{dx}(7x^6 + 1)$ ***[1 mark]*** $= 42x^5$ ***[1 mark]***.

At $x = 1$, $\frac{d^2y}{dx^2} = 42$ ***[1 mark].***

2 a) $\frac{dy}{dx} = 6x^2 - 8x - 4$

[2 marks for all 3 terms correct or 1 mark for 2 terms.]

b) To find the gradient, put $x = 2$ into the answer to part (a):
$6(2^2) - 8(2) - 4 = 24 - 16 - 4 = 4$ ***[1 mark]***.

c) The gradient of the normal is $-1 \div$ the gradient of the tangent $= -1 \div 4 = -\frac{1}{4}$ ***[1 mark]***. At $x = 2$, the y-value is $2(2^3) - 4(2^2) - 4(2) + 12 = 16 - 16 - 8 + 12 = 4$ ***[1 mark]***.
Putting these values into the formula $(y - y_1) = m(x - x_1)$ gives $(y - 4) = -\frac{1}{4}(x - 2) \Rightarrow y = -\frac{1}{4}x + \frac{1}{2} + 4 \Rightarrow$
$y = -\frac{1}{4}x + 4\frac{1}{2}$ ***[1 mark]***.

You could also give your answer in the form $x + 4y = 18$ by multiplying through by 4 to get rid of the fractions.

3 a) $\frac{dy}{dx} = (3 \times mx^{(3-1)}) - (2 \times x^{(2-1)}) + 8(1 \times x^{(1-1)})$

$= 3mx^2 - 2x + 8$

[1 method mark, 1 answer mark]

b) Rearranging the equation of the line parallel to the normal gives the equation: $y = 3 - 4x$, so it has a gradient of -4 ***[1 mark]***. The normal also has gradient -4 because it is parallel to this line ***[1 mark]***. The gradient of the tangent is $-1 \div$ the gradient of the normal $= -1 \div -4 = \frac{1}{4}$ ***[1 mark]***.

c) (i) So you know that when $x = 5$, the gradient

$3mx^2 - 2x + 8 = \frac{1}{4}$. ***[1 mark]***

Now find the value of m:

$m(3 \times 5^2) - (2 \times 5) + 8 = \frac{1}{4}$ ***[1 mark]***

$75m - 2 = \frac{1}{4}$

$m = \frac{9}{4} \times \frac{1}{75} = \frac{9}{300} = \frac{3}{100} = 0.03$ ***[1 mark]***

(ii) When $x = 5$, then:

$y = (\frac{3}{100} \times 5^3) - (5^2) + (8 \times 5) + 2$ ***[1 mark]***

$= \frac{375}{100} - 25 + 40 + 2$

$= \frac{375}{100} + 17 = \frac{2075}{100}$

$= 20.75$ ***[1 mark]***

They've given you all the information you need, in a funny roundabout kinda way. Get comfortable figuring out gradients of normals and tangents and then applying them to curves — otherwise the exam will be very UNcomfortable. You have been warned...

4 a) Rearrange $2x - y = 6$ into an expression for y:
$y = 2x - 6$. ***[1 mark]***. Now find x^2y^2 in terms of x:
$x^2y^2 = x^2(2x - 6)^2 = x^2(4x^2 - 24x + 36)$
$= 4x^4 - 24x^3 + 36x^2$ ***[1 mark]***

b) (i) $\frac{d(4x^4 - 24x^3 + 36x^2)}{dx} = 16x^3 - 72x^2 + 72x$

[2 marks for all three terms, or 1 mark for two]

$= 8(2x^3 - 9x^2 + 9x)$ ***[1 mark]***

So $k = 8$ ***[1 mark]***

(ii) At $x = 1$, $8(2x^3 - 9x^2 + 9x) = 8(2 - 9 + 9)$
$= 16$ ***[1 mark]***.

c) $\frac{d^2W}{dy^2} = 48x^2 - 144x + 72$ ***[1 mark]***

So for $x = 1$: $\frac{d^2W}{dy^2} = 48 - 144 + 72 = -24$ ***[1 mark]***

5 a) $y = 6 + \frac{2x^3 - 6x^2 + 6x}{3} = 6 + \frac{2}{3}x^3 - 2x^2 + 2x$

$\frac{dy}{dx} = 2x^2 - 4x + 2$

[1 mark for each correct term]

b) Stationary points occur when $2x^2 - 4x + 2 = 0$ ***[1 mark]***, i.e. when $x^2 - 2x + 1 = 0$.
Factorising the equation gives: $(x - 1)^2 = 0$.

So stationary point occurs when $x = 1$ ***[1 mark]***.
When $x = 1$:

$y = 6 + \frac{2(1)^3 - 6(1)^2 + 6(1)}{3} = 6\frac{2}{3}$ ***[1 mark]***

So coordinates of the stationary point on the curve are $\left(1, 6\frac{2}{3}\right)$.

6 a) First step, multiply out function to get $y = x^3 - 2x^2 + x$

$\frac{dy}{dx} = 3x^2 - 4x + 1 = 0$ at the stationary point. ***[1 mark]***

Solve using the quadratic formula:

$x = \frac{4 \pm \sqrt{(-4)^2 - (4 \times 3 \times 1)}}{2 \times 3} = \frac{4 \pm 2}{6}$

So solutions are $x = 1$ ***[1 mark]*** and $x = \frac{1}{3}$ ***[1 mark]***.

Substitute these values for x into the original equation for y:

$x = 1 \Rightarrow y = 1(0)^2 = 0$ ***[1 mark]***

$x = \frac{1}{3} \Rightarrow y = \frac{1}{3}\left(-\frac{2}{3}\right)^2 = \frac{1}{3} \times \frac{4}{9} = \frac{4}{27}$ ***[1 mark]***

So the stationary points have coordinates:
$(1, 0)$ and $\left(\frac{1}{3}, \frac{4}{27}\right)$.

Answers

b) $\frac{d^2y}{dx^2} = 6x - 4$ ***[1 mark]***
At $x = 1$, $\frac{d^2y}{dx^2} = 2$ is > 0,
so it's a minimum ***[1 mark]***
At $x = \frac{1}{3}$, $\frac{d^2y}{dx^2} = -2$ is < 0,
so it's a maximum ***[1 mark]***

7 a) Surface area $= [2 \times (d \times x)] + \left[2 \times \left(d \times \frac{x}{2}\right)\right] + \left[x \times \frac{x}{2}\right]$
$= 2dx + \frac{2dx}{2} + \frac{x^2}{2}$ ***[1 mark]***
surface area $= 6$ so $3dx + \frac{x^2}{2} = 6$
$\Rightarrow x^2 + 6dx = 12$ ***[1 mark]***
$d = \frac{12 - x^2}{6x}$ ***[1 mark]***
Volume = width × height × depth = $\frac{x}{2} \times x \times d$
$\Rightarrow V = \frac{x^2}{2} \times \frac{12 - x^2}{6x} = \frac{12x^2 - x^4}{12x} = x - \frac{x^3}{12}$ ***[1 mark]***

b) Differentiate V and then solve for when $\frac{dV}{dx} = 0$:
$\frac{dV}{dx} = 1 - \frac{x^2}{4}$ ***[1 mark for each correct term]***
$1 - \frac{x^2}{4} = 0$ ***[1 mark]***
$\Rightarrow x^2 = 4$
So $x = 2\text{m}$ ***[1 mark]***

c) $\frac{d^2V}{dx^2} = -\frac{x}{2}$ ***[1 mark]***
so when $x = 2$, $\frac{d^2V}{dx^2} = -1$ ***[1 mark]***
$\frac{d^2V}{dx^2}$ is negative, so it's a maximum point. ***[1 mark]***
$x = 2$ at V_{max}, so $V_{max} = 2 - \frac{2^3}{12}$
$V_{max} = \frac{4}{3}\text{ m}^3$ ***[1 mark]***

8 a) $f'(x) = 3x^2 - 6x = 0$ at the stationary points. ***[1 mark]***
$3x^2 - 6x = 0 \Rightarrow x^2 - 2x = 0 \Rightarrow x(x - 2) = 0$
So the stationary points occur at $x = 0$ ***[1 mark]*** and $x = 2$ ***[1 mark]***, which gives:
$y = f(0) = 0 - 0 - 1 = -1$ ***[1 mark]*** and
$y = f(2) = 8 - 12 - 1 = -5$ ***[1 mark]***
So the coordinates are $(0, -1)$ and $(2, -5)$.

b) $f''(x) = 6x - 6$ ***[1 mark]*** so at the stationary points:
$f''(0) = -6$ is negative, so $(0, -1)$ is a maximum ***[1 mark]***.
$f''(2) = 6$ is positive, so $(2, -5)$ is a minimum ***[1 mark]***.

c) $f'(4) = 3(4)^2 - 6(4) = 48 - 24 = 24$, which is positive, so $f(x)$ is increasing. ***[1 mark]***

I love differentiation.

C1 Section 7 — Integration
Warm-up Questions

1) i) Increase the power of x by 1,
ii) divide by the new power,
iii) add a constant.

2) An integral without limits to integrate between.
Because there's more than one right answer.

3) Differentiate your answer, and if you get back the function you integrated in the first place, your answer's right.

4) Check whether there are limits to integrate between. If there are, then it's a definite integral; if not, it's an indefinite integral.

5) a) $2x^5 + C$ b) $\frac{3x^2}{2} + \frac{5x^3}{3} + C$ c) $\frac{3}{4}x^4 + \frac{2}{3}x^3 + C$

6) Integrating gives $y = 3x^2 - 7x + C$; then substitute $x = 1$ and $y = 0$ to find that $C = 4$. So the equation of the curve is $y = 3x^2 - 7x + 4$.

7) a) Integrate to get $y = \frac{3x^4}{4} + 2x + C$. Putting $x = 1$ and $y = 0$ gives $C = -\frac{11}{4}$, and so the required curve is $y = \frac{3x^4}{4} + 2x - \frac{11}{4}$.
b) If the curve has to go through $(1, 2)$ instead of $(1, 0)$, substitute the values $x = 1$ and $y = 2$ to find a different value for C; call this value C_1. Making these substitutions gives $C_1 = -\frac{3}{4}$, and the equation of the new curve is $y = \frac{3x^4}{4} + 2x - \frac{3}{4}$.

8) The area between a curve and the x-axis, starting at the lower limit and going up to the upper limit.

9) a) $\int_0^1 (4x^3 + 3x^2 + 2x + 1)dx$
$= [x^4 + x^3 + x^2 + x]_0^1$
$= 4 - 0 = 4.$

b) $\int_{-3}^{3} (9 - x^2)dx = \left[9x - \frac{x^3}{3}\right]_{-3}^{3}$
$= 18 - (-18) = 36.$

10) a) $A = \int_0^2 (x^3 - 5x^2 + 6x)dx$
$= \left[\frac{x^4}{4} - \frac{5}{3}x^3 + 3x^2\right]_0^2 = \frac{8}{3}$

b) $A = \int_0^2 2x^2 dx + \int_2^6 (12 - 2x)dx$
$= \left[\frac{2}{3}x^3\right]_0^2 + [12x - x^2]_2^6$
$= \frac{16}{3} + 16 = \frac{64}{3}$

c) $A = \int_1^4 (x + 3)dx - \int_1^4 (x^2 - 4x + 7)dx$
$= \left[\frac{x^2}{2} + 3x\right]_1^4 - \left[\frac{x^3}{3} - 2x^2 + 7x\right]_1^4$
$= \frac{33}{2} - 12 = \frac{9}{2}$

Well, that's covered the bog-standard integration skills — but you've got to know how to apply them in all sorts of fiddly ways, so get practising these exam questions...

Exam Questions

1 a) Multiply out the brackets and simplify the terms:
$(5 + 2\sqrt{x})(5 - 2\sqrt{x}) = 25 + 10\sqrt{x} - 10\sqrt{x} - 4x$
$= 25 - 4x$. So $a = 25$ and $b = -4$.
[2 marks: one for each constant]

Answers

b) Integrate your answer from a), treating each term separately:

$\int (25 - 4x)dx = 25x - \left(\frac{4x^2}{2}\right) + C$

$= 25x - 2x^2 + C.$

[2 marks available — 1 for each term. Lose 1 mark if C missing]

Don't forget to add C, don't forget to add C, don't forget to add C. Once, twice, thrice I beg of you, because it's very important.

2 a) Multiply out the brackets in $f'(x)$

$(x - 1)(3x - 1) = 3x^2 - x - 3x + 1$

$= 3x^2 - 4x + 1$ ***[1 mark]***

Now $f(x) = \int (3x^2 - 4x + 1)dx$

$= \frac{3x^3}{3} - \frac{4x^2}{2} + \frac{x}{1} + C$ ***[1 mark]***

$= x^3 - 2x^2 + x + C$ ***[1 mark]***

Input the x and y coordinates to find C:

$10 = 3^3 - 2(3^2) + 3 + C$ ***[1 mark]***

$10 - 27 + 18 - 3 = C$

$C = -2$ ***[1 mark]***

So $f(x) = x^3 - 2x^2 + x - 2$ ***[1 mark]***

b) First calculate the gradient of $f(x)$ when $x = 3$:

$f'(3) = 3(3^2) - (4 \times 3) + 1$

$= 27 - 12 + 1$

$= 16$ ***[1 mark]***

Use the fact that the tangent gradient multiplied by the normal gradient must equal –1 to find the gradient of the normal (n): $16 \times n = -1$ therefore $n = -\frac{1}{16}$ ***[1 mark]***.

Put n and $P(3, 10)$ into the formula for the equation of a line and rearrange until it's in the form $y = \frac{a - x}{b}$:

$y - 10 = -\frac{1}{16}(x - 3)$ ***[1 mark]***

$y = \frac{3}{16} - \frac{x}{16} + 10$

$y = \frac{163 - x}{16}$

So $a = 163$ and $b = 16$. ***[1 mark]***

3 $\int_0^2 (2x - 6x^2 + 1)dx = [x^2 - 2x^3 + x]_0^2$

[1 mark for each term correctly integrated]

$= (2^2 - (2 \times 2^3) + 2) - 0$ ***[1 mark]***

$= -10$ ***[1 mark]***

4 The limits are the x-values when $y = x(x - 1)^2 = 0$.

i.e. the limits are $x = 0$ and $x = 1$ ***[1 mark]***.

Hence, to find the area, calculate:

$\int_0^1 x(x - 1)^2 dx = \int_0^1 (x^3 - 2x^2 + x)\,dx$ ***[1 mark]***

$= \left[\frac{x^4}{4} - \frac{2x^3}{3} + \frac{x^2}{2}\right]_0^1$ ***[1 mark]***

$= \left(\frac{1}{4} - \frac{2}{3} + \frac{1}{2}\right) - (0)$ ***[1 mark]***

$= \frac{3}{12} - \frac{8}{12} + \frac{6}{12}$

$= \frac{1}{12}$ ***[1 mark]***

5 a) $m = \frac{y_2 - y_1}{x_2 - x_1} = \frac{-3 - 0}{1 - -2} = -1$ ***[1 mark]***

$y - y_1 = m(x - x_1)$

$y - -3 = -1(x - 1)$ so $y + 3 = -x + 1$

$y = -x - 2$ ***[1 mark]***

b) Multiply out the brackets and then integrate:

$(x + 2)(x - 2) = x^2 - 4$ ***[1 mark]***

$\int_{-2}^{1} (x^2 - 4)\,dx = \left[\frac{x^3}{3} - 4x\right]_{-2}^{1}$ ***[1 mark]***

$= \left(\frac{1}{3} - 4\right) - \left(-\frac{8}{3} + 8\right)$ ***[1 mark]***

$= -9$ ***[1 mark]***

c) Subtract the area 'above' the line from the area 'above' the curve to leave the area in between. The area 'above' the line is a triangle (where $b = 3$ and $h = -3$), so use the formula for the area of a triangle to calculate it:

Area 'above' line $= \frac{1}{2}bh = \frac{1}{2} \times 3 \times -3 = -4.5$ ***[1 mark]***

So area of shaded region is given by:

$-9 - (-4.5)$ ***[1 mark]*** $= -4.5$ ***[1 mark]***.

You could also have integrated the line $y = -x - 2$ to find the area 'above' the line — you'd have got the same answer.

C1 — Practice Exam One

1 a) The equation has equal roots when the discriminant is equal to zero, so:

$(k + 1)^2 - (4 \times 1 \times 4k) = 0$ ***[1 mark]***

$k^2 + 2k + 1 - 16k = 0$ ***[1 mark]***

$\Rightarrow k^2 - 14k + 1 = 0$, as required ***[1 mark]***.

b) (i) Complete the square:

$k^2 - 14k + 1 = (k - 7)^2 + d$

Expand the right hand side to find d:

$k^2 - 14k + 1 = k^2 - 14k + 49 + d$

$\Rightarrow d = -48$

So $k^2 - 14k + 1 = (k - 7)^2 - 48$.

[2 marks available — 1 mark for –7 term, 1 mark for –48 term]

(ii) You know that $(k - 7)^2 \geq 0$. The minimum value of $k^2 - 14k + 1$ occurs when $(k - 7)^2 = 0$.

The minimum value is therefore –48 ***[1 mark]***.

(iii) $(k - 7)^2 = 0 \Rightarrow k = 7$ ***[1 mark]***.

2 a) First you need to find the gradient of L:

$m = \frac{y_2 - y_1}{x_2 - x_1} = \frac{9 - 6}{-2 - 1} = -1$ ***[1 mark]***.

Now use the formula for the equation of a line:

$y - y_1 = m(x - x_1)$

$y - 6 = -(x - 1)$ ***[1 mark]***

$\Rightarrow y = -x + 7$ ***[1 mark]***.

b) The mid-point is just the average of the end-points:

$M = \left(\frac{1 - 2}{2}, \frac{6 + 9}{2}\right)$ ***[1 mark]*** $= \left(-\frac{1}{2}, \frac{15}{2}\right)$ ***[1 mark]***.

c) The gradient of this new line is $[-1 \div (-1)] = 1$ ***[1 mark]***, so: $y - y_1 = m(x - x_1)$

$y - \frac{15}{2} = \left(x + \frac{1}{2}\right)$ ***[1 mark]***

$2y - 15 = 2x + 1$

$\Rightarrow y = x + 8$ ***[1 mark]***.

Answers

d) Make the equations of L and C equal to each other to find the points of intersection:
$-x + 7 = x^2 + 3x - 5$
$x^2 + 4x - 12 = 0$ ***[1 mark]***
$(x - 2)(x + 6) = 0$
$\Rightarrow x = 2$ ***[1 mark]*** or $x = -6$ ***[1 mark]***.
Sub these back into the equation for L to find the corresponding y-values:
$x = 2 \Rightarrow y = -2 + 7 = 5$ ***[1 mark]***
$x = -6 \Rightarrow y = -(-6) + 7 = 13$ ***[1 mark]***
So the coordinates of intersection are (2, 5) and (–6, 13).

Simultaneous equations — you should be able to do them standing on your head... although don't try that in the exam...

3 a) The circle intersects the x-axis when $y = 0$, i.e.:
$x^2 - 6x - 3 = 0$
Use the quadratic formula to find x:
$x = \frac{6 \pm \sqrt{(-6)^2 - (4 \times 1 \times -3)}}{2 \times 1}$ ***[1 mark]***
$= 3 \pm \frac{\sqrt{48}}{2}$ ***[1 mark]*** $= 3 \pm 2\sqrt{3}$.
So the coordinates of intersection with the x-axis are:
$(3 + 2\sqrt{3}, 0)$ ***[1 mark]*** and $(3 - 2\sqrt{3}, 0)$ ***[1 mark]***.

b) $x^2 - 6x + y^2 = 3$
$(x - 3)^2 - 9 + y^2 = 3$ ***[1 mark]***
$(x - 3)^2 + y^2 = 12$ ***[1 mark]***.

c) $r = \sqrt{12}$ ***[1 mark]*** $= 2\sqrt{3}$ ***[1 mark]***
Q has coordinates (3, 0) ***[1 mark]***.

d) (i) Use the distance formula $d = \sqrt{(x_2 - x_1)^2 + (y_2 - y_1)^2}$:
$d = \sqrt{(5 - 3)^2 + (4 - 0)^2}$ ***[1 mark]***
$= \sqrt{4 + 16} = \sqrt{20}$ ***[1 mark]*** $= 2\sqrt{5}$ ***[1 mark]***.

(ii) Distance from Q to P is $2\sqrt{5}$.
Radius of C is $2\sqrt{3}$.
$2\sqrt{5} > 2\sqrt{3}$, so P lies outside C.
[2 marks available — 1 mark for correct answer, 1 mark for reason]

Remember the process, remember the formulas, and you'll be fine. But the only way to make sure you know it is practice. If you forgot anything, go back to the coordinate geometry section and go through it 'til it sinks in...

4 a) Multiply out the brackets first:
$(5\sqrt{5} + 2\sqrt{3})^2 = (5\sqrt{5} + 2\sqrt{3})(5\sqrt{5} + 2\sqrt{3})$
$= (5\sqrt{5})^2 + 2(5\sqrt{5} \times 2\sqrt{3}) + (2\sqrt{3})^2$
So now you've got three terms to deal with, and they're all a little bit nasty. The first term is:
$(5\sqrt{5})^2 = 5\sqrt{5} \times 5\sqrt{5}$
$= 5 \times 5 \times \sqrt{5} \times \sqrt{5}$
$= 5 \times 5 \times 5$
$= 125$ ***[1 mark]***
The second term is:
$2(5\sqrt{5} \times 2\sqrt{3}) = 2 \times 5 \times 2 \times \sqrt{5} \times \sqrt{3}$
$= 20\sqrt{15}$ (*don't forget,* $\sqrt{5} \times \sqrt{3} = \sqrt{5 \times 3}$) ***[1 mark]***
And the third term is:
$2\sqrt{3} \times 2\sqrt{3} = 2 \times 2 \times \sqrt{3} \times \sqrt{3}$
$= 2 \times 2 \times 3 = 12$ ***[1 mark]***

So all you have to do now is add the three terms together:
$125 + 20\sqrt{15} + 12 = 137 + 20\sqrt{15}$ ***[1 mark]***
So $a = 137$, $b = 20$ and $c = 15$.

b) Rationalising the denominator means getting rid of surds on the bottom line of a fraction.
Multiply top and bottom lines by $\sqrt{5} - 1$:
$\frac{10 \times (\sqrt{5} - 1)}{(\sqrt{5} + 1) \times (\sqrt{5} - 1)}$ ***[1 mark]***
$= \frac{10(\sqrt{5} - 1)}{5 - 1} = \frac{10(\sqrt{5} - 1)}{4} = \frac{5(\sqrt{5} - 1)}{2}$
[1 mark for simplifying top, 1 mark for simplifying bottom]

Mmm, look at all that lovely maths vocab. Rationalising surd denominators — look back at Section 1 if you got stuck with this.

5 a) $(x - 1)(x^2 + x + 1) = 2x^2 - 17$
$x^3 + x^2 + x - x^2 - x - 1 = 2x^2 - 17$ ***[1 mark]***
$\Rightarrow x^3 - 2x^2 + 16 = 0$ ***[1 mark]***.

b) If $(x + 2)$ is a factor then, by the factor theorem, $f(-2)$ will be equal to zero.
Using the equation you found in part a):
$f(-2) = (-2)^3 - 2(-2)^2 + 16$ ***[1 mark]***
$= -8 - 2(4) + 16 = 0$ ***[1 mark]***
$\Rightarrow (x + 2)$ is a factor of $f(x)$.

c) You want to write $f(x)$ in the form $(x + 2)(x^2 + px + q)$.
Multiplying this out gives $x^3 + px^2 + qx + 2x^2 + 2px + 2q$
which simplifies to $x^3 + (p + 2)x^2 + (q + 2p)x + 2q$.
Equate the coefficients in this function with those in your answer to part a) to find p and q:
$x^3 - 2x^2 + 16 = x^3 + (p + 2)x^2 + (q + 2p)x + 2q$

$$\Rightarrow \begin{cases} -2 = p + 2 & \textbf{eqn 1} \\ 0 = q + 2p & \textbf{eqn 2} \\ 16 = 2q & \textbf{eqn 3} \end{cases}$$

eqn 1 $\Rightarrow p = -4$
eqn 3 $\Rightarrow q = 8$
So $x^3 - 2x^2 + 16 = (x + 2)(x^2 - 4x + 8)$.

You could also have used the method from page 26 here — in fact, it probably would have been quicker. It's all about using the method you prefer though.

[3 marks available — 1 mark for working, 1 mark for correct p, 1 mark for correct q]

d) $f(x) = 0$ has at least one root, $x = -2$, as shown in part b).
As shown in part c), $f(x) = (x + 2)(x^2 - 4x + 8)$, so if $f(x) = 0$ then $(x + 2)(x^2 - 4x + 8) = 0$, and so any other roots of $f(x) = 0$ will be the roots of $(x^2 - 4x + 8) = 0$.
The discriminant of $(x^2 - 4x + 8)$ is:
$(-4)^2 - (4 \times 1 \times 8) = -16$ ***[1 mark]***.
$-16 < 0$, so $(x^2 - 4x + 8) = 0$ has no real roots ***[1 mark]***.
Therefore $f(x) = (x + 2)(x^2 - 4x + 8)$ only has one root, $x = -2$ ***[1 mark]***.

e) (i) $\frac{dy}{dx} = f'(x) = 3x^2 - 4x$.
[2 marks available — 1 mark for each correct term]

Answers

(ii) The gradient of the tangent is given by:
$m = f'(1) = 3(1)^2 - 4(1) = -1$ ***[1 mark].***
When $x = 1$, $y = (1)^3 - 2(1)^2 + 16 = 15$ ***[1 mark].***
Now use $y - y_1 = m(x - x_1)$ to find the equation of the tangent:
$y - 15 = -(x - 1)$ ***[1 mark]***
$\Rightarrow y = -x + 16$ ***[1 mark].***

(iii) $f(x)$ is increasing when $f'(x) > 0$.
i.e. when $3x^2 - 4x > 0$. ***[1 mark]***

(iv) The graph of $y = 3x^2 - 4x$ is drawn below.

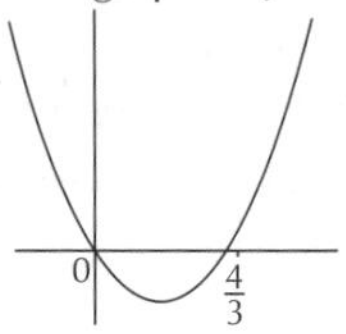

$3x^2 - 4x = 0 \Rightarrow x(3x - 4) = 0$, so the graph crosses the x-axis at $x = 0$ and $x = \frac{4}{3}$.
The graph is positive (i.e. above the x-axis) when $x < 0$ and when $x > \frac{4}{3}$.
Therefore, $3x^2 - 4x > 0$ when $x < 0$ and when $x > \frac{4}{3}$.
[2 marks available — 1 mark for working, 1 mark for correct answer]

What a question that was — a real mash-up of different topics. It's good practice though — the examiners love to mix things up.

6 a) $\frac{dy}{dx} = -2x + 1$ ***[1 mark]***
Stationary points occur when $\frac{dy}{dx} = 0$,
$\Rightarrow -2x + 1 = 0$ ***[1 mark]***
$\Rightarrow x = 0.5$ ***[1 mark].***
$x = 0.5 \Rightarrow y = -(0.5)^2 + 0.5 + 6$
$= -0.25 + 0.5 + 6 = 6.25$ ***[1 mark].***
So there is a stationary point at (0.5, 6.25).

b) (i) Finding the points of intersection:
$-x^2 + x + 6 = x + 2$
$\Rightarrow x^2 = 4$ ***[1 mark]***
$\Rightarrow x = \pm 2$ ***[2 marks].***
Use the equation for L to find the corresponding y-values:
When $x = 2$, $y = 2 + 2 = 4$ ***[1 mark].***
When $x = -2$, $y = -2 + 2 = 0$ ***[1 mark].***
So the coordinates of intersection are (2, 4) and (–2, 0).

(ii)

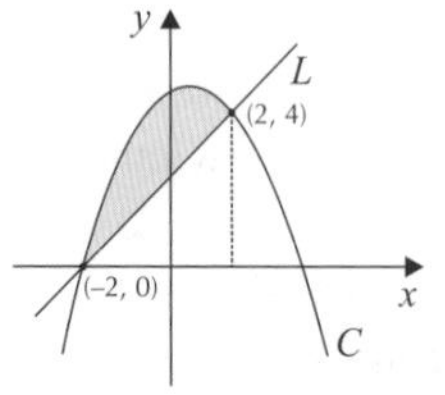

The area between L and C is equal to the area under C between $x = -2$ and $x = 2$ minus the area under L between $x = -2$ and $x = 2$.

The area under C between $x = -2$ and $x = 2$ is given by:
$$\int_{-2}^{2}(-x^2 + x + 6)\mathrm{d}x = \left[-\frac{x^3}{3} + \frac{x^2}{2} + 6x\right]_{-2}^{2} \quad \textbf{\textit{[1 mark]}}$$
$$= \left(-\frac{2^3}{3} + \frac{2^2}{2} + 6(2)\right) - \left(-\frac{(-2)^3}{3} + \frac{(-2)^2}{2} + 6(-2)\right)$$
[1 mark]
$$= \left(-\frac{8}{3} + 14\right) - \left(\frac{8}{3} - 10\right) = \frac{56}{3} \quad \textbf{\textit{[1 mark].}}$$

The area under L between $x = -2$ and $x = 2$ is equal to the area of a triangle of base 4 and height 4 [the vertices of the triangle are (–2, 0), (2, 0) and (2, 4).]
So, the area beneath L is $\frac{1}{2} \times 4 \times 4$ ***[1 mark]***
$= 8$ ***[1 mark].***
The area between C and L is therefore:
$\frac{56}{3} - 8 = \frac{56}{3} - \frac{24}{3} = \frac{32}{3}$ ***[1 mark].***

At the end of the exam, it's tradition to have a comforting cup of tea and vow never to differentiate the equation of a curve ever again. That was just a practice though, so don't celebrate just yet... Oh alright, you can have a cuppa I suppose...

C1 — Practice Exam Two

1 a) Just multiply out the brackets:
$(\sqrt{3} + 1)(\sqrt{3} - 1) = 3 - \sqrt{3} + \sqrt{3} - 1$ ***[1 mark]***
$= 2$ ***[1 mark]***

b) To rationalise the denominator, you want to get rid of the surd on the bottom line of the fraction. To do this, use the difference of two squares — e.g. if the denominator's $\sqrt{a} + b$, you multiply by $\sqrt{a} - b$:
$\frac{\sqrt{3} - 1}{\sqrt{3} + 1} \times \frac{\sqrt{3} - 1}{\sqrt{3} - 1}$ ***[1 mark]***
$= \frac{3 - 2\sqrt{3} + 1}{2}$ ***[1 mark]***
$= 2 - \sqrt{3}$ ***[1 mark]***

c) $x\sqrt{2} = \sqrt{72} - 4\sqrt{8}$
$x\sqrt{2} = \sqrt{36 \times 2} - 4\sqrt{4 \times 2}$
$x\sqrt{2} = 6\sqrt{2} - 8\sqrt{2}$ ***[1 mark]***
$x\sqrt{2} = -2\sqrt{2}$ ***[1 mark]***
$\Rightarrow x = -2$ ***[1 mark]***

2 a) A sketch will definitely help here:

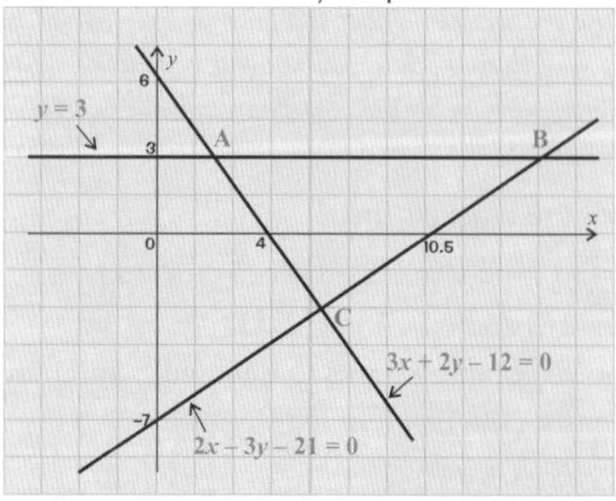

Finding the coordinates of points A, B and C is just a small matter of simultaneous equations. A and B are simple — you already know their y-value is 3, so:
Point A is the point where $y = 3$ and $3x + 2y - 12 = 0$ meet:
$3x + (2 \times 3) - 12 = 0$
$3x + 6 = 12$
$3x = 6$
$x = 2$
Point $A = (2, 3)$ ***[1 mark]***
Point B is the point where $y = 3$ and $2x - 3y - 21 = 0$ meet:

Answers

$2x - (3 \times 3) - 21 = 0$

$2x = 30$

$x = 15$

Point $B = (15, 3)$ ***[1 mark]***

Point C is slightly trickier — you've got to solve a pair of simultaneous equations:

Label the equations: $2x - 3y - 21 = 0$ (a)

$3x + 2y - 12 = 0$ (b)

It's easy to get muddled with all those 3s, 2s and 1s. Just a slip of concentration and then — whoopsy — you've solved 3x – 2y + 12 = 0, and that's no use to anyone...

Multiply (a) by 2 and (b) by 3 to equalise the coefficients of y:

$4x - 6y - 42 = 0$

$9x + 6y - 36 = 0$

Now add together to get rid of y:

$4x + 9x - 6y + 6y - 42 - 36 = 0$ ***[1 mark]***

$13x - 78 = 0$

$13x = 78$

$x = 6$ ***[1 mark]***

Substitute this value of x into (a) or (b):

$(3 \times 6) + 2y - 12 = 0$

$18 - 12 + 2y = 0$

$2y = -6$

$y = -3$

So C has the coordinates (6, –3). ***[1 mark]***

b) To show that the triangle is right-angled, you need to prove that the gradients of two of the lines multiply together to make –1. Looking at the sketch, the right angle looks like it's at C, so use lines BC and AC:

BC: $2x - 3y - 21 = 0$

$3y = 2x - 21$

$y = \frac{2}{3}x - 7$

So the gradient of BC = $\frac{2}{3}$.

AC: $3x + 2y - 12 = 0$

$2y = 12 - 3x$

$y = 6 - \frac{3}{2}x$

So the gradient of AC = $-\frac{3}{2}$ ***[1 mark]***.

$\frac{2}{3} \times -\frac{3}{2} = -1$ ***[1 mark]***

The lines that meet at C are perpendicular, so the triangle must be right-angled.

That gradient rule is proving useful, isn't it — you can find equations of normals and tangents and prove right angles, and it might even be the password for the magical kingdom of Narnia...

c) Use the length formula with $(x_1, y_1) = (2, 3)$, $(x_2, y_2) = (6, -3)$:

$l = \sqrt{(x_2 - x_1)^2 + (y_2 - y_1)^2}$

$= \sqrt{(6-2)^2 + (-3-3)^2}$ ***[1 mark]***

$= \sqrt{16 + 36} = \sqrt{52} = 2\sqrt{13}$ ***[1 mark]***.

3 a) Complete the square by halving the coefficient of x to find the number in the brackets (m):

$x^2 - 7x + 17 = \left(x - \frac{7}{2}\right)^2 + n$ ***[1 mark]***

Now simplify this equation to find n:

$n = x^2 - 7x + 17 - \left[\left(x - \frac{7}{2}\right)^2\right]$

$n = x^2 - 7x + 17 - x^2 + \frac{14x}{2} - \frac{49}{4}$

$n = 17 - \frac{49}{4} = \frac{19}{4}$ ***[1 mark]***

So you can express $x^2 - 7x + 17$ as:

$\left(x - \frac{7}{2}\right)^2 + \frac{19}{4}$ ***[1 mark]***

b) The maximum value of $f(x)$ will be when the denominator is as small as possible — so you want the minimum value of $x^2 - 7x + 17$. Using the completed square above, you can see that the minimum value is $\frac{19}{4}$ because the squared part must be ≥ 0 ***[1 mark]***.

So max value of $f(x)$ is $\frac{1}{19/4} = 1 \times \frac{4}{19} = \frac{4}{19}$ ***[1 mark]***.

You've got to put on your thinking cap for that one — but they give you a big hint by getting you to complete the square first.

c) The mapping is a translation through $\begin{bmatrix} 7/2 \\ 19/4 \end{bmatrix}$, i.e. a shift of $\frac{7}{2}$ to the right and $\frac{19}{4}$ upwards.

[3 marks available — 1 mark for stating that it is a translation, 1 mark for stating amount shifted horizontally, 1 mark for stating amount shifted vertically]

4 a) When $ax^2 + bx + c$ has no real roots, you know that $b^2 - 4ac < 0$.

Here, $a = -j$, $b = 3j$ and $c = 1$ ***[1 mark]***. Therefore

$(3j)^2 - (4 \times -j \times 1) < 0$ ***[1 mark]***

$9j^2 + 4j < 0$ ***[1 mark]***

b) To find the values where $9j^2 + 4j < 0$, you need to start by solving $9j^2 + 4j = 0$:

$j(9j + 4) = 0$, so $j = 0$ ***[1 mark]*** or $9j = -4$

$\Rightarrow j = -\frac{4}{9}$ ***[1 mark]***.

The graph looks like this:

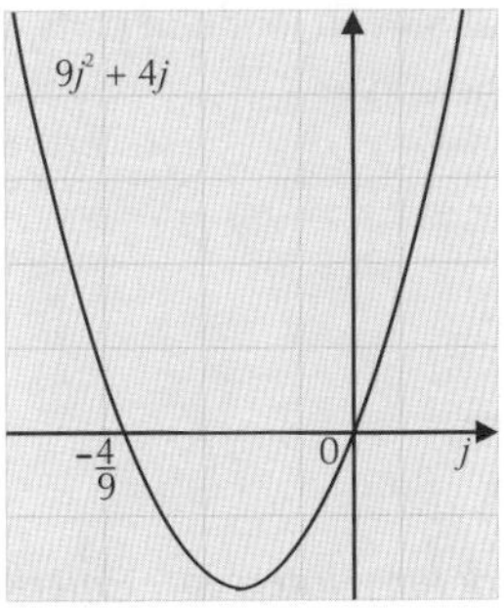

From the graph, you can see that $9j^2 + 4j < 0$ when $-\frac{4}{9} < j < 0$ ***[1 mark]***.

If you made a mistake, do not pass GO, do not collect £200, and move your tiny top hat back to the page on quadratic inequalities...

Answers

5 a) (i) $f'(x) = 3x^2 - 3$.
[2 marks available — 1 mark for each correct term]

(ii) Stationary points occur when $f'(x) = 0$:
$3x^2 - 3 = 0$
$3x^2 = 3$
$x^2 = 1$
$x = \pm 1$.
Plug these back into $y = f(x)$ to find the corresponding y-values:
$x = 1 \Rightarrow y = (1)^3 - 3(1) + 2 = 0$
$x = -1 \Rightarrow y = (-1)^3 - 3(-1) + 2 = 4$
Therefore, stationary points have coordinates (1, 0) and (–1, 4).
[3 marks available — 1 mark for correct working, 1 mark for each correct stationary point]

(iii) Differentiate again to get $f''(x) = 6x$ ***[1 mark]***. Then $f''(1) = 6$, which is positive, so (1, 0) is a minimum ***[1 mark]***, and $f''(-1) = -6$, which is negative, so (–1, 4) is a maximum ***[1 mark]***.

b) (i) Use the remainder theorem:
$f(3) = (3)^3 - 3(3) + 2 = 20$, so the remainder when $f(x)$ is divided by $(x - 3)$ is 20.
[2 marks available — 1 mark for correct working, 1 mark for correct answer]

(ii) Just multiply out the brackets to show that they give the correct equation:
$(x - 1)^2(x + 2) = (x - 1)(x - 1)(x + 2)$ ***[1 mark]***
$= (x^2 - 2x + 1)(x + 2)$
$= x^3 + 2x^2 - 2x^2 - 4x + x + 2$ ***[1 mark]***
$= x^3 - 3x + 2$

c) When $x = 0$, $y = 2$, so the graph intersects the y-axis at (0, 2). From part b)(ii), when $y = 0$, $(x - 1)^2(x + 2) = 0$, and so the graph intersects the x-axis at (–2, 0) and touches the x-axis at (1, 0) [this is a repeated root]. Using this information and the information you found in parts a)(ii) and a)(iii), you can sketch the curve:

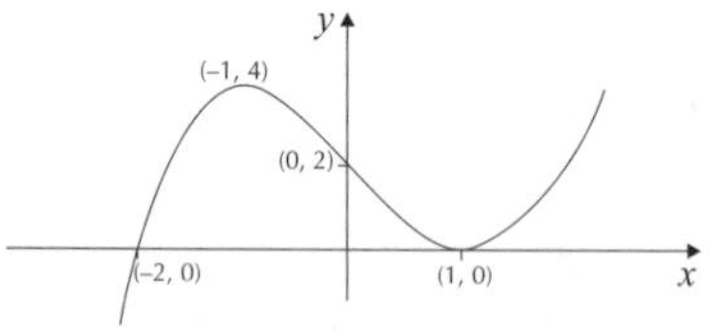

[4 marks available — 1 mark for y-intercept, 1 mark for each x-intercept, 1 mark for correct graph shape]

6 a) $\frac{dy}{dx} = 2x - 4$.
[2 marks available — 1 mark for each correct term]

b) The curve and the line intersect where:
$x^2 - 4x + 6 = -x + 4$
Rearrange and factorise:
$x^2 - 3x + 2 = 0$
$(x - 1)(x - 2) = 0$
$\Rightarrow x = 1$ and $x = 2$
To find the y-coordinates, put the x-values in $y = -x + 4$:
when $x = 1, y = -1 + 4 = 3$
when $x = 2, y = -2 + 4 = 2$
so the two points of intersection are (1,3) and (2,2).
[5 marks available — 1 mark for forming quadratic, 1 mark for each x-value, 1 mark for each y-value]

c) $\int_1^2 (x^2 - 4x + 6)dx = \left[\frac{x^3}{3} - 2x^2 + 6x\right]_1^2$
$= \left(\frac{8}{3} - 8 + 12\right) - \left(\frac{1}{3} - 2 + 6\right)$
$= \frac{7}{3}$
[3 marks available — 1 mark for all terms integrated correctly, 1 mark for correctly substituting in limits, 1 mark for correct final answer]

d) The shaded area is equal to the area under the line $y = -x + 4$ between $x = 1$ and $x = 2$ minus the area under the curve $y = x^2 - 4x + 6$ between $x = 1$ and $x = 2$.
The area under the line $y = -x + 4$ is the area of the trapezium:

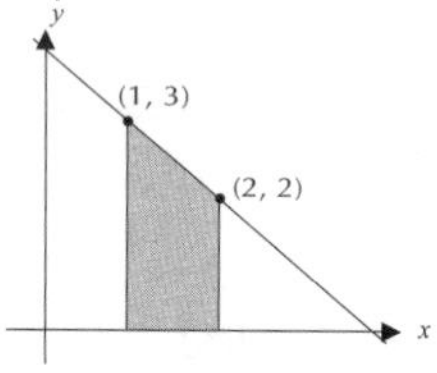

This area is given by: $\left(\frac{2+3}{2}\right) \times 1$ ***[1 mark]*** $= \frac{5}{2}$ ***[1 mark]***.
So the required area is equal to:
$\frac{5}{2} - \frac{7}{3}$ ***[1 mark]*** $= \frac{15}{6} - \frac{14}{6} = \frac{1}{6}$ ***[1 mark]***.

7 a) The centre of the circle must be the midpoint of AB, since AB is a diameter. Midpoint of AB is:
$\left(\frac{2+0}{2}, \frac{1+-5}{2}\right)$ ***[1 mark]*** $= (1, -2)$ ***[1 mark]***
The radius is half of the diameter, so half of length AB. Use Pythagoras' theorem to find the distance:
$AB = \sqrt{(2-0)^2 + (1-(-5))^2}$
$AB = \sqrt{40} = 2\sqrt{10}$ ***[1 mark]***
$\text{Radius} = \frac{2\sqrt{10}}{2} = \sqrt{10}$ ***[1 mark]***

b) The general equation for a circle with centre (a, b) and radius r is: $(x - a)^2 + (y - b)^2 = r^2$.
So for a circle centre (1, –2) and radius $\sqrt{10}$ that gives:
$(x - 1)^2 + (y + 2)^2 = 10$. ***[1 mark]***
Multiply out to get the form given in the question:
$(x - 1)(x - 1) + (y + 2)(y + 2) = 10$ ***[1 mark]***
$x^2 - 2x + 1 + y^2 + 4y + 4 = 10$
$x^2 + y^2 - 2x + 4y - 5 = 0$ ***[1 mark]***

Answers

c) Find the equations of the tangent at A and the normal at C and then solve them simultaneously to find where the lines cross.
The tangent at A is at right angles to the diameter at A.
The diameter at A has the gradient:
$\frac{1 - -5}{2 - 0} = 3$
so the tangent has the gradient $-\frac{1}{3}$ ***[1 mark]***
Put the gradient $-\frac{1}{3}$ and point A (2, 1) into the formula for the equation of a straight line and rearrange:
$y - y_1 = m(x - x_1)$
$y - 1 = -\frac{1}{3}(x - 2)$
$\Rightarrow y = \frac{5}{3} - \frac{x}{3}$ ***[1 mark]***
A normal passes through the centre, so the gradient of the normal through C is the gradient of the line from the centre to C $= \frac{-1 - -2}{4 - 1} = \frac{1}{3}$ ***[1 mark]***
Using the straight line formula at C (4, –1):
$y - -1 = \frac{1}{3}(x - 4)$
$y = \frac{x}{3} - \frac{7}{3}$ ***[1 mark]***
Solve the two equations simultaneously to find the point of intersection:
$\frac{5}{3} - \frac{x}{3} = \frac{x}{3} - \frac{7}{3}$ ***[1 mark]***
$\frac{5}{3} + \frac{7}{3} = \frac{2x}{3}$
$12 = 2x$
$6 = x$ ***[1 mark]***
When $x = 6$, $y = \frac{6}{3} - \frac{7}{3} = -\frac{1}{3}$ ***[1 mark]***
so D has coordinates $\left(6, -\frac{1}{3}\right)$

Oh my giddy goat, we've finished. You'd better go and have a lie down, otherwise all the numbers will fall out of your head and we'd have to start again...

Answers

C2 Section 1 — Algebra and Functions

Warm-up Questions

1) a) x^8 b) a^{15} c) x^6 d) a^8 e) x^4y^3z f) $\frac{b^2c^5}{a}$

2) a) 4 b) 2 c) 8 d) 1 e) $\frac{1}{7}$

Exam Questions

1 a) $27^{\frac{1}{3}} = \sqrt[3]{27}$
$= 3$ ***[1 mark]***

b) $27^{\frac{4}{3}} = (27^{\frac{1}{3}})^4$ ***[1 mark]***
$= 3^4 = 3 \times 3 \times 3 \times 3 = 9 \times 9$
$= 81$ ***[1 mark]***

2 a) (i) A horizontal stretch of scale factor 2 ***[1 mark]***.
(ii) A translation through $\begin{bmatrix}4\\0\end{bmatrix}$ ***[1 mark]***. (i.e. the curve is shifted 4 spaces to the right.)
(iii) A reflection in the x-axis ***[1 mark]***.

b) $y = f(x) + 2$ ***[1 mark]***

3 $10000\sqrt{10} = 10^4 \cdot 10^{\frac{1}{2}}$ ***[1 mark]***
$= 10^{4+\frac{1}{2}}$ ***[1 mark]***
$= 10^{\frac{9}{2}}$
so $k = \frac{9}{2}$ ***[1 mark]***

4 $2x^4 - 32x^2 = 2x^2(x^2 - 16)$
$= 2x^2(x + 4)(x - 4)$
[3 marks available in total — 1 mark for each correct factor]

5 $\frac{x + 5x^3}{\sqrt{x}} = x^{-\frac{1}{2}}(x + 5x^3)$ ***[1 mark]***
$= x^{\frac{1}{2}} + 5x^{\frac{5}{2}}$ ***[1 mark]***

C2 Section 2 — Sequences and Series

Warm-up Questions

1) a) nth term = $4n - 2$ b) nth term = $0.5n - 0.3$
c) nth term = $-3n + 24$ d) nth term = $-6n + 82$

2) $a_{k+1} = a_k + 5,\ a_1 = 32$

3) Last term (l) = 15

4) Common difference (d) = 0.75

5) $S_8 = 168$

6) First work out n: $a = 5,\ l = 65,\ d = 3$
so, $65 = 5 + 3(n - 1)$
so, $n = 21$
Now use: $S_{21} = 21 \times \frac{(5 + 65)}{2}$
so, $S_{21} = 735$
They're not gonna hand it to you on a plate — you often have to work out a and d from the question. Luckily it's pretty simple. Phew.

7) a) Common difference (d) = 4
b) 15th term = 63 c) $S_{10} = 250$

8) First, work out d:
We know that the 7th term is 36 and the 10th is 30.
So the difference in 3 'moves' is –6 i.e. $d = -2$.
This gives the expression for the nth term as $-2n + 50$.
So, $a = 48$ (i.e. when $n = 1$)
and, $l = 40$ ($n = 5$). Now find the sum of the series:
$S_5 = 5 \times \frac{(48 + 40)}{2}$
so $S_5 = 220$

9) a) $S_{20} = 610$ b) $S_{10} = 205$

10) a) $a = 2,\ r = -3$
You find r by putting the information you're given into the formula for u_2: $u_2 = u_1 \times r$, so $-6 = 2 \times r \Rightarrow r = -3$.
10th term, $u_{10} = ar^9$
$= 2 \times (-3)^9 = -39366$

b) $S_{10} = \frac{2(1 - (-3)^{10})}{1 - (-3)} = \frac{1 - (-3)^{10}}{2} = -29524$

11) a) $a = 2,\ r = 4$, so $S_{12} = \frac{2(4^{12} - 1)}{4 - 1} = 11{,}184{,}810$

b) $a = 30,\ r = ½$, so $S_{12} = \frac{30\left(1 - \left[\frac{1}{2}\right]^{12}\right)}{1 - \frac{1}{2}} = 59.985$ (to 3 d.p.)

12) a) $r = 2$, so series is divergent.
b) $r = 1/3$, so series is convergent.
c) $r = 1/3$, so series is convergent.
d) $r = 1/4$, so series is convergent.

13) a) r = 2nd term ÷ 1st term
$r = 12 \div 24 = ½$

b) 7th term = ar^6
$= 24 \times (½)^6$
$= 0.375$ (or $\frac{3}{8}$)

c) $S_{10} = \frac{24\left(1 - \left[\frac{1}{2}\right]^{10}\right)}{1 - \frac{1}{2}} = 47.953$ (to 3 d.p.)

d) $S_\infty = \frac{a}{1 - r} = \frac{24}{1 - \frac{1}{2}} = 48$

14) $a = 2,\ r = 3$
You need $ar^{n-1} = 1458$, i.e. $2 \times 3^{n-1} = 1458$, i.e. $3^{n-1} = 729$.
Using trial and error:
$3^4 = 81;\ 3^5 = 243;\ 3^6 = 729$.
So $n - 1 = 6$, i.e. $n = 7$, the 7th term = 1458.

15) 1 5 10 10 5 1

16) $1 + 12x + 66x^2 + 220x^3$

17) $29120x^4$

18) $(2 + 3x)^5 = 2^5\left(1 + \frac{3}{2}x\right)^5$
$= 2^5\left[1 + \frac{5}{1}\left(\frac{3}{2}x\right) + \frac{5 \times 4}{1 \times 2}\left(\frac{3}{2}x\right)^2 + \ldots\right]$
x^2 term is $2^5 \times \frac{5 \times 4}{1 \times 2}\left(\frac{3}{2}x\right)^2$
so coefficient is $2^5 \times \frac{5 \times 4}{1 \times 2} \times \frac{3^2}{2^2} = 720$

Answers

Exam Questions

1 a) $h_2 = h_{1+1} = 2 \times 5 + 2 = 12$ ***[1 mark]***

$h_3 = h_{2+1} = 2 \times 12 + 2 = 26$ ***[1 mark]***

$h_4 = 2h_3 + 2 = 54$ ***[1 mark]***

b) $\sum_{r=3}^{6} h_r = h_3 + h_4 + h_5 + h_6$

$h_5 = 2h_4 + 2 = 110$ ***[1 mark]***

$h_6 = 2(110) + 2 = 222$

so $\sum_{r=3}^{6} h_r = 26 + 54 + 110 + 222$ ***[1 mark]***

$= 412$ ***[1 mark]***

Nothing hard here, just pop the numbers in. Pop, pop, pop...

2 a) $a_2 = 3k + 11$ ***[1 mark]***

$a_3 = 3a_2 + 11$

$= 3(3k + 11) + 11 = 9k + 33 + 11$

$= 9k + 44$ ***[1 mark]***

$a_4 = 3a_3 + 11$

$= 3(9k + 44) + 11$

$= 27k + 143$ ***[1 mark]***

b) $\sum_{r=1}^{4} a_r = k + (3k + 11) + (9k + 44) + (27k + 143)$

$= 40k + 198$ ***[1 mark]***

$40k + 198 = 278$

$40k = 278 - 198 = 80$ ***[1 mark]***

$k = 2$ ***[1 mark]***

3 a) $\left(\frac{1}{2x} + \frac{x}{2}\right)^3 = \left(\frac{1}{2x}\right)^3 + 3\left(\frac{1}{2x}\right)^2\left(\frac{x}{2}\right) + 3\left(\frac{1}{2x}\right)\left(\frac{x}{2}\right)^2 + \left(\frac{x}{2}\right)^3$

$= \frac{1}{8x^3} + \frac{3}{8x} + \frac{3x}{8} + \frac{x^3}{8}$

[3 marks available — 1 mark for correct use of formula, 1 mark for correct coefficients, 1 mark for correct simplified final answer].

b) $(2 + x^2)\left(\frac{1}{2x} + \frac{x}{2}\right)^3 = (2 + x^2)\left(\frac{1}{8x^3} + \frac{3}{8x} + \frac{3x}{8} + \frac{x^3}{8}\right)$.

The only parts of this multiplication which will give an x term are $2\left(\frac{3x}{8}\right)$ and $x^2\left(\frac{3}{8x}\right)$. ***[1 mark]***

So, x term will be: $2\left(\frac{3x}{8}\right) + x^2\left(\frac{3}{8x}\right) = x\left(\frac{6}{8} + \frac{3}{8}\right)$ ***[1 mark]***

and the coefficient of x is $\frac{9}{8}$ ***[1 mark].***

4 a) The ratio = 1.3, which is > 1, so the sequence is divergent. ***[1 mark]***

b) $u_3 = 12 \times 1.3^2 = 20.28$ ***[1 mark]***

$u_{10} = 12 \times 1.3^9 = 127.25$ ***[1 mark]***

5 $(4 + 3x)^{10}$

$$= 4^{10}\left[1 + \frac{10}{1}\left(\frac{3}{4}x\right) + \frac{10 \times 9}{1 \times 2}\left(\frac{3}{4}x\right)^2 + \frac{10 \times 9 \times 8}{1 \times 2 \times 3}\left(\frac{3}{4}x\right)^3 + \frac{10 \times 9 \times 8 \times 7}{1 \times 2 \times 3 \times 4}\left(\frac{3}{4}x\right)^4 + \ldots\right]$$

So the x coefficient $= 4^{10} \times \frac{10}{1} \times \frac{3}{4} = 7864320$ ***[1 mark]***

x^2 coefficient $= 4^{10} \times \frac{90}{2} \times \frac{9}{16} = 26542080$ ***[1 mark]***

x^3 coefficient $= 4^{10} \times \frac{720}{6} \times \frac{27}{64} = 53084160$ ***[1 mark]***

x^4 coefficient $= 4^{10} \times \frac{5040}{24} \times \frac{81}{256} = 69672960$ ***[1 mark]***

No problems there... as long as you've got binomial expansion straight in your head. If it's still a tangle of factors, powers and garden gnomes, go back and sort it out — you'll be glad you did.

6 a) Use $u_n = ar^{n-1}$:

$u_2 = ar = 250$

$u_4 = ar^3 = 10$.

You now have a pair of simultaneous equations. Multiply the first equation by r^2 to get $ar^3 = 250r^2$, then you can solve them:

$ar^3 = 250r^2$

$ar^3 = 10$

$\Rightarrow 250r^2 = 10$

$r^2 = \frac{10}{250} = \frac{1}{25}$

So $r = \pm\frac{1}{5}$.

[4 marks available — 1 mark for forming a pair of simultaneous equations, 1 mark for working, 1 mark for each correct r value.]

b) (i) Use that $ar = 250$ (from part a)):

$a = \frac{250}{r} = \frac{250}{(1/5)} = 1250$ ***[1 mark].***

(ii) Use $u_n = ar^{n-1}$:

$u_8 = ar^7 = 1250\left(\frac{1}{5}\right)^7$ ***[1 mark]*** $= \frac{2}{125}$ ***[1 mark].***

(iii) Use $S_n = \frac{a(1 - r^n)}{1 - r}$:

$S_8 = \frac{1250\left(1 - \left(\frac{1}{5}\right)^8\right)}{1 - \left(\frac{1}{5}\right)}$ ***[1 mark]***

$S_8 = 1562.5$ (1 d.p.) ***[1 mark].***

7 a) $S_\infty = \frac{a}{1 - r} = \frac{20}{1 - \frac{3}{4}} = \frac{20}{\frac{1}{4}} = 80$

[2 marks available — 1 mark for formula, 1 mark for correct answer]

b) $u_{15} = ar^{14} = 20 \times \left(\frac{3}{4}\right)^{14} = 0.356$ (to 3 sig. fig.)

[2 marks available — 1 mark for formula, 1 mark for correct answer]

8 a) $u_1 = 10$

$u_2 = au_1 + b = 10a + b = -1$

$u_3 = au_2 + b = -a + b = \frac{6}{5}$

You now have a pair of simultaneous equations:

$10a + b = -1$ **(eqn. 1)** ***[1 mark]***

$-a + b = \frac{6}{5}$ **(eqn. 2)** ***[1 mark]***

Eliminate b by subtracting **eqn. 2** from **eqn. 1**:

$10a + b + a - b = -1 - \frac{6}{5}$ ***[1 mark]***

$11a = -\frac{11}{5}$

$\Rightarrow a = -\frac{1}{5}$ ***[1 mark]***

And so (from **eqn. 2**):

$b = \frac{6}{5} + a = \frac{6}{5} - \frac{1}{5} = 1$ ***[1 mark].***

Answers

b) First find the value of u_4:

$u_4 = au_3 + b = -\frac{1}{5}u_3 + 1$

$= -\frac{1}{5}\left(\frac{6}{5}\right) + 1 = \frac{19}{25}$ ***[1 mark]***

So $u_5 = -\frac{1}{5}u_4 + 1 = -\frac{1}{5}\left(\frac{19}{25}\right) + 1 = \frac{106}{125}$ ***[1 mark].***

c) The sequence is defined by the function $f(u_n) = -\frac{1}{5}u_n + 1$.

Setting $L = f(L)$ gives $L = -\frac{1}{5}L + 1$ ***[1 mark].***

This rearranges to $\frac{6}{5}L = 1$, and so $L = \frac{5}{6}$ ***[1 mark].***

9 a) $a_{31} = 22 + (31 - 1)(-1.1)$ ***[1 mark]***

$= 22 + 30(-1.1)$

$= 22 - 33 = -11$ ***[1 mark]***

b) $a_k = 0$

$a_1 + (k-1)d = 0$

$22 + (k-1) \times -1.1 = 0$ ***[1 mark]***

$k - 1 = \frac{-22}{-1.1} = \frac{220}{11} = 20$

$k = 20 + 1 = 21$ ***[1 mark]***

c) We want to find the first value of n for which $S_n < 0$.

Using the formula for sum of a series:

$S_n = \frac{n}{2}[2 \times 22 + (n-1)(-1.1)] < 0$ ***[1 mark]***

$S_n = \frac{n}{2}(44 - 1.1n + 1.1) < 0$

$\frac{n}{2}(45.1 - 1.1n) < 0$ ***[1 mark]***

$\frac{n}{2}(45.1 - 1.1n) = 0$

$\Rightarrow \frac{n}{2} = 0$ or $45.1 - 1.1n = 0$

$\Rightarrow n = 0$ or $n = 41$ ***[1 mark]***

The coefficient of n^2 is negative so graph is n-shaped.
Need to find negative part, so $n < 0$ or $n > 41$.
Since n cannot be negative then $n > 41$.
Now we just want the first (i.e. lowest) value of n for which this is true, which is $n = 42$. ***[1 mark]***

Take a look back at the stuff on quadratic inequalities (on page 23) if you're a bit unsure about what just happened.

10 a) $S_\infty = \frac{a}{1-r}$ and $u_2 = ar$ ***[1 mark]***

So $36 = \frac{a}{1-r}$ i.e. $36 - 36r = a$ ***[1 mark]***

and $5 = ar$. ***[1 mark]***

Substituting for a gives: $5 = (36 - 36r)r = 36r - 36r^2$

i.e. $36r^2 - 36r + 5 = 0$ ***[1 mark]***

b) Factorising gives: $(6r - 1)(6r - 5) = 0$

So $r = \frac{1}{6}$ or $r = \frac{5}{6}$. ***[1 mark for each correct value]***

If $r = \frac{1}{6}$ and $ar = 5$ then $\frac{a}{6} = 5$ i.e. $a = 30$

If $r = \frac{5}{6}$ and $ar = 5$ then $\frac{5a}{6} = 5$ i.e. $a = 6$

[1 mark for each correct value]

11 a) $(1 + 3x)^5 =$

$1 + 5(3x) + \frac{5 \cdot 4}{1 \cdot 2}(3x)^2 + \frac{5 \cdot 4 \cdot 3}{1 \cdot 2 \cdot 3}(3x)^3 + \frac{5 \cdot 4 \cdot 3 \cdot 2}{1 \cdot 2 \cdot 3 \cdot 4}(3x)^4 + (3x)^5$.

This simplifies to:

$1 + 15x + 90x^2 + 270x^3 + 405x^4 + 243x^5$.

[6 marks available — 1 mark for using binomial expansion, 1 mark for 15x, 1 mark for 90x², 1 mark for 270x³, 1 mark for 405x⁴, 1 mark for 243x⁵]

b) When you multiply

$1 + 15x + 90x^2 + 270x^3 + 405x^4 + 243x^5$

by $(1 + x)$, the only terms which are going to multiply together to give an x^2 term are $(1 \times 90x^2)$ ***[1 mark]*** and $(x \times 15x)$ ***[1 mark].***

So the x^2 term will be $(1 \times 90x^2) + (x \times 15x)$

$= (90 + 15)x^2 = 105x^2$ ***[1 mark].***

So the coefficient of x^2 is 105, as required.

C2 Section 3 — Trigonometry Warm-up Questions

1) $\tan x = \frac{\sin x}{\cos x}$; $\cos^2 x = 1 - \sin^2 x$

2) Sine Rule: $\frac{x}{\sin X} = \frac{y}{\sin Y} = \frac{z}{\sin Z}$

Cosine Rule: $x^2 = y^2 + z^2 - 2yz\cos X$

Area: $\frac{1}{2}xy\sin Z$

3) (a) B = 125°, a = 3.66 m, c = 3.10 m, area is 4.64 m²

(b) r = 20.05 km, P = 1.49°, Q = 168.51°

4) Freda's angles are 22.3°, 49.5°, 108.2°

5) One triangle: c = 4.98, C = 72.07°, B = 72.93°
Other possible triangle: c = 3.22, C = 37.93°, B = 107.07°

6) $\cos 30° = \frac{\sqrt{3}}{2}$, $\sin 30° = \frac{1}{2}$, $\tan 30° = \frac{1}{\sqrt{3}}$

$\cos 45° = \frac{1}{\sqrt{2}}$, $\sin 45° = \frac{1}{\sqrt{2}}$, $\tan 45° = 1$

$\cos 60° = \frac{1}{2}$, $\sin 60° = \frac{\sqrt{3}}{2}$, $\tan 60° = \sqrt{3}$

7)

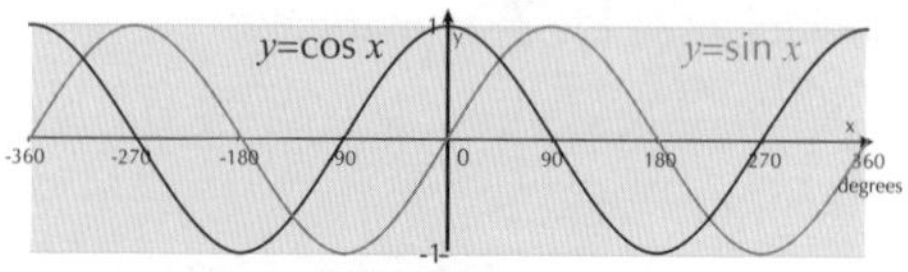

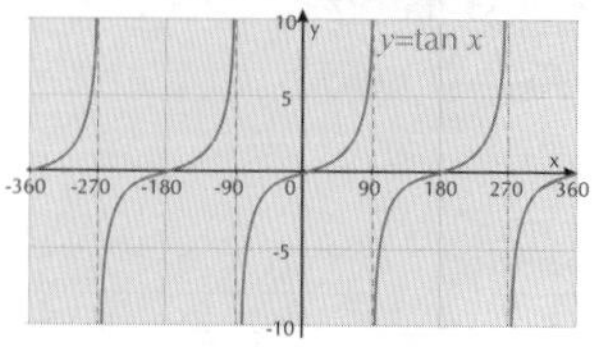

8) (a)

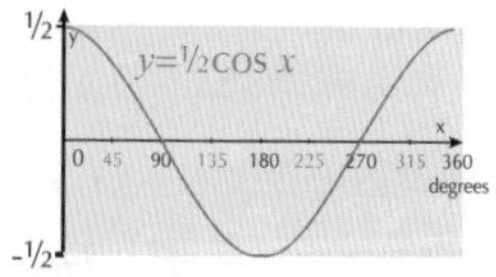

Answers

(b)

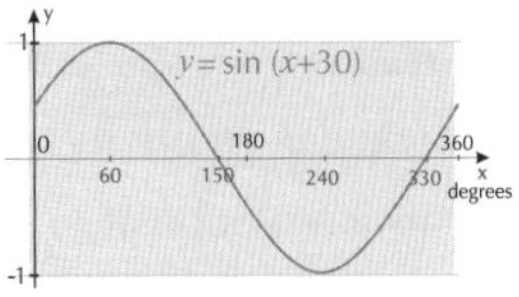

(c)

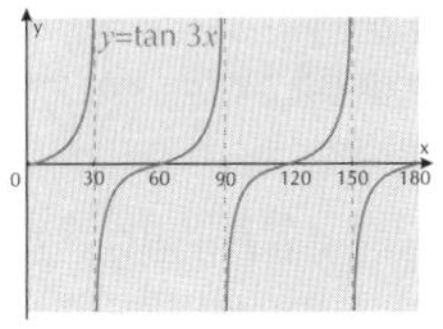

9) (a) (i) $\theta = 240°, 300°$.

(ii) $\theta = 135°, 315°$.

(iii) $\theta = 135°, 225°$.

(b) (i) $\theta = 33.0°, 57.0°, 123.0°, 147.0°, -33.0°, -57.0°, -123.0°, -147.0°$

(ii) $\theta = -17.5°, 127.5°$

(iii) $\theta = 179.8°$

10) $x = 70.5°, 120°, 240°, 289.5°$.

11) $x = -30°$

12) $(\sin y + \cos y)^2 + (\cos y - \sin y)^2$

$\equiv (\sin^2 y + 2\sin y\cos y + \cos^2 y) +$

$(\cos^2 y - 2\cos y\sin y + \sin^2 y)$

$\equiv 2(\sin^2 y + \cos^2 y) \equiv 2$

13) $\dfrac{\sin^4 x + \sin^2 x\cos^2 x}{\cos^2 x - 1} \equiv -1$

LHS:

$\dfrac{\sin^2 x(\sin^2 x + \cos^2 x)}{(1 - \sin^2 x) - 1}$

$\equiv \dfrac{\sin^2 x}{-\sin^2 x} \equiv -1 \equiv \text{RHS}$

Exam Questions

1 a) Using the cosine rule:

$a^2 = b^2 + c^2 - 2bc\cos A$

If XY is a, then angle $A = 180° - 100° = 80°$.

$XY^2 = 150^2 + 250^2 - (2 \times 150 \times 250 \times \cos 80°)$

$XY^2 = 71976.3867$

$XY = \sqrt{71976.3867} = 268.28$ m (to 2 d.p.)

= 268 m to the nearest m.

[2 marks available — 1 mark for correct substitution into cosine rule formula, and 1 mark for correct answer.]

b) Using the sine rule:

$\dfrac{a}{\sin A} = \dfrac{b}{\sin B}$, so $\dfrac{250}{\sin\theta} = \dfrac{268.2842 \text{ (from (a))}}{\sin 80°}$.

Rearranging gives:

$\dfrac{\sin\theta}{\sin 80°} = \dfrac{250}{268.2842} = 0.93$ to 2 d.p.

[3 marks available — 1 mark for correct substitution into sine rule formula, 1 mark for rearrangement into the correct form, and 1 mark for correct final answer.]

2 a) Using the cosine rule with ΔAMC:

$a^2 = b^2 + c^2 - 2bc\cos A$

$2.30^2 = 2.20^2 + 2.20^2 - (2 \times 2.20 \times 2.20 \times \cos\theta)$

$\cos\theta = \dfrac{5.29 - 4.84 - 4.84}{-9.68} = 0.4535...$

$\theta = \cos^{-1} 0.4535... = 1.10$ rad to 3 s.f.

[2 marks available — 1 mark for correct substitution into cosine rule formula, and 1 mark for correct answer in radians.]

...or, you could divide it into 2 right-angled triangles and then use: $\theta = 2\sin^{-1}\frac{1.15}{2.20} = 1.10$ rad. Whatever works.

b) Arc length $S = r\theta = 2.20 \times 1.10$ ***[1 mark]***

= 2.42 m ***[1 mark]***.

Perimeter of slab = 2.42 + 1.5 + 1.5 = 5.42 m

[1 mark].

c) Area of slab =

Area ΔABC + Area ΔAMC – Area sector AMC.

Area of the triangles can be found using $\frac{1}{2}ab\sin C$.

Area of sector can be found using $\frac{1}{2}r^2\theta$.

So area of slab = $\left(\frac{1}{2} \times 1.50 \times 1.50 \times \sin 1.75\right) +$

$\left(\frac{1}{2} \times 2.20 \times 2.20 \times \sin 1.10\right) - \left(\frac{1}{2} \times 2.20^2 \times 1.10\right)$

$= 1.107 + 2.157 - 2.662 = 0.602$ m² to 3 s.f.

[5 marks available — 1 mark for correct calculation of each of the three shapes, 1 mark for combining the three shapes in the correct way, 1 mark for correct final answer.]

3 a) Area of silver triangle = $\frac{1}{2}ab\sin\theta$

$= \frac{1}{2} \times 5 \times 8 \times \sin\frac{\pi}{4} = \frac{20}{\sqrt{2}}$ cm²

Area of whole sector = $\frac{1}{2}r^2\theta$

$= \frac{1}{2} \times 8^2 \times \frac{\pi}{4} = 8\pi$ cm².

Area of glass piece = $8\pi - \frac{20}{\sqrt{2}} = 11.0$ cm² (3 s.f.)

[3 marks available — 1 mark for area of triangle, 1 mark for area of sector, 1 mark for correct final answer.]

b) First, find the length of the arc AB:

$AB = 8 \times \frac{\pi}{4} = 2\pi$ cm ***[1 mark]***

Next, find the length of AX using the cosine rule:

$AX^2 = 5^2 + 8^2 - 2(5 \times 8 \times \cos\frac{\pi}{4})$ ***[1 mark]***

$AX = \sqrt{89 - \frac{80}{\sqrt{2}}} = 5.6948...$ cm ***[1 mark]***

So, perimeter of glass shape = $2\pi + 5.6948... + (8 - 5)$

= 15.0 cm (3 s.f.) ***[1 mark]***

4 a) (i)

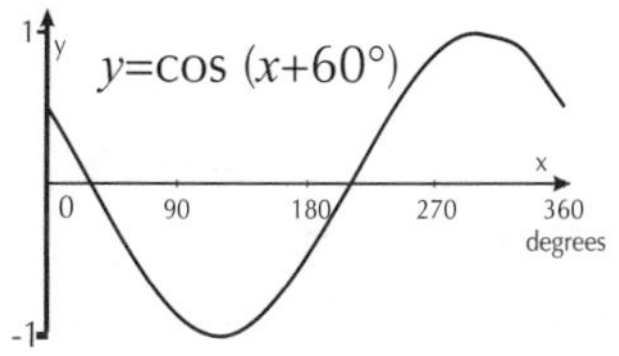

[2 marks available — 1 mark for correct shape of cos x graph, 1 mark for shift of 60° to the left.]

Answers

(ii) E.g. the graph of $y = \cos(x + 60°)$ cuts the x-axis at 30° and 210°, so for $0 \le x \le 360°$, $\cos(x + 60°) = 0$ when $x = 30°$ and 210°.

[2 marks available — 1 mark for each correct solution.]

b)

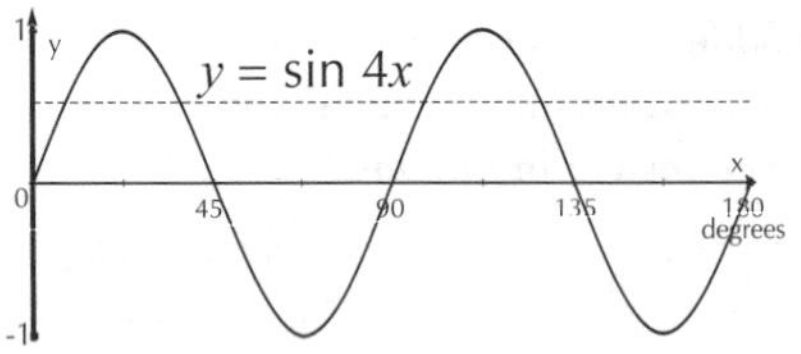

[2 marks available — 1 mark for correct y = sin x graph, 1 mark for 2 repetitions of the sine wave between 0 and 180°.]

c) E.g. $\sin 4x = 0.5$

$4x = 30°$

$x = 7.5°$ is one solution. ***[1 mark]***

The graph in (b) shows there are 4 solutions between 0 and 180°, which, by the symmetry of the graph, lie 7.5° from where the graph cuts the x-axis, as follows:

$x = 45° - 7.5° = 37.5°$. ***[1 mark]***

$x = 90° + 7.5° = 97.5°$. ***[1 mark]***

$x = 135° - 7.5° = 127.5°$. ***[1 mark]***

5 a) $3\cos x = 2\sin x$, and $\tan x = \frac{\sin x}{\cos x}$,

You need to substitute tan in somewhere, so look at how you can rearrange to get sin/cos in the equation...

Divide through by $\cos x$ to give:

$3\frac{\cos x}{\cos x} = 2\frac{\sin x}{\cos x}$

$\Rightarrow 3 = 2\tan x$

$\Rightarrow \tan x = \frac{3}{2}$ (or = 1.5).

[2 marks available — 1 mark for correct substitution of tan x, 1 mark for correct final answer.]

b) Using $\tan x = 1.5$

$x = 56.3°$ ***[1 mark]***

and a 2nd solution can be found from

$x = 180° + 56.3° = 236.3°$ ***[1 mark]***.

Don't forget the other solution! Either use the CAST diagram or sketch a graph to help.

6 a) $\tan\left(x + \frac{\pi}{6}\right) = \sqrt{3}$

$x + \frac{\pi}{6} = \frac{\pi}{3}$ ***[1 mark]***

Tan is positive in the 3rd quadrant of our good friend the CAST diagram, so we can work out the other solution as follows...

2nd solution between 0 and 2π can be found from:

$x + \frac{\pi}{6} = \pi + \frac{\pi}{3} = \frac{4}{3}\pi$ ***[1 mark]***. So subtracting $\frac{\pi}{6}$ from each solution gives: $x = \frac{\pi}{6}$ ***[1 mark]***, $x = \frac{7\pi}{6}$ ***[1 mark]***

b) For this one, it's helpful to change the range — so look for solutions in the range $-\frac{\pi}{4} \le x - \frac{\pi}{4} \le 2\pi - \frac{\pi}{4}$.

$2\cos\left(x - \frac{\pi}{4}\right) = \sqrt{3}$

so $\cos\left(x - \frac{\pi}{4}\right) = \frac{\sqrt{3}}{2}$.

Solving this gives $x - \frac{\pi}{4} = \frac{\pi}{6}$, which is in the range — so it's a solution ***[1 mark]***. From the symmetry of the cos graph there's another solution at $2\pi - \frac{\pi}{6} = \frac{11\pi}{6}$. But this is outside the range for $x - \frac{\pi}{4}$, so you can ignore it ***[1 mark]***. Using symmetry again, there's also a solution at $-\frac{\pi}{6}$ — and this one is in your range. So solutions for $x - \frac{\pi}{4}$ are $-\frac{\pi}{6}$ and $\frac{\pi}{6}$ ***[1 mark]*** $\Rightarrow x = \frac{\pi}{12}$ and $\frac{5\pi}{12}$. ***[1 mark]***

You might find it useful to sketch the graph for this one — or you could use the CAST diagram if you prefer.

c) $\sin 2x = -\frac{1}{2}$, so look for solutions in the range $0 \le 2x \le 4\pi$ ***[1 mark]***.

It's easier to see what's going on by drawing a graph for this one:

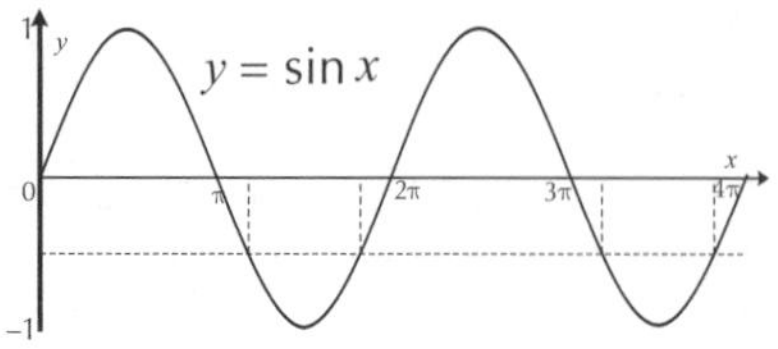

The graph shows there are 4 solutions between 0 and 4π. Putting $\sin 2x = -½$ into your calculator gives you the solution $2x = -\frac{\pi}{6}$, but this is outside the range ***[1 mark]***. From the graph, you can see that the solutions within the range occur at $\pi + \frac{\pi}{6}$, $2\pi - \frac{\pi}{6}$, $3\pi + \frac{\pi}{6}$ and $4\pi - \frac{\pi}{6}$ ***[1 mark]***, so $2x = \frac{7\pi}{6}, \frac{11\pi}{6}, \frac{19\pi}{6}$ and $\frac{23\pi}{6}$ ***[1 mark]***.

Dividing by 2 gives:

$x = \frac{7\pi}{12}, \frac{11\pi}{12}, \frac{19\pi}{12}$ and $\frac{23\pi}{12}$ ***[2 marks for all 4 correct, 1 mark for 2 correct]***

7 a) $2(1 - \cos x) = 3\sin^2 x$, and $\sin^2 x = 1 - \cos^2 x$.

$\Rightarrow 2(1 - \cos x) = 3(1 - \cos^2 x)$ ***[1 mark]***.

You need to get the whole equation into either sin or cos to get something useful at the end, so get used to spotting places to use the trig identities.

$\Rightarrow 2 - 2\cos x = 3 - 3\cos^2 x$

$\Rightarrow 3\cos^2 x - 2\cos x - 1 = 0$ ***[1 mark]***.

b) From (a), the equation can be written as:

$3\cos^2 x - 2\cos x - 1 = 0$

Now this looks suspiciously like a quadratic equation, which can be factorised...

$(3\cos x + 1)(\cos x - 1) = 0$

$\Rightarrow \cos x = -\frac{1}{3}$ ***[1 mark]*** or $\cos x = 1$ ***[1 mark]***

For $\cos x = -\frac{1}{3}$

$x = 109.5°$ (to 1 d.p.) ***[1 mark]***,

and a 2nd solution can be found from

$x = (360° - 109.5°) = 250.5°$ ***[1 mark]***.

For $\cos x = 1$

$x = 0°$ ***[1 mark]*** and 360° ***[1 mark]***.

8 $2 - \sin x = 2\cos^2 x$, and $\cos^2 x = 1 - \sin^2 x$

$\Rightarrow 2 - \sin x = 2(1 - \sin^2 x)$

$\Rightarrow 2 - \sin x = 2 - 2\sin^2 x$

$\Rightarrow 2\sin^2 x - \sin x = 0$

Now simply factorise and all will become clear...

$\sin x(2\sin x - 1) = 0 \Rightarrow \sin x = 0$ or $\sin x = \frac{1}{2}$.
For $\sin x = 0$, $x = 0$, π and 2π.
For $\sin x = \frac{1}{2}$, $x = \frac{\pi}{6}$ and $\pi - \frac{\pi}{6} = \frac{5\pi}{6}$.
[6 marks available — 1 mark for correct substitution using trig identity, 1 mark for factorising quadratic in sin x, 1 mark for finding correct values of sin x, 1 mark for all three solutions when sin x = 0, 1 mark for each of the other 2 correct solutions.]

9 a) $(1 + 2\cos x)(3\tan^2 x - 1) = 0$
$\Rightarrow 1 + 2\cos x = 0 \Rightarrow \cos x = -\frac{1}{2}$.
OR:
$3\tan^2 x - 1 = 0 \Rightarrow \tan^2 x = \frac{1}{3} \Rightarrow \tan x = \frac{1}{\sqrt{3}}$ or $-\frac{1}{\sqrt{3}}$.
For $\cos x = -\frac{1}{2}$
$x = \frac{2\pi}{3}$ and $-\frac{2\pi}{3}$.
Drawing the cos x graph helps you find the second one here, and don't forget the limits are $-\pi \le x \le \pi$.
For $\tan x = \frac{1}{\sqrt{3}}$
$x = \frac{\pi}{6}$ and $-\pi + \frac{\pi}{6} = -\frac{5\pi}{6}$.
For $\tan x = -\frac{1}{\sqrt{3}}$
$x = -\frac{\pi}{6}$ and $-\frac{\pi}{6} + \pi = \frac{5\pi}{6}$.
Again, look at the graph of tan x if you're unsure.
[6 marks available — 1 mark for each correct solution.]

b) $3\cos^2 x = \sin^2 x \Rightarrow 3 = \frac{\sin^2 x}{\cos^2 x}$
$\Rightarrow 3 = \tan^2 x \Rightarrow \tan x = \pm\sqrt{3}$
For $\tan x = \sqrt{3}$, $x = \frac{\pi}{3}$ and $\frac{\pi}{3} - \pi = -\frac{2\pi}{3}$.
For $\tan x = -\sqrt{3}$, $x = -\frac{\pi}{3}$ and $-\frac{\pi}{3} + \pi = \frac{2\pi}{3}$.
[4 marks available — 1 mark for each correct solution.]

C2 Section 4 — Exponentials and Logarithms

Warm-up Questions

1) a) $3^3 = 27$ so $\log_3 27 = 3$
b) To get fractions you need negative powers
$3^{-3} = 1/27$
$\log_3 (1/27) = -3$
c) Logs are subtracted so divide
$\log_3 18 - \log_3 2 = \log_3 (18 \div 2)$
$= \log_3 9$
$= 2$ $(3^2 = 9)$

2) a) Logs are added so you multiply — remember $2\log 5 = \log 5^2$.
$\log 3 + 2\log 5 = \log (3 \times 5^2)$
$= \log 75$
b) Logs are subtracted so you divide and the power half means square root
$\frac{1}{2}\log 36 - \log 3 = \log (36^{\frac{1}{2}} \div 3)$
$= \log (6 \div 3)$
$= \log 2$
c) Logs are subtracted so you divide and the power quarter means fourth root
$\log 2 - \frac{1}{4}\log 16 = \log (2 \div 16^{\frac{1}{4}})$
$= \log (2 \div 2)$
$= \log 1 = 0$

3) This only looks tricky because of the algebra, just remember the laws: $\log_b (x^2 - 1) - \log_b (x - 1) = \log_b \{(x^2 - 1)/(x - 1)\}$
Then use the difference of two squares:
$(x^2 - 1) = (x - 1)(x + 1)$ and cancel to get
$\log_b (x^2 - 1) - \log_b (x - 1) = \log_b (x + 1)$

4) a) Filling in the answers is just a case of using the calculator

x	−3	−2	−1	0	1	2	3
y	0.0156	0.0625	0.25	1	4	16	64

Check that it agrees with what we know about the graphs. It goes through the common point (0, 1), and it follows the standard shape.

b) Then you just need to draw the graph, and use a scale that's just right.

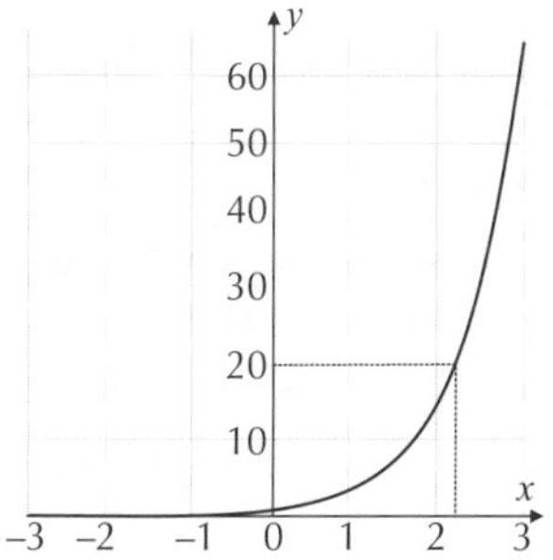

c) The question tells you to use the graph to get your answer, so you'll need to include the construction lines, but check the answer with a calculator.
$x = \log 20 / \log 4 = 2.16$, but you can't justify this accuracy if your graph's not up to it, so 2.2 is a good estimate.

5) a) $x = \log_{10}240 / \log_{10}10 = \log_{10}240 = 2.380$
b) $x = 10^{5.3} = 199526.2... = 200000$ (to 3 s.f.)
c) $2x + 1 = \log_{10}1500 = 3.176$, so $2x = 2.176$, so $x = 1.088$
d) $(x - 1)\log 4 = \log 200$, so $x - 1 = \log 200 / \log 4 = 3.822$, so $x = 4.822$

6) First solve for $1.5^P > 1\,000\,000$
$P \times \log_{10}1.5 > \log_{10}1\,000\,000$,
so $P > (\log_{10}1\,000\,000) / (\log_{10}1.5)$, $P > 34.07$.
We need the next biggest integer, so this will be $P = 35$.

Exam Questions

1 a) (i) $\log_a 20 - 2\log_a 2$
$= \log_a 20 - \log_a (2^2)$ ***[1 mark]***
$= \log_a (20 \div 2^2)$ ***[1 mark]***
$= \log_a 5$ ***[1 mark]***

Answers

(ii) $\frac{1}{2}\log_a 16 + \frac{1}{3}\log_a 27$
$= \log_a (16^{\frac{1}{2}}) + \log_a (27^{\frac{1}{3}})$ ***[1 mark]***
$= \log_a (16^{\frac{1}{2}} \times 27^{\frac{1}{3}})$ ***[1 mark]***
$= \log_a (4 \times 3) = \log_a 12$ ***[1 mark]***

b) (i) $\log_2 64 = 6$ ***[1 mark]*** (since $2^6 = 64$)
(ii) $2\log_3 9 = \log_3 (9^2) = \log_3 81$ ***[1 mark]***
$\log_3 81 = 4$ ***[1 mark]*** (since $3^4 = 81$)

c) (i) $\log_6 25 = \frac{\log 25}{\log 6} = 1.7965$ to 4 d.p. ***[1 mark]***
(ii) $\log_3 10 + \log_3 2 = \log_3 (10 \times 2) = \log_3 20$ ***[1 mark]***
$\log_3 20 = \frac{\log 20}{\log 3} = 2.7268$ to 4 d.p. ***[1 mark]***

2 a) $2^x = 9$, so taking logs of both sides gives
$\log 2^x = \log 9$
This is usually the first step in getting x on its own — then you can use your trusty log laws...
$\Rightarrow x \log 2 = \log 9$
$\Rightarrow x = \frac{\log 9}{\log 2} = 3.17$ to 2 d.p.
[3 marks available — 1 mark for taking logs of both sides, 1 mark for x log 2 = log 9, and 1 mark for correct final answer.]

b) $2^{2x} = (2^x)^2$ (from the power laws) ***[1 mark]***,
so let $y = 2^x$ and $y^2 = 2^{2x}$. This gives a quadratic in y:
$y^2 - 13y + 36 = 0$
Now the big question is — will it factorise? You betcha...
$(y - 9)(y - 4) = 0$, so $y = 9$ or $y = 4$, that is,
$\Rightarrow 2^x = 9$ ***[1 mark]*** or $2^x = 4$ ***[1 mark]***
From (a), for $2^x = 9$, $x = 3.17$ to 2 d.p. ***[1 mark]***
and for $2^x = 4$, $x = 2$ (since $2^2 = 4$) ***[1 mark]***.

3 $\log_7 (y + 3) + \log_7 (2y + 1) = 1$
$\Rightarrow \log_7 ((y + 3)(2y + 1)) = 1$
To remove the $\log_7$, do 7 to the power of each side:
$(y + 3)(2y + 1) = 7^1 = 7$
Multiply out, rearrange, and re-factorise:
$2y^2 + 7y + 3 = 7$
$\Rightarrow 2y^2 + 7y - 4 = 0$
$\Rightarrow (2y - 1)(y + 4) = 0$
$\Rightarrow y = \frac{1}{2}$ or $y = -4$,
but since $y > 0$, $y = \frac{1}{2}$ is the only solution.
[5 marks available — 1 mark for combining the two logs, 1 mark for 7 to the power of each side, 1 mark for the correct factorisation of the quadratic, 1 mark for correct solutions and 1 mark for stating that only y = ½ is a valid solution.]

4 a) $\log_3 x = -\frac{1}{2}$, so do 3 to the power of each side to remove the log:
$x = 3^{-\frac{1}{2}}$ ***[1 mark]***
$\Rightarrow x = \frac{1}{3^{\frac{1}{2}}}$ ***[1 mark]*** $\Rightarrow x = \frac{1}{\sqrt{3}}$ ***[1 mark]***.

b) $2\log_3 x = -4$
$\Rightarrow \log_3 x = -2$, and 3 to the power of each side gives:
$x = 3^{-2}$ ***[1 mark]***
$\Rightarrow x = \frac{1}{9}$ ***[1 mark]***

5 a) $6^{(3x + 2)} = 9$, so taking logs of both sides gives:
$(3x + 2)\log 6 = \log 9$ ***[1 mark]***
$\Rightarrow 3x + 2 = \frac{\log 9}{\log 6} = 1.2262...$ ***[1 mark]***
$\Rightarrow x = (1.2262... - 2) \div 3 = -0.258$ to 3 s.f. ***[1 mark]***

b) $3^{(y^2 - 4)} = 7^{(y + 2)}$, so taking logs of both sides gives:
$(y^2 - 4)\log 3 = (y + 2)\log 7$ ***[1 mark]***
$\Rightarrow \frac{(y^2 - 4)}{(y + 2)} = \frac{\log 7}{\log 3} = 1.7712...$ ***[1 mark]***
The top of the fraction is a 'difference of two squares' so it will simplify as follows...
$\frac{(y - 2)(y + 2)}{(y + 2)} = 1.7712...$ ***[1 mark]***
$\Rightarrow y - 2 = 1.7712...$ ***[1 mark]*** $\Rightarrow y = 3.77$ to 3 s.f. ***[1 mark]***

6 a) $\log_4 p - \log_4 q = \frac{1}{2}$, so using the log laws:
$\log_4 (\frac{p}{q}) = \frac{1}{2}$
Doing 4 to the power of both sides gives:
$\frac{p}{q} = 4^{\frac{1}{2}} = \sqrt{4} = 2$
$\Rightarrow p = 2q$
[3 marks available — 1 mark for combining the two logs, 1 mark for 4 to the power of each side, 1 mark for the correct final working.]

b) Since $p = 2q$ (from (a)), the equation can be written:
$\log_2 (2q) + \log_2 q = 7$ ***[1 mark]***
This simplifies to:
$\log_2 (2q^2) = 7$ ***[1 mark]***
Doing 2 to the power of both sides gives:
$2q^2 = 2^7 = 128$ ***[1 mark]***
$\Rightarrow q^2 = 64$, $\Rightarrow q = 8$ (since p and q are positive) ***[1 mark]***
$p = 2q \Rightarrow p = 16$ ***[1 mark]***

C2 Section 5 — Differentiation Warm-up Questions

1) a) $\frac{dy}{dx} = 4x^3 + \frac{1}{2\sqrt{x}}$
b) $\frac{dy}{dx} = -\frac{14}{x^3} + \frac{3}{2\sqrt{x^3}} + 36x^2$

2) $\frac{dy}{dx} = 3x^2 - \frac{3}{x^2}$; this is zero at (1, 4) and (–1, –4).
$\frac{d^2y}{dx^2} = 6x + \frac{6}{x^3}$; at $x = 1$ this is positive so (1, 4) is a minimum; at $x = -1$ this is negative, so (–1, –4) is a maximum.

3) The tangent and normal must go through (16, 6).
Differentiate to find $\frac{dy}{dx} = \frac{3}{2}\sqrt{x} - 3$, so gradient at (16, 6) is 3.
Therefore tangent can be written $y_T = 3x + c_T$;

Answers

putting $x = 16$ and $y = 6$ gives $6 = 3 \times 16 + c_T$, so $c_T = -42$, and the equation of the tangent is $y_T = 3x - 42$.

The gradient of the normal must be $-\frac{1}{3}$, so the equation of the normal is $y_N = -\frac{1}{3}x + c_N$

Substituting in the coordinates of the point (16, 6) gives $6 = -\frac{16}{3} + c_N \Rightarrow c_N = \frac{34}{3}$; so the normal is $y_N = -\frac{1}{3}x + \frac{34}{3} = \frac{1}{3}(34 - x)$.

Exam Questions

1 a) Rewrite all the terms as powers of x:

$y = x^7 + \frac{2}{x^3} = x^7 + 2x^{-3}$ **[1 mark]**

and then differentiate each term:

$\frac{dy}{dx} = 7x^6 + (-3)2x^{-4}$

$= 7x^6 - \frac{6}{x^4}$ **[1 mark]**

b) This is a second-order derivative — just differentiate the answer for part a):

$\frac{d^2y}{dx^2} = \frac{d}{dx}(7x^6 - 6x^{-4})$ **[1 mark]**

$= (7 \times 6)x^5 - (6 \times -4)x^{-5}$

$= 42x^5 + \frac{24}{x^5}$ **[1 mark]**

2 a) Rewrite the expression in powers of x, so it becomes:

$\frac{x^2 + 3x^{\frac{3}{2}}}{x^{\frac{1}{2}}}$ **[1 mark]**

Then divide the top of the fraction by the bottom:

$\frac{x^2}{x^{\frac{1}{2}}} + \frac{3x^{\frac{3}{2}}}{x^{\frac{1}{2}}}$

$= x^{\frac{3}{2}} + 3x$

So $p = \frac{3}{2}$ **[1 mark]** and $q = 1$ **[1 mark]**.

b) Use your answer to part a) to rewrite the equation as:

$y = 3x^3 + 5 + x^{\frac{3}{2}} + 3x$ **[1 mark]**.

Then differentiate each term to give:

$\frac{dy}{dx} = 9x^2 + \frac{3}{2}x^{\frac{1}{2}} + 3$ **[1 mark for each correct term]**.

3 a) Rewrite the expression in powers of x, so it becomes $x^{-\frac{1}{2}} + x^{-1}$. Then differentiate to get

$\frac{dy}{dx} = -\frac{1}{2}x^{-\frac{3}{2}} - x^{-2}$ **[1 mark for each correct term]**.

Putting $x = 4$ into the derivative gives:

$-\frac{1}{2}4^{-\frac{3}{2}} - 4^{-2} = -\frac{1}{2}(\sqrt{4})^{-3} - \frac{1}{4^2}$

$= -\frac{1}{2} \cdot \frac{1}{2^3} - \frac{1}{16} = -\frac{1}{2} \cdot \frac{1}{8} - \frac{1}{16}$

$= -\frac{1}{16} - \frac{1}{16} = -\frac{1}{8}$

[1 method mark, 1 answer mark]

b) Gradient of the normal to the curve is:

$-1 \div -\frac{1}{8} = 8$ **[1 mark]**.

Use $y - y_1 = m(x - x_1)$ **[1 mark]**:

$y - \frac{3}{4} = 8(x - 4)$ **[1 mark]** $\Rightarrow y - \frac{3}{4} = 8x - 32$

$\Rightarrow y = 8x - \frac{125}{4}$ **[1 mark]**.

4 a) Rewrite the expression in powers of x, so it becomes $y = x + x^{-2}$. Then differentiate to get $\frac{dy}{dx} = 1 - 2x^{-3}$ **[1 mark]**. Make this equal to zero to find stationary point: $1 - 2x^{-3} = 0$ **[1 mark]** $\Rightarrow 1 = 2x^{-3}$

$\Rightarrow 2 = x^3 \Rightarrow x = \sqrt[3]{2}$ **[1 mark]**.

b) Differentiate again: $\frac{d^2y}{dx^2} = 6x^{-4}$ **[1 mark]**. Plug in $x = \sqrt[3]{2}$:

$\frac{d^2y}{dx^2} = 6(\sqrt[3]{2})^{-4}$ **[1 mark]**.

This is positive, so the stationary point is a minimum **[1 mark]**.

Knowing the Power Laws really well will have helped you a lot with that question. If you struggled with them, take a look at p. 57 to refresh your memory.

5 a) Find the value of x that gives the minimum value of y — the stationary point of curve y.

Differentiate, and then solve $\frac{dy}{dx} = 0$:

$\frac{dy}{dx} = \frac{1}{\sqrt{x}} - \frac{27}{x^2}$ **[1 mark for each term]**

$\frac{1}{\sqrt{x}} - \frac{27}{x^2} = 0$ **[1 mark]** $\Rightarrow \frac{1}{\sqrt{x}} = \frac{27}{x^2}$

$x^2 \div x^{1/2} = 27 \Rightarrow x^{3/2} = 27$ **[1 mark]**

$x = \sqrt[3]{27^2} = 9.$ **[1 mark]**

So, 9 miles per hour gives the minimum coal consumption.

b) $\frac{d^2y}{dx^2} = \frac{54}{x^3} - \frac{1}{2\sqrt{x^3}}$ **[1 mark]**.

At the stationary point $x = 9$,

so $\frac{54}{9^3} - \frac{1}{2\sqrt{9^3}} = 0.05555...$ which is positive,

therefore the stationary point is a minimum **[1 mark]**.

c) $y = 2\sqrt{9} + \frac{27}{9} = 9$ **[1 mark]**.

C2 Section 6 — Integration

Warm-up Questions

1) a) $\int_1^2 \left(\frac{8}{x^5} + \frac{3}{\sqrt{x}}\right)dx = \left[-\frac{2}{x^4} + 6\sqrt{x}\right]_1^2$

$= \left(-\frac{2}{16} + 6\sqrt{2}\right) - (-2 + 6) = -\frac{33}{8} + 6\sqrt{2}$

b) $\int_1^6 \frac{3}{x^2}\,dx = \left[\frac{-3}{x}\right]_1^6 = -\frac{1}{2} - (-3) = \frac{5}{2}$

2) $\int_1^8 y\,dx = \int_1^8 x^{-\frac{1}{3}}dx = \left[\frac{3}{2}x^{\frac{2}{3}}\right]_1^8$

$= \left(\frac{3}{2} \times 8^{\frac{2}{3}}\right) - \left(\frac{3}{2} \times 1^{\frac{2}{3}}\right) = \left(\frac{3}{2} \times 4\right) - \left(\frac{3}{2} \times 1\right) = \frac{9}{2}$

3) a) $h = \frac{(3 - 0)}{3} = 1$

$x_0 = 0$: $y_0 = \sqrt{9} = 3$

$x_1 = 1$: $y_1 = \sqrt{8} = 2.8284$

$x_2 = 2$: $y_2 = \sqrt{5} = 2.2361$

$x_3 = 3$: $y_3 = \sqrt{0} = 0$

Answers

$$\int_a^b y\,dx \approx \frac{1}{2}[(3+0)+2(2.8284+2.2361)]$$
$= 6.5645 \approx 6.56$

b) $h = \frac{(1.2-0.2)}{5} = 0.2$

$x_0 = 0.2$: $y_0 = 0.2^{0.04} = 0.93765$
$x_1 = 0.4$: $y_1 = 0.4^{0.16} = 0.86363$
$x_2 = 0.6$: $y_2 = 0.6^{0.36} = 0.83202$
$x_3 = 0.8$: $y_3 = 0.8^{0.64} = 0.86692$
$x_4 = 1$: $y_4 = 1^1 = 1$
$x_5 = 1.2$: $y_5 = 1.2^{1.44} = 1.30023$

$$\int_a^b y\,dx \approx \frac{0.2}{2}\left[\begin{matrix}(0.93765+1.30023)\\+2(0.86363+0.83202+0.86692+1)\end{matrix}\right]$$
$= 0.1 \times 9.36302 \approx 0.936$

Exam Questions

1 a) $f(x) = \frac{x^{\frac{1}{2}}}{\frac{1}{2}} + 4x - \frac{5x^4}{4} + C$

and then simplify each term further if possible...

$= 2\sqrt{x} + 4x - \frac{5x^4}{4} + C$

[3 marks available — 1 mark for each term. Lose 1 mark if C missing or terms not simplified, e.g. ÷½ not converted to ×2. Note — you don't need to put surds in for it to be simplified — indices are fine.]

b) First rewrite everything in terms of powers of x:

$f'(x) = 2x + 3x^{-2}$

Now you can integrate each term (don't forget to add C):

$f(x) = \frac{2x^2}{2} + \frac{3x^{-1}}{-1} + C$

Then simplify each term:

$f(x) = x^2 - \frac{3}{x} + C$

[2 marks available — 1 mark for each term. Lose 1 mark if C missing or terms not simplified.]

c) Following the same process as in part b):

$f'(x) = 6x^2 - \frac{1}{3}x^{-\frac{1}{2}}$

$f(x) = \frac{6x^3}{3} + \frac{1}{3}(x^{\frac{1}{2}} \div \frac{1}{2}) + C$

$f(x) = 2x^3 + \frac{2}{3}\sqrt{x} + C$

[2 marks available — 1 mark for each term. Lose 1 mark if C missing or terms not simplified.]

2 a) Multiply out the brackets and simplify the terms:

$(5 + 2\sqrt{x})^2 = (5 + 2\sqrt{x})(5 + 2\sqrt{x})$
$= 25 + 10\sqrt{x} + 10\sqrt{x} + 4x$
$= 25 + 20\sqrt{x} + 4x$

So $a = 25$, $b = 20$ and $c = 4$

[3 marks: one for each constant]

b) Integrate your answer from a), treating each term separately:

$\int (25 + 20\sqrt{x} + 4x)dx = 25x + \left(20x^{\frac{3}{2}} \div \frac{3}{2}\right) + \left(\frac{4x^2}{2}\right) + C$

$= 25x + \frac{40\sqrt{x^3}}{3} + 2x^2 + C$

[3 marks available – 1 for each term. Lose 1 mark if C missing or answers not simplified (surds not necessary)]

3 $\int_2^7 (2x - 6x^2 + \sqrt{x})dx = \left[x^2 - 2x^3 + \frac{2\sqrt{x^3}}{3}\right]_2^7$

[1 mark for each correct term]

$= \left(7^2 - (2 \times 7^3) + \frac{2\sqrt{7^3}}{3}\right) - \left(2^2 - (2 \times 2^3) + \frac{2\sqrt{2^3}}{3}\right)$

[1 mark]

$= -624.6531605 - (-10.11438192)$
$= -614.5387786 = -614.5388$ to 4 d.p. ***[1 mark]***

4 To find f(x) you integrate f ′(x), but it helps to write all terms in powers of x, so $5\sqrt{x} = 5x^{\frac{1}{2}}$ and $\frac{6}{x^2} = 6x^{-2}$ ***[1 mark]***

Now integrate each term:

$\int (2x + 5x^{\frac{1}{2}} + 6x^{-2})dx = \frac{2x^2}{2} + \left(5x^{\frac{3}{2}} \div \frac{3}{2}\right) + \left(\frac{6x^{-1}}{-1}\right) + C$

$f(x) = x^2 + \frac{10\sqrt{x^3}}{3} - \frac{6}{x} + C$

[2 marks for correct terms, 1 mark for +C]

You've been given a point on the curve so you can calculate the value of C:

If $y = 7$ when $x = 3$, then

$3^2 - \frac{6}{3} + \frac{10\sqrt{3^3}}{3} + C = 7$ ***[1 mark]***

$9 - 2 + 10\sqrt{3} + C = 7$
$7 + 10\sqrt{3} + C = 7$
$C = -10\sqrt{3}$
$f(x) = x^2 - \frac{6}{x} + \frac{10\sqrt{x^3}}{3} - 10\sqrt{3}$ ***[1 mark]***

5 a) Rearrange the terms so each is written as a power of x, showing your working:

$\frac{1}{\sqrt{36x}} = \frac{1}{\sqrt{36}\sqrt{x}} = \frac{1}{6} \times \frac{1}{\sqrt{x}} = \frac{1}{6}x^{-\frac{1}{2}}$ ***[1 mark]***

$2\left(\sqrt{\frac{1}{x^3}}\right) = 2\left(\frac{1}{x^3}\right)^{\frac{1}{2}} = 2(x^{-3})^{\frac{1}{2}}$
$= 2(x^{(-3 \times \frac{1}{2})}) = 2x^{-\frac{3}{2}}$ ***[1 mark]***

This shows that f ′(x) = $\frac{1}{6}x^{-\frac{1}{2}} - 2x^{-\frac{3}{2}}$ — so $A = \frac{1}{6}$ and $B = 2$ ***[1 mark]***

b) Integrate f ′(x) to find f (x):

$f(x) = \left(\frac{1}{6} \times x^{\frac{1}{2}} \div \frac{1}{2}\right) - \left(2x^{-\frac{1}{2}} \div -\frac{1}{2}\right) + C$ ***[1 mark]***

$= \frac{1}{3}x^{\frac{1}{2}} + \left(-2 \div -\frac{1}{2}\right)\left(\frac{1}{\sqrt{x}}\right) + C$

$= \frac{\sqrt{x}}{3} + \frac{4}{\sqrt{x}} + C$ ***[1 mark]***

Now use the coordinates (1, 7) to find the value of C:

$7 = \frac{\sqrt{1}}{3} + \frac{4}{\sqrt{1}} + C$ ***[1 mark]***

$7 - \frac{1}{3} - 4 = C$

$C = \frac{8}{3}$

So $y = \frac{\sqrt{x}}{3} + \frac{4}{\sqrt{x}} + \frac{8}{3}$ ***[1 mark]***

Well, aren't we having lovely integrating fun. Keep toddling through, and it'll be time for tea and biscuits in no time.

Answers

6 a) The tangent at (1, 2) has the same gradient as the curve at that point, so use $f'(x)$ to calculate the gradient:

$f'(1) = 1^3 - 2$ **[1 mark]**

$= -1$ **[1 mark]**

Put this into the straight-line equation $y - y_1 = m(x - x_1)$:

$y - 2 = -1(x - 1)$ **[1 mark]**

$y = -x + 1 + 2$

$y = -x + 3$ **[1 mark]**

No need to go off on a tangent here — just find the gradient and then find the equation. Boom. Done. And move swiftly on...

b) $f(x) = \int\left(x^3 - \frac{2}{x^2}\right)dx = \int(x^3 - 2x^{-2})dx$ **[1 mark]**

$= \frac{x^4}{4} - 2\frac{x^{-1}}{-1} + C = \frac{x^4}{4} + 2x^{-1} + C$ **[1 mark]**

$= \frac{x^4}{4} + \frac{2}{x} + C$

Now use the coordinates (1, 2) to find the value of C:

$2 = \frac{1^4}{4} + \frac{2}{1} + C$ **[1 mark]**

$2 - \frac{1}{4} - 2 = C$

$C = -\frac{1}{4}$ **[1 mark]**

So $f(x) = \frac{x^4}{4} + \frac{2}{x} - \frac{1}{4}$ **[1 mark]**

7 a) (i) $h = \frac{8-2}{3} = 2$ **[1 mark]**

$x_0 = 2$ $y_0 = \sqrt{(3 \times 2^3)} + \frac{2}{\sqrt{2}} = 6.31319$

$x_1 = 4$ $y_1 = \sqrt{(3 \times 4^3)} + \frac{2}{\sqrt{4}} = 14.85641$

$x_2 = 6$ $y_2 = \sqrt{(3 \times 6^3)} + \frac{2}{\sqrt{6}} = 26.27234$

$x_3 = 8$ $y_3 = \sqrt{(3 \times 8^3)} + \frac{2}{\sqrt{8}} = 39.89894$ **[1 mark]**

$\int_2^8 y\,dx \approx \frac{2}{2}[6.31319 + 2(14.85641 + 26.27234) + 39.89894]$ **[1 mark]**

$= 128.46963 \approx 128.47$ to 2 d.p. **[1 mark]**

(ii) $h = \frac{5-1}{4} = 1$ **[1 mark]**

$x_0 = 1$ $y_0 = \frac{1^3 - 2}{4} = -0.25$

$x_1 = 2$ $y_1 = \frac{2^3 - 2}{4} = 1.5$

$x_2 = 3$ $y_2 = \frac{3^3 - 2}{4} = 6.25$

$x_3 = 4$ $y_3 = \frac{4^3 - 2}{4} = 15.5$

$x_4 = 5$ $y_4 = \frac{5^3 - 2}{4} = 30.75$ **[1 mark]**

$\int_1^5 y\,dx \approx \frac{1}{2}[-0.25 + 2(1.5 + 6.25 + 15.5) + 30.75]$ **[1 mark]** ≈ 38.5 **[1 mark]**

b) Increase the number of intervals calculated. **[1 mark]**

8 $n = 5$, $h = \frac{4 - 1.5}{5} = 0.5$ **[1 mark]**

$x_1 = 2.0$, $x_2 = 2.5$, $x_3 = 3.0$ **[1 mark]**

$y_0 = 2.8182$ **[1 mark]**, $y_3 = 6.1716$ **[1 mark]**,

$y_4 = 7.1364$ **[1 mark]**

$\int_{1.5}^4 y\,dx \approx \frac{0.5}{2}[2.8182 + 2(4 + 5.1216 + 6.1716 + 7.1364) + 8]$

[1 mark] $= 13.91935 = 13.9$ to 3 s.f. **[1 mark]**

C2 Practice Exam 1

1 a) $a^1 = a$, so $\log_a a = 1$. **[1 mark]**

b) (i) If $f(x) = 4^{2x}$, then $4^x = f\left(\frac{x}{2}\right)$, which is a horizontal stretch, scale factor 2. **[1 mark]**

(ii) $4^{(2x+1)} = 4^{2\left(x+\frac{1}{2}\right)} = f\left(x + \frac{1}{2}\right)$, which is a translation through $\begin{bmatrix} -\frac{1}{2} \\ 0 \end{bmatrix}$. i.e. 'horizontal shift of $\frac{1}{2}$ to the left'.

[1 mark]

c) Take logs of both sides and use laws of logs:

$\log 4^{(2x+1)} = \log 9$ **[1 mark]**

$\Rightarrow (2x + 1)\log 4 = \log 9$ **[1 mark]**

$\Rightarrow 2x + 1 = \log 9 \div \log 4$

$\Rightarrow x = \frac{1}{2}\left(\frac{\log 9}{\log 4} - 1\right) = 0.292$ (3 s.f.) **[1 mark]**

d) Use the laws of logs:

$\log_a x = \log_a 8 + 2(\log_a 4 - \log_a 2) - 1$

$\Rightarrow \log_a x = \log_a 8 + \log_a 16 - \log_a 4 - 1$ **[1 mark]**

$\Rightarrow \log_a x = \log_a(8 \times 16 \div 4) - 1 = \log_a 32 - 1$ **[1 mark]**

Take exponentials of both sides:

$x = a^{\log_a 32 - 1}$

$\Rightarrow x = a^{\log_a 32}a^{-1}$ **[1 mark]**

$\Rightarrow x = 32a^{-1} = \frac{32}{a}$. **[1 mark]**

2 a) Use the formula for sector area:

Area $= \frac{1}{2}r^2\theta = \frac{1}{2} \times (4)^2 \times 0.2$ **[1 mark]**

$= 1.6$ cm^2 **[1 mark]**.

b) Use the formula for arc length:

Arc length $= r\theta = 4 \times 0.2$ **[1 mark]**

$= 0.8$ cm **[1 mark]**

So the perimeter of the whole sector is:

$4 + 4 + 0.8 = 8.8$ cm **[1 mark]**.

3 a) Work out the terms one by one:

$x_1 = 6$

$x_2 = 3x_1 - 4 = (3 \times 6) - 4 = 14$

$x_3 = 3x_2 - 4 = (3 \times 14) - 4 = 38$ **[1 mark]**

$x_4 = 3x_3 - 4 = (3 \times 38) - 4 = 110$

$x_4 = 110$ **[1 mark]**

b) (i) Use the information given and the general expression for the nth term in a sequence: $u_n = a + (n - 1)d$ to formulate expressions for the 3rd and 7th term:

$u_3 = a + 2d = 9$

$u_7 = a + 6d = 33$ **[1 mark]**

Now solve them as simultaneous equations:

$u_7 - u_3 = a + 6d - a - 2d = 33 - 9$

$4d = 24$

$d = 6$ **[1 mark]**

$a + 2d = 9$

$a + (2 \times 6) = 9$

$a + 12 = 9$

$a = 9 - 12 = -3$ **[1 mark]**

So the first term is –3 and the common difference is 6.

Answers

(ii) $S_n = \frac{n}{2}[2a + (n-1)d]$

So $S_{12} = \frac{12}{2}[(2 \times -3) + 6(12-1)]$ ***[1 mark]***

$= 6[-6 + 66]$ ***[1 mark]***

$= 6 \times 60$

So $S_{12} = 360$ ***[1 mark]***

(iii) $(6n + 1)$ is the rule for finding the nth term, so jot down the first few terms:

$u_1 = (6 \times 1) + 1 = 7$

$u_2 = (6 \times 2) + 1 = 13$

$u_3 = (6 \times 3) + 1 = 19$

So you can see that $d = 6$ and $a = 7$ ***[1 mark]***

It's the same as the sequence before, except the first term is 7 instead of −3. So...

$\sum_{1}^{12}(6n + 1) = S_{12} = \frac{12}{2}[(2 \times 7) + (11 \times 6)]$

$= 6(14 + 66) = 6 \times 80 = 480$ ***[1 mark]***

You could note that each term is 10 more than the equivalent term in part (ii), so just add 10 × 12 to your answer for part (ii).

4 a) Use the formula for the sum to infinity of a series, $S_\infty = \frac{a}{1-r}$.

You are told that the sum to infinity is 10 times the first term, so:

$10a = \frac{a}{1-r}$

$10 = \frac{1}{1-r}$

$10 - 10r = 1$

$10r = 9$

$\Rightarrow r = 0.9$

[3 marks available — 1 mark for correct use of formula, 1 mark for correct working, 1 mark for final answer]

b) Now use the formula for the nth term of a series, $u_n = ar^{n-1}$. You are told that the third term is 81, so:

$u_3 = ar^{3-1}$

$\Rightarrow 81 = ar^2$ ***[1 mark]***

$81 = a \times 0.9^2$

$a = 81 \div 0.9^2 = 100$ ***[1 mark]***.

c) Finally, use the formula for the sum of n terms of a series, $S_n = \frac{a(1-r^n)}{1-r}$:

$S_7 = \frac{100(1-0.9^7)}{1-0.9}$ ***[1 mark]*** $= 521.7$ (1 d.p.) ***[1 mark]***.

5 a) You can use the coefficients from Pascal's Triangle to make things a bit quicker:

$\left(1 + \frac{3}{x^3}\right)^4 = 1 + 4\left(\frac{3}{x^3}\right) + 6\left(\frac{3}{x^3}\right)^2 + 4\left(\frac{3}{x^3}\right)^3 + \left(\frac{3}{x^3}\right)^4$

$= 1 + \frac{12}{x^3} + \frac{54}{x^6} + \frac{108}{x^9} + \frac{81}{x^{12}}$.

[5 marks available — 1 mark for coefficients from Pascal's triangle, 1 mark for each correct final term (except 1)]

b) (i) $y = 1 + \frac{12}{x^3} + \frac{54}{x^6} + \frac{108}{x^9} + \frac{81}{x^{12}} =$

$1 + 12x^{-3} + 54x^{-6} + 108x^{-9} + 81x^{-12}$, so:

$\frac{dy}{dx} = -36x^{-4} - 324x^{-7} - 972x^{-10} - 972x^{-13}$

$= -\frac{36}{x^4} - \frac{324}{x^7} - \frac{972}{x^{10}} - \frac{972}{x^{13}}$.

[4 marks available — 1 for each correct final term]

(ii) $\int_{-1}^{1}\left(1 + \frac{3}{x^3}\right)^4 dx =$

$\int_{-1}^{1}(1 + 12x^{-3} + 54x^{-6} + 108x^{-9} + 81x^{-12})dx$ ***[1 mark]***

$= \left[x - 6x^{-2} - \frac{54}{5}x^{-5} - \frac{108}{8}x^{-8} - \frac{81}{11}x^{-11}\right]_{-1}^{1}$ ***[1 mark]***

$= \left[1 - 6 - \frac{54}{5} - \frac{108}{8} - \frac{81}{11}\right]$

$- \left[-1 - 6 - \frac{54}{5(-1)^5} - \frac{108}{8(-1)^8} - \frac{81}{11(-1)^{11}}\right]$ ***[1 mark]***

$= 2 - \frac{108}{5} - \frac{162}{11} = -34.3$ (3 s.f.). ***[1 mark]***

6 a) $\frac{x^2 + 2x}{\sqrt{x}} = x^{-\frac{1}{2}}(x^2 + 2x) = x^{\frac{3}{2}} + 2x^{\frac{1}{2}}$ ***[1 mark for each correct term]***, so $m = \frac{3}{2}$ and $n = \frac{1}{2}$.

b) From part a) above, $y = x^{\frac{3}{2}} + 2x^{\frac{1}{2}} + 3x^3 - x$.

Differentiating this gives

$\frac{dy}{dx} = \frac{3}{2}x^{\frac{1}{2}} + \left(2 \cdot \frac{1}{2}\right)x^{-\frac{1}{2}} + (3 \cdot 3)x^2 - 1$

$= \frac{3}{2}x^{\frac{1}{2}} + x^{-\frac{1}{2}} + 9x^2 - 1$

[4 marks available — 1 mark for each correct term]

c) Plug $x = 1$ into $\frac{dy}{dx}$ to find m, the gradient of the tangent to the curve at $x = 1$:

$m = \frac{3}{2}(1)^{\frac{1}{2}} + (1)^{-\frac{1}{2}} + 9(1)^2 - 1 = \frac{21}{2}$ ***[1 mark]***.

Plug $x = 1$ into the equation for y to find the corresponding y-value:

$y = \frac{1^2 + 2(1)}{\sqrt{1}} + 3(1)^3 - 1 = 5$ ***[1 mark]***.

Now use the equation of a straight line, $y - y_1 = m(x - x_1)$ to find the equation of the tangent:

$y - 5 = \frac{21}{2}(x - 1)$ ***[1 mark]***

$2y - 10 = 21x - 21$

$2y = 21x - 11$ — as required ***[1 mark]***.

Answers

7 a) Use the formula:

$$\int_a^b y\,dx \approx \frac{h}{2}[y_0 + 2(y_1 + y_2 + ...y_{n-1}) + y_n]$$

Remember that n is the number of strips (in this case 3), and h is the width of each strip:

$h = \frac{b-a}{n} = \frac{3-0}{3} = 1$ ***[1 mark]***

Work out each y value:

$x_0 = 0$ $y_0 = \sqrt{(3+0)} = \sqrt{3} = 1.732$ (3 d.p.)

$x_1 = 1$ $y_1 = \sqrt{(3+1)} = \sqrt{4} = 2$

$x_2 = 2$ $y_2 = \sqrt{(3+4)} = \sqrt{7} = 2.646$ (3 d.p.)

$x_3 = 3$ $y_3 = \sqrt{(3+9)} = \sqrt{12} = 3.464$ (3 d.p.) ***[1 mark]***

And put all these values into the formula:

$\int_0^2 \sqrt{3+x^2}\,dx \approx \frac{1}{2}[1.732 + 2(2 + 2.646) + 3.464]$ ***[1 mark]***

$= 7.24$ (3 s.f.) ***[1 mark]***

b) Improve your estimate by using more strips / more ordinates. ***[1 mark]***

8 a) First, use the cosine rule, $a^2 = b^2 + c^2 - 2bc\cos A$, with $a = x$, $b = 4$, $c = 6$ and $A = 30$:

$x^2 = 4^2 + 6^2 - (2 \times 4 \times 6)\cos 30$ ***[1 mark]***

$\Rightarrow x^2 = 16 + 36 - 48\cos 30 = 10.43$

$\Rightarrow x = 3.23$ cm (3 s.f.). ***[1 mark]***

b) Now use the sine rule $\frac{a}{\sin A} = \frac{b}{\sin B} = \frac{c}{\sin C}$ with B = β and C = α:

$\frac{3.23}{\sin 30} = \frac{4}{\sin\beta} = \frac{6}{\sin\alpha}$ ***[1 mark]***

$\Rightarrow \sin\beta = \frac{4\sin 30}{3.23} = 0.6192$

$\Rightarrow \beta = 38.3°$ (3 s.f.) ***[1 mark]***.

Angles in a triangle sum to 180°, so

$\alpha = 180° - 30° - 38.26° = 111.7°$ (3 s.f.) ***[1 mark]***

c) Finally, use the area formula:

Area $= \frac{1}{2}bc\sin 30 = \frac{1}{2} \times 4 \times 6 \times \sin 30$ ***[1 mark]***

$= 6$ cm^2 ***[1 mark]***.

9 a) Use the trig identity $\tan\theta \equiv \frac{\sin\theta}{\cos\theta}$:

$\tan^2\theta + \frac{\tan\theta}{\cos\theta} = 1 \Rightarrow \frac{\sin^2\theta}{\cos^2\theta} + \frac{\sin\theta}{\cos^2\theta} = 1$ ***[1 mark]***

Put over a common denominator: $\frac{\sin^2\theta + \sin\theta}{\cos^2\theta} = 1$

$\Rightarrow \sin^2\theta + \sin\theta = \cos^2\theta$

Now use the identity $\cos^2\theta \equiv 1 - \sin^2\theta$ to give:

$\sin^2\theta + \sin\theta = 1 - \sin^2\theta$ ***[1 mark]***

$\Rightarrow 2\sin^2\theta + \sin\theta - 1 = 0$ ***[1 mark]***.

b) Factorising the quadratic from a) gives:

$(2\sin\theta - 1)(\sin\theta + 1) = 0$ ***[1 mark]***

$\Rightarrow \sin\theta = \frac{1}{2}$ or $\sin\theta = -1$ ***[1 mark]***

$\sin\theta = \frac{1}{2} \Rightarrow \theta = 30°$ ***[1 mark]*** or

$\theta = (180° - 30°) = 150°$ ***[1 mark]***

(from the CAST diagram or graph of $\sin\theta$)

$\sin\theta = -1 \Rightarrow \theta = 270°$. ***[1 mark]***

So, solutions are $\theta = 30°$, 150° and 270°

Don't forget that θ has to be between 0° and 360° — that last value of θ will come up as −90° on your calculator, so you have to work out θ = 360° − 90° = 270°.

C2 Practice Exam 2

1 a) Rearrange according to Laws of Indices (if you've forgotten, they've got their very own page in Section 1):

$36^{-\frac{1}{2}} = \frac{1}{36^{\frac{1}{2}}} = \frac{1}{\sqrt{36}}$ ***[1 mark]***

$= \frac{1}{6}$ ***[1 mark]***

b) First simplify the surd on the bottom of the fraction:

$\sqrt[m]{x^n} = x^{\frac{n}{m}}$ so $\sqrt{x^4} = x^2$ ***[1 mark]***

Then rewrite the entire expression so that you're only multiplying things (remember that $\div x^n = \times x^{-n}$):

$\frac{x^6 \times x^3}{x^2} \div x^{\frac{1}{2}} = x^6 \times x^3 \times x^{-2} \times x^{-\frac{1}{2}}$ ***[1 mark]***

Finally, add the powers together, because $x^m \times x^n = x^{m+n}$:

$= x^{6+3-2-\frac{1}{2}} = x^{\frac{13}{2}}$ ***[1 mark]***

Lots of laws to remember there. Make sure you don't multiply powers when you should be adding, and vice versa.

c) From part b), the required integral is:

$\int_0^1 (x^{\frac{13}{2}})dx = \left[\frac{2}{15}x^{\frac{15}{2}}\right]_0^1$ ***[1 mark]*** $= \left(\frac{2}{15}\right) - (0)$ ***[1 mark]***

$= \frac{2}{15}$ ***[1 mark]***.

2 a) In the laws of logs: $\log_a a = 1$. So $\log_3 3 = 1$. ***[1 mark]***

b) Use the laws of logs to rewrite the expression:

$\log_a 4 + 3\log_a 2 = \log_a(4 \times 2^3) = \log_a 32$. ***[1 mark]***

Therefore $\log_a \alpha = \log_a 32$ so $\alpha = 32$. ***[1 mark]***

c) Take logs of both sides and use laws of logs:

$0.5^{2x} = 0.3 \Rightarrow \log(0.5^{2x}) = \log 0.3$

$\Rightarrow 2x\log 0.5 = \log 0.3$ ***[1 mark]***

$\Rightarrow 2x = \log 0.3 \div \log 0.5 = 1.736...$ ***[1 mark]***

$\Rightarrow x = 0.868$ (3 s.f.) ***[1 mark]***

3 a) Using the binomial expansion formula:

$(1+x)^n =$

$1 + \frac{n}{1}x + \frac{n(n-1)}{1 \times 2}x^2 + \frac{n(n-1)(n-2)}{1 \times 2 \times 3}x^3 + ... + x^n$

Expand the expression $(1 + ax)^{10}$ into this form:

$1 + \frac{10}{1}(ax) + \frac{10 \times 9}{1 \times 2}(ax)^2 + \frac{10 \times 9 \times 8}{1 \times 2 \times 3}(ax)^3 + ...$ ***[1 mark]***

Then simplify each coefficient:

$(1 + ax)^{10} = 1 + 10ax + 45a^2x^2 + 120a^3x^3 + ...$ ***[1 mark]***

b) *Pay attention to the number at the front of the bracket. If it's not a 1, you have to rearrange everything so it is a 1. Fiddly but important.*

First take a factor of 2 to get it in the form $(1 + ax)^n$:

$(2+3x)^5 = \left[2\left(1 + \frac{3}{2}x\right)\right]^5 = 2^5\left(1 + \frac{3}{2}x\right)^5 = 32\left(1 + \frac{3}{2}x\right)^5$

Now expand:

$32\left[1 + \frac{5}{1}\left(\frac{3}{2}x\right) + \frac{5 \times 4}{1 \times 2}\left(\frac{3}{2}x\right)^2 + ...\right]$ ***[1 mark]***

You only need the x^2 term, so simplify that one:

$32 \times \frac{20}{2} \times \left(\frac{3}{2}\right)^2 \times x^2 = 720x^2$

So the coefficient of x^2 is 720 ***[1 mark]***

Answers

c) Look back to the x^2 term from part a) — $45a^2x^2$. This is equal to $720x^2$ so just rearrange the formula to find a:

$45a^2 = 720$

$a^2 = 16$

$a = \pm 4$ ***[1 mark]***

And remember that part a) tells you that $a > 0$, so $a = 4$. ***[1 mark]***

4 a)

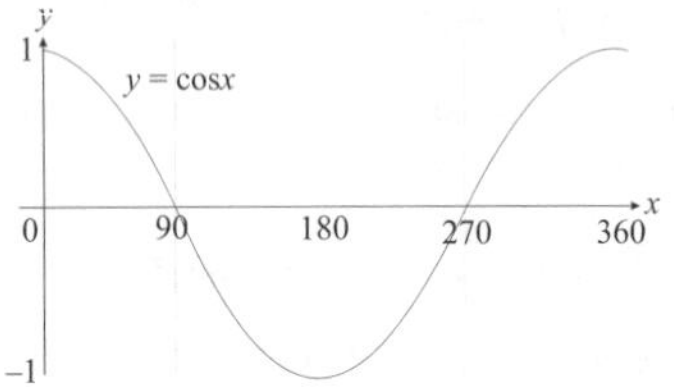

[1 mark]

b) Remember that $\sin^2 x + \cos^2 x = 1$ so with a little rearranging you can replace the $\sin^2$ with a $1 - \cos^2$:

$2[1 - \cos^2 x] = 1 + \cos x$ ***[1 mark]***

Now multiply out the bracket and rearrange:

$2[1 - \cos^2 x] = 1 + \cos x$

$\Rightarrow 2 - 2\cos^2 x = 1 + \cos x$ ***[1 mark]***

$\Rightarrow 2\cos^2 x + \cos x - 1 = 0$ ***[1 mark]***

c) $2\cos^2 x + \cos x - 1 = 0$ factorises to give:

$(2\cos x - 1)(\cos x + 1) = 0$

so $\cos x = \frac{1}{2}$ or $\cos x = -1$ ***[1 mark]***

By taking the inverse cosine of these you get:

$x = \cos^{-1}\left(\frac{1}{2}\right) = 60°$ ***[1 mark]***

$x = \cos^{-1}(-1) = 180°$ ***[1 mark]***

The graph from part a) shows there are three possible solutions, as the curve intersects the line $y = \frac{1}{2}$ twice. The other solution is $x = 360° - 60° = 300°$. ***[1 mark]***

So the three solutions are:

$x = 60°$, $x = 180°$ and $x = 300°$.

d) (i) A shift of 60° to the right maps $\cos x$ to $\cos(x - 60°)$

A shift of 1 up maps $\cos x$ to $1 + \cos x$

Putting these together gives $f(x) = 1 + \cos(x - 60°)$.

[2 marks available — 1 mark for vertical shift, 1 mark for horizontal shift]

(ii)

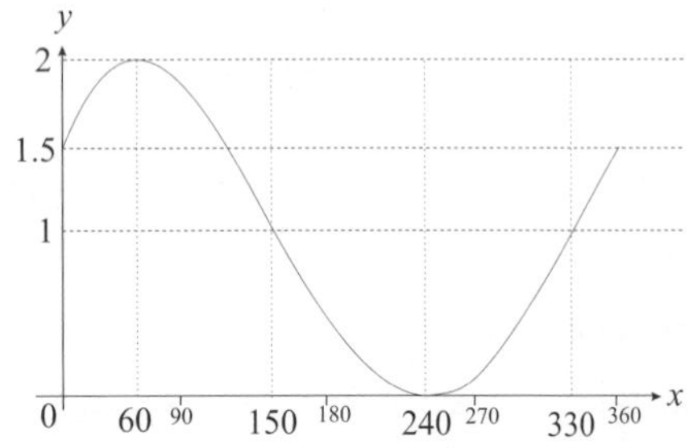

[3 marks available — 1 mark for shape, 1 mark for shift right 60°, 1 mark for shift up 1]

(iii) Stretch in y-direction ***[1 mark]*** scale factor $\frac{1}{2}$ ***[1 mark]***.

5 a) First put the two known terms into the formula for the nth term of a geometric series: $u_n = ar^{n-1}$:

$u_3 = ar^2 = \frac{5}{2}$ $\qquad u_6 = ar^5 = \frac{5}{16}$

Divide the expression for u_6 by the expression for u_3 to get an expression just containing r and solve it:

$\frac{ar^5}{ar^2} = \frac{5}{16} \div \frac{5}{2} \Rightarrow r^3 = \frac{5}{16} \times \frac{2}{5} = \frac{10}{80}$ ***[1 mark]***

$r^3 = \frac{1}{8}$ $\qquad r = \sqrt[3]{\frac{1}{8}}$ $\qquad r = \frac{1}{2}$ ***[1 mark]***

Put this value back into the expression for u_3 to find a:

$a\left(\frac{1}{2}\right)^2 = \frac{5}{2} \Rightarrow \frac{a}{4} = \frac{5}{2} \Rightarrow a = 10$ ***[1 mark]***

The nth term is $u_n = ar^{n-1}$, where $r = \frac{1}{2}$ and $a = 10$

$u_n = 10 \times \left(\frac{1}{2}\right)^{n-1} = 10 \times \frac{1}{2^{n-1}} = \frac{10}{2^{n-1}}$ ***[1 mark]***

b) This asks you to find the sum of the first ten terms, so input the known values of n, a and r (from part a)) into the formula for the sum of the first n terms:

$S_n = \frac{a(1 - r^n)}{1 - r}$, $S_{10} = \frac{10\left(1 - \left(\frac{1}{2}\right)^{10}\right)}{1 - \left(\frac{1}{2}\right)}$ ***[1 mark]***

$= 10 \times 2 \times \left(1 - \frac{1}{2^{10}}\right)$

$= 20\left(1 - \frac{1}{1024}\right) = 20 \times \frac{1023}{1024} = 5 \times \frac{1023}{256}$ ***[1 mark]***

$= \frac{5115}{256}$ ***[1 mark]***

c) Just pop the known values of a and r into the formula for the sum to infinity:

$S_\infty = \frac{a}{1 - r} = \frac{10}{1 - \left(\frac{1}{2}\right)}$ ***[1 mark]***

$= 10 \div \frac{1}{2} = 10 \times 2$

$= 20$ ***[1 mark]***

6 a) Arc length $= r\theta$, so the arc length of the sector $= 0.5r$ ***[1 mark]***.

The perimeter of the sector is then $0.5r + 2r = 2.5r = \frac{5}{2}r$ ***[1 mark]***.

Sector area $= \frac{1}{2}r^2\theta = \frac{1}{2} \times r^2 \times 0.5 = \frac{1}{4}r^2$ ***[1 mark]***.

The area and perimeter are equal, so:

$\frac{5}{2}r = \frac{1}{4}r^2$ ***[1 mark]***. Multiply throughout by 4 to get:

$10r = r^2$

$r^2 - 10r = 0$

$r(r - 10) = 0$ ***[1 mark]***

$\Rightarrow r = 0$ or $r = 10$.

r is a radius, so assume $r > 0$, and so $r = 10$ ***[1 mark]***.

b) (i) Use the cosine rule, $a^2 = b^2 + c^2 - 2bc\cos A$:

$x^2 = 10^2 + 10^2 - (2 \times 10 \times 10)\cos 0.5$ ***[1 mark]***

$x^2 = 24.48$

$\Rightarrow x = 4.95$ cm (3 s.f.) ***[1 mark]***

Answers

(ii) Use the formula for the area of a triangle,

Area = $\frac{1}{2}ab\sin C$:

Area of $OAB = (\frac{1}{2} \times 10 \times 10)\sin 0.5$ ***[1 mark]***

$= 23.97$ cm^2 ***[1 mark]***.

Area of sector from part a) = $\frac{1}{2}r^2\theta = \frac{1}{2} \times 10^2 \times 0.5$ ***[1 mark]***

$= 25$ cm^2 ***[1 mark]***.

So the area of the shaded region is:

$25 - 23.97 = 1.03$ cm^2 (3 s.f.) ***[1 mark]***.

7 a) $y = \frac{1}{x^2} + x^2 = x^{-2} + x^2$

$\frac{dy}{dx} = -2x^{-3} + 2x$ ***[1 mark]*** $= -\frac{2}{x^3} + 2x$.

The curve is stationary when $\frac{dy}{dx} = 0$, i.e. when

$-\frac{2}{x^3} + 2x = 0$ ***[1 mark]***

$\Rightarrow \frac{1}{x^3} = x$

$x^4 = 1$ ***[1 mark]***

$\Rightarrow x = 1$ or $x = -1$ ***[1 mark]***.

Substitute into original equation to find y-coordinates:

$\frac{1}{(-1)^2} + (-1)^2 = \frac{1}{1^2} + 1^2 = 2$ ***[1 mark]***

So coordinates of stationary points are:

(–1, 2) and (1, 2).

b) The gradient of the tangent at $x = 2$ is equal to $\frac{dy}{dx}$ at $x = 2$.

i.e. gradient $= -\frac{2}{2^3} + 2(2) = \frac{15}{4}$ ***[1 mark]***.

When $x = 2$, $y = \frac{1}{2^2} + 2^2 = \frac{17}{4}$ ***[1 mark]***.

So the equation of the tangent is:

$y - \frac{17}{4} = \frac{15}{4}(x - 2)$ ***[1 mark]***

$4y - 17 = 15x - 30$

$\Rightarrow y = \frac{1}{4}(15x - 13)$ ***[1 mark]***

c) The required area is given by:

$\int_1^2 (x^{-2} + x^2)dx = \left[\frac{x^{-1}}{-1} + \frac{x^3}{3}\right]_1^2$ ***[1 mark]*** $= \left[-\frac{1}{x} + \frac{x^3}{3}\right]_1^2$

$= \left(-\frac{1}{2} + \frac{8}{3}\right) - \left(-1 + \frac{1}{3}\right)$ ***[1 mark]*** $= \frac{17}{6}$ ***[1 mark]***.

8 a) Look up the trapezium formula in the nice formula booklet they give you:

$\int_a^b y\,dx \approx \frac{h}{2}[y_0 + 2(y_1 + y_2 + \ldots y_{n-1}) + y_n]$

and remember that n is the number of intervals (in this case 4), and h is the width of each strip:

$h = \frac{b-a}{n} = \frac{2-0}{4} = 0.5$ ***[1 mark]***

Work out each y value:

$x_0 = 0$ $\quad y_0 = 2^{0^2} = 2^0 = 1$

$x_1 = 0.5$ $\quad y_1 = 2^{0.5^2} = 2^{0.25} = 1.189$ (3 d.p.)

$x_2 = 1$ $\quad y_2 = 2^{1^2} = 2^1 = 2$

$x_3 = 1.5$ $\quad y_3 = 2^{1.5^2} = 2^{2.25} = 4.757$ (3 d.p)

$x_4 = 2$ $\quad y_4 = 2^{2^2} = 2^4 = 16$ ***[1 mark]***

And put all these values into the formula:

$\int_0^2 2^{x^2}dx \approx \frac{0.5}{2}[1 + 2(1.189 + 2 + 4.757) + 16]$ ***[1 mark]***

$= \frac{1}{4}(17 + 15.892)$

$= 8.22$ (3 s.f.) ***[1 mark]***

b) Look at the diagram of the curve — it's U-shaped.
A trapezium on each strip goes higher than the curve and so has a greater area than that under the curve. ***[1 mark]***
So the trapezium rule gives an overestimate for the area.
[1 mark]

Core blimey — that's both Core modules finished (assuming you did them in order). But don't put your feet up for too long — there's more maths joy to come...

Answers

S1 Section 1 — Numerical Measures

Warm-up Questions

1) $\Sigma f = 16$, $\Sigma fx = 22$, so mean = 22 ÷ 16 = 1.375
Median position = 17 ÷ 2 = 8.5, so median = 1
Mode = 0.

2)

Speed	mid-class value x	Number of cars f	fx
30 - 34	32	12 (12)	384
35 - 39	37	37 (49)	1369
40 - 44	42	9	378
45 - 50	47.5	2	95
	Totals	60	2226

Estimated mean = 2226 ÷ 60 = 37.1 mph
Median position is 61 ÷ 2 = 30.5.
This is in class 35 - 39.
Easy eh? It doesn't hurt to double-check your mid-class values though.

3) Put the 20 items of data in order:
1, 4, 5, 5, 5, 5, 5, 6, 6, 7, 7, 8, 10, 10, 12, 15, 20, 20, 30, 50
Then the median position is 10.5, and since the 10th and the 11th items are both 7, the median = 7.
Lower quartile position = 5.5, so lower quartile = 5.
Upper quartile position = 15.5,
so upper quartile = (12 + 15) ÷ 2 = 13.5.

4) $$\text{Mean} = \frac{11+12+14+17+21+23+27}{7}$$
$$= \frac{125}{7} = 17.9 \text{ to 3 sig. fig.}$$
$$\text{s.d.} = \sqrt{\frac{11^2+12^2+14^2+17^2+21^2+23^2+27^2}{7} - \left(\frac{125}{7}\right)^2}$$
$$= \sqrt{30.98} = 5.57 \text{ to 3 sig. fig.}$$
Just numbers and a formula. Simple.

5)

Score	Mid-class value, x	x^2	f	fx	fx^2
100 - 106	103	10609	6	618	63654
107 - 113	110	12100	11	1210	133100
114 - 120	117	13689	22	2574	301158
121 - 127	124	15376	9	1116	138384
128 - 134	131	17161	2	262	34322
		Totals	50 (= Σf)	5780 (= Σfx)	670618 (= Σfx^2)

$$\text{Mean} = \frac{5780}{50} = 115.6$$
$$\text{Variance} = \frac{670\,618}{50} - 115.6^2 = 49$$

6) a) $\sum fx = 8 + 14 + 6 + 0 + 0 + 0 + 14 = 42$
$\sum f = 8 + 7 + 2 + 0 + 0 + 0 + 2 = 19$
$$\text{So mean} = \bar{x} = \frac{\sum fx}{\sum f} = \frac{42}{19} = 2.21 \text{(to 2 d.p.)}$$
Median position = (19 + 1) ÷ 2 = 10,
so the median = 2.
Mode = 1.

b) Range = maximum value – minimum value = 7 – 1 = 6
To find the interquartile range, you need to find the upper and lower quartiles.
First, the lower quartile (Q_1).
Since 19 ÷ 4 = 4.75 is not a whole number, the lower quartile position is rounded up to 5. So $Q_1 = 1$.
Now find the upper quartile (Q_3).
Since 19 × 3 ÷ 4 = 14.25 is not a whole number, the upper quartile position is rounded up to 15. So $Q_3 = 2$.
So the interquartile range = 2 – 1 = 1.
To find the variance, add some more rows to the table:

Number of cars owned, x	1	2	3	4	5	6	7
Frequency, f	8	7	2	0	0	0	2
x^2	1	4	9	16	25	36	49
fx^2	8	28	18	0	0	0	98

$$\text{Variance} = \frac{\sum fx^2}{\sum f} - \bar{x}^2 = \frac{8+28+18+98}{19} - \left(\frac{42}{19}\right)^2$$
$$= \frac{152}{19} - \left(\frac{42}{19}\right)^2 = \frac{1124}{361} = 3.11 \text{(to 2 d.p.)}.$$
$$\text{Standard deviation} = \sqrt{\text{variance}} = \sqrt{\frac{1124}{361}}$$
$$= 1.76 \text{ (to 2 d.p.)}$$

7) Let $y = x - 20$.
Then
$\bar{y} = \bar{x} - 20$ or $\bar{x} = \bar{y} + 20$
$\sum y = 125$ and $\sum y^2 = 221$
So $\bar{y} = \frac{125}{100} = 1.25$ and $\bar{x} = 1.25 + 20 = 21.25$
Variance of $y = \frac{221}{100} - 1.25^2 = 0.6475$,
and so s.d. of $y = 0.805$ to 3 sig. fig.
Therefore s.d. of $x = 0.805$ to 3 sig. fig.
If you got in a muddle, look back at stuff about linear scaling.

8)

Time	Mid-class x	$y = x - 35.5$	f	fy	fy^2
30 - 33	31.5	-4	3	-12	48
34 - 37	35.5	0	6	0	0
38 - 41	39.5	4	7	28	112
42 - 45	43.5	8	4	32	256
		Totals	20 (= Σf)	48 (= Σfy)	416 (= Σfy^2)

$$\bar{y} = \frac{48}{20} = 2.4$$
So $\bar{x} = \bar{y} + 35.5 = 2.4 + 35.5 = 37.9$ minutes
Variance of $y = \frac{416}{20} - 2.4^2 = 15.04$,
and so s.d. of $y = 3.88$ minutes, to 3 sig. fig.
But s.d. of x = s.d. of y,
and so s.d. of $x = 3.88$ minutes, to 3 sig. fig.
Even if you did the linear scaling differently, your answer should be the same.

Answers

Exam Questions

1 a) Let $y = x - 30$.

$\overline{y} = \frac{228}{19} = 12$ and so $\underline{\overline{x} = \overline{y} + 30 = 42}$ ***[1 mark]***

Variance of $y = \frac{3040}{19} - 12^2 = 16$ ***[1 mark]***,

and so s.d. of $y = 4$.

But s.d. of x = s.d. of y, and so $\underline{\text{s.d. of } x = 4}$ ***[1 mark]***

b) $\overline{x} = \frac{\sum x}{19} = 42$

And so $\sum x = 42 \times 19 = \underline{798}$ ***[1 mark]***

Variance of $x = \frac{\sum x^2}{19} - \overline{x}^2 = \frac{\sum x^2}{19} - 42^2 = 16$ ***[1 mark]***

And so $\sum x^2 = (16 + 42^2) \times 19 = \underline{33\,820}$ ***[1 mark]***

A bit harder...

c) New $\sum x = 798 + 32 = 830$ ***[1 mark]***

So new $\bar{x} = \frac{830}{20} = \underline{41.5}$ ***[1 mark]***

New $\sum x^2 = 33\,820 + 32^2 = 34\,844$ ***[1 mark]***

So new variance $= \frac{34\,844}{20} - 41.5^2 = 19.95$

and new $\underline{\text{s.d.} = 4.47}$ to 3 sig. fig. ***[1 mark]***

2 a) (i) Worker A's times in order: 2, 3, 4, 4, 5, 5, 5, 7, 10, 12
Median position = 5.5, so median = 5 minutes
[1 mark]
Worker B's times in order: 3, 4, 6, 7, 8, 8, 9, 9, 10, 11
Median position = 5.5, so median = 8 minutes
[1 mark]

(ii) For Worker A:
Lower quartile position = 3,
so lower quartile = 4 minutes ***[1 mark]***
Upper quartile position = 8,
so upper quartile = 7 minutes ***[1 mark]***

For Worker B:
Lower quartile position = 3,
so lower quartile = 6 minutes ***[1 mark]***
Upper quartile position = 8,
so upper quartile = 9 minutes ***[1 mark]***

b) Various statements could be made,
e.g. the times for Worker B are longer than those for Worker A, on average.
The IQR for both workers is the same (3 minutes) — generally they both work with the same consistency.
The range for Worker A (10 minutes) is larger than that for Worker B (8 minutes). Worker A had a few items he/she could iron very quickly and a few which took a long time.
[1 mark for any sensible answer]

c) Worker A would be better to employ. The median time is shorter than for Worker B, and the upper quartile is shorter than the median of Worker B. Worker A would generally iron more items in a given time than worker B.
[1 mark for any sensible answer]

Don't be put off by these questions — you just have to show you understand what the data is telling you.

3 a) $\overline{a} = \frac{60.3}{20} = 3.015$ g ***[1 mark]***

b) Variance of brand A $= \frac{219}{20} - 3.015^2$ ***[1 mark]***

$= 1.860$ g^2 ***[1 mark]***

So s.d. of brand A = 1.36 g to 3 sig. fig. ***[1 mark]***

c) E.g. Brand A chocolate drops are heavier on average than brand B. Brand B chocolate drops are much closer to their mean weight than brand A.
[1 mark for each of 2 sensible statements]

"Mmm, chocolate drops" does not count as a sensible statement...

d) Mean of A and B $= \frac{\sum a + \sum b}{50} = \frac{60.3 + (30 \times 2.95)}{50}$

$= 2.976$ g ***[1 mark]***

$\frac{\sum b^2}{30} - 2.95^2 = 1$, and so $\sum b^2 = 291.075$ ***[1 mark]***

Variance of A and $B = \frac{\sum a^2 + \sum b^2}{50} - 2.976^2$

$= \frac{219 + 291.075}{50} - 2.976^2$

$= 1.3449$ ***[1 mark]***

So s.d. $= \sqrt{1.3449} = 1.16$ g to 3 sig. fig. ***[1 mark]***

Work through each step carefully so you don't make silly mistakes and lose any lovely marks.

4 a)

No. of Hits, x	12	13	14	15	16	17	18	19	20	21	22	23	24	25
Frequency, f	1	2	3	2	1	1	0	0	0	0	3	0	0	1
fx	12	26	42	30	16	17	0	0	0	0	66	0	0	25

$\sum fx = 234$ ***[1 mark]*** and $\sum f = 14$,

so $\overline{x} = \frac{234}{14} = 16.7$ (to 3 sig.fig) ***[1 mark]***

Total number of people = 14

Median position = (14 + 1) ÷ 2 = 7.5 ***[1 mark]***,

so median = average of 7th and 8th values = <u>15 hits</u>
[1 mark].

Modes = <u>14 hits and 22 hits</u> ***[1 mark]***

b) The mode is least appropriate ***[1 mark]***, since there is no unique value, and one of these modes is a long way from the bulk of the data ***[1 mark for any suitable reason]***.

Answers

S1 Section 2 — Probability

Warm-up Questions

1) You need a way to record all the possible scores. You could do this with a tree diagram, but a table will do just as well:

		Dice					
		1	2	3	4	5	6
Coin	H	2	4	6	8	10	12
	T	5	6	7	8	9	10

a) There are 12 outcomes in total, and 9 of these are more than 5, so P(score > 5) = 9/12 = 3/4

b) There are 6 outcomes which have a tail showing, and 3 of these are even, so P(even score given that you throw a tail) = 3/6 = 1/2

Hmm, a bit fiddly but not too bad.

2) a) There are 30 workers altogether, and 12 drink coffee, so:
$P(\text{drinks coffee}) = \frac{12}{30} = \frac{2}{5}$

b) 7 of the workers drink milky tea without sugar, so:
$P(\text{drinks milky tea without sugar}) = \frac{7}{30}$

c) There are 18 + 7 – 4 = 21 workers who either drink tea or take only sugar, so:
$P(\text{drinks tea or takes only sugar}) = \frac{21}{30} = \frac{7}{10}$

3) a) 20% of the people eat chips, and 10% of these is 2% — so 2% eat both chips and sausages.
Now you can draw the Venn diagram:

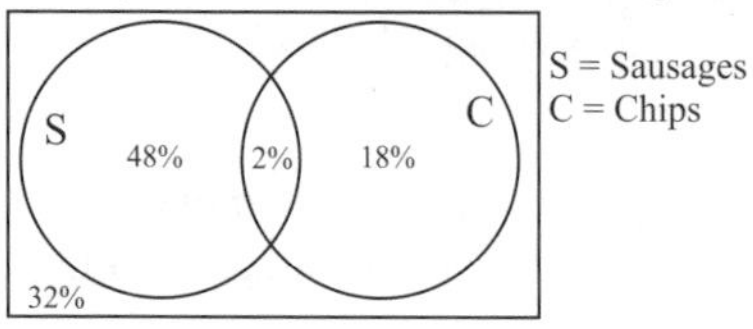

By reading the numbers in the appropriate sets from the diagram you can see...

b) 18% eat chips but not sausages.

c) 18% + 48% = 66% eat chips or sausages, but not both.

These questions do make you work up an appetite... mmm, sausages...

4) a) There are 36 – 18 – 8 = 10 people whose surnames begin with C. So the probability that the first person to leave the room has a surname beginning with C is $\frac{10}{36} = \frac{5}{18}$. Then the probability that the second person to leave the room has a surname beginning with C is $\frac{9}{35}$.

This means the probability of both people having surnames beginning with C is $\frac{5}{18} \times \frac{9}{35} = \frac{1}{14}$.

Since there is only <u>one way</u> for both people to have "C-surnames" (i.e. first person has a "C-surname", then second person also has a "C-surname"), this must be the total probability.

b) The probability that the first person has an "A-surname", the second person has a "B-surname" and the third has a "C-surname" is $\frac{18}{36} \times \frac{8}{35} \times \frac{10}{34} = \frac{1440}{42\,840} = \frac{4}{119}$.
But there are 6 ways to "arrange" these 3 surnames (since you'd have 3 choices for the first name, 2 choices for the second and only 1 choice for the third — and 3 × 2 × 1 = 6). So the total probability that the three people have surnames beginning with different letters is $6 \times \frac{4}{119} = \frac{24}{119}$.

If you find it easier to use a tree diagram, then use one.

5) a)

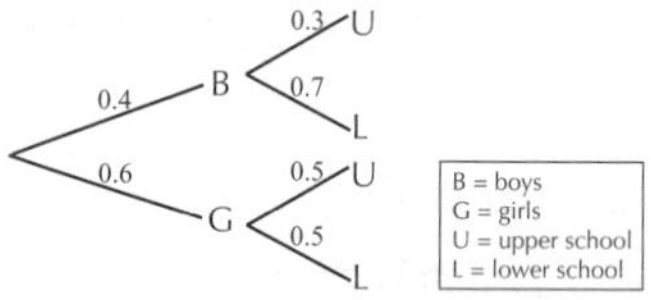

b) Choosing an upper school pupil means either 'boy and upper' or 'girl and upper'.
P(boy and upper) = 0.4 × 0.3 = 0.12.
P(girl and upper) = 0.6 × 0.5 = 0.30.
So P(Upper) = 0.12 + 0.30 = 0.42.

6) Draw a tree diagram:

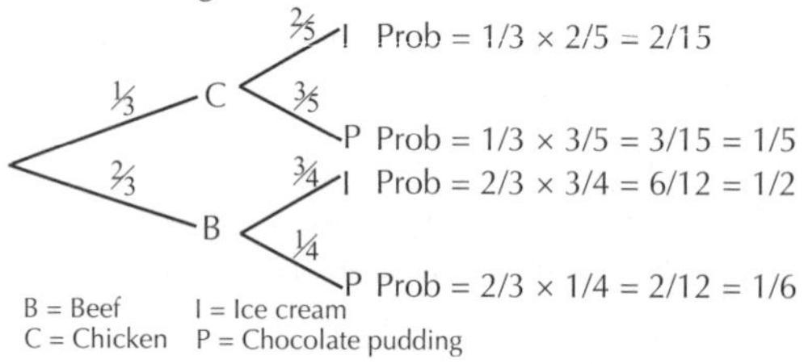

a) P(chicken or ice cream but not both)
$= P(C \cap P) + P(B \cap I) = 1/5 + 1/2 = 7/10$

b) $P(\text{ice cream}) = P(C \cap I) + P(B \cap I) = 2/15 + 1/2 = 19/30$

c) $P(\text{chicken}|\text{ice cream}) = P(C \cap I) \div P(I)$
$= (2/15) \div (19/30) = 4/19$

You aren't asked to draw a tree diagram, but it makes it a lot easier if you do.

d) P(chicken) = 1/3, and from part c),
P(chicken|ice cream) = 4/19.
Since P(chicken) ≠ P(chicken|ice cream), the events 'eats chicken' and 'eats ice cream' are not independent.

Or you could say that:
P(chicken and ice cream) = P(chicken | ice cream) × P(ice cream) = 4/19 × 19/30 = 2/15 (using parts b) and c)).
This does not equal P(chicken) × P(ice cream) = 1/3 × 19/30 = 19/90, so the two events are not independent.

Answers

Exam Questions

1 Most questions become easier to answer if you draw a picture. In this case, that means a Venn diagram.

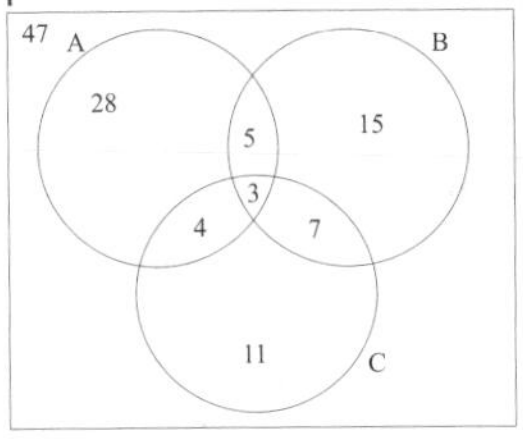

a) Add up the numbers in all the circles to get 73 people out of 120 buy at least 1 type of soap ***[1 mark]***. So the probability = 73/120 ***[1 mark]***

b) Add up the numbers in the intersections to get 5 + 3 + 4 + 7 = 19, meaning that 19 people buy at least two soaps ***[1 mark]***, so the probability a person buys at least two types = 19/120 ***[1 mark]***.

c) 28 + 11 + 15 = 54 people buy only 1 soap ***[1 mark]***, and of these 15 buy soap B ***[1 mark]***.

So probability of a person who only buys one type of soap buying type B is 15/54 = 5/18 ***[1 mark]***

2 Once again... a picture will help. Here you have a sequence of events, so you need a tree diagram.

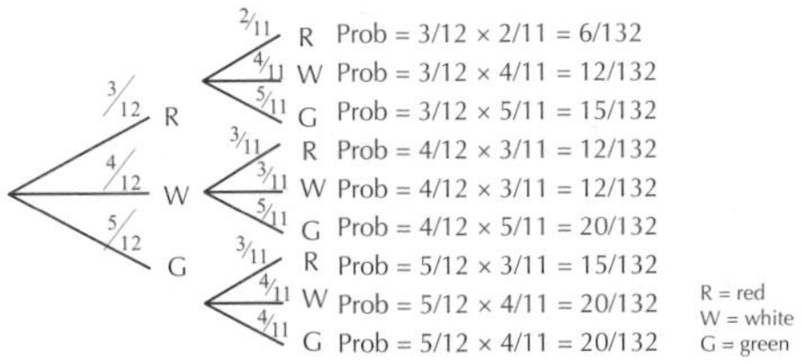

a) The second counter is green means one of three outcomes 'red then green' or 'white then green' or 'green then green'. So P(2nd is green) = 15/132 + 20/132 + 20/132 ***[1 mark]*** = 55/132 = 5/12 ***[1 mark]***

b) For both to be red there's only one outcome: 'red then red' ***[1 mark]***. P(both red) = 6/132 = 1/22 ***[1 mark]***

c) 'Both same colour' is the complementary event of 'not both same colour'. So P(not same colour) = 1 – P(both same colour) ***[1 mark]***. Both same colour is either (R and R) or (W and W) or (G and G).
P(not same colour) = 1 – [6/132 + 12/132 + 20/132] ***[1 mark]*** = 1 – 38/132 = 94/132 = 47/66 ***[1 mark]***
(Alternatively, 1 mark for showing P(RW or RG or WR or WG or GR or GW), 1 mark for adding the 6 correct probabilities and 1 mark for the correct answer.)

Ooh, that was a long one. Shouldn't be too tricky though — especially with a nice, clear tree diagram.

3 a) (i) J and K are independent, so
$P(J \cap K) = P(J) \times P(K) = 0.7 \times 0.1 = 0.07$ ***[1 mark]***

(ii) $P(J \cup K) = P(J) + P(K) - P(J \cap K)$ ***[1 mark]***
$= 0.7 + 0.1 - 0.07 = 0.73$ ***[1 mark]***

b) Drawing a quick Venn Diagram often helps:

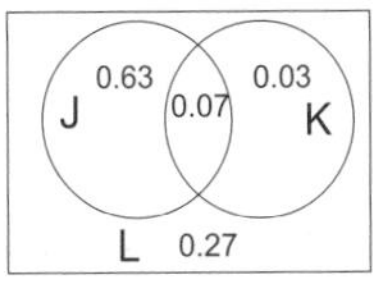

$P(L|K') = P(L \cap K') \div P(K')$
Now $L \cap K' = L$ — think about it — all of L is contained in K', so $L \cap K'$ (the 'bits in both L and K') are just the bits in L.
Therefore $P(L \cap K') = P(L) = 1 - P(K \cup J) = 1 - 0.73 = 0.27$ ***[1 mark]***
$P(K') = 1 - P(K) = 1 - 0.1 = 0.9$ ***[1 mark]***
And so $P(L|K') = 0.27 \div 0.9 = 0.3$ ***[1 mark]***

That was a bit complicated — you just need to put your thinking cap on and DON'T PANIC.

4 Draw a tree diagram:

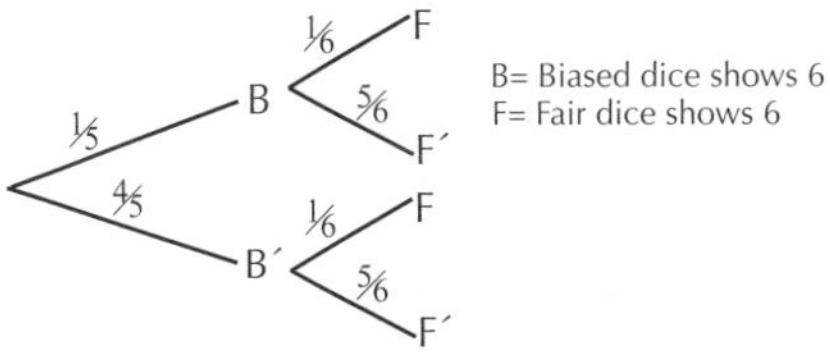

a) $P(B') = 0.8$ ***[1 mark]***

b) Either at least one of the dice shows a 6 or neither of them do, so these are complementary events. Call F the event 'the fair dice shows a 6'.
Then $P(F \cup B) = 1 - P(F' \cap B')$ ***[1 mark]***
$= 1 - (4/5 \times 5/6) = 1 - 2/3 = 1/3$ ***[1 mark]***

c) P(exactly one 6 | at least one 6)
= P(exactly one 6 ∩ at least one 6) ÷ P(at least one 6).
The next step might be a bit easier to get your head round if you draw a Venn diagram:

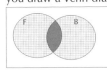

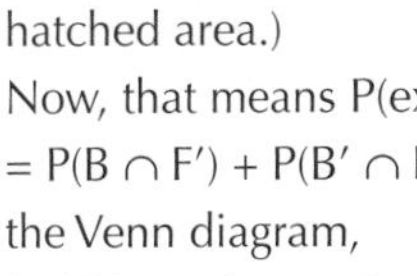

'exactly one 6' ∩ 'at least one 6' = 'exactly one 6'
(Look at the diagram — 'exactly one 6' is the cross-hatched area, and 'at least one 6' is the cross-hatched area plus the grey bit. So the bit in common to both is just the cross-hatched area.)
Now, that means P(exactly one 6 ∩ at least one 6)
$= P(B \cap F') + P(B' \cap F)$ — this is the cross-hatched area in the Venn diagram,
i.e. P(exactly one 6 ∩ at least one 6)
$= (1/5 \times 5/6) + (4/5 \times 1/6) = 9/30 = 3/10$
(using the fact that B and F are independent) ***[1 mark]***
P(at least one 6) = 1/3 (from b)).

And all of this means P(exactly one 6 | at least one 6)
= 3/10 ÷ 1/3 ***[1 mark]*** = 9/10 ***[1 mark]***

Blauuurgh — the noise of a mind boggling. Part c is difficult to get your head round, but it's just a matter of remembering the right conditional probability formula, breaking it down into separate parts and working through it step by step. Yay.

Answers

S1 Section 3 — Binomial Distribution

Warm-up Questions

1) a) There are 10 trials altogether, and you need to know how many ways there are to arrange 6 'successes' and 4 'failures'. This is $\binom{10}{6} = \frac{10!}{6! \times 4!} = \frac{3\,628\,800}{720 \times 24} = 210$.

It doesn't matter if you think of heads as 'success' or 'failure' — you'll get the same answer either way.

This is because $\binom{10}{6} = \binom{10}{4} = \frac{10!}{6! \times 4!}$.

b) There are $\binom{7}{4} = \frac{7!}{4! \times 3!} = \frac{5040}{24 \times 6} = 35$ different orders.

You'd be a while counting all these on your fingers.

2) a) $P(5 \text{ heads}) = \binom{10}{5} \times 0.5^5 \times 0.5^5$
$= \frac{10!}{5!5!} \times 0.5^{10} = 0.246$ (to 3 sig.fig.).

b) $P(9 \text{ heads}) = \binom{10}{9} \times 0.5^9 \times 0.5$
$= \frac{10!}{9!1!} \times 0.5^{10} = 0.00977$ (to 3 sig.fig.).

3) a) Binomial — there are a fixed number of independent trials (30) with two possible results ('prime' / 'not prime'), a constant probability of success, and the random variable is the total number of successes.

b) Binomial — there are a fixed number of independent trials (however many students are in the class) with two possible results ('heads' / 'tails'), a constant probability of success, and the random variable is the total number of successes.

c) Not binomial — the probability of being dealt an ace changes each time, since the total number of cards decreases as each card is dealt.

d) Not binomial — the number of trials is not fixed.

It's weird to have to write actual sentences in a maths exam, but be ready for it.

4) a) Use tables with $n = 10$ and $p = 0.5$.
If X represents the number of heads, then:
$P(X \geq 5) = 1 - P(X < 5) = 1 - P(X \leq 4)$
$= 1 - 0.3770 = 0.6230$

b) $P(X \geq 9) = 1 - P(X < 9) = 1 - P(X \leq 8)$
$= 1 - 0.9893 = 0.0107$

You do have to be prepared to monkey around with the numbers the tables give you.

5) a) You can't use tables here (because they don't include $p = 0.27$), so you have to use the probability function.

$P(X = 4) = \binom{14}{4} \times 0.27^4 \times (1 - 0.27)^{10}$
$= 0.229$ (to 3 sig.fig.)

b) $P(X < 2) = P(X = 0) + P(X = 1)$
$= \binom{14}{0} \times 0.27^0 \times (1 - 0.27)^{14}$
$+ \binom{14}{1} \times 0.27^1 \times (1 - 0.27)^{13}$
$= 0.012204... + 0.063195...$
$= 0.0754$ (to 3 sig.fig.)

c) $P(5 < X \leq 8) = P(X = 6) + P(X = 7) + P(X = 8)$
$= \binom{14}{6} \times 0.27^6 \times (1 - 0.27)^8$
$+ \binom{14}{7} \times 0.27^7 \times (1 - 0.27)^7$
$+ \binom{14}{8} \times 0.27^8 \times (1 - 0.27)^6$
$= 0.093825... + 0.039660... + 0.012835...$
$= 0.146$ (to 3 sig.fig.)

6) For parts a)-c), use tables with $n = 25$ and $p = 0.15$.

a) $P(X \leq 3) = 0.4711$

b) $P(X \leq 7) = 0.9745$

c) $P(X \leq 15) = 1.0000$

For parts d)-f), define a new random variable $T \sim B(15, 0.35)$. Then use tables with $n = 15$ and $p = 0.35$.

d) $P(Y \leq 3) = P(T \geq 12) = 1 - P(T < 12)$
$= 1 - P(T \leq 11) = 1 - 0.9995 = 0.0005$

e) $P(Y \leq 7) = P(T \geq 8) = 1 - P(T < 8)$
$= 1 - P(T \leq 7) = 1 - 0.8868 = 0.1132$

f) $P(Y \leq 15) = 1$ (since 15 is the maximum possible value).

These last few parts (where you can't use the tables without a bit of messing around first) are quite awkward, so make sure you get lots of practice.

7) From tables:

a) $P(X \leq 15) = 0.9997$

b) $P(X < 4) = P(X \leq 3) = 0.1302$

c) $P(X > 7) = 1 - P(X \leq 7) = 1 - 0.0639 = 0.9361$

For parts d)-f) where $X \sim B(n, p)$ with $p > 0.5$, define a new random variable $Y \sim B(n, q)$, where $q = 1 - p$. Then use tables.

d) Define $Y \sim B(50, 0.2)$. Then $P(X \geq 40) = P(Y \leq 10) = 0.5836$

e) Define $Y \sim B(30, 0.3)$. Then $P(X = 20) = P(Y = 10)$
$= P(Y \leq 10) - P(Y \leq 9) = 0.7304 - 0.5888 = 0.1416$

f) Define $Y \sim B(10, 0.25)$. Then $P(X = 7) = P(Y = 3)$
$= P(Y \leq 3) - P(Y \leq 2) = 0.7759 - 0.5256 = 0.2503$

Or you could use: $P(X = 7) = \binom{10}{7} 0.75^7 \times 0.25^3$

8) a) mean $= 20 \times 0.4 = 8$; variance $= 20 \times 0.4 \times 0.6 = 4.8$

b) mean $= 40 \times 0.15 = 6$; variance $= 40 \times 0.15 \times 0.85 = 5.1$

c) mean $= 25 \times 0.45 = 11.25$;
variance $= 25 \times 0.45 \times 0.55 = 6.1875$

d) mean $= 50 \times 0.8 = 40$; variance $= 50 \times 0.8 \times 0.2 = 8$

e) mean $= 30 \times 0.7 = 21$; variance $= 30 \times 0.7 \times 0.3 = 6.3$

f) mean $= 45 \times 0.012 = 0.54$;
variance $= 45 \times 0.012 \times 0.988 = 0.53352$

Answers

Exam Questions

1 a) (i) Define a new random variable $Y \sim B(12, 0.4)$.
Then $P(X < 8) = P(Y > 4) = 1 - P(Y \leq 4)$
$= 1 - 0.4382$ ***[1 mark]*** $= 0.5618$ ***[1 mark]***

(ii) $P(X = 5) = P(Y = 7) = P(Y \leq 7) - P(Y \leq 6)$ ***[1 mark]***
$= 0.9427 - 0.8418 = 0.1009$ ***[1 mark]***

Or you could use the probability function for part (ii):

$$P(X = 5) = \binom{12}{5} \times 0.6^5 \times 0.4^7 = 0.1009$$

(iii) $P(3 < X \leq 7) = P(X$ is greater than 3 and less than or equal to 7$) = P(Y$ is less than 9 and greater than or equal to 5$) = P(5 \leq Y < 9) = P(5 \leq Y \leq 8)$ ***[1 mark]***
$= P(Y \leq 8) - P(Y \leq 4)$ ***[1 mark]*** $= 0.9847 - 0.4382$
$= 0.5465$ ***[1 mark]***

b) (i) $P(Y = 4) = \frac{11!}{4!7!} \times 0.8^4 \times 0.2^7$ ***[1 mark]***
$= 0.00173$ (to 3 sig. fig.) ***[1 mark]***

(ii) $E(Y) = 11 \times 0.8 = 8.8$ ***[1 mark]***

(iii) $Var(Y) = 11 \times 0.8 \times 0.2 = 1.76$ ***[1 mark]***

2 a) (i) Let X represent the number of apples that contain a maggot. Then $X \sim B(40, 0.15)$ ***[1 mark]***.
$P(X < 6) = P(X \leq 5) = 0.4325$ ***[1 mark]***

(ii) $P(X > 2) = 1 - P(X \leq 2)$ ***[1 mark]***
$= 1 - 0.0486 = 0.9514$ ***[1 mark]***

(iii) $P(X = 12) = P(X \leq 12) - P(X \leq 11)$ ***[1 mark]***
$= 0.9957 - 0.9880 = 0.0077$ ***[1 mark]***

Or you could use the probability function for part (iii):

$$P(X = 12) = \binom{40}{12} \times 0.15^{12} \times 0.85^{28} = 0.0077$$

b) The probability that a crate contains more than 2 apples with maggots is 0.9514 (from part a) (ii)).
So define a random variable Y, where Y is the number of crates that contain more than 2 apples with maggots.
Then $Y \sim B(3, 0.9514)$ ***[1 mark]***.
You need to find $P(Y = 2) + P(Y = 3)$. This is:

$$\binom{3}{2} \times 0.9514^2 \times (1 - 0.9514)$$
$$+ \binom{3}{3} \times 0.9514^3 \times (1 - 0.9514)^0 \text{ [1 mark]}$$
$$= 0.1320 + 0.8612 = 0.993 \text{ (to 3 d.p.) [1 mark]}$$

3 a) (i) The probability of Simon being able to solve each crossword needs to remain the same ***[1 mark]***, and all the outcomes need to be independent (i.e. Simon solving or not solving a puzzle one day should not affect whether he will be able to solve it on another day) ***[1 mark]***.

(ii) The total number of puzzles he solves (or the number he fails to solve) ***[1 mark]***.

b) $P(X = 4) = p^4 \times (1 - p)^{14} \times \frac{18!}{4!14!}$ ***[1 mark]***

$P(X = 5) = p^5 \times (1 - p)^{13} \times \frac{18!}{5!13!}$ ***[1 mark]***

So $p^4 \times (1 - p)^{14} \times \frac{18!}{4!14!} = p^5 \times (1 - p)^{13} \times \frac{18!}{5!13!}$ ***[1 mark]***

Dividing by p^4, $(1 - p)^{13}$ and $\frac{18!}{4!13!}$ gives:

$\frac{1 - p}{14} = \frac{p}{5}$ ***[1 mark]***, or $5 = 19p$.

This means $p = \frac{5}{19}$ ***[1 mark]***.

S1 Section 4 — Normal Distribution

Warm-up Questions

1) Use the Z-tables:

a) $P(Z < 0.84) = 0.79955$

b) $P(Z < 2.95) = 0.99841$

c) $P(Z > 0.68) = 1 - P(Z \leq 0.68) = 1 - 0.75175 = 0.24825$

d) $P(Z \geq 1.55) = 1 - P(Z < 1.55)$
$= 1 - P(Z \leq 1.55) = 1 - 0.93943 = 0.06057$

e) $P(Z < -2.10) = P(Z > 2.10) = 1 - P(Z \leq 2.10)$
$= 1 - 0.98214 = 0.01786$

f) $P(Z \leq -0.01) = P(Z \geq 0.01)$
$= 1 - P(Z < 0.01) = 1 - 0.50399 = 0.49601$

g) $P(Z > 0.10) = 1 - P(Z \leq 0.10) = 1 - 0.53983 = 0.46017$

h) $P(Z \leq 0.64) = 0.73891$

i) $P(Z > 0.23) = 1 - P(Z \leq 0.23) = 1 - 0.59095 = 0.40905$

j) $P(0.10 < Z \leq 0.50) = P(Z \leq 0.50) - P(Z \leq 0.10)$
$= 0.69146 - 0.53983 = 0.15163$

k) $P(-0.62 \leq Z < 1.10) = P(Z < 1.10) - P(Z < -0.62)$
$= P(Z < 1.10) - P(Z > 0.62) = P(Z < 1.10) - (1 - P(Z \leq 0.62))$
$= 0.86433 - (1 - 0.73237) = 0.59670$

l) $P(-0.99 < Z \leq -0.74) = P(Z \leq -0.74) - P(Z \leq -0.99)$
$= P(Z \geq 0.74) - P(Z \geq 0.99)$
$= (1 - P(Z < 0.74)) - (1 - P(Z < 0.99))$
$= (1 - 0.77035) - (1 - 0.83891) = 0.06856$

I know... these all get a bit fiddly. As always — take it nice and slow, and double-check each step as you do it. And remember that, since a normal distribution is continuous, $P(Z < z) = P(Z \leq z)$ and $P(Z > z) = P(Z \geq z)$ for any value of z.

2) a) If $P(Z < z) = 0.91309$, then from the Z-table, $z = 1.36$.

b) If $P(Z < z) = 0.58706$, then from the Z-table, $z = 0.22$.

c) If $P(Z > z) = 0.03593$, then $P(Z \leq z) = 0.96407$.
From the Z-table, $z = 1.80$.

d) If $P(Z < z) = 0.99$, then from the percentage-points table, $z = 2.3263$.

e) If $P(Z \leq z) = 0.40129$, then z must be negative (and so won't be in the Z-table).
But this means $P(Z < -z) = 1 - 0.40129 = 0.59871$.
Using the Z-table, $-z = 0.25$, so $z = -0.25$.

It's getting a bit trickier here, with all the z and −z business. If you need to draw a graph here to make it a bit clearer what's going on, then draw one.

Answers

f) If $P(Z \geq z) = 0.995$, then $P(Z \leq -z) = 0.995$ as well.
From the percentage points table, $-z = 2.5758$.
So $z = -2.5758$.

When you've answered a question like this, always ask yourself whether your answer looks 'about right'. Here, you need a number that Z is very very likely to be greater than... so your answer is going to be negative, and it's going to be pretty big.
So z = −2.5758 looks about right.

3) a) $P(X < 55) = P\left(Z < \frac{55-50}{\sqrt{16}}\right) = P(Z < 1.25) = 0.89435$

b) $P(X < 42) = P\left(Z < \frac{42-50}{\sqrt{16}}\right) = P(Z < -2)$
$= P(Z > 2) = 1 - P(Z \leq 2) = 1 - 0.97725 = 0.02275$

c) $P(X > 56) = P\left(Z > \frac{56-50}{\sqrt{16}}\right) = P(Z > 1.5)$
$= 1 - P(Z \leq 1.5) = 1 - 0.93319 = 0.06681$

d) $P(47 < X < 57) = P(X < 57) - P(X \leq 47)$
$= P(Z < 1.75) - P(Z \leq -0.75)$
$= 0.95994 - P(Z \geq 0.75)$
$= 0.95994 - (1 - P(Z < 0.75))$
$= 0.95994 - (1 - 0.77337) = 0.73331$

4) a) $P(X < 0) = P\left(Z < \frac{0-5}{7}\right) = P(Z < -0.71) = P(Z > 0.71)$
$= 1 - P(Z \leq 0.71) = 1 - 0.76115 = 0.23885$

b) $P(X < 1) = P\left(Z < \frac{1-5}{7}\right) = P(Z < -0.57) = P(Z > 0.57)$
$= 1 - P(Z \leq 0.57) = 1 - 0.71566 = 0.28434$

c) $P(X > 7) = P\left(Z > \frac{7-5}{7}\right) = P(Z > 0.29)$
$= 1 - P(Z \leq 0.29) = 1 - 0.61409 = 0.38591$

d) $P(2 < X < 4) = P(X < 4) - P(X \leq 2)$
$= P(Z < -0.14) - P(Z \leq -0.43)$
$= P(Z > 0.14) - P(Z \geq 0.43)$
$= (1 - P(Z \leq 0.14)) - (1 - P(Z < 0.43))$
$= (1 - 0.55567) - (1 - 0.66640)) = 0.11073$

5) $P(X < 8) = 0.89251$ means $P\left(Z < \frac{8-\mu}{\sqrt{10}}\right) = 0.89251$.

From tables, $\frac{8-\mu}{\sqrt{10}} = 1.24$.

So $\mu = 8 - 1.24 \times \sqrt{10} = 4.08$ (to 3 sig. fig.).

6) $P(X > 221) = 0.30854$ means $P\left(Z > \frac{221-\mu}{8}\right) = 0.30854$,

or $P\left(Z \leq \frac{221-\mu}{8}\right) = 1 - 0.30854 = 0.69146$.

From tables, $\frac{221-\mu}{8} = 0.5$.

So $\mu = 221 - 8 \times 0.5 = 217$

7) $P(X < 13) = 0.6$ means $P\left(Z < \frac{13-11}{\sigma}\right) = 0.6$.

From the percentage points table, $\frac{13-11}{\sigma} = 0.2533$.

So $\sigma = \frac{2}{0.2533} = 7.90$ (to 3 sig.fig.).

8) $P(X \leq 110) = 0.96784$ means
$P\left(Z < \frac{110-108}{\sigma}\right) = 0.96784$, or $P\left(Z < \frac{2}{\sigma}\right) = 0.96784$.

From tables, $\frac{2}{\sigma} = 1.85$.

So $\sigma = \frac{2}{1.85} = 1.08$ (to 3 sig.fig.).

9) $P(X < 15.2) = 0.97831$ means $P\left(Z < \frac{15.2-\mu}{\sigma}\right) = 0.97831$.

From tables, $\frac{15.2-\mu}{\sigma} = 2.02$, or $\underline{2.02\sigma + \mu = 15.2}$.

$P(X > 14.8) = 0.10565$ means $P\left(Z > \frac{14.8-\mu}{\sigma}\right) = 0.10565$,

or $P\left(Z \leq \frac{14.8-\mu}{\sigma}\right) = 1 - 0.10565 = 0.89435$.

From tables, $\frac{14.8-\mu}{\sigma} = 1.25$, or $\underline{1.25\sigma + \mu = 14.8}$.

Solving the underlined simultaneous equations

gives $\sigma = 0.52$ and $\mu = 14.15$ (to 2 d.p.).

10) If X represents the mass of an item, then $X \sim N(55, 4.4^2)$.

a) $P(X < 55) = P\left(Z < \frac{55-55}{4.4}\right) = P(Z < 0) = 0.5$.

b) $P(X < 50) = P\left(Z < \frac{50-55}{4.4}\right) = P(Z < -1.14)$
$= P(Z > 1.14) = 1 - P(Z \leq 1.14)$
$= 1 - 0.87286 = 0.12714$

c) $P(X > 60) = P\left(Z > \frac{60-55}{4.4}\right) = P(Z > 1.14)$
$= 1 - P(Z \leq 1.14) = 1 - 0.87286 = 0.12714$

You could have done c) without all that working using the fact that a normal distribution is symmetrical about its mean. Since here the mean is 55, P(X > 60) = P(X < 50), and this is what you worked out in part b).

11) If X represents the mass of an egg, then $X \sim N(1.4, 0.3^2)$.

a) $P(X < 1) = P\left(Z < \frac{1-1.4}{0.3}\right) = P(Z < -1.33)$
$= P(Z > 1.33) = 1 - P(Z \leq 1.33)$
$= 1 - 0.90824 = 0.09176$

b) $P(X > 1.5) = P\left(Z > \frac{1.5-1.4}{0.3}\right) = P(Z > 0.33)$
$= 1 - P(Z \leq 0.33)$
$= 1 - 0.62930 = 0.37070$

c) $P(1.3 < X < 1.6) = P(X < 1.6) - P(X < 1.3)$
$= P\left(Z < \frac{1.6-1.4}{0.3}\right) - P\left(Z < \frac{1.3-1.4}{0.3}\right)$
$= P(Z < 0.67) - P(Z < -0.33)$
$= 0.74857 - (1 - P(Z < 0.33))$
$= 0.74857 - (1 - 0.62930) = 0.37787$

Answers

12) a) Here X ~ N(80, 15). You need to use your percentage points table for these.
If $P(X > a) = 0.01$, then $P\left(Z > \frac{a - 80}{\sqrt{15}}\right) = 0.01$,
i.e. $P\left(Z \le \frac{a - 80}{\sqrt{15}}\right) = 0.99$.
So $\frac{a - 80}{\sqrt{15}} = 2.3263$ (using the table).
Rearrange this to get $a = 80 + 2.3263 \times \sqrt{15} = 89.01$

b) $|X - 80| < b$ means that X is 'within b' of 80,
i.e. $80 - b < X < 80 + b$.
Since 80 is the mean of X, and since a normal distribution is symmetrical,
$P(80 - b < X < 80 + b) = 0.8$ means that
$P(80 < X < 80 + b) = 0.4$
i.e. $P\left(\frac{80 - 80}{\sqrt{15}} < Z < \frac{80 + b - 80}{\sqrt{15}}\right) = 0.4$
i.e. $P\left(0 < Z < \frac{b}{\sqrt{15}}\right) = 0.4$
This means that $P\left(Z < \frac{b}{\sqrt{15}}\right) - P(Z \le 0) = 0.4$
i.e. $P\left(Z < \frac{b}{\sqrt{15}}\right) - 0.5 = 0.4$, or $P\left(Z < \frac{b}{\sqrt{15}}\right) = 0.9$
Use your percentage points table to find that
$\frac{b}{\sqrt{15}} = 1.2816$, or $b = 1.2816 \times \sqrt{15} = 4.964$ (to 3 d.p.)

An elephant never forgets... that a normal distribution is symmetrical. It's often very useful for figuring out these sorts of questions, and elephants love calculating probability distributions more than they love peanuts.
Sketches are also very useful (and elephants love those too). For example, the statement $P(|X - 80| < b) = 0.8$ can be summarised on a diagram like this:

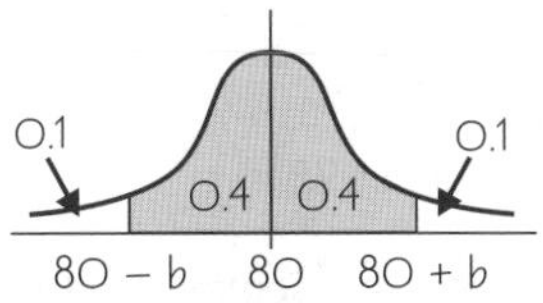

From here, it's just one small elephant hop to the equation $P(X < 80 + b) = P\left(Z < \frac{b}{\sqrt{15}}\right) = 0.9$, and with that, you're nearly at the answer.

Exam Questions

1 a) Let X represent the exam marks. Then $X \sim N(50, 30^2)$.
$P(X \ge 41) = P\left(Z \ge \frac{41 - 50}{30}\right)$
$= P\left(Z \ge -\frac{9}{30}\right) = P(Z \ge -0.3)$ ***[1 mark]***
$= P(Z \le 0.3) = 0.61791$ ***[1 mark]***
So $0.61791 \times 1000 = 618$ is the expected number who passed the exam ***[1 mark]***.

b) If a is the mark needed for a distinction, then:
$P(X \ge a) = 0.1$ means $P\left(Z \le \frac{a - 50}{30}\right) = 0.9$ ***[1 mark]***.
From the percentage points table, $\frac{a - 50}{30} = 1.2816$ ***[1 mark]***,
or $a = 88.4$ ***[1 mark]***.

2 Assume that the lives of the batteries are distributed as: $N(\mu, \sigma^2)$. Then $P(X < 20) = 0.4$ and $P(X < 30) = 0.8$.
[1 mark]
Transform these 2 equations to get:
$P\left[Z < \frac{20 - \mu}{\sigma}\right] = 0.4$ and $P\left[Z < \frac{30 - \mu}{\sigma}\right] = 0.8$ ***[1 mark]***
Now you need to look up 0.6 and 0.8 in your percentage points table to get:
$\frac{20 - \mu}{\sigma} = -0.2533$ ***[1 mark]*** and $\frac{30 - \mu}{\sigma} = 0.8416$ ***[1 mark]***
You need to look up $p = 0.6$ since $p = 0.4$ isn't in the table. Then make the value from the table negative.
Now rewrite these as:
$20 - \mu = -0.2533\sigma$ and $30 - \mu = 0.8416\sigma$. ***[1 mark]***
Subtract these two equations to get:
$10 = (0.8416 + 0.2533)\sigma$
i.e. $\sigma = \frac{10}{0.8416 + 0.2533} = 9.1333$ ***[1 mark]***
Now use this value of σ in one of the equations above:
$\mu = 20 + 0.2533 \times 9.1333 = 22.31$ ***[1 mark]***
So $X \sim N(22.31, 9.13^2)$ i.e. $X \sim N(22.31, 83.4)$
This might feel a bit repetitive but if you're comfortable doing probability distribution questions in lots of ever-so-very-slightly different ways then the exam'll be a doddle.

3 a) $X \sim N(120, 25^2)$.
$P(X > 145) = P\left(Z > \frac{145 - 120}{25}\right) = P(Z > 1)$ ***[1 mark]***
$= 1 - P(Z \le 1)$ ***[1 mark]***
$= 1 - 0.84134 = 0.15866$ ***[1 mark]***.

b) $P(120 < X < j) = 0.46407$ means
$P(X < j) - P(X \le 120) = 0.46407$ ***[1 mark]***.
$P(X < j) - P(X \le 120)$
$= P\left(Z < \frac{j - 120}{25}\right) - P\left(Z \le \frac{120 - 120}{25}\right)$ ***[1 mark]***
$= P\left(Z < \frac{j - 120}{25}\right) - 0.5 = 0.46407$,
or $P\left(Z < \frac{j - 120}{25}\right) = 0.96407$ ***[1 mark]***
From tables, $\frac{j - 120}{25} = 1.80$,
or $j = 1.8 \times 25 + 120 = 165$ ***[1 mark]***

Answers

4 a) For a normal distribution, mean = median, so median = 12 inches ***[1 mark]***.

b) Let X represent the base diameters. Then $P(X > 13) = 0.05$ ***[1 mark]***. Since $X \sim N(12, \sigma^2)$, this means:

$P(X < 13) = P\left(Z < \frac{13-12}{\sigma}\right) = 0.95$ ***[1 mark]***.

From the percentage points table, $\frac{1}{\sigma} = 1.6449$ ***[1 mark]***.

This means $\sigma = \frac{1}{1.6449} = 0.61$ (to 2 sig. fig.) ***[1 mark]***.

c) $P(X < 10.8) = P\left(Z < \frac{10.8-12}{0.61}\right) = P(Z < -1.97)$ ***[1 mark]***
$= 1 - P(Z \leq 1.97) = 1 - 0.97558 = 0.02442$ ***[1 mark]***

So you would expect $0.02442 \times 100 \approx 2$ pizza bases to be discarded ***[1 mark]***.

d) P(at least 1 base too small) = 1 – P(no bases too small).
P(base not too small) = 1 – 0.02442 = 0.97558.
P(no bases too small) = $0.97558^3 = 0.9285$ ***[1 mark]***
P(at least 1 base too small) = 1 – 0.9285 ***[1 mark]***
= 0.0715 ***[1 mark]***.

5 a) Let X represent the volume of compost in a bag. Then $X \sim N(50, 0.4^2)$.

$P(X < 49) = P\left(Z < \frac{49-50}{0.4}\right) = P(Z < -2.50)$ ***[1 mark]***
$= 1 - P(Z \leq 2.50)$ ***[1 mark]***
$= 1 - 0.99379 = 0.00621$ ***[1 mark]***

b) $P(X > 50.5) = P\left(Z > \frac{50.5-50}{0.4}\right) = P(Z > 1.25)$ ***[1 mark]***
$= 1 - P(Z \leq 1.25)$ ***[1 mark]***
$= 1 - 0.89435 = 0.10565$ ***[1 mark]***

So in 1000 bags, 0.10565×1000 ***[1 mark]***
≈ 106 bags ***[1 mark]*** (approximately) would be expected to contain more than 50.5 litres of compost.

c) $P(Y < 74) = 0.10$ means $P\left(Z < \frac{74-75}{\sigma}\right) = 0.1$.

To use the percentage points table, rewrite this as
$P\left(Z < \frac{75-74}{\sigma}\right) = 0.9$ ***[1 mark]***,

which gives $\frac{1}{\sigma} = 1.2816$ ***[1 mark]***.

So, $\sigma = \frac{1}{1.2816} = 0.78$ litres ***[1 mark]***.

S1 Section 5 — Estimation

Warm-up Questions

1) a) All the members of the tennis club.
b) By using a random sample.

2) A simple random sample is one where every member of the population has an equal chance of being in the sample and each selection is independent.

3) a) Yes
b) No — it contains unknown parameter σ.
c) No — it contains unknown parameter μ.
d) Yes

There's no excuse for getting these ones wrong. You've just got to look for any unknown parameters — if you find one, it's not a statistic.

4) The sample mean is an unbiased estimate of the population mean — this is: $\frac{\sum x}{n} = \frac{80.5}{10} = 8.05$

An unbiased estimate of the population variance is:

$$\frac{n}{n-1}\left[\frac{\sum x^2}{n} - \left(\frac{\sum x}{n}\right)^2\right] = \frac{10}{9}\left[\frac{653.13}{10} - \left(\frac{80.5}{10}\right)^2\right] = 0.567 \text{ (to 3 d.p.)}.$$

5) a) Standard error of sample mean $= \frac{\sigma}{\sqrt{n}}$
$= \frac{5}{\sqrt{15}} = 1.29$ (to 3 sig.fig.)

b) Because the population follows a normal distribution, the sampling distribution of the sample mean is also normal.
$\overline{X} \sim N\left(7.2, \frac{25}{15}\right) = N\left(7.2, \frac{5}{3}\right)$

c) $P(\overline{X} < 7.5) = P\left(Z < \frac{7.5-7.2}{\sqrt{5/3}}\right)$
$= P(Z < 0.23) = 0.59095$

6) a) Because the sample size is large, the Central Limit Theorem tells you that the sampling distribution of the sample mean will be approximately normal.
$\overline{X} \sim N\left(18, \frac{4^2}{80}\right) = N(18, 0.2)$ (approximately)

b) $P(\overline{X} > 19) \approx P\left(Z > \frac{19-18}{\sqrt{0.2}}\right) = P(Z > 2.24)$
$= 1 - P(Z \leq 2.24)$
$= 1 - 0.98745$
$= 0.01255$

7) This is a normal distribution with a known standard deviation, so a 99% confidence interval is given by:
$\left(\overline{X} - z\frac{\sigma}{\sqrt{n}}, \overline{X} + z\frac{\sigma}{\sqrt{n}}\right)$, where $z = 2.5758$.
Since the sample mean is 18.2, the confidence interval is:
$\left(18.2 - 2.5758 \times \frac{0.4}{\sqrt{25}}, 18.2 + 2.5758 \times \frac{0.4}{\sqrt{25}}\right)$
$= (17.99, 18.41)$ (to 2 d.p.).

Answers

8) Here, the population itself may not be normally distributed, but the Central Limit Theorem tells you that the sampling distribution of the sample mean will still be approximately normal — meaning a 95% confidence interval is given by:

$\left(\overline{X} - z\frac{\sigma}{\sqrt{n}}, \overline{X} + z\frac{\sigma}{\sqrt{n}}\right)$, where $z = 1.96$.

Since the sample mean is 33.8, the confidence interval is:

$$\left(33.8 - 1.96 \times \frac{3}{\sqrt{120}}, 33.8 + 1.96 \times \frac{3}{\sqrt{120}}\right)$$
$$= (33.26, 34.34) \text{ (to 2 d.p.)}.$$

9) You're going to need an unbiased estimate of the population standard deviation — but first work out an unbiased estimate of the population variance.

$$s^2 = \frac{n}{n-1}\left[\frac{\sum x^2}{n} - \left(\frac{\sum x}{n}\right)^2\right] = \frac{100}{99}\left[\frac{122.1}{100} - \left(\frac{104}{100}\right)^2\right]$$
$$= 0.140808...$$

So an unbiased estimate of the population standard deviation is $s = \sqrt{0.140808...}$.

Now you can find an unbiased estimate of the sample mean's standard error — this is:

$$\frac{s}{\sqrt{n}} = \frac{\sqrt{0.140808...}}{\sqrt{100}} = 0.0375 \text{ (to 3 sig. fig.)}.$$

10) Sample mean $= \frac{\sum x}{n} = \frac{24.2}{80} = 0.3025$

Unbiased estimate of population variance is:

$$s^2 = \frac{n}{n-1}\left[\frac{\sum x^2}{n} - \left(\frac{\sum x}{n}\right)^2\right]$$
$$= \frac{80}{79}\left[\frac{41.3}{80} - \left(\frac{24.2}{80}\right)^2\right] = 0.43012...$$

And so an unbiased estimate of the population standard deviation is the square root of this, which is $s = 0.656$ (to 3 d.p.).

Because the sample size is large, you can assume that the sampling distribution of $\overline{X}$ is approximately normal and your estimate of the population standard deviation is reasonably good, which means that a 95% confidence interval for the population mean is given by:

$\left(\overline{x} - z\frac{s}{\sqrt{n}}, \overline{x} + z\frac{s}{\sqrt{n}}\right)$, where $z = 1.96$

$$= \left(0.3025 - 1.96 \times \frac{0.656}{\sqrt{80}}, 0.3025 + 1.96 \times \frac{0.656}{\sqrt{80}}\right)$$
$$= (0.16, 0.45) \text{ (to 2 d.p.)}.$$

Exam Questions

1 a) You need to use mid-interval values. So add some extra rows to your table.

Height, h (cm)	0-40	40-80	80-100	100-140	140-160
Frequency, f	5	8	11	19	7
Mid-interval value, x	20	60	90	120	150
fx	100	480	990	2280	1050
x^2	400	3600	8100	14400	22500
fx^2	2000	28 800	89 100	273 600	157 500

$\sum f = 50, \sum fx = 4900, \sum fx^2 = 551000$.

This gives an (unbiased) estimate of the mean of:

$\frac{\sum fx}{\sum f} = \frac{4900}{50} = 98$ cm ***[1 mark]***

b) For an unbiased estimate of the standard deviation, first find an unbiased estimate of the variance. This is:

$\frac{n}{n-1}\left[\frac{\sum fx^2}{\sum f} - \left(\frac{\sum fx}{\sum f}\right)^2\right] = \frac{50}{49}\left[\frac{551000}{50} - \left(\frac{4900}{50}\right)^2\right]$ ***[1 mark]***

$= 1444.9$ ***[1 mark]***

And so an unbiased estimate of the population standard deviation is: $\sqrt{1444.9} = 38.0$ cm (to 1 d.p.) ***[1 mark]***.

2 a) The mean weight of the frogs in the sample is:

$\frac{3840}{30} = 128$ g ***[1 mark]***

b) The standard error of the sample mean weight is:

$\frac{\sigma}{\sqrt{n}} = \frac{8.5}{\sqrt{30}}$ ***[1 mark]*** $= 1.55$ g (to 2 d.p.) ***[1 mark]***

c) A 95% confidence interval is given by:

$\left(\overline{x} - z\frac{\sigma}{\sqrt{n}}, \overline{x} + z\frac{\sigma}{\sqrt{n}}\right)$ ***[1 mark]***

$= \left(128 - 1.96 \times \frac{8.5}{\sqrt{30}}, 128 + 1.96 \times \frac{8.5}{\sqrt{30}}\right)$ ***[1 mark]***

$= (125.0, 131.0)$ (to 1d.p.) ***[1 mark]***.

d) The weights of the frogs are normally distributed, so if the random variable Y represents the weights of the individual frogs, then $Y \sim N(128, 8.5^2)$ *(using the estimated mean)* ***[1 mark]***. If you transform this to the standard normal distribution (by subtracting the mean and dividing by the standard deviation), then you need to find z with:

$P\left(-z \le \frac{Y - 128}{8.5} \le z\right) = 0.99$,

i.e. $P(-z \le Z \le z) = 0.99$, or $P(Z \le z) = 0.995$

Using the normal percentage-points table, this gives $z = 2.5758$ ***[1 mark]***.

Since $-2.5758 \le \frac{Y - 128}{8.5} \le 2.5758$,

the range for Y you need is:

$(128 - 2.5758 \times 8.5, 128 + 2.5758 \times 8.5)$

$= (106.1, 149.9)$ ***[1 mark]***

3 a) $s^2 = \frac{n}{n-1}\left[\frac{\sum x^2}{n} - \left(\frac{\sum x}{n}\right)^2\right]$

$= \frac{60}{59}\left[\frac{620}{60} - \left(\frac{184}{60}\right)^2\right]$ ***[1 mark]***

$= 0.94463...$ ***[1 mark]***

So an unbiased estimate of the sample mean's standard error is:

$\frac{s}{\sqrt{n}} = \frac{\sqrt{0.94463...}}{\sqrt{60}}$ ***[1 mark]*** $= 0.125$ (to 3 sig. fig.) ***[1 mark]***

Answers

b) Since the sample size is large, the estimate of the standard error should be reasonably good ***[1 mark]***, and the sampling distribution of the sample mean should be approximately normal ***[1 mark]***. This means that a 98% confidence interval for the population mean is given by:

$\left(\overline{x} - z\frac{s}{\sqrt{n}}, \overline{x} + z\frac{s}{\sqrt{n}}\right)$, where $z = 2.3263$ ***[1 mark]***

$= \left(\frac{184}{60} - 2.3263 \times 0.125, \frac{184}{60} + 2.3263 \times 0.125\right)$ ***[1 mark]***

$= (2.78, 3.36)$ (to 2 d.p.). ***[1 mark]***

c) Because part of this confidence interval is below 3 ***[1 mark]***, these results do not provide evidence that the average distance covered in 10 minutes is above 3 metres ***[1 mark]*** (although they do not provide evidence that the manager is wrong either).

4 a) $s^2 = \frac{\sum(x - \overline{x})^2}{n - 1} = \frac{1382}{39} = 35.4$ (to 3 sig. fig.) ***[1 mark]***

b) The population is modelled by a normal distribution and so the sampling distribution of the sample mean will also be normal ***[1 mark]***, so a 99% confidence interval for the mean is given by:

$\left(\overline{x} - z\frac{s}{\sqrt{n}}, \overline{x} + z\frac{s}{\sqrt{n}}\right)$, where $z = 2.5758$ ***[1 mark]***

$= \left(\frac{1676}{40} - 2.5758 \times \frac{\sqrt{1382/39}}{\sqrt{40}}, \frac{1676}{40} + 2.5758 \times \frac{\sqrt{1382/39}}{\sqrt{40}}\right)$ ***[1 mark]***

$= (39.5, 44.3)$ (to 3 sig.fig.) ***[1 mark]***

c) Since this confidence interval does not include 46 minutes, these results do not support the claim of Ravi's manager ***[1 mark]***. Since 42 minutes is within the confidence interval, these results provide no reason to doubt Ravi's claim ***[1 mark]***.

S1 Section 6 — Correlation and Regression

Warm-up Questions

1) a)

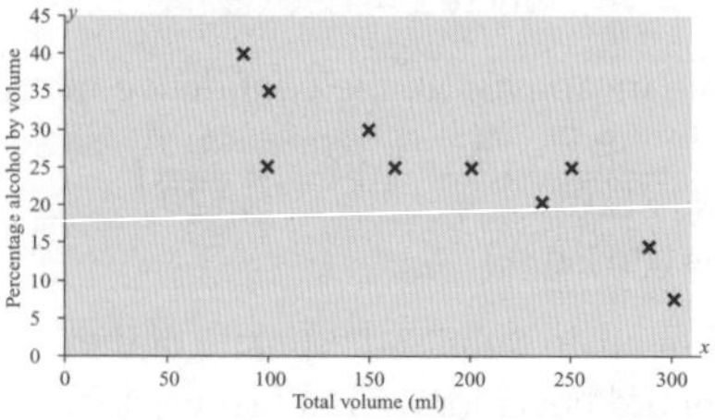

b) First you need to find these values:

$\sum x = 1880$, $\sum y = 247$, $\sum x^2 = 410400$, $\sum y^2 = 6899$ and $\sum xy = 40600$.

Then put these values into the PMCC formula:

$$\frac{40600 - \frac{[1880][247]}{10}}{\sqrt{\left(410400 - \frac{[1880]^2}{10}\right)\left(6899 - \frac{[247]^2}{10}\right)}}$$

$$= \frac{40600 - 46436}{\sqrt{(410400 - 353440)(6899 - 6100.9)}}$$

$$= \frac{-5836}{\sqrt{56960 \times 798.1}} = \frac{-5836}{6742.3865} = -0.866 \text{ (to 3 sig.fig.)}$$

c) The PMCC tells you that there is a strong negative correlation between drink volume and alcohol concentration — cocktails with smaller volumes tend to have higher concentrations of alcohol.

Don't panic about that nasty ol' PMCC equation. You need to know how to USE it, but they give you the formula in the exam, so you don't need to REMEMBER it. Hurrah.

2) a) **Independent**: the annual number of sunny days
Dependent: the annual number of volleyball-related injuries

b) **Independent**: the annual number of rainy days
Dependent: the annual number of Monopoly-related injuries

c) **Independent**: a person's disposable income
Dependent: a person's spending on luxuries

d) **Independent**: the number of cups of tea drunk per day
Dependent: the number of trips to the loo per day

e) **Independent**: the number of festival tickets sold
Dependent: the number of pairs of Wellington boots bought

3) a) (i) $S_{rr} = 26816.78 - \frac{517.4^2}{10} = 46.504$

(ii) $S_{rw} = 57045.5 - \frac{517.4 \times 1099}{10} = 183.24$

b) $b = \frac{S_{rw}}{S_{rr}} = \frac{183.24}{46.504} = 3.94$

c) $a = \overline{w} - b\overline{r}$, where $\overline{w} = \frac{\sum w}{10} = 109.9$

and $\overline{r} = \frac{\sum r}{10} = 51.74$

So $a = 109.9 - 3.94 \times 51.74 = -94.0$

Answers

d) The equation of the regression line is: $w = 3.94r - 94.0$

e) When $r = 60$, the regression line gives an estimate for w of:
$w = 3.94 \times 60 - 94.0 = 142.4$ g

f) This estimate might not be very reliable because it uses an r-value from outside the range of the original data.
It is extrapolation.

You'll be given the equations for finding a regression line — but you still need to know how to use them, otherwise the formula booklet will just be a blur of incomprehensible squiggles. Oh, and you need to practise USING them of course...

4) Substitute expressions for w and r into the regression line's equation from Question 3: $8Q = 3.94(P + 5) - 94.0$
Now rearrange to get an equation of the form $Q = a + bP$:
$8Q = 3.94P + (3.94 \times 5 - 94.0)$
i.e. $Q = 0.493P - 9.29$.

Exam Questions

1) a)

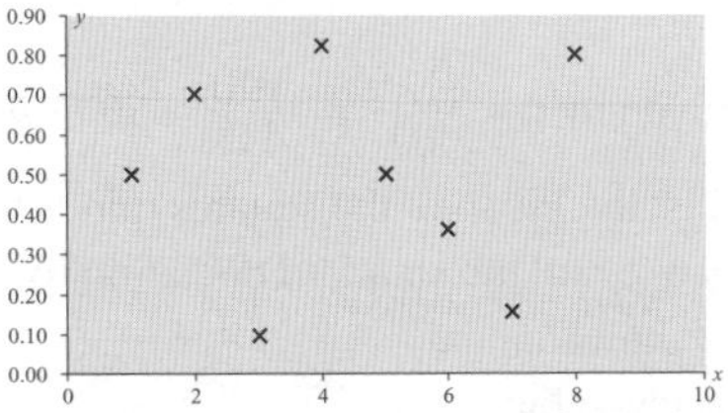

[2 marks for all points plotted correctly, or 1 mark if at least 3 points are plotted correctly.]

Aren't scatter diagrams pretty... Just make sure you're not so distracted by their artistic elegance that you forget to be accurate and lose easy marks.

b) You need to work out these sums:

$\sum x = 36, \sum y = 3.94,$
$\sum x^2 = 204, \sum y^2 = 2.4676, \sum xy = 17.66$

Then:

$$S_{xx} = \sum x^2 - \frac{(\sum x)^2}{n} = 204 - \frac{36^2}{8} = 42$$

$$S_{yy} = \sum y^2 - \frac{(\sum y)^2}{n} = 2.4676 - \frac{3.94^2}{8} = 0.52715$$

$$S_{xy} = \sum xy - \frac{(\sum x)(\sum y)}{n} = 17.66 - \frac{36 \times 3.94}{8} = -0.07$$

[3 marks available — 1 for each correct term]

This means:

$$r = \frac{S_{xy}}{\sqrt{S_{xx}S_{yy}}} = \frac{-0.07}{\sqrt{42 \times 0.52715}} = -0.015 \text{ (to 3 d.p.)}.$$

[1 mark]

c) This very small value for the correlation coefficient tells you that there appears to be only a very weak linear relationship (a negative correlation) between the two variables (or perhaps no linear relationship at all) **[1 mark].**

2 a)

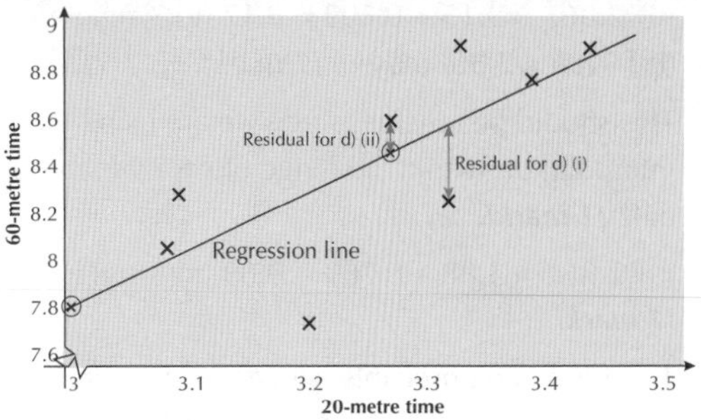

[2 marks for all points plotted correctly, or 1 mark if at least 3 points are plotted correctly.]

b) It's best to make a table like this one, first:

									Totals
20-metre time, x	3.39	3.2	3.09	3.32	3.33	3.27	3.44	3.08	26.12
60-metre time, y	8.78	7.73	8.28	8.25	8.91	8.59	8.9	8.05	67.49
x^2	11.4921	10.24	9.5481	11.0224	11.0889	10.6929	11.8336	9.4864	85.4044
xy	29.7642	24.736	25.5852	27.39	29.6703	28.0893	30.616	24.794	220.645

[2 marks for at least three correct sums, or 1 mark if one total found correctly.]

Then: $S_{xy} = 220.645 - \frac{26.12 \times 67.49}{8} = 0.29015$
[1 mark]

$S_{xx} = 85.4044 - \frac{26.12^2}{8} = 0.1226$
[1 mark]

Then the gradient b is given by:

$$b = \frac{S_{xy}}{S_{xx}} = \frac{0.29015}{0.1226} = 2.3666$$

[1 mark]

And the intercept a is given by:

$$a = \bar{y} - b\bar{x} = \frac{\sum y}{n} - b\frac{\sum x}{n}$$
$$= \frac{67.49}{8} - 2.3666 \times \frac{26.12}{8} = 0.709$$

[1 mark]

So the regression line has equation: $y = 2.367x + 0.709$
[1 mark]

To plot the line, find two points that the line passes through. A regression line always passes through $(\bar{x}, \bar{y})$, which here is (3.27, 8.44). Then put $x = 3$ (say) to find that the line also passes through (3, 7.81).
Now plot these points (in circles) on your scatter diagram, and draw the regression line through them
[1 mark for plotting the line correctly].

Hmm, lots of fiddly things to calculate there. Remember, you get marks for method as well as correct answers, so take it step by step and show all your workings. And don't fall into the old trap of using one of your data points to plot the regression line, since there's no guarantee that any of your data points will satisfy the regression equation (and so be on the regression line).

Answers

c) (i) $y = 2.367 \times 3.15 + 0.709 = 8.17$ (to 3 sig. fig.), (8.16 if $b = 2.3666$ used) ***[1 mark]***

This should be reliable, since we are using interpolation within the range of x for which we have data ***[1 mark].***

(ii) $y = 2.367 \times 3.88 + 0.709 = 9.89$ (to 3 sig. fig.) ***[1 mark]***

This could be unreliable, since we are extrapolating beyond the range of the data ***[1 mark].***

d) (i) residual = $8.25 - (2.367 \times 3.32 + 0.709)$ $= -0.317$ (3 sig. fig.), (-0.316 if $b = 2.3666$ used)

[1 mark for calculation, 1 mark for plotting residual correctly (see diagram)]

(ii) residual = $8.59 - (2.367 \times 3.27 + 0.709)$ $= 0.141$ (3 sig. fig.), (0.142 if $b = 2.3666$ used)

[1 mark for calculation, 1 mark for plotting residual correctly (see diagram)]

3 a) Put the values into the correct PMCC formula:

$$\text{PMCC} = \frac{S_{xy}}{\sqrt{S_{xx}S_{yy}}} = \frac{12666}{\sqrt{310880 \times 788.95}} = \frac{12666}{15661.059} = 0.809$$

[1 mark for correctly substituting the values into the PMCC formula, and 1 mark for the correct final answer.]

b) There is a strong positive correlation between the miles cycled in the morning and calories consumed for lunch. Generally, the further they have cycled, the more they eat ***[1 mark].***

c) 0.809 ***[1 mark]***

Remember — the PMCC won't be affected if you multiply all the variables by a constant, so changing the data from miles to km doesn't change it.

4 a) (i) At $x = 12.5$, $y = 211.599 + (9.602 \times 12.5) = 331.624$
(ii) At $x = 14.7$, $y = 211.599 + (9.602 \times 14.7) = 352.748$
[1 mark for each value of y correctly calculated]

b) Using the equation
'Residual = Observed y-value – Estimated y-value':
At $x = 12.5$: Residual = 332.5 – 331.624 = 0.876 ***[1 mark]***
At $x = 14.7$: Residual = 352.1 – 352.748 = –0.648 ***[1 mark]***

And that's the end of that — Section 6 done and dusted.

S1 — Practice Exam One

1) a) The events are mutually exclusive, so they can't both happen. Hence $P(A \cap B) = 0$ ***[1 mark].***

b) For mutually exclusive events, $P(A \cup B) = P(A) + P(B)$ ***[1 mark].*** So $P(A \cup B) = 0.3 + 0.4 = 0.7$ ***[1 mark].***

c) The probability that neither event happens is equal to $1 - P(A \cup B)$ ***[1 mark]*** $= 1 - 0.7 = 0.3$ ***[1 mark].***

d) $P(A \mid B)$ is the probability of A, given that B has already happened. But since A and B are mutually exclusive, they can't both happen — so $P(A \mid B)$ must equal zero.
You can use the formula for conditional probability to get the same answer:

$$P(A \mid B) = \frac{P(A \cap B)}{P(B)}.$$

[1 mark for the correct answer, and 1 mark for a reasonable explanation.]

There'll be some probability formulas in your exam formula booklet — but that's no excuse for not learning all this stuff properly.

2) a) The mean (μ) is given by $\mu = \frac{\sum x}{10} = \frac{500}{10} = 50$ ***[1 mark].***

The variance (σ^2) is 'the mean of the squares minus the square of the mean'. And the standard deviation (σ) is just the square root of the variance.
So the variance is given by

$$\sigma^2 = \frac{\sum x^2}{10} - 50^2 = \frac{25\,622}{10} - 2500 = 62.2 \quad \textbf{[1 mark]}$$

So $\sigma = \sqrt{62.2} = 7.89$ (to 3 sig. fig.) ***[1 mark].***

b) (i) The mean will be unchanged ***[1 mark]***, because the new value is equal to the original mean ***[1 mark].***

(ii) The standard deviation will decrease ***[1 mark].*** This is because the standard deviation measures the deviation of values from the mean. So by adding a new value that's equal to the mean, you're not adding to the total deviation from the mean, but as you have an extra reading, you now have to divide by 11 (not 10) when you work out the variance ***[1 mark].***

Understanding what the standard deviation actually is can help you get your head round questions like this.

3) a) A normal distribution is symmetrical, so the median (i.e. the value that 50% of values fall below) is the same as the mean ***[1 mark].*** So the median = 93 °C ***[1 mark].***

It's an easy question really, but not 100% obvious when you first look at it.

Answers

b) P(X ≥ 95) = 0.2.
Transform this to a statement about the standard normal distribution Z by subtracting the mean (93) and dividing by the standard deviation (σ) ***[1 mark]***:

$P\left(Z > \frac{95-93}{\sigma}\right) = 0.2$, i.e. $P\left(Z \leq \frac{2}{\sigma}\right) = 0.8$ ***[1 mark].***

Using the percentage points table: $\frac{2}{\sigma} = 0.8416$ ***[1 mark].***

So $\sigma = \frac{2}{0.8416} = 2.3764... = 2.38$ (to 3 sig.fig.) ***[1 mark].***

c) You need to find P(X < 88). This equals:

$P\left(Z < \frac{88-93}{2.3764}\right) = P(Z < -2.10) = 1 - P(Z < 2.10)$

[1 mark]. From tables, P(Z < 2.10) = 0.98214.
So P(X < 88) = 1 – 0.98214 = 0.0179 (to 3 sig. fig.) ***[1 mark].***

d) You need to find a value d with $P(93 - d \leq T \leq 93 + d) = 0.99$. Because of the symmetry, you can say that you need to find d with $P(T \leq 93 + d) = 0.995$ ***[1 mark]***, or

$P\left(Z \leq \frac{93+d-93}{2.3764}\right) = P\left(Z \leq \frac{d}{2.3764}\right) = 0.995$ ***[1 mark].***

Using the percentage points table, this gives:

$\frac{d}{2.3764} = 2.5758$ ***[1 mark]***,

or $d = 6.12$ (to 3 sig. fig.) ***[1 mark].***

With a normal distribution, if you transform the variable to Z, then chances are you're on the right track.

4) a) Because the cards are not being replaced, the probability of choosing a picture card does not remain constant, which means X cannot follow a binomial distribution ***[1 mark].***

b) (i) Since the cards are now being replaced after each pick, Y will follow a binomial distribution.

Since $\frac{12}{52} = \frac{3}{13}$, $Y \sim B(3, \frac{3}{13})$ ***[1 mark].***

$P(Y = 2) = \binom{3}{2} \times \left(\frac{3}{13}\right)^2 \times \frac{10}{13}$ ***[1 mark]***

$= \frac{270}{13^3} = \frac{270}{2197}$

$= 0.123$ (to 3 sig.fig.) ***[1 mark].***

(ii) $E(Y) = 3 \times \frac{3}{13} = \frac{9}{13} = 0.692$ (to 3 sig.fig.) ***[1 mark].***

(iii) $Var(Y) = 3 \times \frac{3}{13} \times \frac{10}{13} = \frac{90}{169} = 0.533$ (to 3 sig.fig.) ***[1 mark]***

c) The probability of picking a red card is always 0.5.
So the probability of any student picking exactly 3 red cards in 4 picks is:

$P(\text{pick 3 red cards}) = \binom{4}{3} \times 0.5^3 \times (1 - 0.5)$

$= 4 \times 0.5^4 = 0.25$ ***[1 mark]***

This means that Q follows a binomial distribution: $Q \sim B(20, 0.25)$ ***[1 mark].***

You need to find $P(2 \leq Q \leq 8)$. $p = 0.25$ and $n = 20$ are in binomial tables, so:

$P(2 \leq Q \leq 8) = P(Q \leq 8) - P(Q < 2)$

$= P(Q \leq 8) - P(Q \leq 1)$ ***[1 mark]***

$= 0.9591 - 0.0243 = 0.935$ (to 3 sig. fig.) ***[1 mark]***

You could do this question without tables. But it would take a while, because you'd need to find 7 individual probabilities, and add them together. That's the beauty of tables — they can tell you lots of information very quickly.

5) a) (i) It's best to add an extra row to the table:

Number of days (x)	1	2	3	4	5	6	7	8
Frequency (f)	5	3	1	0	0	0	0	1
fx	5	6	3	0	0	0	0	8

Then $\sum f = 10$ and $\sum fx = 22$ ***[1 mark]***, which gives

$\bar{x} = \frac{\sum fx}{\sum f} = \frac{22}{10} = 2.2$ ***[1 mark]***

(ii) There are 10 values, meaning the median is in position (10 + 1) ÷ 2 = 5.5, so you need to take the mean of the 5th and 6th values ***[1 mark].*** Since the 5th value is 1 and the 6th is 2, this will be the midpoint of 1 and 2. So the median = 1.5 ***[1 mark].***

b) No, the range is not a very appropriate measure because the outlier of 8 ***[1 mark]*** will make it unrepresentative of the dataset as a whole (since all but one of the data values are either 1, 2 or 3) ***[1 mark].***

There are usually loads of numbers in these 'data analysis' questions. So take your time, do things carefully, and watch that your fingers don't accidentally hit the wrong calculator buttons.

6) a) (i) $\bar{x} = \frac{2246.3}{80} = 28.07875 = 28.1$ (to 3 sig.fig.) ***[1 mark]***

(ii) $s^2 = \frac{n}{n-1}\left[\frac{\sum x^2}{n} - \left(\frac{\sum x}{n}\right)^2\right]$

$= \frac{80}{79}\left[\frac{63885.4}{80} - \left(\frac{2246.3}{80}\right)^2\right]$ ***[1 mark]***

$= 10.27979... = 10.3$ (to 3 sig.fig.) ***[1 mark]***

b) Since n is large, the confidence interval will be:

$\left(\bar{x} - z\frac{s}{\sqrt{n}}, \bar{x} + z\frac{s}{\sqrt{n}}\right)$, where $z = 1.96$ ***[1 mark]***

$= \left(28.079 - 1.96 \times \frac{\sqrt{10.280}}{\sqrt{80}}, 28.079 + 1.96 \times \frac{\sqrt{10.280}}{\sqrt{80}}\right)$

[1 mark for each correct expression]

$= (27.4, 28.8)$ (to 3 sig. fig.) ***[1 mark]***

Answers

c) These results do provide evidence to help justify this claim ***[1 mark]***, since all of the confidence interval is above 27 cm ***[1 mark]***.

Confidence-interval questions... if you can do one, then you should be able to have a pretty good crack at them all.

7) a) (i) Call C the event 'has had a crash' and G the event 'wears glasses'.

Then $P(C') = 1 - \frac{9}{30} = \frac{21}{30} = \frac{7}{10}$ ***[1 mark]***.

(ii) You don't have to draw a tree diagram, but it helps to make things clearer.

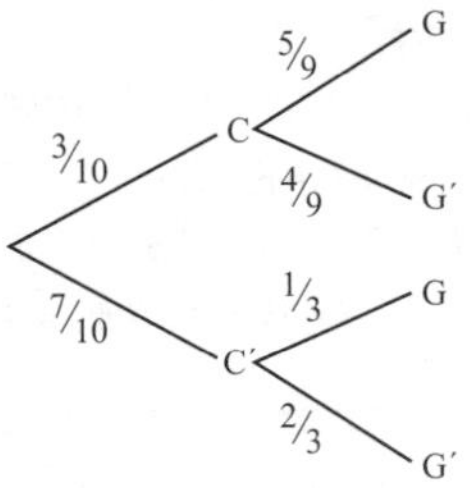

The easiest way is to work out the probabilities of the two branches ending in G by multiplying along each branch, and then adding the results.

$P(G \cap C) = \frac{3}{10} \times \frac{5}{9} = \frac{15}{90} = \frac{1}{6}$ ***[1 mark]***,

and $P(G \cap C') = \frac{7}{10} \times \frac{1}{3} = \frac{7}{30}$ ***[1 mark]***.

Adding these together you get:

$$P(G) = \frac{1}{6} + \frac{7}{30} = \frac{5+7}{30} = \frac{12}{30} = \frac{2}{5} \text{ [1 mark]}.$$

b) You need to find: $P(C \mid G) = \frac{P(C \cap G)}{P(G)}$ ***[1 mark]***.

You've just worked out $P(C \cap G)$ and P(G), so

$P(C \mid G) = \frac{1}{6} \div \frac{2}{5}$ ***[1 mark]***

$= \frac{1}{6} \times \frac{5}{2} = \frac{5}{12}$ ***[1 mark]***.

c) There are various possible 'arrangements' of the three older people (i.e. 35 or over) and the one younger person (i.e. less than 35) if you record the order in which they are selected. The probability of <u>one</u> of these arrangements is

$\frac{13}{30} \times \frac{12}{29} \times \frac{11}{28} \times \frac{17}{27} = \frac{29\,172}{657\,720} = \frac{2431}{54\,810}$ ***[1 mark]***.

Since there are 4 possible arrangements of the four people, the probability of <u>any</u> of these occurring must be:

$4 \times \frac{2431}{54\,810}$ ***[1mark]*** $= \frac{9724}{54\,810} = \frac{4862}{27\,405}$

$= 0.177$ (to 3 sig.fig.) ***[1mark]***

You didn't need to draw a tree diagram in this question, but it does help to be able to 'look at a picture' of the information in the question, rather than just read it.

8) a) $S_{xx} = \sum x^2 - \frac{(\sum x)^2}{n}$

$= 847 - \frac{79^2}{8} = 66.875 = 66.9$ (to 3 sig. fig.) ***[1 mark]***

$S_{yy} = \sum y^2 - \frac{(\sum y)^2}{n}$

$= 45\,884 - \frac{600^2}{8} = 884$ ***[1 mark]***

$S_{xy} = \sum xy - \frac{\sum x \sum y}{n}$

$= 6143 - \frac{79 \times 600}{8} = 218$ ***[1 mark]***

b) $r = \frac{S_{xy}}{\sqrt{S_{xx} \times S_{yy}}} = \frac{218}{\sqrt{66.875 \times 884}}$ ***[1 mark]***

$= 0.897$ (to 3 sig. fig.) ***[1 mark]***

c) The amount of revision (x) ***[1 mark]***, since the exam mark depends on the amount of revision done, not the other way around ***[1 mark]***.

The explanatory variable is also known as the independent variable (and the response variable as the dependent variable). The dependent variable depends on the value of the independent variable.

d) The correlation coefficient is very close to 1, so if the points were plotted on a scatter graph, they would all lie close to a straight line ***[1 mark]***.

e) $b = \frac{S_{xy}}{S_{xx}} = \frac{218}{66.875} = 3.26$ (to 3 sig. fig.) ***[1 mark]***

$a = \bar{y} - b\bar{x} = \frac{600}{8} - 3.26 \times \frac{79}{8}$ ***[1 mark]***

$= 42.8$ (to 3 sig. fig.) ***[1 mark]***

So $y = 42.8 + 3.26x$ ***[1 mark]***.

f) Using the regression line to estimate the mark:
$y = 42.8 + 3.26 \times 8 = 68.88 \approx 69$ marks ***[1 mark]***.

g) This should be a fairly reliable estimate, since it is an interpolation between two known values / This estimate may be unreliable, since this student may not fit the pattern generated by the others ***[1 mark for any sensible comment]***.

Answers

S1 — Practice Exam Two

1) a) (i) There are 40 sweets altogether, and 6 + 8 + 2 = 16 of them have a soft centre.

So P(soft centre) = $\frac{16}{40} = \frac{2}{5}$ ***[1 mark].***

(ii) There are 5 + 2 = 7 sweets made with white chocolate, and so 33 are not made with white chocolate ***[1 mark].***

So P(not made with white chocolate) = $\frac{33}{40}$ ***[1 mark].***

(iii) There are 24 + 19 – 11 = 32 sweets either with a hard centre or made with milk chocolate ***[1 mark].*** So

P(hard centre or milk chocolate) = $\frac{32}{40} = \frac{4}{5}$ ***[1 mark].***

(iv) Of the 14 plain chocolates, 8 have a hard centre ***[1 mark].***

So P(hard centre | plain chocolate) = $\frac{8}{14} = \frac{4}{7}$ ***[1 mark].***

b) P(plain then milk then white) =

$\frac{14}{40} \times \frac{19}{39} \times \frac{7}{38} = \frac{1862}{59\,280} = \frac{49}{1560}$ ***[1 mark].***

But there are 3 × 2 × 1 = 6 ways to select one sweet made with each type of chocolate ***[1 mark].***

So P(plain, milk, white in any order)

$= 6 \times \frac{49}{1560}$ ***[1 mark]*** $= \frac{49}{260}$ ***[1 mark]***

As long as you didn't get distracted by the mention of chocolate, then that question should have been a fairly easy start.

2) a) $T \sim N(132, 40^2)$. So

$P(T > 160) = P\left(Z > \frac{160-132}{40}\right) = P(Z > 0.7)$ ***[1 mark]***

$= 1 - P(Z \leq 0.7)$ ***[1 mark]***

$= 1 - 0.75804$

$= 0.24196 = 0.242$ (to 3 sig. fig.) ***[1 mark]***

b) $P(60 \leq T \leq 90) = P(T \leq 90) - P(T < 60)$ ***[1 mark]***

$= P\left(Z \leq \frac{90-132}{40}\right) - P\left(Z \leq \frac{60-132}{40}\right)$

$= P(Z \leq -1.05) - P(Z \leq -1.8)$ ***[1 mark]***

$= (1 - P(Z \leq 1.05))$

$\quad - (1 - P(Z \leq 1.8))$ ***[1 mark]***

$= (1 - 0.85314) - (1 - 0.96407)$

$= 0.11093 = 0.111$ (to 3 sig. fig.) ***[1 mark]***

c) $P(T = 135) = 0$ ***[1 mark]***

You should have got lots of practice with the normal distribution by now, so hopefully it feels like a good friend.

3) a) Mean = $\frac{\sum x}{n} = \frac{66.5}{12}$ ***[1 mark]***

= 5.54 (or £5540) (to 3 sig. fig.) ***[1 mark].***

Variance $= \frac{\sum x^2}{n} - \left(\frac{\sum x}{n}\right)^2 = \frac{390.97}{12} - \left(\frac{66.5}{12}\right)^2$ ***[1 mark]***

= 1.87 (to 3 sig. fig.) ***[1 mark].***

b) The ordered list of the 12 data points is:
3.8, 4.1, 4.2, 4.6, 4.9, 5.5, 5.8, 5.9, 6.0, 6.2, 6.4, 9.1.

The position of the median is $\frac{1}{2}(n+1) = \frac{1}{2}(12+1) = 6.5$, so take the average of the 6th and 7th values.

So the median Q_2 is: $\frac{1}{2}(5.5 + 5.8) = 5.65$ ***[1 mark].***

Since 12 ÷ 4 = 3, the lower quartile is the average of the 3rd and 4th values.

So the lower quartile Q_1 is: $\frac{1}{2}(4.2 + 4.6) = 4.4$ ***[1 mark].***

Since 12 ÷ 4 × 3 = 9, the upper quartile is the average of the 9th and 10th values.

So the upper quartile Q_3 is: $\frac{1}{2}(6.0 + 6.2) = 6.1$ ***[1 mark].***

c) You need to use the mid-class values. It's best to add some extra rows to the table.

Sales (£'000s)	4-5	5-6	6-7	7-8	8-9	9-10
Number of weeks, f	1	2	2	3	4	1
Mid-class value, y	4.5	5.5	6.5	7.5	8.5	9.5
$f \times y$	4.5	11	13	22.5	34	9.5
y^2	20.25	30.25	42.25	56.25	72.25	90.25
$f \times y^2$	20.25	60.5	84.5	168.75	289	90.25

(i) $\bar{x} = \frac{\sum fy}{\sum f} = \frac{94.5}{13}$ ***[1 mark]*** = 7.27 (to 3 sig.fig.) ***[1 mark]***

(ii) First use the table to calculate

$\sum fy^2 = 713.25$ ***[1 mark]***

Then you can use the formula for the variance:

Variance $= \frac{\sum fy^2}{\sum f} - \bar{x}^2 = \frac{713.25}{13} - \left(\frac{94.5}{13}\right)^2$ ***[1 mark]***

= 2.02 (to 3 sig fig.) ***[1 mark]***

Numbers, numbers everywhere.

4) a)

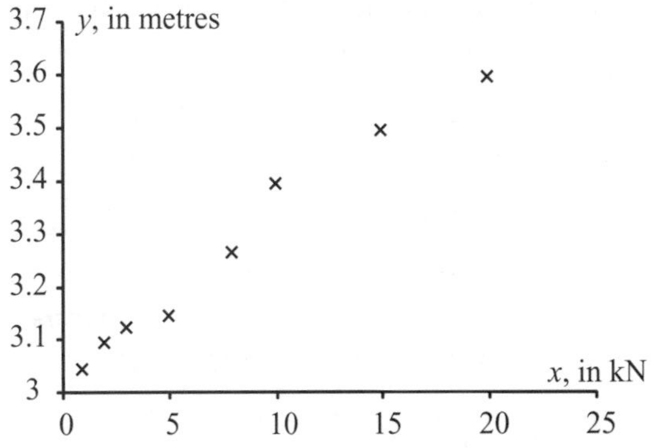

[2 marks for all points correctly plotted, or 1 mark if at least 4 points are correctly plotted.]

b) $S_{xx} = \sum x^2 - \frac{(\sum x)^2}{n} = 828 - \frac{64^2}{8} = 316$ ***[1 mark]***

$S_{xy} = \sum xy - \frac{\sum x \sum y}{n} = 219.05 - \frac{64 \times 26.2}{8} = 9.45$
[1 mark].

Answers

c) If $y = a + bx$, then
$b = \frac{S_{xy}}{S_{xx}} = \frac{9.45}{316} = 0.029905... = 0.0299$ (to 3 sig.fig.).
[1 mark].
And $a = \bar{y} - b\bar{x}$
$= \frac{26.2}{8} - 0.0299 \times \frac{64}{8}$ ***[1 mark]***
$= 3.0358 = 3.04$ (to 3 sig. fig.) ***[1 mark]***.

So the equation of the regression line is:
$y = 3.04 + 0.0299x$ ***[1 mark]***.

All these formulas for the regression line coefficients will be on your formula sheet.

d) a represents the length of the cable when it is not under tension (i.e. when $x = 0$) ***[1 mark]***.
b represents the extra extension of the cable when the tension is increased by 1 kN ***[1 mark]***.

e) $y = 3.04 + 0.0299x$, so when $x = 30$,
$y = 3.04 + 0.0299 \times 30 = 3.937$ ***[1 mark]***
$= 3.94$ metres (to 3 sig. fig.) ***[1 mark]***
(or 3.93 m if $a = 3.0358$ used)

f) This estimate may be unreliable as it involves extrapolating beyond the range of the experimental data ***[1 mark]***.

5) a) This is $\bar{x} = \frac{1344}{36} = 37.3$ minutes (to 3 sig.fig.) ***[1 mark]***.

b) First you need an unbiased estimate of the population variance. This is:

$$s^2 = \frac{n}{n-1}\left[\frac{\sum x^2}{n} - \left(\frac{\sum x}{n}\right)^2\right]$$

$$= \frac{36}{35}\left[\frac{51983}{36} - \left(\frac{1344}{36}\right)^2\right] \textbf{\textit{[1 mark]}} = 51.628... \textbf{\textit{[1 mark]}}$$

So $s = \sqrt{51.628...} = 7.1853... = 7.19$ (to 3 sig.fig.) ***[1 mark]***.

So an estimate of the standard error of the sample mean is:
$\frac{s}{\sqrt{n}} = \frac{7.1853...}{\sqrt{36}}$ ***[1 mark]***
$= 1.1975... = 1.20$ (to 3 sig.fig.) ***[1 mark]***.

c) The confidence interval will be given by:

$$\left(\bar{x} - z\frac{s}{\sqrt{n}}, \bar{x} + z\frac{s}{\sqrt{n}}\right)$$

$$= \left(\frac{1344}{36} - 2.5758 \times 1.198, \frac{1344}{36} + 2.5758 \times 1.198\right)$$ ***[1 mark]***

$= (34.2, 40.4)$ (to 3 sig. fig.) ***[1 mark]***

6) a) $P(X < 7.5) = P\left(\frac{X-8}{\sqrt{1.2}} < \frac{7.5-8}{\sqrt{1.2}}\right)$
$= P(Z < -0.46)$ ***[1 mark]***
$= 1 - P(Z \leq 0.46)$ ***[1 mark]***
$= 1 - 0.67724$
$= 0.32276 = 0.323$ (to 3 sig. fig.) ***[1 mark]***

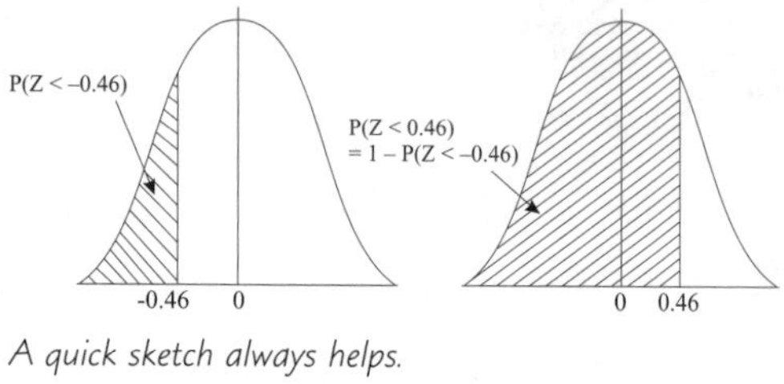

A quick sketch always helps.

b) You need to find the probability that the duration is less than 7 minutes or more than 9 minutes.
This is $P(X > 9) + P(X < 7)$.
By symmetry, these are equal, so:
$P(X < 7) + P(X > 9) = 2 \times P\left(Z < \frac{7-8}{\sqrt{1.2}}\right)$ ***[1 mark]***
$= 2 \times P(Z < -0.91)$ ***[1 mark]***
$= 2 \times (1 - P(Z \leq 0.91))$ ***[1 mark]***
$= 2 \times (1 - 0.81859)$
$= 0.36282 = 0.363$ (to 3 sig. fig.) ***[1 mark]***

These do definitely get a bit tricky. Just have a good think about what you need to find out before you launch into a load of working out. Getting loads of practice at this kind of question is dead helpful — it gets you used to working with those tables.

c) You need to find d where: $P(X > d) = 0.01$.
$P(X > d) = 0.01$, so $P\left(Z \leq \frac{d-8}{\sqrt{1.2}}\right) = 0.99$ ***[1 mark]***.

Using the percentage points table: $\frac{d-8}{\sqrt{1.2}} = 2.3263$ ***[1 mark]***.

So $d = \sqrt{1.2} \times 2.3263 + 8$ ***[1 mark]***,
i.e. $d = 10.5$ minutes (to 3 sig. fig.) ***[1 mark]***

These 'standardise the normal variable' questions get everywhere. They look hard, but you soon get used to them. And once you get your head round the basic idea, you'll be able to do pretty much anything they ask you.

7) a) (i) $P(X > 1) = 1 - P(X \leq 1)$.
From tables, $P(X \leq 1) = 0.3917$ ***[1 mark]***.
So $P(X > 1) = 1 - 0.3917 = 0.6083$
$= 0.608$ (to 3 sig. fig.) ***[1 mark]***.

(ii) $P(2 < X \leq 6) = P(X \leq 6) - P(X \leq 2)$ ***[1 mark]***
$= 0.9976 - 0.6769$ ***[1 mark]*** $= 0.3207$
$= 0.321$ (to 3 sig. fig.) ***[1 mark]***

b) If you consider all 3 students, then there are 60 independent events with a 0.1 probability each time of 'success'. So if y is the total number of golden tickets they find, then $Y \sim B(60, 0.1)$ ***[1 mark]***.

$P(Y = 3) = \binom{60}{3} \times 0.1^3 \times 0.9^{57}$ ***[1 mark]***
$= \frac{60 \times 59 \times 58}{3 \times 2 \times 1} \times 0.1^3 \times 0.9^{57}$
$= 0.0844$ (to 3 sig.fig.) ***[1 mark]***.

c) (i) $T \sim B(10, 0.1)$.
Then $E(T) = np = 10 \times 0.1 = 1$ ***[1 mark]***, and
$Var(T) = np(1-p) = 10 \times 0.1 \times 0.9 = 0.9$ ***[1 mark]***.

(ii) The actual values of 0.2 for the mean and 0.19 for the variance are a long way from the theoretical values of 1 and 0.9 ***[1 mark]***, and so one of the binomial assumptions does not seem to be true. If 10% of the chocolate bars really do contain a golden ticket, then it seems unlikely that these samples of 10 bars can be genuinely random ***[1 mark]***.

Don't rush in when you first start a question — think it through carefully, otherwise you might start running in the wrong direction.

Answers

M1 Section 1 — Kinematics

Warm-up Questions

1) $u = 3;\ v = 9;\ a = a;\ s = s;\ t = 2$. Use $s = \frac{1}{2}(u + v)t$
$s = \frac{1}{2}(3 + 9) \times 2 \qquad s = \frac{1}{2}(12) \times 2 = 12$ m

2) 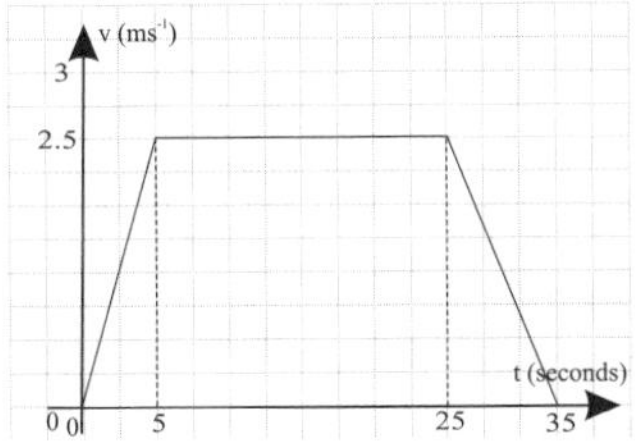

distance = (5 × 2.5) ÷ 2 + (20 × 2.5) + (10 × 2.5) ÷ 2
= 68.75 m

3) velocity = area under (t, a) graph

a) $t = 3$, velocity = (3 × 5) ÷ 2 = 7.5 ms^{-1}

b) $t = 5$, velocity = 7.5 + (2 × 5) = 17.5 ms^{-1}

c) $t = 6$, velocity = 17.5 + (1 × 5) ÷ 2 = 20 ms^{-1}

4) Displacement = (15 × 0.25) – (10 × 0.75) = –3.75 km
Time taken = 1 hour
Average velocity = –3.75 kmh^{-1} (i.e. 3.75 kmh^{-1} south)

5) $(3\mathbf{i} + 7\mathbf{j}) + 2 \times (-2\mathbf{i} + 2\mathbf{j}) - 3 \times (\mathbf{i} - 3\mathbf{j})$
$= (3 - 4 - 3)\mathbf{i} + (7 + 4 + 9)\mathbf{j} = -4\mathbf{i} + 20\mathbf{j}$

6) 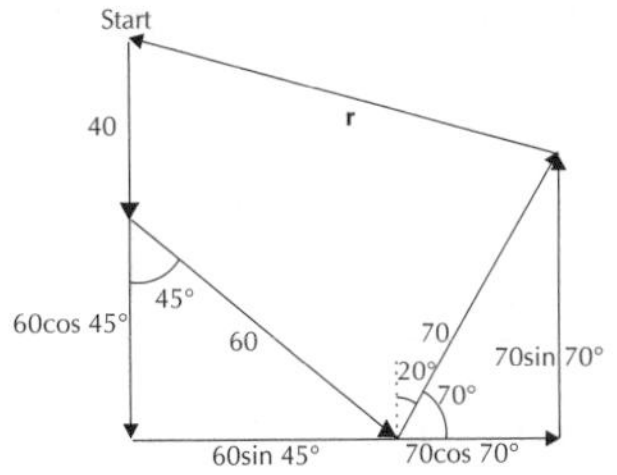

Resolving East: 60sin45° + 70cos70° = 66.4 miles
Resolving North: –40 – 60cos45° + 70sin70°
= –16.6 miles

Magnitude of $\mathbf{r} = \sqrt{66.4^2 + 16.6^2} = 68.4$ miles

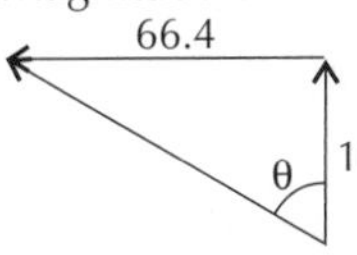

Direction = $\theta = \tan^{-1}\left(\frac{66.4}{16.6}\right) = 76.0°$

Bearing is 360° – 76.0° = 284°

Top-top-tip: When answering questions concerning vectors or forces always draw a diagram. I promise it'll make it simpler.

Exam Questions

1 a) Using $u = v - at$
$u = 17 - (9.8 \times 1.2)$
So, $u = 5.24$ ms^{-1}

[3 marks available in total]:
- ***1 mark for using appropriate equation***
- ***1 mark for correct workings***
- ***1 mark for correct value of u***

These kind of questions are trivial if you've memorised all those constant acceleration equations. If you haven't, you know what to do now (turn to p. 158).

b) Using $s = ut + \frac{1}{2}at^2$
$s = (17 \times 2.1) + \frac{1}{2}(9.8 \times 2.1^2)$
So, $s = 57.31$
$h = \frac{s}{14} = 4.09$ m (3 s.f.)

[4 marks available in total]:
- ***1 mark for using appropriate equation***
- ***1 mark for correct value of s***
- ***1 mark for correct workings***
- ***1 mark for correct value of h***

2 a) 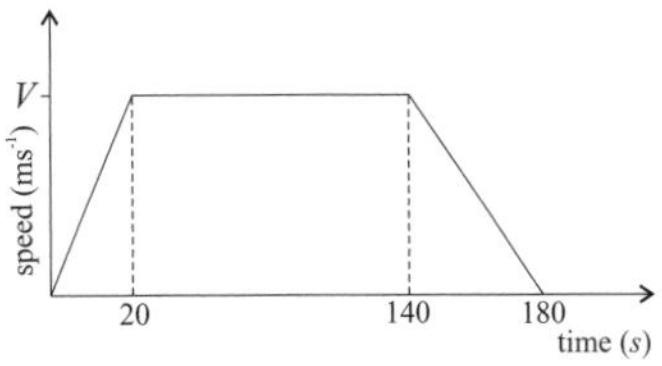

[3 marks available in total]:
- ***1 mark for the correct shape***
- ***1 mark for the correct times***
- ***1 mark for correctly marking V on the vertical axis***

b) Area under graph (area of trapezium) = distance
$\frac{1}{2}(120 + 180)V = 2100$
$V = \frac{2100}{150} = 14$ ms^{-1}

[3 marks available in total]:
- ***1 mark for using area under graph = distance***
- ***1 mark for correct workings***
- ***1 mark for correct value of V***

The area could also be worked out in other ways, say, using two triangles and a rectangle. Best to use whatever's easiest for you.

c) Distance = area under graph
= ½ × 40 × 14 = 280 m

[2 marks available in total]:
- ***1 mark for using area under graph = distance***
- ***1 mark for correct value***

d) Using $a = \frac{v - u}{t}$.
Acceleration period:
$t = 20,\ v = 14,\ u = 0$
$a = (14 - 0) \div 20 = 0.7$ ms^{-2}
Deceleration period:
$t = 40,\ v = 0,\ u = 14$
$a = (0 - 14) \div 40 = -0.35$ ms^{-2}

Answers

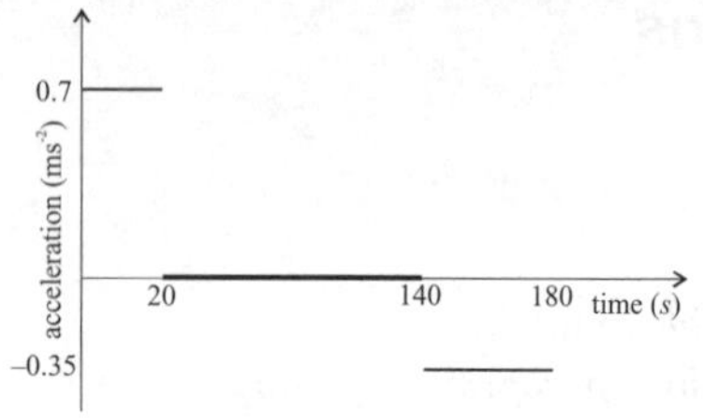

[3 marks available in total]:
- ***1 mark for correct shape***
- ***1 mark for correct value of acceleration (0.7 ms⁻²)***
- ***1 mark for correct value of deceleration (0.35 ms⁻²)***

Yep, sometimes you're not given the values you need to label a graph and you have to work them out yourself. All good practice.

3 a) $u = u$; $v = 20$; $a = 9.8$; $s = 8$
Using $v^2 = u^2 + 2as$
$20^2 = u^2 + (2 \times 9.8 \times 8)$
$u = \sqrt{400 - 156.8}$
So, $u = 15.6$ ms⁻¹ (3 s.f.)

[3 marks available in total]:
- ***1 mark for using appropriate equation***
- ***1 mark for correct workings***
- ***1 mark for correct value of u***

b) Using $v = u + at$
$20 = -15.59 + 9.8t$
So, $9.8t = 35.59$
hence $t = 3.63$ s (3 s.f.)

[3 marks available in total]:
- ***1 mark for using appropriate equation***
- ***1 mark for correct workings***
- ***1 mark for correct value of t***

Constant use of those constant acceleration equations... See, I wasn't fooling around — learn them.

4

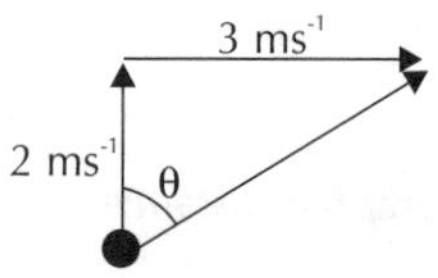

Magnitude = $\sqrt{2^2 + 3^2} = \sqrt{13} = 3.61$ ms⁻¹
$\theta = \tan^{-1}\left(\frac{3}{2}\right) = 56.3°$
So angle to river bank is 90° – 56.3° = 33.7°

[3 marks available in total]:
- ***1 mark for calculating the magnitude of the velocity***
- ***1 mark for correct workings***
- ***1 mark for calculating the angle from the bank***

5 For the first particle:
$\mathbf{s}_A = \mathbf{s}_O + \mathbf{v}t$
$= (\mathbf{i} + 2\mathbf{j}) + 8(3\mathbf{i} + \mathbf{j}) = (\mathbf{i} + 2\mathbf{j}) + (24\mathbf{i} + 8\mathbf{j}) = (25\mathbf{i} + 10\mathbf{j})$ m
For the second particle:
$\mathbf{s}_B = \mathbf{s}_A + \mathbf{v}t = (25\mathbf{i} + 10\mathbf{j}) + 5(-4\mathbf{i} + 2\mathbf{j})$
$= (25\mathbf{i} + 10\mathbf{j}) + (-20\mathbf{i} + 10\mathbf{j}) = (5\mathbf{i} + 20\mathbf{j})$ m

[4 marks available in total]:
- ***1 mark for correct workings for $\mathbf{s}_A$***
- ***1 mark for correct value of $\mathbf{s}_A$***
- ***1 mark for correct workings for $\mathbf{s}_B$***
- ***1 mark for correct value of $\mathbf{s}_B$***

Argh. My head hurts with the sheer quantity of **i***s and* **j***s in those workings. Better to take it a step at a time (it's good for your health).*

6 a) Speed = $\sqrt{7^2 + (-3)^2} = \sqrt{58}$
= 7.62 ms⁻¹

[2 marks available in total]:
- ***1 mark for correct workings***
- ***1 mark for correct value***

*A bit of a classic here, and easy if you remember that for a vector (x***i*** + y***j***) then the magnitude = $\sqrt{x^2 + y^2}$.*

b) Angle from horizontal = $\tan^{-1}\left(\frac{-3}{7}\right)$
= –23.2° (i.e. 23° south of east)
The bearing is measured from north
so, bearing = 90 + 23.2 = 113°

[3 marks available in total]:
- ***1 mark for correct workings***
- ***1 mark for calculating angle***
- ***1 mark for correct bearing***

Bearings are measured from the north, because that's where Polar bears live, and where the Be[a]ring Strait is — maybe...

c) Position at $t = 4$:
$(\mathbf{i} + 5\mathbf{j}) + 4(7\mathbf{i} - 3\mathbf{j}) = (29\mathbf{i} - 7\mathbf{j})$ m
Displacement to 15**i**:
$15\mathbf{i} - (29\mathbf{i} - 7\mathbf{j}) = (-14\mathbf{i} + 7\mathbf{j})$ m
Velocity = $\frac{\mathbf{s}}{t} = (-14\mathbf{i} + 7\mathbf{j})/3.5 = (-4\mathbf{i} + 2\mathbf{j})$
So, $a = -4$ and $b = 2$

[4 marks available in total]:
- ***1 mark for calculating position at t = 4***
- ***1 mark for calculating displacement to 15i***
- ***1 mark for correct workings***
- ***1 mark for correct values of a and b***

7 a) Call acceleration $\mathbf{a} = A\mathbf{i} + B\mathbf{j}$. Using $\mathbf{s} = \mathbf{u}t + \frac{1}{2}\mathbf{a}t^2$:
$(6\mathbf{i} + 11\mathbf{j}) - (\mathbf{i} + \mathbf{j}) = 10(3\mathbf{i} - \mathbf{j}) + (\frac{1}{2} \times (A\mathbf{i} + B\mathbf{j}) \times 10^2)$
$\Rightarrow 5\mathbf{i} + 10\mathbf{j} = 30\mathbf{i} - 10\mathbf{j} + 50(A\mathbf{i} + B\mathbf{j})$
$\Rightarrow -25\mathbf{i} + 20\mathbf{j} = 50(A\mathbf{i} + B\mathbf{j})$
$\Rightarrow A\mathbf{i} + B\mathbf{j} = (-25 \div 50)\mathbf{i} + (20 \div 50)\mathbf{j} = -0.5\mathbf{i} + 0.4\mathbf{j}$ ms⁻².

[3 marks available in total]:
- ***1 mark for using appropriate equation***
- ***1 mark for correct workings***
- ***1 mark for correct value of a***

b) Use $\mathbf{v} = \mathbf{u} + \mathbf{a}t$ ***[1 mark]***:
$\mathbf{v} = (3\mathbf{i} - \mathbf{j}) + 6(-0.5\mathbf{i} + 0.4\mathbf{j})$ ***[1 mark]***
$= (3 - 3)\mathbf{i} + (-1 + 2.4)\mathbf{j}$
$= 1.4\mathbf{j}$ ***[1 mark]***
i.e. the particle is moving parallel to **j**.

Answers

M1 Section 2 — Statics and Forces

Warm-up Questions

1) a) 

Small point mass, no air resistance, no wind, released from rest, constant acceleration due to gravity.

b) 

Small point mass, no air resistance, no wind, released from rest, constant acceleration due to gravity.

c) 

Same assumptions as in a) and b), although it might not be safe to ignore wind if outside as table tennis balls are very light.

You need to get familiar with modelling and all the terminology used in M1 — it's going to be really tricky to figure out M1 questions if you're not.

2) 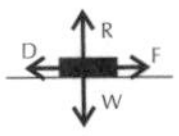

Assumptions: Point mass, one point of contact with ground, constant driving force D from engine, constant friction, F, includes road resistance and air resistance, acceleration = 0 as it's moving at 25mph (in a straight line).

3) a) 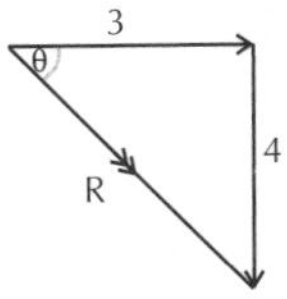

$R = \sqrt{4^2 + 3^3} = 5\text{ N}$

$\tan\theta = \frac{4}{3}$

$\theta = 53.1^\circ$ below the horizontal

b) 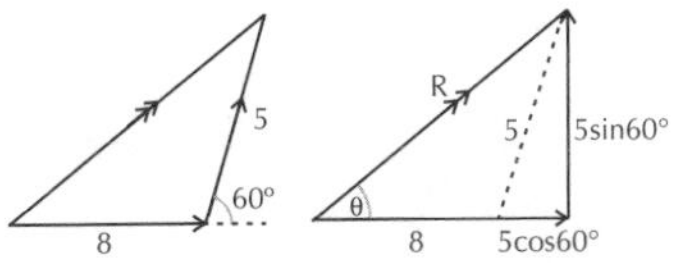

$R = \sqrt{(8 + 5\cos 60^\circ)^2 + (5\sin 60^\circ)^2} = 11.4\text{ N}$

$\tan\theta = \dfrac{5\sin 60^\circ}{8 + 5\cos 60^\circ} = 0.412...$

So $\theta = 22.4^\circ$ above the horizontal (3 s.f.)

c) Total force up $= 6 - 4\sin 10^\circ - 10\sin 20^\circ = 1.885$ N

Total force left $= 10\cos 20^\circ - 4\cos 10^\circ = 5.458$ N

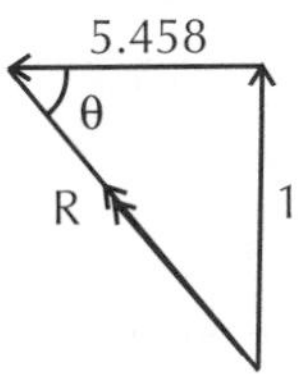

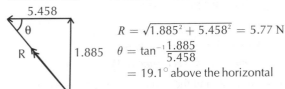

$R = \sqrt{1.885^2 + 5.458^2} = 5.77\text{ N}$

$\theta = \tan^{-1}\frac{1.885}{5.458}$

$= 19.1^\circ$ above the horizontal

4) 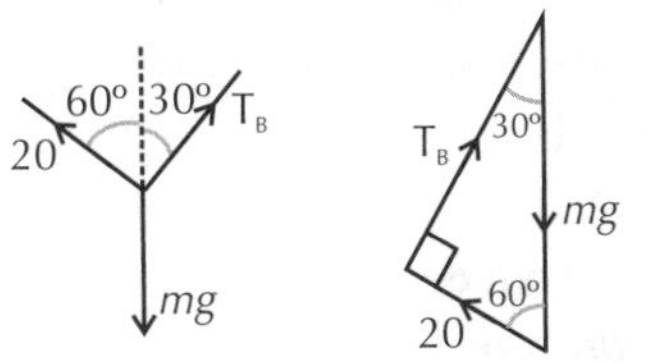

a) $\tan 30^\circ = \dfrac{20}{T_B}$

$T_B = \dfrac{20}{\tan 30^\circ}$

$= 34.6\text{ N}$

b) $\sin 30^\circ = \dfrac{20}{mg}$

$mg = \dfrac{20}{\sin 30^\circ}$

$m = 4.08\text{ kg}$

5) 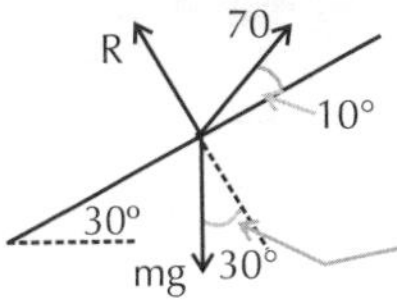

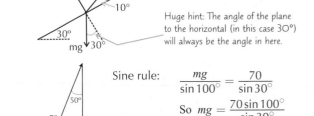

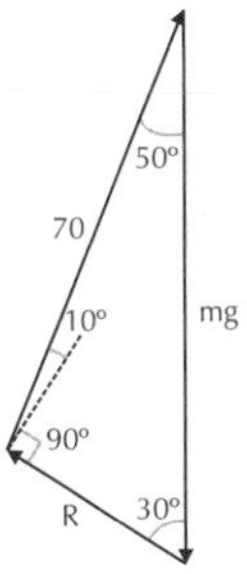

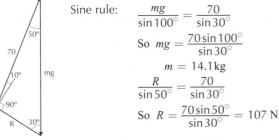

Sine rule: $\dfrac{mg}{\sin 100^\circ} = \dfrac{70}{\sin 30^\circ}$

So $mg = \dfrac{70\sin 100^\circ}{\sin 30^\circ}$

$m = 14.1\text{kg}$

$\dfrac{R}{\sin 50^\circ} = \dfrac{70}{\sin 30^\circ}$

So $R = \dfrac{70\sin 50^\circ}{\sin 30^\circ} = 107\text{ N}$

Yet another example of how triangles might just save your ~~life~~ mark for the M1 module. Although you could also have solved it by resolving forces parallel and perpendicular to the slope if that floats your boat.

6) 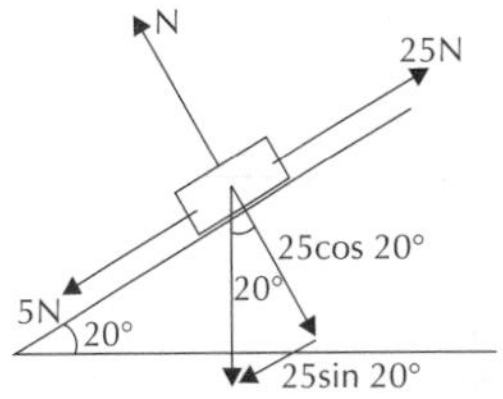

Force perpendicular to the slope: N = 25cos20° = 23.5 N

Force parallel to the slope: 25 – 25sin20° – 5 = 11.4 N.
So the resultant force is 11.4 N up the slope.

7) a) 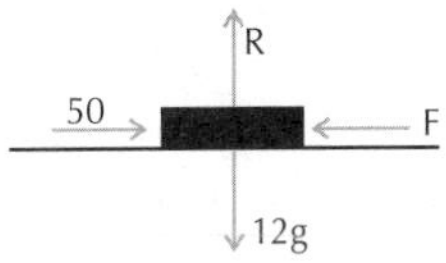

Resolve vertically: $R = 12g$

Use formula: $F \le \mu R$

$F \le \frac{1}{2}(12g)$

$F \le 58.8\text{ N}$

50 N isn't big enough to overcome friction — so it has no overall motion.

b) Force would have to be > 58.8 N

Exam Questions

1 Resolve horizontally: 0 + 5cos30° = 4.330 N
Resolve vertically: 4 – 5sin30° = 1.5 N

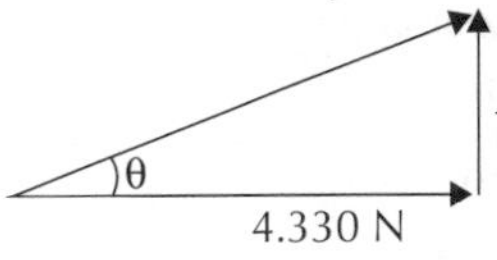

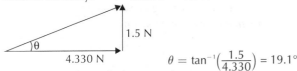

$\theta = \tan^{-1}\left(\frac{1.5}{4.330}\right) = 19.1^\circ$

i.e. θ = 19.1° above the horizontal

Magnitude = $\sqrt{1.5^2 + 4.330^2} = 4.58$ N

[4 marks available in total]:
- ***1 mark for resolving horizontally***
- ***1 mark for resolving vertically***
- ***1 mark for calculating the direction***
- ***1 mark for calculating the magnitude***

Triangles — how do I love thee? Let me count the ways...
...one big way, really. They're just super useful when resolving things.

Answers

2 a) 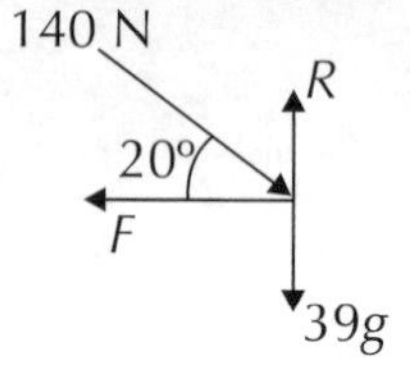

[2 marks available in total]:
- ***1 mark for drawing 4 correct arrows***
- ***1 mark for correctly labelling the arrows***

b) Resolve vertically:
R = 39g + 140sin20° = 430 N
Resolve horizontally:
F = 140cos20° = 132 N

[4 marks available in total]:
- ***1 mark for resolving vertically***
- ***1 mark for resolving horizontally***
- ***1 mark for correct reaction magnitude***
- ***1 mark for correct friction magnitude***

No need to panic if friction is involved — it's just another thing to consider when resolving (and an extra arrow to draw).

3 a) $\mathbf{R} = \mathbf{P} + \mathbf{Q} = (2\mathbf{i} - 11\mathbf{j}) + (7\mathbf{i} + 5\mathbf{j}) = (9\mathbf{i} - 6\mathbf{j})$

[2 marks available in total]:
- ***1 mark for correct workings***
- ***1 mark for correct resultant***

b) $|\mathbf{R}| = \sqrt{9^2 + (-6)^2} = 10.8$ N

[2 marks available in total]:
- ***1 mark for correct workings***
- ***1 mark for correct magnitude***

4 a) 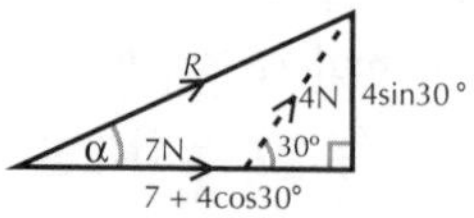

Resolve vertically:
4sin30° = 2
Resolve horizontally:
7 + 4cos30° = 10.464...

$R = \sqrt{2^2 + (7 + 4\cos 30°)^2}$
= 10.7 N (3 s.f.)

[4 marks available in total]:
- ***1 mark for diagram***
- ***1 mark for resolving vertically***
- ***1 mark for resolving horizontally***
- ***1 mark for correct magnitude***

b) $\tan\alpha = \dfrac{2}{7 + 4\cos 30} = 0.191...$
α = 10.8° (3 s.f.)

[2 marks available in total]:
- ***1 mark for correct workings***
- ***1 mark for correct value of α***

5 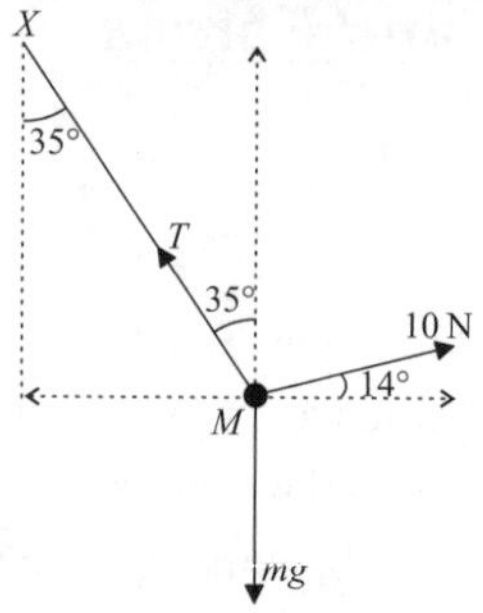

Resolving vertically: mg = 10sin14° + Tcos35°
Need to find T, the tension in the rod.
Resolving horizontally: Tsin35° = 10cos14°
So, $T = \dfrac{10\cos 14°}{\sin 35°} = 16.92$ N
So mg = 10sin14° + 16.92cos35° = 16.28 N
Therefore, the mass of $M = \dfrac{16.28}{g} = 1.66$ kg (3 s.f.)

[4 marks available in total]:
- ***1 mark for resolving vertically***
- ***1 mark for resolving horizontally***
- ***1 mark for correct value of T***
- ***1 mark for correct mass of M***

I think this question is sort of fun, but then I also think rods are sort of pretty... Anyway, a good clear diagram here will simplify matters no end.

6 a) 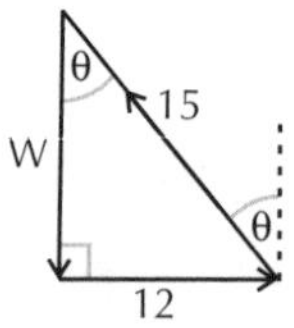

$\sin\theta = \dfrac{12}{15}$, so, θ = 53.1°

[2 marks available in total]:
- ***1 mark for correct workings***
- ***1 mark for the correct value of θ***

b) $15^2 = W^2 + 12^2$
$W = \sqrt{15^2 - 12^2} = 9$ N

[2 marks available in total]:
- ***1 mark for correct workings***
- ***1 mark for correct value of W***

c) Remove W and the particle moves in the opposite direction to W, i.e. upwards. This resultant of the two remaining forces is 9 N upwards, because the particle was in equilibrium beforehand.

[2 marks available in total]:
- ***1 mark for correct magnitude***
- ***1 mark for correct direction***

The word 'state' in an exam question means that you shouldn't need to do any extra calculation to answer it.

Answers

7

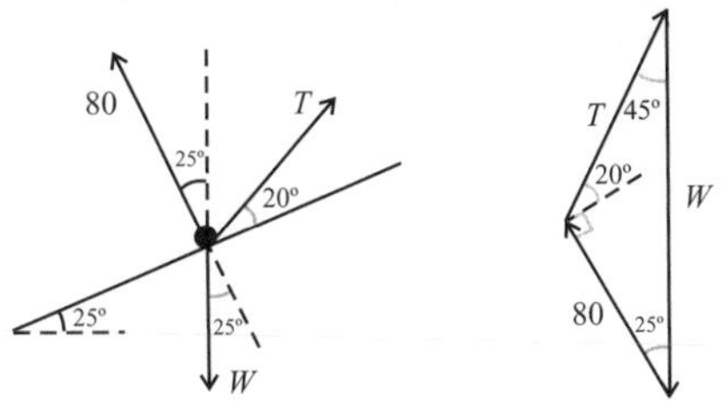

a) $\frac{T}{\sin 25^\circ} = \frac{80}{\sin 45^\circ}$

So, $T = \frac{80 \sin 25^\circ}{\sin 45^\circ}$ = 47.8 N

[3 marks available in total]:
- ***1 mark for diagram***
- ***1 mark for correct workings***
- ***1 mark for correct value of T***

b) $\frac{W}{\sin 110^\circ} = \frac{80}{\sin 45^\circ}$

So, $W = \frac{80 \sin 110^\circ}{\sin 45^\circ}$ = 106 N

[2 marks available in total]:
- ***1 mark for correct workings***
- ***1 mark for correct value of W***

Do you remember the sine rule? If not, go and revise it (try p. 74) — a bit of trigonometry never hurt anyone, and it'll serve you well in M1. You can solve pretty much anything with triangles...

8 a) Resolving horizontally:

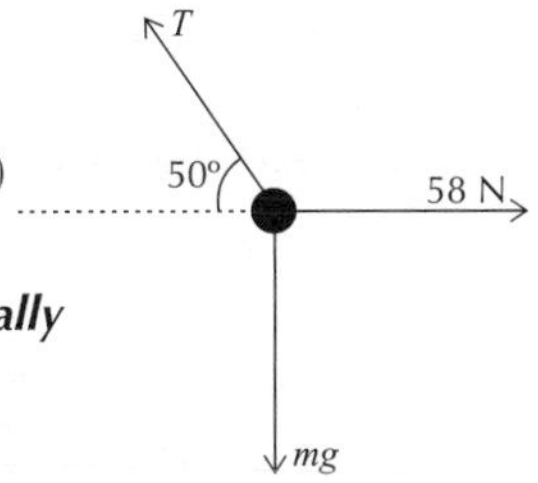

$T\cos 50^\circ = 58$ N

so, $T = \frac{58}{\cos 50^\circ}$ = 90.2 N (3 s.f.)

[3 marks available in total]:
- ***1 mark for resolving horizontally***
- ***1 mark for correct workings***
- ***1 mark for correct value of T***

b) Resolving vertically:

$mg = T\sin 50^\circ = 90.23\sin 50^\circ$

so, $mg = 69.12...$ and $m = 7.05$ kg (3 s.f.)

[3 marks available in total]:
- ***1 mark for resolving vertically***
- ***1 mark for correct workings***
- ***1 mark for correct value of m***

9 Resolve horizontally: $S\cos 40^\circ = F$

Resolve vertically: $R = 2g + S\sin 40^\circ$

It's limiting friction so $F = \mu R$

So, $S\cos 40^\circ = \frac{3}{10}(2g + S\sin 40^\circ)$

$S\cos 40^\circ = 0.6g + 0.3S\sin 40^\circ$

$S\cos 40^\circ - 0.3S\sin 40^\circ = 0.6g$

$S(\cos 40^\circ - 0.3\sin 40^\circ) = 0.6g$

$S = 10.3$ N

[4 marks available in total]:
- ***1 mark for resolving horizontally***
- ***1 mark for resolving vertically***
- ***1 mark for correct workings***
- ***1 mark for correct value of S***

A ring on a rod — it might be a car on a road, or a sled on snow... it's all the same mathematically.

M1 Section 3 — Dynamics
Warm-up Questions

1) a) $(5 \times 3) + (4 \times 1) = (5 \times 2) + (4 \times v)$

$19 = 10 + 4v$

$v = 2.25$ ms^{-1} to the right

b) $(5 \times 3) + (4 \times 1) = 9v$

$19 = 9v$

$v = 2.11$ ms^{-1} to the right (3 s.f.)

c) $(5 \times 3) + (4 \times -2) = (5 \times -v) + (4 \times 3)$

$7 = -5v + 12$

$5v = 5$

$v = 1$ ms^{-1} to the left

d) $(m \times 6) + (8 \times 2) = (m \times 2) + (8 \times 4)$

$6m + 16 = 2m + 32$

$4m = 16$

$m = 4$ kg

Collision questions have me bouncing off the ceiling... Be careful with your directions (positive and negative) and it'll all be okay.

2) Resolve horizontally: $F_{net} = ma$

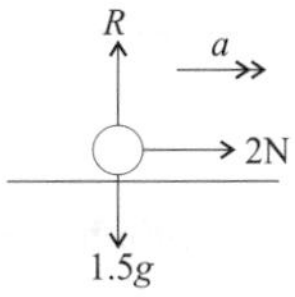

$2 = 1.5a$ so $a = 1\frac{1}{3}$ ms^{-2}

$v = u + at$ $v = 0 + (1\frac{1}{3} \times 3) = 4$ ms^{-1}

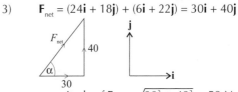

I hope that bit of resolving was simple enough (if not you might want to go revise).

3) $\mathbf{F}_{net} = (24\mathbf{i} + 18\mathbf{j}) + (6\mathbf{i} + 22\mathbf{j}) = 30\mathbf{i} + 40\mathbf{j}$

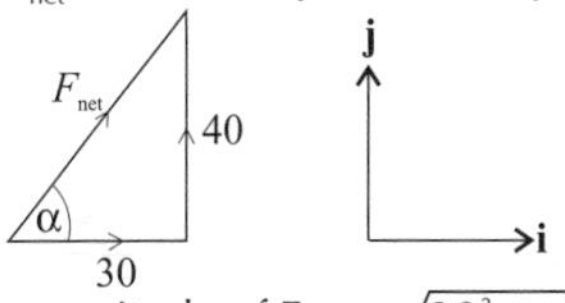

magnitude of $\mathbf{F}_{net} = \sqrt{30^2 + 40^2} = 50$ N

$F_{net} = ma$, so $50 = 8a$, which gives $a = 6.25$ ms^{-2}

$\tan\alpha = \frac{40}{30}$, i.e. $\alpha = 53.1^\circ$

$s = ut + \frac{1}{2}at^2$, $s = 0 \times 3 + \frac{1}{2} \times 6.25 \times 3^2 = 28.1$ m

4) Resolve horizontally:

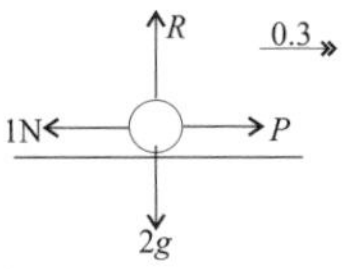

$F_{net} = ma$ $P - 1 = 2 \times 0.3$

So, P = 1.6 N

Resolve vertically: $R = 2g$

Limiting friction: $F = \mu R$, so $1 = \mu \times 2g$,

which gives $\mu = 0.05$ (to 2 d.p.)

5) Resolving in ↖ direction:

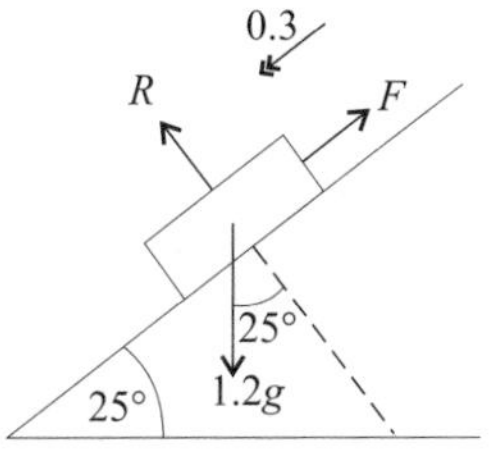

$F_{net} = ma$

$R - 1.2g\cos 25^\circ = 1.2 \times 0$

$R = 1.2g\cos 25^\circ$

$R = 10.66$ N

Resolving in ↙ direction:

$F_{net} = ma$

$1.2g\sin 25^\circ - F = 1.2 \times 0.3$

So, $F = 1.2g\sin 25^\circ - 1.2 \times 0.3 = 4.610$ N

Answers

Limiting friction, so:
$F = \mu R$
$4.610 = \mu \times 10.66$
$\mu = 0.43$ (to 2 d.p.)
Assumptions:
i) brick slides down line of greatest slope
ii) acceleration is constant
iii) no air resistance
iv) point mass / particle

6) Resolving in ↖ direction:

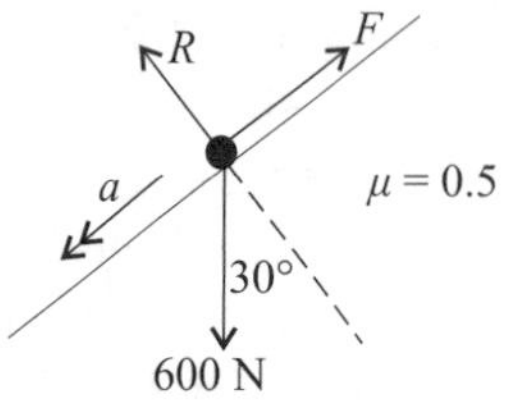

$F_{net} = ma$
$R - 600\cos30° = \left(\frac{600}{g}\right) \times 0$
$R = 600\cos30°$
Sliding, so $F = \mu R$
$F = 0.5 \times 600\cos30° = 259.8$
Resolving in ↙ direction:
$600\sin30° - F = \left(\frac{600}{g}\right)a$
$600\sin30° - 259.8 = 61.22 \times a$
$a = 0.6566 \text{ ms}^{-2}$

$u = 0$, $s = 20$, $a = 0.6566$, $v = ?$

$v^2 = u^2 + 2as$
$v^2 = 0^2 + 2 \times 0.6566 \times 20$
$v = 5.12 \text{ ms}^{-1}$ (3 s.f.)

7) Taking tractor and trailer together (and calling the resistance force on the trailer R):

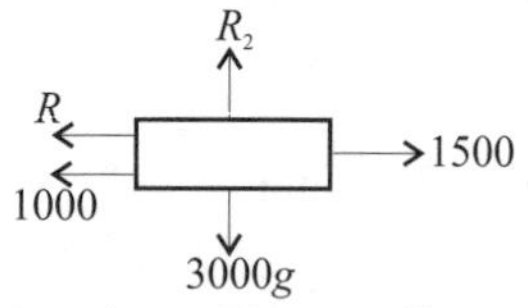

Resolving horizontally: $F_{net} = ma$
$1500 - R - 1000 = 3000 \times 0$
$R = 500$ N
For trailer alone:

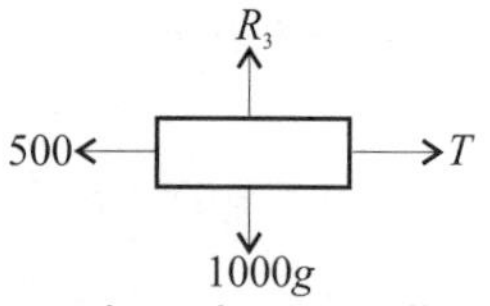

Resolving horizontally:
$F_{net} = ma$
$T - 500 = 1000 \times 0$
$T = 500$ N

T could be found instead by looking at the horizontal forces acting on the tractor alone.

8) Resolving downwards for A:

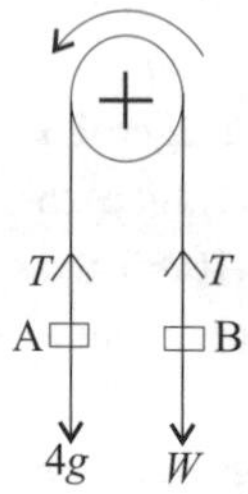

$F_{net} = ma$
$4g - T = 4 \times 1.2$
$T = 4g - 4.8$ ①
Resolving upwards for B:
$F_{net} = ma$
$T - W = \frac{W}{g} \times 1.2$ ②

Sub ① into ② :
$(4g - 4.8) - W = \frac{W}{g}(1.2)$
$4g - 4.8 = W(1 + \frac{1.2}{g})$
So $W = 30.6$ N

Here the particles are connected over a pulley, rather than in a straight line, but the key is still resolving — downwards and upwards instead of horizontally and vertically. No need for any panic then. Phew.

Exam Questions

1 a) Constant velocity, so, $a = 0$
Resolve horizontally:

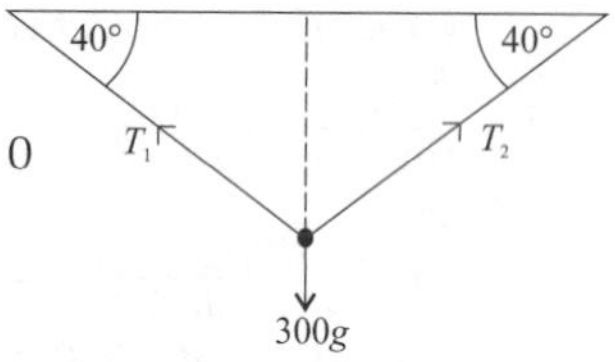

$F_{net} = ma$
$T_2\cos40° - T_1\cos40° = 300 \times 0$
$T_2\cos40° = T_1\cos40°$
$T_2 = T_1$
Resolve vertically:
$F_{net} = ma$
$T_1\sin40° + T_2\sin40° - 300g = 300 \times 0$
Let $T_1 = T_2 = T$: $2T\sin40° = 300g$
$T = 2290$ N (to 3 s.f.)

[4 marks available in total]:
- ***1 mark for resolving horizontally***
- ***1 mark for resolving vertically***
- ***1 mark for substituting T_1 or T_2***
- ***1 mark for the correct value of T***

More triangles, that's what I like to see...

b) Resolve horizontally: $F_{net} = ma$
$T_2\cos40° - T_1\cos40° = 300 \times 0.4$
$T_2 - T_1 = 156.65$ N ①
Resolve vertically:
$F_{net} = ma$
$T_1\sin40° + T_2\sin40° - 300g = 300 \times 0$
$T_1\sin40° + T_2\sin40° = 300g$
So $T_1 + T_2 = 4573.83$ N ②
from ①: $T_2 = T_1 + 156.65$
into ②: $T_1 + (T_1 + 156.65) = 4573.83$
so $2T_1 = 4417.18$
$T_1 = 2210$ N (to 3 s.f.)
and, $T_2 = 2370$ N (to 3 s.f.)

[6 marks available in total]:
- ***1 mark for resolving horizontally***
- ***1 mark for finding ①***
- ***1 mark for resolving vertically***
- ***1 mark for finding ②***
- ***1 mark for correct value of T_1***
- ***1 mark for correct value of T_2***

c) E.g. cables are inextensible, particle is considered as a point mass, there's no air resistance.

[2 marks available in total]:
- ***1 mark each for any 2 relevant assumptions.***

If you got these right, I will make the assumption that you've done some revision...

Answers

2 a) Resolving in ↗ direction:

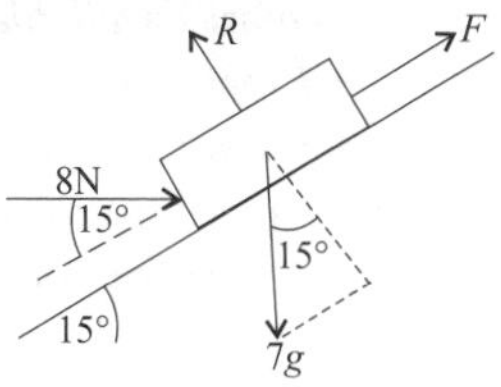

$F_{net} = ma$

$8\cos15° + F - 7g\sin15° = 7 \times 0$

$F = 7g\sin15° - 8\cos15°$

$F = 10.03$ N

Resolving in ↖ direction:

$F_{net} = ma$

$R - 8\sin15° - 7g\cos15° = 7 \times 0$

$R = 8\sin15° + 7g\cos15° = 68.33$ N

Limiting friction:

$F = \mu R$, i.e. $10.03 = \mu \times 68.33$,

which gives $\mu = 0.15$ (2 d.p.)

[5 marks available in total]:
- ***1 mark for resolving in ↗ direction***
- ***1 mark for correct value of F_{net} in ↗ direction***
- ***1 mark for resolving in ↖ direction***
- ***1 mark correct value of R***
- ***1 mark for correct value of μ***

b) 8 N removed:

Resolving in ↙ direction:

$7g\sin15° - F = 7a$ ①

Resolving in ↖ direction:

$R - 7g\cos15° = 7 \times 0$

$R = 7g\cos15° = 66.26$ N

$F = \mu R$

$F = 0.147 \times 66.26 = 9.74$ N

①: $7g\sin15° - 9.74 = 7a$

$a = \frac{8.01}{7} = 1.14$ ms^{-2} (to 2 d.p.)

$s = 3;\ u = 0;\ a = 1.14;\ t = ?$ $\quad s = ut + \frac{1}{2}at^2$

$3 = 0 + \frac{1}{2} \times 1.14 \times t^2 \quad t = \sqrt{\frac{6}{1.14}} = 2.3$ s (to 2 s.f.)

[7 marks available in total]:
- ***1 mark for resolving in ↙ direction***
- ***1 mark for resolving in ↖ direction***
- ***1 mark for correct value of R***
- ***1 mark for correct value of F***
- ***1 mark for correct value of a***
- ***1 mark for appropriate calculation method for t***
- ***1 mark for correct value of t***

3 a) Considering the car and the caravan together:

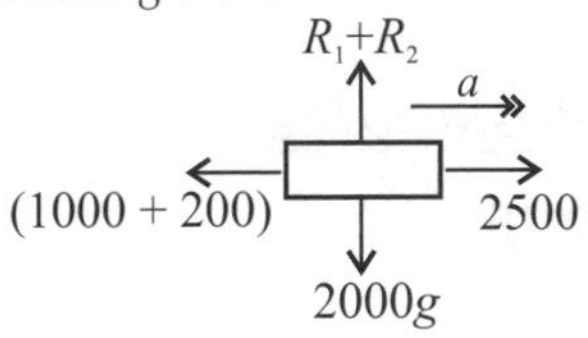

Resolving horizontally:

$F_{net} = ma$

$2500 - 1200 = 2000a$

$a = 0.65$ ms^{-2}

[3 marks available in total]:
- ***1 mark for resolving horizontally***
- ***1 mark for correct workings***
- ***1 mark for correct value of a***

b) Either: *Caravan*

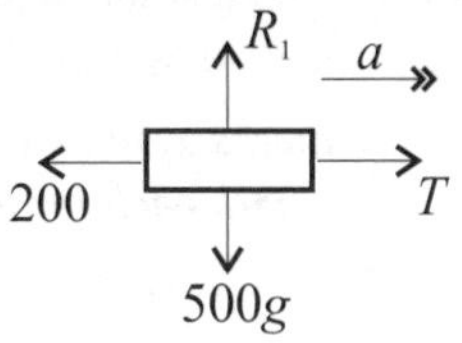

Resolving horizontally:

$F_{net} = ma$

$T - 200 = 500 \times 0.65$

$T = 525$ N

[2 marks available in total]:
- ***1 mark for resolving horizontally***
- ***1 mark for correct value of T***

Or: *Car*

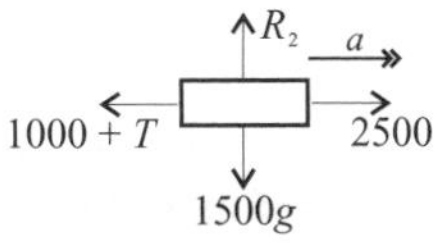

Resolving horizontally:

$F_{net} = ma$

$2500 - (1000 + T) = 1500 \times 0.65$

$2500 - 1000 - T = 975$

$1500 - 975 = T$

$T = 525$ N

[2 marks available in total]:
- ***1 mark for resolving horizontally***
- ***1 mark for correct value of T***

Two different methods, one correct answer. At the end of the day, it doesn't matter which you use (but show your diagrams and workings), although it's certainly a bonus if you manage to pick the simpler way and save a bit of time in the exam.

4 Before / After

$(0.8 \times 4) + (1.2 \times 2) = (0.8 \times 2.5) + 1.2v$

$3.2 + 2.4 = 2.0 + 1.2v$

$v = 3$ ms^{-1}

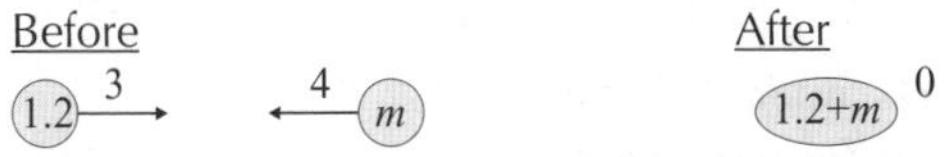

$(1.2 \times 3) + (m \times -4) = (1.2 + m) \times 0$

$3.6 = 4m$

$m = 0.9$ kg

[4 marks available in total]:
- ***1 mark for using conservation of momentum***
- ***1 mark for correct value of v***
- ***1 mark for correct workings***
- ***1 mark for correct value of m***

Diagrams are handy for collision questions too, partly because they make the question clearer for you, but they also make it easier for the examiner to see how you're going about answering the question.

5 a)

$(4000 \times 2.5) + (1000 \times 0) = 5000v$

$v = 2$ ms^{-1}

[2 marks available in total]:
- ***1 mark for using conservation of momentum***
- ***1 mark for correct value of v***

Answers

b) E.g. no resistance (e.g. friction) to motion; wagons can be modelled as particles.

[2 marks available in total]:
- ***1 mark each for any valid assumptions***

If you've forgotten what 'any valid assumptions' might be, I recommend you have a look at the table on p.168 again...

6 a) $(8\mathbf{i} - 3\mathbf{j}) = (x\mathbf{i} + y\mathbf{j}) + (5\mathbf{i} + \mathbf{j})$

So, $x\mathbf{i} + y\mathbf{j} = (8\mathrm{i} - 3\mathrm{j}) - (5\mathrm{i} + \mathrm{j})$, so $x = 3$ and $y = -4$

[2 marks available in total]:
- ***1 mark for correct value of x***
- ***1 mark for correct value of y***

b) Magnitude of resultant force $= \sqrt{8^2 + (-3)^2} = \sqrt{73}$

Using $F = ma$:

$\sqrt{73} = 2.5a$, so $a = 3.42$ ms^{-2} (to 3 s.f.)

[3 marks available in total]:
- ***1 mark using F = ma***
- ***1 mark for correct workings***
- ***1 mark for correct value of a***

7 a) Resolving forces acting on A:

$7g - T = 7a$

Resolving forces acting on B:

$T - 3g = 3a$, so $T = 3a + 3g$

Substituting T:

$7g - 3a - 3g = 7a$, so $4g = 10a$

hence $a = 3.92$ ms^{-2}

Using $t = \frac{(v - u)}{a}$:

$t = (5.9 - 0) \div 3.92$

So, $t = 1.51$ s (to 3 s.f.)

[4 marks available in total]:
- ***1 mark for resolving forces***
- ***1 mark for correct value of a***
- ***1 mark for correct workings***
- ***1 mark for correct value of t***

b) Using $s = \frac{v^2 - u^2}{2a}$

$s = (5.9^2 - 0^2) \div (2 \times 3.92)$

$s = 4.44$ m (to 3 s.f.)

[2 marks available in total]:
- ***1 mark for correct workings***
- ***1 mark for correct value of s***

c) When A hits the ground, speed of A = speed of B = 5.9 ms^{-1}

B will then continue to rise, momentarily stop and then fall freely under gravity. String will be taut again when displacement of B = 0.

So, $a = -9.8$, $s = 0$, $u = 5.9$

Using $s = ut + \frac{1}{2}at^2$:

$0 = 5.9t + \frac{1}{2}(-9.8)t^2 = 5.9t - 4.9t^2$

Solve for t:

$4.9t^2 = 5.9t$, so $t(4.9t - 5.9) = 0$

and so $t = 0$ s or $t = 5.9 \div 4.9 = 1.20$ s

So the string becomes taut again at $t = 1.20$ s (to 3 s.f.)

[4 marks available in total]:
- ***1 mark for using $s = ut + \frac{1}{2}at^2$***
- ***1 mark for correct workings***
- ***1 mark for solving for t***
- ***1 mark for correct value of T***

8 a) Total momentum $= m\begin{pmatrix} u \\ 1 \end{pmatrix} + 5\begin{pmatrix} 0 \\ -5 \end{pmatrix}$ ***[1 mark]***

b) (i) Using answer to part a) and conservation of momentum:

$m\begin{pmatrix} u \\ 1 \end{pmatrix} + 5\begin{pmatrix} 0 \\ -5 \end{pmatrix} = \begin{pmatrix} mu \\ m - 25 \end{pmatrix} = (5 + m)\begin{pmatrix} 2 \\ -1 \end{pmatrix}$ ***[1 mark]***

Considering the "vertical" component only:

$m - 25 = -5 - m$ ***[1 mark]***

$\Rightarrow 2m = 20 \Rightarrow m = 10$ ***[1 mark]***

(ii) First, find u using $\begin{pmatrix} mu \\ m - 25 \end{pmatrix} = (5 + m)\begin{pmatrix} 2 \\ -1 \end{pmatrix}$

Considering the "horizontal" component:

$mu = 2(5 + m) \Rightarrow 10u = 2(5 + 10)$ ***[1 mark]***

$\Rightarrow 10u = 30 \Rightarrow u = 3$ ***[1 mark]***

Find the speed using Pythagoras:

$\sqrt{3^2 + 1^2} = \sqrt{10} = 3.16$ ms^{-1} (3 s.f.) ***[1 mark]***

9 a) Using conservation of momentum:

$m\begin{pmatrix} 8 \\ 4 \end{pmatrix} + 3m\begin{pmatrix} 0 \\ 0 \end{pmatrix} = m\begin{pmatrix} v_P \\ 1 \end{pmatrix} + 3m\begin{pmatrix} 1 \\ v_Q \end{pmatrix}$ ***[1 mark]***. As there is a factor of m in every term, they can be cancelled: $\begin{pmatrix} 8 \\ 4 \end{pmatrix} = \begin{pmatrix} v_P \\ 1 \end{pmatrix} + 3\begin{pmatrix} 1 \\ v_Q \end{pmatrix}$.

Considering the "horizontal" and "vertical" components separately:

$8 = v_P + 3 \Rightarrow v_P = 5$ ***[1 mark]***

$4 = 1 + 3v_Q \Rightarrow v_Q = 1$ ***[1 mark]***

b) Now you have the velocities of P and Q as $\begin{pmatrix} 5 \\ 1 \end{pmatrix}$ ms^{-1} and $\begin{pmatrix} 1 \\ 1 \end{pmatrix}$ ms^{-1} respectively, use Pythagoras to find the speeds: ***[1 mark]***

Speed of $P = \sqrt{5^2 + 1^2} = \sqrt{26} = 5.10$ ms^{-1} (3 s.f.) ***[1 mark]***

Speed of $Q = \sqrt{1^2 + 1^2} = \sqrt{2} = 1.41$ ms^{-1} (3 s.f.) ***[1 mark]***

M1 Section 4 — Projectiles
Warm-up Questions

1)

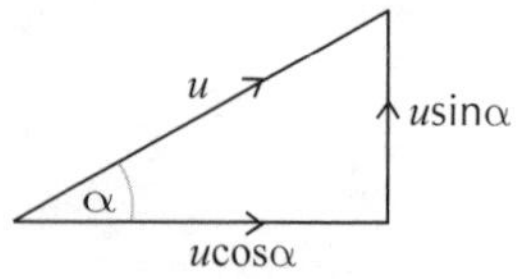

So, parallel to the horizontal, the initial velocity is $u\cos\alpha$.

2) Resolving horizontally (taking right as +ve):

$u = 120$; $s = 60$; $a = 0$; $t = ?$

$s = ut + \frac{1}{2}at^2$

$60 = 120t + \frac{1}{2} \times 0 \times t^2$

$t = 0.5$ s

Answers

Resolving vertically (taking down as +ve):
$u = 0$; $s = ?$; $a = 9.8$; $t = 0.5$
$s = ut + \frac{1}{2}at^2$
$= (0 \times 0.5) + (0.5 \times 9.8 \times 0.5^2)$
$= 1.23$ m (to 3 s.f.)

3) Resolving vertically (taking up as +ve):
$u = 22\sin\alpha$; $a = -9.8$; $t = 4$; $s = 0$
s = 0 because the ball lands at the same vertical level it started at.
$s = ut + \frac{1}{2}at^2$
$0 = 22\sin\alpha \times 4 + (0.5 \times -9.8 \times 4^2)$
Rearranging: $\sin\alpha = \frac{78.4}{88}$
$\Rightarrow \alpha = 63.0°$ (3 s.f.)
There are other ways to answer this question — you could use $v = u + at$ and use $t = 2$, which is the time taken to reach the highest point, when $v = 0$. I like my way though.

4) The minimum speed of the particle occurs when the particle is at the highest point of its flight, when the vertical component of its velocity is zero. (The horizontal component of its velocity is the same throughout the particle's motion.) When the vertical component of velocity is zero, the particle's speed is equal to the horizontal component of velocity.
So, minimum speed = 20cos20° = 18.8 ms^{-1} (3 s.f.).

Exam Questions

1 Resolving horizontally, taking right as +ve :
$u = 20\cos 30°$; $s = 30$; $a = 0$; $t = ?$

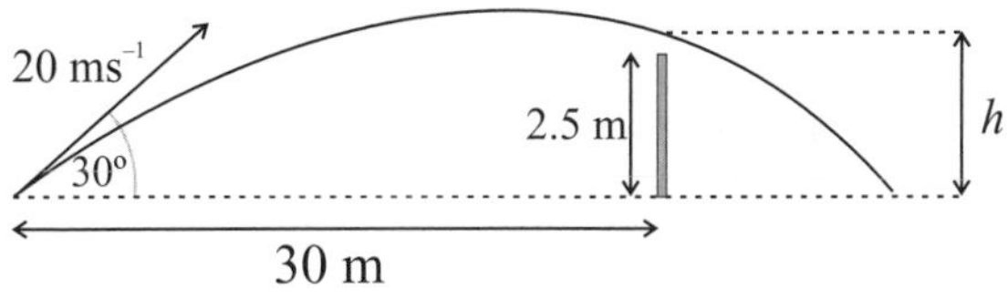

$s = ut + \frac{1}{2}at^2$ ***[1 mark]***
$30 = (20\cos 30° \times t)$
$t = 1.732$ s ***[1 mark]***
Resolving vertically, taking up as +ve:
$s = h$; $u = 20\sin 30°$; $t = 1.732$; $a = -9.8$
$s = ut + \frac{1}{2}at^2$ ***[1 mark]***
$h = (20\sin 30° \times 1.732) + (\frac{1}{2} \times -9.8 \times 1.732^2)$
$= 2.62$ m (to 3 s.f.) ***[1 mark]***
Therefore the ball goes over the crossbar. ***[1 mark]***
Assumptions: e.g. ball is a point mass/no air or wind resistance/no spin on the ball ***[1 mark]***

That was always my problem when I was taking free kicks — I didn't model the flight of the ball properly before kicking it, so no wonder I never scored.

2 a) $\tan\alpha = \frac{3}{4} \Rightarrow \sin\alpha = \frac{3}{5}$ and $\cos\alpha = \frac{4}{5}$ ***[1 mark]***
Resolving vertically, taking down as +ve:
$u = u_y = 15\sin\alpha = 9$; ***[1 mark]***
$s = 11$; $a = 9.8$; $t = ?$
$s = ut + \frac{1}{2}at^2$ ***[1 mark]***
$11 = 9t + 4.9t^2$ ***[1 mark]***
Use the quadratic formula to find:
$t = 0.839$ s (3 s.f.) ***[1 mark]***

b) Resolving horizontally, taking right as +ve:
$u = u_x = 15\cos\alpha = 15 \times \frac{4}{5} = 12$; ***[1 mark]***
$s = ?$; $t = 0.8390$ s
$OB = 12 \times 0.8390$ ***[1 mark]***
$= 10.07$ m
So stone misses H by $10.07 - 9 = 1.07$ m (3 s.f.) ***[1 mark]***

c) Resolving vertically, taking down as +ve:
$u = u_y = 9$; $s = 11$;
$a = 9.8$; $v = v_y$
$v^2 = u^2 + 2as$ ***[1 mark]***
$v_y^2 = 9^2 + (2 \times 9.8 \times 11) = 296.6$ ***[1 mark]***
Resolving horizontally, taking right as +ve:
$v_x = u_x = 12$ ***[1 mark]***.
So speed of landing = $\sqrt{v_x^2 + v_y^2}$
$= \sqrt{12^2 + 296.6}$ ***[1 mark]*** $= 21.0$ ms^{-1} (3 s.f.) ***[1 mark]***

d)

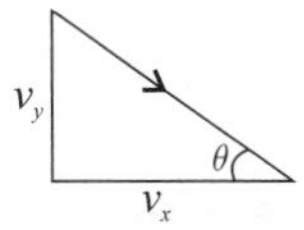

The angle, θ, that the direction of motion makes with the horizontal can be found using trigonometry: $\tan\theta = \frac{v_y}{v_x}$.
From part c), $v_y = \sqrt{296.6} = 17.22$ and $v_x = 12$, so:
$\tan\theta = \frac{17.22}{12}$ ***[1 mark]*** $= 1.435$
$\Rightarrow \theta = 55.1°$ (3 s.f.) ***[1 mark]***.

What a lovely way to finish off M1. If there was anything you struggled with then take a look back at the relevant pages, then its on to the practice exams.

M1 Practice Exam 1

1 a) With uniform motion questions, always start by writing down the data you have and the data you need.
$u = 15$, $v = 40$, $a = ?$, $t = 4$, $s = ?$
You need to find a, so '$v = u + at$' is the equation you need.
Rearrange to make a the subject and substitute in:
$a = \frac{v - u}{t} = \frac{40 - 15}{4} = 6.25$ ms^{-2}
[2 marks available in total]:
- ***1 mark for using '$v = u + at$' or equivalent***
- ***1 mark for correct value of a***

Answers

b) $u = 40$, $v = 26$, $a = -2.8$, $t = ?$, $s = ?$

You need t, so it's '$v = u + at$'. Rearrange and substitute:

$$t = \frac{v - u}{a} = \frac{26 - 40}{-2.8} = 5 \text{ s}$$

[2 marks available in total]:

- ***1 mark for using '$v = u + at$' or equivalent***
- ***1 mark for correct value of t***

c) You've now got all the other quantities except the two distances, so you can use any of the formulas with s in.

I'm going for '$s = \frac{1}{2}(u + v)t$' because it's nice and easy:

A to B: $s = \frac{1}{2}(u + v)t = \frac{1}{2}(15 + 40) \times 4 = 110$ m

B to C: $s = \frac{1}{2}(u + v)t = \frac{1}{2}(40 + 26) \times 5 = 165$ m

AC = 275 m

[3 marks available in total]:

- ***1 mark for using '$s = \frac{1}{2}(u + v)t$' or equivalent***
- ***1 mark for either intermediate distance correct***
- ***1 mark for the correct value of AC***

You REALLY need to know those equations.

2 a)

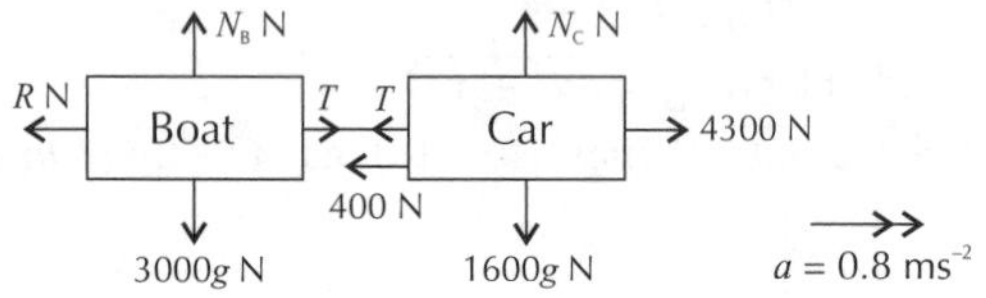

(i) Resolve horizontal forces acting on the car only:

$F_{net} = ma$ ***[1 mark]***

$\Rightarrow 4300 - 400 - T = 1600 \times 0.8$ ***[1 mark]***

$\Rightarrow T = 4300 - 400 - (1600 \times 0.8) = 2620$ N ***[1 mark]***.

(ii) Resolve horizontal forces acting on the boat only:

$F_{net} = ma$ ***[1 mark]***

$T - R = 3000 \times 0.8$ ***[1 mark]***

$\Rightarrow R = 2620 - (3000 \times 0.8) = 220$ N ***[1 mark]***.

b) (i) $F_{net} = ma \Rightarrow -220 = 3000a$ ***[1 mark]***

$\Rightarrow a = -220 \div 3000 = -0.07333$ ms^{-2} ***[1 mark]***.

So the boat's deceleration is 0.0733 ms^{-2} (3 s.f.).

(ii) Use $v^2 = u^2 + 2as$: ***[1 mark]***

$0 = 7^2 + (2 \times -0.07333 \times s)$ ***[1 mark]***

$\Rightarrow s = -49 \div (2 \times -0.07333) = 334$ m (3 s.f.) ***[1 mark]***.

3 a) For collision questions, always decide at the start (and label) which direction you're going to use as positive.

Diagram:

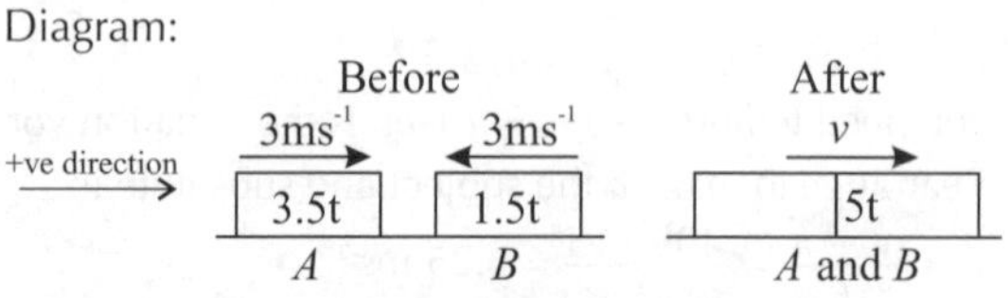

Use Conservation of Momentum to solve:

'total momentum of A and B before collision = momentum of combined trucks after collision'

$(3.5 \times 3) + (1.5 \times -3) = 5v$ ***[1 mark]***

So, $v = 6 \div 5 = 1.2$ ms^{-1} in the positive direction.

After the collision, the combined trucks move at 1.2 ms^{-1} ***[1 mark]*** in the same direction as A was moving before the collision ***[1 mark]***.

b)

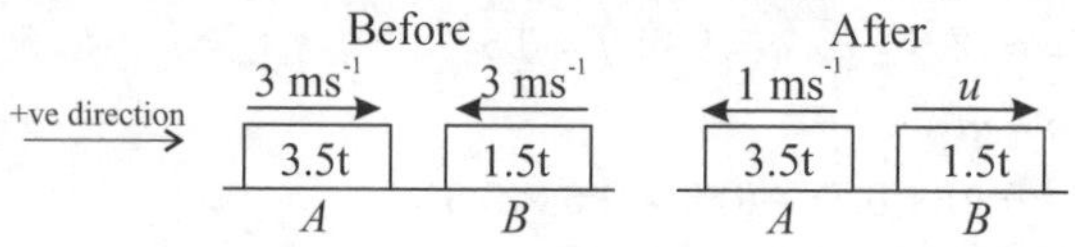

As in part a), use Conservation of Momentum to solve:

$(3.5 \times 3) + (1.5 \times -3) = (3.5 \times -1) + (1.5 \times u)$ ***[1 mark]***

So $u = 9.5 \div 1.5 = 6.33$ ms^{-1} (3 s.f.) ***[1 mark]*** in the same direction A was moving before the collision ***[1 mark]***.

Not sure why, but I find these questions very satisfying when I've solved them. Maybe I'm just a freak.

4 a) $u = 0$, $v = ?$, $a = 1.5$, $t = 8$, $s = ?$

So you're going to need '$v = u + at$'

$v = 0 + (1.5 \times 8) = 12$ ms^{-1}

Greatest speed = 12 ms^{-1}

[2 marks available in total]:

- ***1 mark for using '$v = u + at$' or equivalent***
- ***1 mark for correct value of speed***

b) Use the information from part (a) to mark when the cyclist stops accelerating and travels at a constant speed:

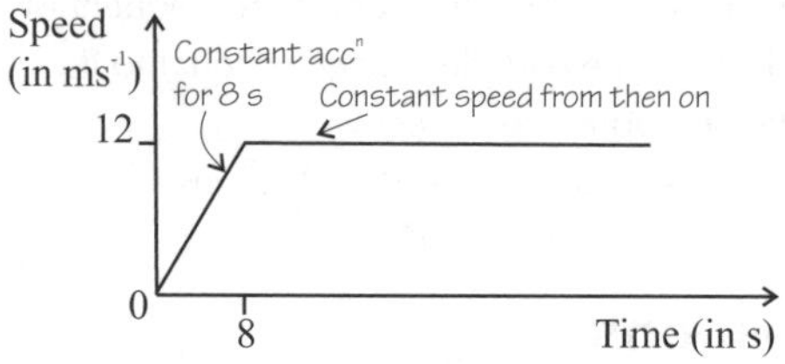

[2 marks available in total]:

- ***1 mark for correct shape***
- ***1 mark for correct numbers***

c)

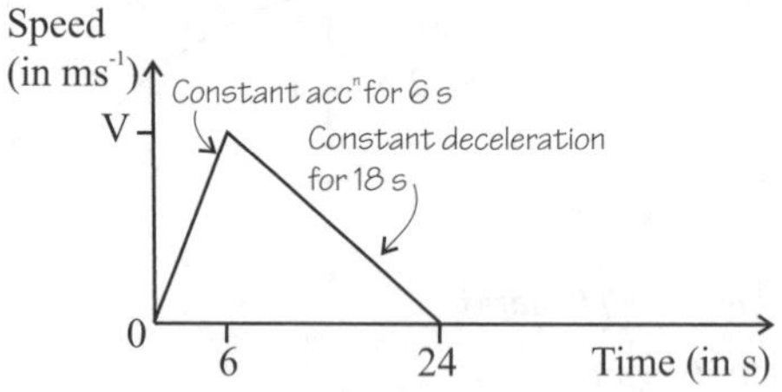

[2 marks available in total]:

- ***1 mark for correct shape***
- ***1 mark for correct numbers***

d) First work out how long the cyclist was travelling at a constant speed:

total time until overtaking = 24 s,

so cyclist travelled at a constant speed for 24 – 8 = 16 s.

Total distance travelled = area under speed-time graph.

The area is split into two simple shapes:

area of triangle = $\frac{1}{2} \times 12 \times 8 = 48$

area of rectangle = $12 \times 16 = 192$

Distance travelled = 48 + 192 = 240 m

[3 marks available in total]:

- ***1 mark for using 'distance = area under graph' or equivalent***
- ***1 mark for correct working***
- ***1 mark for the correct value for distance travelled***

Answers

e) The cycle and car travel the same distance, so you can use the result from part (d). Again, the distance travelled is the area under the graph, so you can work back from there. This time the area under the graph is a simple triangle, so it's a bit easier.

distance = area under graph = $240 = \frac{1}{2} \times 24 \times V$

So, $V = 20$ ms^{-1}

[3 marks available in total]:
- ***1 mark for 'distance = area under graph' or equivalent***
- ***1 mark for recognising both cycle and car travelled the same distance***
- ***1 mark for correct value of V***

This is as easy as Mechanics gets, so there's no excuse for getting questions like this wrong. If you found this hard, get it learnt before the exam. These questions never vary much — if you can do one, you can do them all.

5 a) Here's a nice little diagram to help you out:

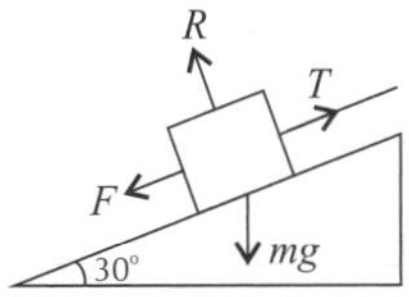

where T is the tension in the string, F is the friction between the block and the surface, mg is the block's weight and R is the reaction force of the surface acting on the block.

Resolving perpendicular to the slope:

$R - mg\cos 30 = 0$ ***[1 mark]***

$R = 5 \times 9.8 \times \cos 30 = 42.435... = 42.4$ N (3 s.f.) ***[1 mark]***

b) Resolve forces parallel to slope:

$T = F + mg\sin 30$ ***[1 mark]***

$F = \mu R = 42.44\mu$

$= 42.44 \times 0.3 = 12.73$ ***[1 mark]***

$\Rightarrow T = 12.73 + (5 \times 9.8 \times \sin 30)$

$= 37.23$ N (4 s.f.) $= 37.2$ N (3 s.f.) ***[1 mark]***

c) The block now slides down the slope, so friction acts up the slope, opposing the motion:

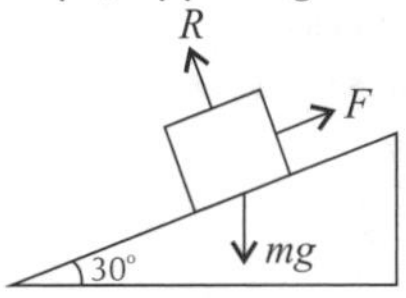

Again, resolving parallel to the slope:

$F_{net} = mg\sin 30 - F$ (taking down the slope as positive)

$= (5 \times 9.8 \times \sin 30) - F$ ***[1 mark]***

$= 11.77$ N (4 s.f.) $= 11.8$ N (3 s.f.) ***[1 mark]***

d) $F_{net} = ma$ ***[1 mark]***

$\Rightarrow a = 11.77 \div 5 = 2.35$ ms^{-2} (3 s.f.) ***[1 mark]***

So the block accelerates down the slope at a rate of 2.35 ms^{-2}.

6 a) The forces are given in terms of components, so to find the resultant force you just add them as vectors:

$\mathbf{F} = \mathbf{F}_1 + \mathbf{F}_2 = (3\mathbf{i} + 2\mathbf{j})$ N $+ (2\mathbf{i} - \mathbf{j})$ N $= (5\mathbf{i} + \mathbf{j})$ N

To find the angle of the resultant force, it's just trig as usual...

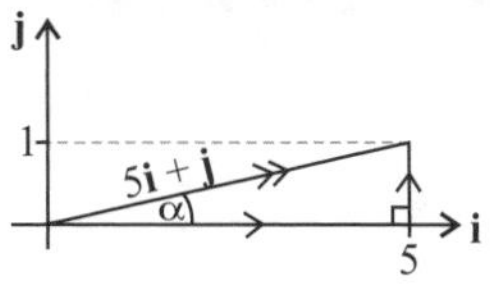

angle $= \alpha = \tan^{-1}\left(\frac{1}{5}\right) = 11.3°$

[3 marks available in total]:
- ***1 mark for adding vectors to get resultant force***
- ***1 mark for using tan⁻¹ or equivalent correctly***
- ***1 mark for correct answer***

b) There are a couple of ways of doing this, but both involve Pythagoras and '$F = ma$'. First way:

Find the magnitude of the resultant force:

$|\mathbf{F}| = \sqrt{(5^2 + 1^2)} = \sqrt{26} = 5.10$ N

Then use Newton's second law — $F = ma$:

Rearrange to give: $a = \frac{F}{m} = \frac{5.10}{0.5} = 10.2$ ms^{-2}

(Second way uses $\mathbf{F} = m\mathbf{a}$ to find vector form of accⁿ, then Pythagoras to find magnitude. Either way is fine.)

Magnitude of acceleration = 10.2 ms^{-2}

[4 marks available in total]:
- ***1 mark for using 'F = ma'***
- ***1 for correct calculation of <u>either</u> magnitude of force <u>or</u> vector form of acceleration (whichever method is used)***
- ***1 mark for correct workings***
- ***1 for correct value of a***

Here we go again — "split vector into components, use Pythagoras to find magnitude, ya de ya de ya". Yawn...

7 a) $\mathbf{u} = (40\mathbf{i} + 108\mathbf{j})$

$\mathbf{a} = (4\mathbf{i} + 12\mathbf{j})$

$t = \frac{1}{2}$

Use $\mathbf{s} = \mathbf{u}t + \frac{1}{2}\mathbf{a}t^2$: ***[1 mark]***

$\mathbf{s} = \frac{1}{2}(40\mathbf{i} + 108\mathbf{j}) + \left[\frac{1}{2} \times \left(\frac{1}{2}\right)^2 \times (4\mathbf{i} + 12\mathbf{j})\right]$ ***[1 mark]***

$\mathbf{s} = (20.5\mathbf{i} + 55.5\mathbf{j})$ km ***[1 mark]***

Make sure you keep track of the units you're using — you have to convert the time from minutes to hours to match the units of the velocity and acceleration.

b) $\mathbf{u} = (60\mathbf{i} + 120\mathbf{j})$

$t = \frac{1}{4}$

$\mathbf{s} = (20.5\mathbf{i} + 55.5\mathbf{j}) - (2.5\mathbf{i} + 25.5\mathbf{j}) = (18\mathbf{i} + 30\mathbf{j})$

Use $\mathbf{s} = \mathbf{u}t + \frac{1}{2}\mathbf{a}t^2$: ***[1 mark]***

$(18\mathbf{i} + 30\mathbf{j}) = \frac{1}{4}(60\mathbf{i} + 120\mathbf{j}) + \frac{1}{2} \times \left(\frac{1}{4}\right)^2 \times \mathbf{a}$ ***[1 mark]***

$\Rightarrow \frac{1}{32}\mathbf{a} = (18\mathbf{i} + 30\mathbf{j}) - (15\mathbf{i} + 30\mathbf{j})$

$\Rightarrow \mathbf{a} = 32(3\mathbf{i} + 0\mathbf{j}) = 96\mathbf{i}$ ***[1 mark]***

So the magnitude of the acceleration is 96 kmh^{-2} ***[1 mark]***.

Answers

c) $\mathbf{u} = (30\mathbf{i} - 40\mathbf{j})$
$\mathbf{a} = (5\mathbf{i} + 8\mathbf{j})$
$\mathbf{v} = \mathbf{v}$
$t = ?$
$\mathbf{s} = ?$
Use $\mathbf{v} = \mathbf{u} + \mathbf{a}t$ to find the time that R is moving parallel to $\mathbf{i}$ (the unit vector in the direction of east): ***[1 mark]***
$\mathbf{v} = (30\mathbf{i} - 40\mathbf{j}) + t(5\mathbf{i} + 8\mathbf{j})$ ***[1 mark]***
When R is moving parallel to $\mathbf{i}$, the $\mathbf{j}$-component of its velocity is zero, so:
$0 = -40 + 8t \Rightarrow t = 5$. ***[1 mark]***
Now use $\mathbf{s} = \mathbf{u}t + \frac{1}{2}\mathbf{a}t^2$ to find the position of R at this time: ***[1 mark]***
$\mathbf{s} = 5(30\mathbf{i} - 40\mathbf{j}) + \frac{1}{2} \times 25 \times (5\mathbf{i} + 8\mathbf{j})$ ***[1 mark]***
$\mathbf{s} = (150\mathbf{i} - 200\mathbf{j}) + (62.5\mathbf{i} + 100\mathbf{j})$
$\Rightarrow \mathbf{s} = (212.5\mathbf{i} - 100\mathbf{j})$ km ***[1 mark]***.

8 a) Consider motion vertically, taking up as positive:
$u = 10\sin 20$, $a = -9.8$, $v = 0$.
Use $v^2 = u^2 + 2as$ ***[1 mark]***:
$0 = (10\sin 20)^2 - 19.6s$ ***[1 mark]***
$\Rightarrow s = (10\sin 20)^2 \div 19.6 = 0.597$ m.
The frisbee is thrown from 1 m above the ground, so the maximum height reached is $1 + 0.597 = 1.60$ m (3 s.f.)
[1 mark]

b) Again consider vertical motion:
$u = 10\sin 20$, $a = -9.8$, $s = -1$, $t = ?$.
Use $s = ut + \frac{1}{2}at^2$ ***[1 mark]***:
$-1 = (10\sin 20)t - 4.9t^2$ ***[1 mark]***
$\Rightarrow 4.9t^2 - (10\sin 20)t - 1 = 0$.
Use the quadratic formula to find t:
$$t = \frac{10\sin 20 + \sqrt{(-10\sin 20)^2 + (4 \times 4.9 \times 1)}}{9.8}$$ ***[1 mark]***
$= 0.91986... = 0.920$ s (3 s.f.) ***[1 mark]***

You don't need to worry about the other value of t that the formula gives you as it will be negative, and you can't have a negative time.

c) Consider horizontal motion:
$u = 10\cos 20$, $a = 0$, $t = 0.9199$, $s = ?$:
Use $s = ut + \frac{1}{2}at^2$ ***[1 mark]***:
$s = 10\cos 20 \times 0.9199$ ***[1 mark]*** $= 8.64$ m (3 s.f.) ***[1 mark]***

M1 Practice Exam 2

1 a) Resolve along x-axis:
$P\sin 40° - 20\cos(105-90)° = 0$
so, $P = \frac{20\cos 15°}{\sin 40°} = 30.054... = 30.1$ N (3 s.f.)
[2 marks available in total]:
- ***1 mark for correct workings***
- ***1 mark for correct value of P***

Something simple to get you started. I bet you'd never have guessed it would involve resolving forces.

b) Resolve along y-axis:
$R = 30.05\cos 40° + 35 - 20\sin 15°$
$= 52.8$ N (3 s.f.) up the y-axis
[3 marks available in total]:
- ***1 mark for correct workings***
- ***1 mark for correct value of R***
- ***1 mark for correct direction***

2 a) Using $F_{net} = ma$ and resolving vertically:
$34\,000\,000 - (1\,400\,000 \times 9.8) = (1\,400\,000 \times a)$
So, $a = \frac{34\,000\,000 - 13\,720\,000}{1\,400\,000}$
$= 14.485... = 14.5$ ms^{-2} (3 s.f.)
[2 marks available in total]:
- ***1 mark for resolving vertically***
- ***1 mark for correct value of a***

b) Again, using $F_{net} = ma$ and resolving vertically:
$34\,000\,000 - 13\,720\,000 - R = 1\,400\,000 \times 12$
so, $R = 34\,000\,000 - 13\,720\,000 - 16\,800\,000$
$= 3\,480\,000$ N
[2 marks available in total]:
- ***1 mark for resolving vertically***
- ***1 mark correct value of R***

c) $u = 0$; $a = 12$; $s = 20\,000$; $t = ?$
Using $s = ut + \frac{1}{2}at^2$:
$20\,000 = (0 \times t) + \frac{1}{2}(12 \times t^2)$
so, $t^2 = 20\,000 \div 6$ and $t = 58$ s (to the nearest second)
[2 marks available in total]:
- ***1 mark for using '$s = ut + \frac{1}{2}at^2$'***
- ***1 mark for correct value of t***

d) Use $v^2 = u^2 + 2as$ and the values from c) to find the speed at 20 km:
$v^2 = 0^2 + 2(12 \times 20\,000)$
$\Rightarrow v^2 = 480\,000$ ***[1 mark]***
The acceleration of the rocket is now $-g$, so using $v^2 = u^2 + 2as$:
$0 = 480\,000 + (2 \times -9.8 \times s)$ ***[1 mark]***
At its maximum height, the rocket's velocity will be 0 ms^{-1}.
So $s = 24\,500$ m (3 s.f.) ***[1 mark]***
Add to initial height:
$20\,000 + 24\,500 = 44\,500$ m $= 44.5$ km (3 s.f.) ***[1 mark]***

You could have used other motion equations here, but it's best to use one which includes the values that you're given in the question, not values that you've found yourself.

e) e.g. g is constant/R is constant up to 20 km/the force from the engines is constant/the mass of the rocket is constant.
[1 mark for any valid assumption]

Answers

3 a)

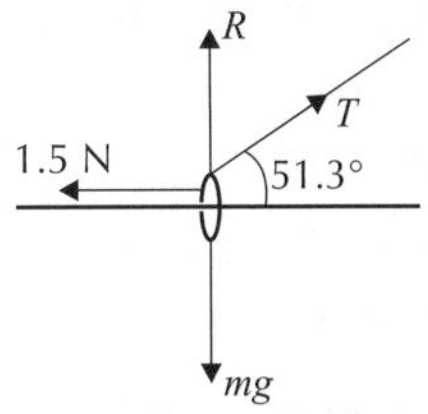

[2 marks available in total]:

- ***1 mark for correct force arrows***
- ***1 mark for correctly labelling the forces***

b) Resolving horizontally:

$T\cos 51.3° - 1.5 = 0$ so $T = \frac{1.5}{\cos 51.3} = 2.399$ N

Resolving vertically upwards:

$R + T\sin 51.3° - mg = 0$

$R = \frac{F}{\mu} = 1.5 \div 0.6 = 2.5$ N

so $mg = 2.5 + 2.399\sin 51.3°$

and $m = 0.45$ kg (to 2 s.f.)

[6 marks available in total]:

- ***1 mark for resolving horizontally***
- ***1 mark for correct value of T***
- ***1 mark for resolving vertically***
- ***1 mark for using F = μR***
- ***1 mark for correct workings***
- ***1 mark for correct value of m***

4 a)

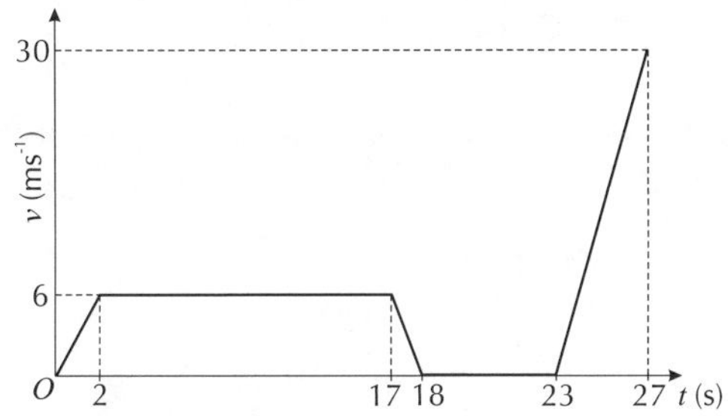

[2 marks available in total]:

- ***1 mark for correct shape***
- ***1 mark for correct numbers***

b) a = gradient of graph

$t = 0 - 2$ s: $a = 6 \div 2 = 3$ ms^{-2}

$t = 17 - 18$ s: $a = -6 \div 1 = -6$ ms^{-2}

$t = 23 - 27$ s: $a = 30 \div 4 = 7.5$ ms^{-2}

Greatest acceleration is 7.5 ms^{-2}.

[3 marks available in total]:

- ***1 mark for finding gradients or equivalent***
- ***1 mark for correct workings***
- ***1 mark for correct statement of greatest acceleration***

c) Distance = area under the graph

$t = 0 - 2$ s: $s = \frac{1}{2}(6)2 = 6$ m (triangle: area = $\frac{1}{2}bh$)

$t = 2 - 17$ s: $s = 6 \times 15 = 90$ m (rectangle)

$t = 17 - 18$ s: $s = \frac{1}{2}(6)1 = 3$ m (triangle)

$t = 18 - 23$ s: $s = 0 \times 5 = 0$ m (rectangle)

$t = 23 - 27$ s: $s = \frac{1}{2}(30)4 = 60$ m (triangle)

Total distance = 6 + 90 + 3 + 0 + 60 = 159 m

[3 marks available in total]:

- ***1 mark for using distance = area under graph***
- ***1 mark for correct workings***
- ***1 mark for correct value of total distance***

I don't really like roller coasters, but they do have some great maths involved. I'd give up writing books to go and design them, unless I had to test them. I feel sick thinking about that.

5 a) Consider vertical motion, taking up as positive:

$a = -g$, $u = U\sin\alpha$, $v = 0$, $s = ?$:

Use $v^2 = u^2 + 2as$ ***[1 mark]***:

$0 = U^2\sin^2\alpha - 2gs$ ***[1 mark]***

$\Rightarrow s = \frac{U^2\sin^2\alpha}{2g}$ ***[1 mark]***.

The frog is initially 0.5 m above the ground, so the maximum height, h, it reaches is:

$h = \frac{1}{2} + \frac{U^2\sin^2\alpha}{2g} = \frac{g + U^2\sin^2\alpha}{2g}$ m, as required ***[1 mark]***.

b) Still considering vertical motion, use $v = u + at$ ***[1 mark]***:

$0 = U\sin\alpha - gt$ ***[1 mark]*** $\Rightarrow t = \frac{U\sin\alpha}{g}$ s ***[1 mark]***.

c) You first need to find the horizontal and vertical components of the frog's motion when it lands.

Vertically:

$u = U\sin\alpha$, $a = -g$, $s = -0.5$, $v = v_V$.

Use $v^2 = u^2 + 2as$ ***[1 mark]***:

$v_V^2 = U^2\sin^2\alpha + g$ ***[1 mark]***.

Horizontally, $v_H = u_H = U\cos\alpha$, as acceleration is zero ***[1 mark]***.

So, $V = \sqrt{v_V^2 + v_H^2} = \sqrt{U^2\sin^2\alpha + g + U^2\cos^2\alpha}$ ***[1 mark]***

$= \sqrt{U^2(\sin^2\alpha + \cos^2\alpha) + g}$

$= \sqrt{U^2 + g}$ ms^{-1}, as required ***[1 mark]***.

6 a) Call the new particle R.

Use conservation of momentum:

$m_P\mathbf{u}_P + m_Q\mathbf{u}_Q = (m_P + m_Q)\mathbf{v}_R$ ***[1 mark]***

$5\begin{pmatrix}3\\-9\end{pmatrix} + 10\begin{pmatrix}0\\0\end{pmatrix} = 15\mathbf{v}_R$ ***[1 mark]***

$\Rightarrow \begin{pmatrix}15\\-45\end{pmatrix} = 15\mathbf{v}_R$

$\Rightarrow \mathbf{v}_R = \begin{pmatrix}1\\-3\end{pmatrix}$ ms^{-1} ***[1 mark]***.

b) Again, use conservation of momentum:

$m_P\mathbf{u}_P + m_Q\mathbf{u}_Q = m_P\mathbf{v}_P + m_Q\mathbf{v}_Q$ ***[1 mark]***

$5\begin{pmatrix}3\\-9\end{pmatrix} + 10\begin{pmatrix}0\\0\end{pmatrix} = 5\begin{pmatrix}2\\0\end{pmatrix} + 10\mathbf{v}_Q$ ***[1 mark]***

$\Rightarrow \begin{pmatrix}15\\-45\end{pmatrix} - \begin{pmatrix}10\\0\end{pmatrix} = 10\mathbf{v}_Q$

$\Rightarrow \begin{pmatrix}5\\-45\end{pmatrix} = 10\mathbf{v}_Q$

$\Rightarrow \mathbf{v}_Q = \begin{pmatrix}0.5\\-4.5\end{pmatrix}$ ***[1 mark]***.

7 a) The pirate ship is travelling with constant speed, so use $\mathbf{s} = \mathbf{u}t$ with $\mathbf{u} = (9\mathbf{i} + 8\mathbf{j})$ and $\mathbf{s} = (63\mathbf{i} + 56\mathbf{j})$:

$(63\mathbf{i} + 56\mathbf{j}) = t(9\mathbf{i} + 8\mathbf{j})$ ***[1 mark]***

$\Rightarrow t = 7$ hours ***[1 mark]***.

So time of arrival is 1350 + 0700 = 2050 hours ***[1 mark]***.

Answers

b) $t = 7$

$\mathbf{s} = (63\mathbf{i} + 56\mathbf{j}) - (-18\mathbf{i} + 6\mathbf{j}) = (81\mathbf{i} + 50\mathbf{j})$ ***[1 mark]***

$\mathbf{u} = \mathbf{0}$ (the naval ship is at rest at 1350)

$\mathbf{a} = ?$

Use $\mathbf{s} = \mathbf{u}t + \frac{1}{2}\mathbf{a}t^2$: ***[1 mark]***

$(81\mathbf{i} + 50\mathbf{j}) = \frac{1}{2} \times 49 \times \mathbf{a}$ ***[1 mark]***

$\Rightarrow \mathbf{a} = \frac{2}{49}(81\mathbf{i} + 50\mathbf{j})$

$\Rightarrow \mathbf{a} = (3.31\mathbf{i} + 2.04\mathbf{j})$ kmh^{-2} (3 s.f.) ***[1 mark]***

Argh, mateys, we've been caught by those naval scallywags... If only we'd studied M1 harder we might've escaped capture. Learn from those scurvy pirates — revise. Disclaimer: CGP does not condone piracy.

8 a) 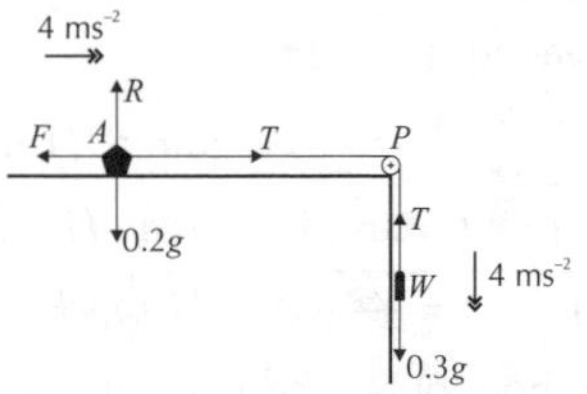

If mass of $A = 0.2$ kg, mass of $W = 1.5 \times 0.2 = 0.3$ kg

For W, $F_{net} = ma$:

$0.3g - T = 0.3(4)$

$T = 2.94 - 1.2 = 1.74$

For A, $F_{net} = ma$:

$T - F = 4(0.2) = 0.8$

$F = \mu R$,

resolving vertically: $R = 0.2g = 1.96$

so $F = 1.96\mu$,

so $T - 1.96\mu = 0.8$,

$1.74 - 1.96\mu = 0.8$

$1.96\mu = 0.94$

so $\mu = 0.47959... = 0.480$ (3 s.f.)

[5 marks available in total]:

- ***1 mark for using correct masses of A and W***
- ***1 mark for using 'F = ma'***
- ***1 mark for correct value of T***
- ***1 mark for correct value of R***
- ***1 mark for correct value of μ***

b) Find speed of A when string goes slack (this will be the same as the speed of W when it hits the floor):

$u = 0$; $v = ?$; $a = 4$; $s = h$

Use $v^2 = u^2 + 2as$:

$v^2 = 0^2 + 2(4 \times h)$ ***[1 mark]***

$v^2 = 8h \Rightarrow v = \sqrt{8h}$ ***[1 mark]***

Find the acceleration of A once the string has gone slack by resolving forces horizontally:

$F_{net} = ma \quad \Rightarrow \quad -F = 0.2a$, ***[1 mark]***

where $F = \mu R = 1.96\mu = 0.94$ (from part a) ***[1 mark]***

So, $-0.94 = 0.2a \Rightarrow a = -4.7$ ms^{-2} ***[1 mark]***

For the motion of A between the point where the string goes slack and P:

$u = \sqrt{8h}$; $v = 3$; $a = -4.7$; $s = \frac{7}{4}h - h = \frac{3}{4}h$

Use $v^2 = u^2 + 2as$:

$9 = 8h - (2 \times 4.7 \times 0.75h) = 0.95h$ ***[1 mark]***

So $h = 9.47$ m (3 s.f.) ***[1 mark]***

Tricky. And a little bit sticky.

c) E.g. string is inextensible — so A and W have the same acceleration when W is falling.

[1 mark for correct explanation]

9 a) Draw a triangle of the velocities, and use the fact that the banks are parallel to find one of the interior angles:

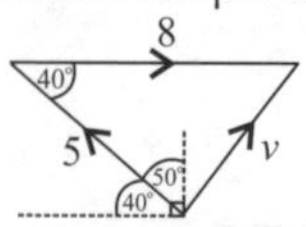

[1 mark]

Use the cosine rule to find the magnitude of the resultant velocity: $v^2 = 8^2 + 5^2 - (2 \times 8 \times 5)\cos 40$ ***[2 marks]***

$v^2 = 27.716$

$\Rightarrow v = 5.26$ ms^{-1} (3 s.f.) ***[1 mark]***

b) The boat is modelled as a particle. ***[1 mark]***

A simple question to finish off. Well, you've earned it. That's M1 in the bag, so practise, practise and practise some more until the exam. If the next thing you have to do is the exam, then good luck — Mechanics loves you.

Answers

D1 Section 1 — Algorithms

Warm-up Questions

1) a) Input: raw ingredients (e.g. vegetables, water etc.)
Output: vegetable soup.
b) Input: starting point (Leicester Square)
Output: final destination (the Albert Hall)
c) Input: components (e.g. shelves, screws etc.)
Output: finished TV cabinet *(in theory — I'm not much good at flat-pack)*

2)

x	y
17	56
~~8~~	~~112~~
~~4~~	~~224~~
~~2~~	~~448~~
1	896
Total	952

So 17 × 56 = 952.

3) Diamond-shaped boxes are used for decisions — they'll ask a question.

4) $a = 16$

n	b	Output	$n = a$?
1	16	1	No
2	8	2	No
3	$5\frac{1}{3}$		No
4	4	4	No
5	$3\frac{1}{5}$		No
6	$2\frac{2}{3}$		No
7	$2\frac{2}{7}$		No
8	2	8	No
9	$1\frac{7}{9}$		No
10	$1\frac{3}{5}$		No
11	$1\frac{5}{11}$		No
12	$1\frac{1}{3}$		No
13	$1\frac{3}{13}$		No
14	$1\frac{1}{7}$		No
15	$1\frac{1}{15}$		No
16	1	16	Yes

So the factors of 16 are 1, 2, 4, 8 and 16.

5) 72, 57, 64, 54, 68, 71 swap
57, 72, 64, 54, 68, 71 swap
57, 64, 72, 54, 68, 71 swap
57, 64, 54, 72, 68, 71 swap
57, 64, 54, 68, 72, 71 swap
57, 64, 54, 68, 71, 72 end of first pass.
At the end of the second pass, the list is:
57, 54, 64, 68, 71, 72.
At the end of the third pass, the list is:
54, 57, 64, 68, 71, 72.
There are no swaps on the fourth pass, so the list is in order.
It took 4 passes to get the list in order.

6) The maximum number of comparisons is 9 + 8 + 7 + 6 + 5 + 4 + 3 + 2 + 1 = 45 (or ½ × 9 × 10 = 45).

7)
First pass: 21, 11, 23, 19, 28, 26 swap
Second pass: 11, 21, 23, 19, 28, 26 no swap
Third pass: 11, 21, 23, 19, 28, 26 swap
11, 21, 19, 23, 28, 26 swap
11, 19, 21, 23, 28, 26 no swap
Fourth pass: 11, 19, 21, 23, 28, 26 no swap
Fifth pass: 11, 19, 21, 23, 28, 26 swap
11, 19, 21, 23, 26, 28 no swap

8) There are 7 numbers, so you want 7/2 = 3.5 = 3 subsets:
Subset 1: 101 113 84
Subset 2: 98 87
Subset 3: 79 108
Sorting these subsets gives:
Subset 1: 84 101 113
Subset 2: 87 98
Subset 3: 79 108
Putting this list back together gives: 84, 87, 79, 101, 98, 108, 113. Now divide the list up into 3/2 = 1.5 = 1 subset and use a shuttle sort to sort the numbers:
First pass: 84, 87, 79, 101, 98, 108, 113 no swap
Second pass: 84, 87, 79, 101, 98, 108, 113 swap
84, 79, 87, 101, 98, 108, 113 swap
Third pass: 79, 84, 87, 101, 98, 108, 113 no swap
Fourth pass: 79, 84, 87, 101, 98, 108, 113 swap
79, 84, 87, 98, 101, 108, 113 no swap
Fifth pass: 79, 84, 87, 98, 101, 108, 113 no swap
Sixth pass: 79, 84, 87, 98, 101, 108, 113 no swap
The shuttle sort is completed, so the list is in order.

9) Use the first item (0.8) as the pivot. The list becomes:
0.7, 0.5, 0.4, 0.1, 0.8, 1.2, 1.0.
Use 0.7 as the pivot for the first list, and 1.2 as the pivot for the second list. The list becomes:
0.5 0.4, 0.1, 0.7, 0.8, 1.0, 1.2.
1.0 is in a list on its own, so it's in the right place. In the other list, use 0.5 as the pivot, so the list becomes:
0.4, 0.1, 0.5, 0.7, 0.8, 1.0, 1.2.
Use 0.4 for the pivot in the remaining list, so the final (ordered) list is:
0.1, 0.4, 0.5, 0.7, 0.8, 1.0, 1.2.
If you'd used different pivots, your working would look different.

Exam Questions

1 a) Use the first item (77) as the pivot. The list becomes:
77, 83, 96, 105, 78, 89, 112, 80, 98, 94 ***[1 mark]***
Use 83 as the next pivot. The list becomes:
77, 78, 80, 83, 96, 105, 89, 112, 98, 94 ***[1 mark]***
Use 78 as the pivot in the first list, and 96 as the pivot in the second list. The list becomes:
77, 78, 80, 83, 89, 94, 96, 105, 112, 98 ***[1 mark]***
80 is in a list on its own, so it's in the correct place.
Use 89 and 105 as the pivots. The list becomes:
77, 78, 80, 83, 89, 94, 96, 98, 105, 112 ***[1 mark]***
94, 98 and 112 are in lists on their own, so they are in the correct places. The list is now in order.
[1 mark for stating all pivots used]

Answers

b) (i) After the first pass, the final (6th) number in the list will be in the correct position. ***[1 mark]***

(ii) There are 6 items, so the maximum number of passes is 6 ***[1 mark]***. The maximum number of swaps is 5 + 4 + 3 + 2 + 1 = 15 ***[1 mark]***.

2 a)

First pass:	1.3, 0.8, 1.8, 0.5, 1.2, 0.2, 0.9	swap
Second pass:	0.8, 1.3, 1.8, 0.5, 1.2, 0.2, 0.9	no swap
Third pass:	0.8, 1.3, 1.8, 0.5, 1.2, 0.2, 0.9	swap
	0.8, 1.3, 0.5, 1.8, 1.2, 0.2, 0.9	swap
	0.8, 0.5, 1.3, 1.8, 1.2, 0.2, 0.9	swap
		[1 mark]
Fourth pass:	0.5, 0.8, 1.3, 1.8, 1.2, 0.2, 0.9	swap
	0.5, 0.8, 1.3, 1.2, 1.8, 0.2, 0.9	swap
	0.5, 0.8, 1.2, 1.3, 1.8, 0.2, 0.9	no swap
		[1 mark]
Fifth pass:	0.5, 0.8, 1.2, 1.3, 1.8, 0.2, 0.9	swap
	0.5, 0.8, 1.2, 1.3, 0.2, 1.8, 0.9	swap
	0.5, 0.8, 1.2, 0.2, 1.3, 1.8, 0.9	swap
	0.5, 0.8, 0.2, 1.2, 1.3, 1.8, 0.9	swap
	0.5, 0.2, 0.8, 1.2, 1.3, 1.8, 0.9	swap
		[1 mark]
Sixth pass:	0.2, 0.5, 0.8, 1.2, 1.3, 1.8, 0.9	swap
	0.2, 0.5, 0.8, 1.2, 1.3, 0.9, 1.8	swap
	0.2, 0.5, 0.8, 1.2, 0.9, 1.3, 1.8	swap
	0.2, 0.5, 0.8, 0.9, 1.2, 1.3, 1.8	no swap
		[1 mark]

There are 7 numbers, so there must be 7 – 1 = 6 passes ***[1 mark]***.

b) On the first pass, there was 1 comparison and 1 swap ***[1 mark]***.

3 a)

N	C	D	Output	$N = A$?
1	8	12	1	No
2	4	6	2	No
3	$2\frac{2}{3}$	4		No
4	2	3	4	No
5	$1\frac{3}{5}$	$2\frac{2}{5}$		No
6	$1\frac{1}{3}$	2		No
7	$1\frac{1}{7}$	$1\frac{5}{7}$		No
8	1	$1\frac{1}{2}$		Yes

The results are 1, 2 and 4.

[3 marks available in total:
- ***1 mark for correct values of C;***
- ***1 mark for correct values of D;***
- ***1 mark for correct outputs (there should be 3 outputs)]***

b) (i) This algorithm produces the common factors of the inputs. ***[1 mark]***

(ii) The output would be 1 ***[1 mark]***, as 19 and 25 have no common factors. ***[1 mark]***

4 Use the first name (Mark) as the pivot. The list becomes:
Dan, Adam, James, Helen, Mark, Stella, Robert ***[1 mark]***
Use Dan as the pivot for the first list and Stella as the pivot for the second list. The list becomes:
Adam, Dan, James, Helen, Mark, Robert, Stella ***[1 mark]***
Adam and Robert are in lists on their own, so they are in the correct place. Use James as the pivot for the other list. The list becomes:
Adam, Dan, Helen, James, Mark, Robert, Stella ***[1 mark]***
Helen is in a list on her own, so the list is in order
[1 mark for stating all pivots used].

5 a) On the first pass, there were 4 comparisons ***[1 mark]*** and 3 swaps ***[1 mark]***.

You can work this out by calculating how many subsets are needed — 8/2 = 4 subsets, each with 2 numbers in, so there will be 4 comparisons. You have to see how many numbers have swapped places — in this case, 71 and 59 have swapped, 63 and 60 have swapped and 72 and 55 have swapped, so there were 3 swaps.

b) On the second pass, you need 4/2 = 2 subsets:
First subset: 54 60 68 63
Second subset: 59 55 71 72 ***[1 mark]***
Putting these in order gives:
First subset: 54 60 63 68
Second subset: 55 59 71 72 ***[1 mark]***
Putting the list back together gives:
54, 55, 60, 59, 63, 71, 68, 72 ***[1 mark]***.
You now need 2/2 = 1 subset — this is just the list above.
Sort this using a shuttle sort:

First pass:	54, 55, 60, 59, 63, 71, 68, 72	no swap
Second pass:	54, 55, 60, 59, 63, 71, 68, 72	no swap
Third pass:	54, 55, 60, 59, 63, 71, 68, 72	swap
	54, 55, 59, 60, 63, 71, 68, 72	no swap
Fourth pass:	54, 55, 59, 60, 63, 71, 68, 72	no swap
Fifth pass:	54, 55, 59, 60, 63, 71, 68, 72	no swap
Sixth pass:	54, 55, 59, 60, 63, 71, 68, 72	swap
	54, 55, 59, 60, 63, 68, 71, 72	no swap
Seventh pass:	54, 55, 59, 60, 63, 68, 71, 72	no swap

[2 marks for correct shuttle sort, or 1 mark for correct method but with mistakes made]

D1 Section 2 — Algorithms on Graphs
Warm-up Questions

1) a) A graph which has a number associated with each edge.

b) A graph in which one or more of the edges have a direction associated with them.

c) A connected graph with no cycles.

d) A subgraph which contains all the vertices of the original graph and is also a tree.

2) Yes

3) E.g. ADBC

4) E.g. BCDB

Answers

5) E.g.

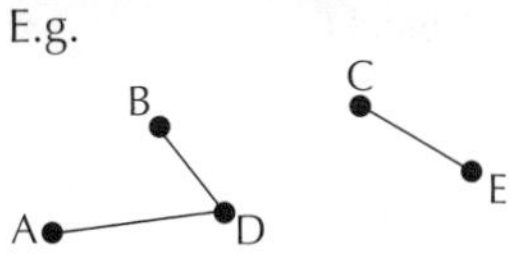

6) A = 1, B = 2, C = 3, D = 3, E = 1
Sum of degrees = double number of edges

7)

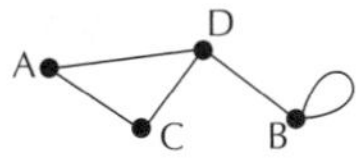

8) a) ABDEF. Shortest route A to F is 9.
Trace back from F to A on the final diagram on page 215.
The weight of each edge you go along must equal the difference in the final values of the vertices at each end of the edge.

b)

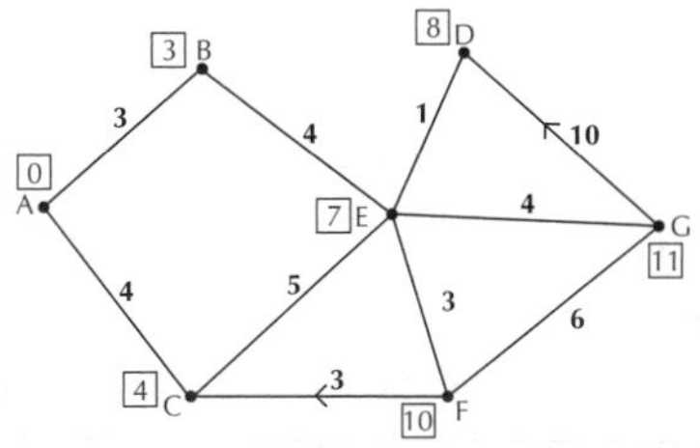

ABEG (11)

Exam Questions

1 a) DG (22) – add; EF (24) – add; BE (25) – add; DE (25) – add; EG (26) – don't add; AB (26) – add; BD (27) – don't add; BF (28) – don't add; FH (28) – add; CF (30) – add; AC (32) – don't add; GH (33) – don't add. Vertices of equal length can be considered in either order. ***[3 marks available — 1 mark for edges in correct order, 2 marks for all added edges correct. Lose 1 mark for each error.]***

b) 22 + 24 + 25 + 25 + 26 + 28 + 30 = £180 ***[1 mark]***

c)

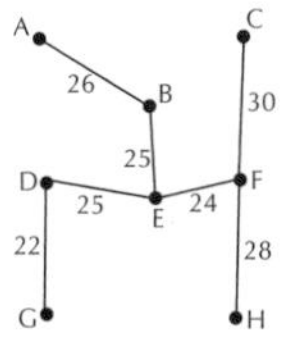

[2 marks for correct edges. Lose 1 mark for each error.]

d) FC / CF.
Order edges added: EF, EB, ED, DG (or ED, DG, EB), BA, FH, FC.
[2 marks available. 1 mark for correct edge, 1 mark for evidence that Prim's algorithm has been applied.]

e) E.g. Prim's algorithm can be applied to data in matrix form; you don't have to check for cycles using Prim's; the tree grows in a connected way using Prim's.
[2 marks — 1 mark for each.]

2 a) Bipartite graph ***[1 mark]***

b) 1 (e.g. AD) ***[1 mark]***

c) 8 (2 × number of edges) ***[1 mark]***

d) The sum of the orders is double the number of edges, so is always even ***[1 mark]***. There are 5 vertices, and the sum of 5 odd numbers is always odd ***[1 mark]***.

3 a)

	A	B	C	D	E
A	–	14	22	21	18
B	14	–	19	21	20
C	22	19	–	21	15
D	21	21	21	–	24
E	18	20	15	24	–

Order edges added: AB, AE, EC, AD/BD/CD
(You only need one of AD, BD and CD — they're interchangeable.)

[3 marks available — 2 marks for edges in correct order (1 mark if one error). 1 mark for correct use of matrix.]

Each time you circle a number, write down which edge it represents by reading the row and column labels. Don't leave it until the end, or you'll have forgotten the order you added them in. Doh.

b) E.g.

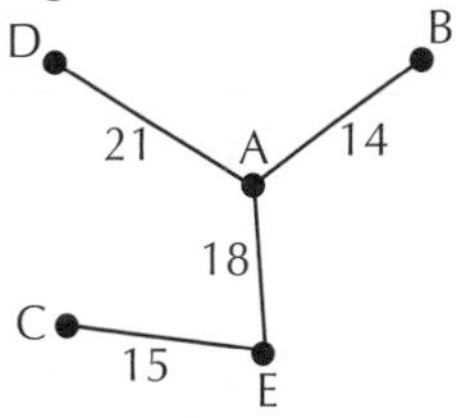

D may be connected to B or C instead of A ***[1 mark]***.
weight = 68 ***[1 mark]***.

c) 2 (using any of the alternatives AD, BD and CD) ***[1 mark]***.

4 a) 8 (no. of vertices – 1) ***[1 mark]***

b)

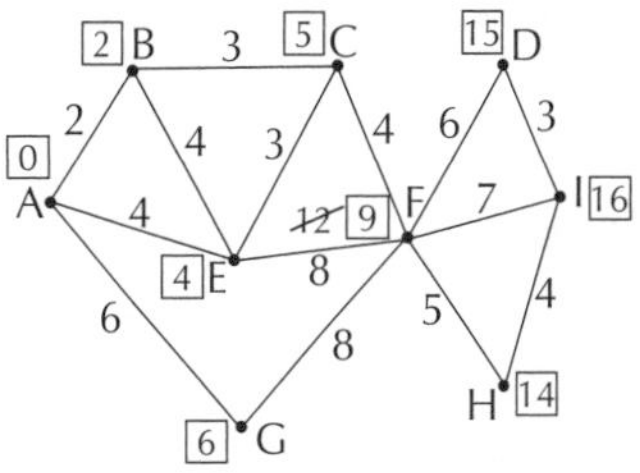

Fastest route = ABCFI, 16 minutes

[6 marks available — 1 mark for route, 1 mark for 16 minutes, 4 marks for all vertices correctly completed in diagram, lose 1 mark for each error.]

Find the fastest route by tracing back from the final destination. You know if a path is on the route because its weight is the difference between the final values at either end of it.

c) $6 + x + 4 < 16$, $x < 6$
Fastest route from A to G is 6. Fastest route from H to I is 4. Total route AGHI must be less than 16 minutes.
[2 marks available — 1 mark for $6 + x + 4$ as new route length, 1 mark for solving inequality for x.]

Answers

D1 Section 3 — The Route Inspection Problem
Warm-up Questions

1) a) Eulerian
   b) semi-Eulerian
   c) semi-Eulerian
   d) neither

Eulerian. It's my new favourite word. In fact, I think I'll name my first-born child "Eulerian".

2) A, B, F, J
   AB + FJ, AF + BJ, AJ + BF

3) a) It's Eulerian, so length = weight of network = 36.
   b) It's semi-Eulerian, so length = 31 (weight of network) + 4 (distance AB) = 35
   c) 4 odd vertices: A, B, D, F
   AB + DF = 7 + 8 = 15
   AD + BF = 5 + 3 = 8 (minimum)
   AF + BD = 4 + 7 = 11
   length = 42 (weight of network) + 8 = 50

4) a) 36, any vertex.
   b) 31, start and end at A and B
   c) BF is shortest distance between odd vertices, so start and end at A and D. Length = 42 + 3 = 45.

Exam Questions

1 a) Logo A = semi-Eulerian ***[1 mark]***
   Logo B = neither ***[1 mark]***
  b) (i) Logo A = 0 ***[1 mark]***
   Logo B = once ***[1 mark]***
   (ii) B or D ***[1 mark]***
  c) Logo A = 1 ***[1 mark]***
   Logo B = 2 ***[1 mark]***

2 a) There are odd vertices (A, D, I and J) ***[1 mark]***

So it's not Eulerian. And there are four odd vertices, so it's not even semi-Eulerian.

  b) Odd vertices are A, D, I, J ***[1 mark]***
   Pairings: AD + IJ = 180 + 100 = 280
   AI + DJ = 440 + 310 = 750
   AJ + ID = 490 + 340 = 830
   [1 mark for pairings, 1 mark for lengths]
   minimum pairing = AD + IJ ***[1 mark]***
   route length = 2740 + 280
   = 3020 ***[1 mark]*** m ***[1 mark]***

The question says that you must start and end at K — but as you've made the graph effectively Eulerian, it doesn't actually matter which vertex you start and finish at.

  c) (i) IJ is minimum distance between odd vertices ***[1 mark]***
   Length of route = 2740 + 100 = 2840 m ***[1 mark]***
   (ii) A or D ***[1 mark]***

3 a) Odd vertices are B, G, M, L ***[1 mark]***
   Pairings: BL + GM = 41 + 26 = 67
   BG + LM = 30 + 29 = 59
   BM + GL = 30 + 28 = 58
   [1 mark for pairings, 1 mark for times]
   minimum pairing = BM + GL ***[1 mark]***
   route time = 336 + 58 = 394 ***[1 mark]*** mins ***[1 mark]***
  b) (i) GM is minimum ***[1 mark]*** so end at L ***[1 mark]***
   (ii) 336 + 26 = 362 minutes ***[1 mark]***

4 a) Odd vertices are A, D, E, F ***[1 mark]***
   Pairings: AD + EF = 12 + 15 = 27
   AE + DF = 12 + 21 = 33
   AF + DE = 18 + 6 = 24
   [1 mark for pairings, 1 mark for distances]
   minimum pairing = AF + DE ***[1 mark]***
   route length = 106 + 24 = 130 ***[1 mark]*** miles ***[1 mark]***
  b) Example route = ABFGEFBACBEDECDA ***[1 mark]***
   so 5 times ***[1 mark]***.
   Alternatively, you could say that C has four edges connected to it, so must be passed through twice. E has six edges connected to it (including the extra pass along DE) so must be passed through three times ***[1 mark]***, so 2 + 3 = 5 times past an ice-cream shop ***[1 mark]***.
  c) 106 × 2 ***[1 mark]*** = 212 miles ***[1 mark]***

You've effectively doubled the edges and made the graph Eulerian. You have to traverse it twice, so the distance is just double the network's weight.

D1 Section 4 — Travelling Salesperson Problem
Warm-up Questions

1) a) E.g. ABCDEA, ACEBDA, AEBDCA

Your routes must start and end at A and contain B, C, D and E exactly once each.

  b) From A: ACEBDA (weight = 31)
   From B: BECADB (weight = 31)
   From C: CEBADC (weight = 30)
   From D: DCEBAD (weight = 30)
   From E: ECADBE (weight = 31)
   Best upper bound = 30
  c) Deleting A:

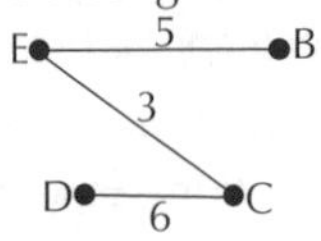

$x = 4$, $y = 7$, $W = 14$, Lower bound = 25

Answers

Deleting B:

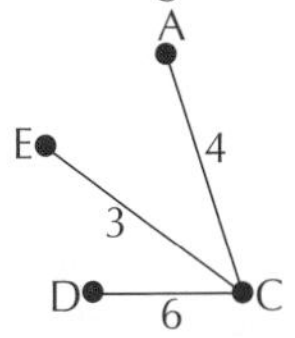

$x = 5$, $y = 7$, $W = 13$, Lower bound = 25

Deleting C:

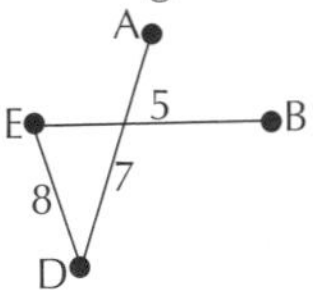

$x = 3$, $y = 4$, $W = 20$, Lower bound = 27

Deleting D:

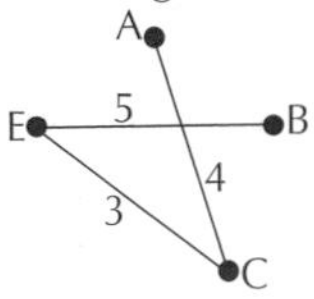

$x = 6$, $y = 7$, $W = 12$, Lower bound = 25

Deleting E:

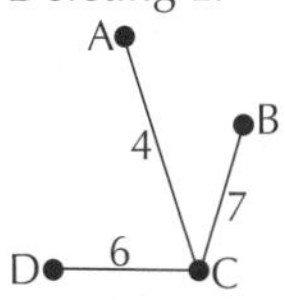

$x = 3$, $y = 5$, $W = 17$, Lower bound = 25

Best lower bound = 27

d) $27 \leq$ optimum solution ≤ 30

2) a) FCEBADF, upper bound = 38

b) Deleting B:

	(A)	B	(C)	(D)	(E)	(F)
A	—	4	6	5	8	12
B	4	—	14	22	6	11
C	(6)	14	—	18	3	5
D	(5)	22	18	—	13	15
E	8	6	(3)	13	—	20
F	12	11	(5)	15	20	—

$x = 4$, $y = 6$, $W = 6 + 5 + 3 + 5 = 19$

Lower bound = 29

3)

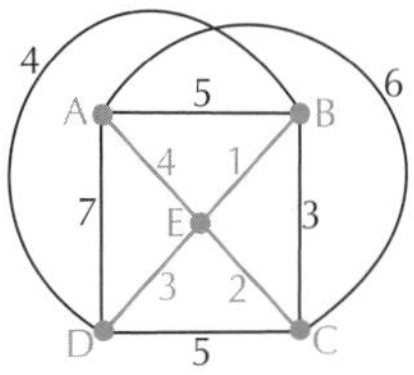

Exam Questions

1 a) (i) To complete the graph, you need to add edges between A and C, A and D, C and F and D and F — so you need to add 4 edges ***[1 mark]***.

(ii) The shortest distance between A and C is 43 (ABC), the shortest distance between A and D is 41 (AED), the shortest distance between C and F is 62 (CEF) and the shortest distance between D and F is 49 (DEF) ***[2 marks for all distances correct, 1 mark for up to 3 distances correct]***. Adding these edges to the diagram:

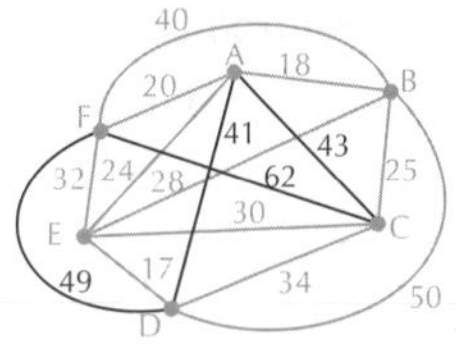

[1 mark]

b) DE (17) – add; AB (18) – add; AF (20) – add; AE (24) – add; BC (25) – add. That's all vertices joined, so don't need to consider any more. ***[3 marks available — 1 mark for edges in correct order, 2 marks for all added edges correct, otherwise 1 mark for all but one edge correct.]***

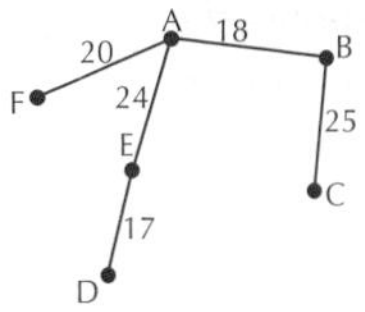

[1 mark]

Weight = 17 + 18 + 20 + 24 + 25 = 104 ***[1 mark]***

c) Weight of reduced network = 104 – 25 = 79 ***[1 mark]***
Shortest edges incident to C = 25 + 30 = 55 ***[1 mark]***
Lower bound = 79 + 55 = 134 ***[1 mark]***

d) ABCEDFA ***[1 mark]***
Weight = 18 + 25 + 30 + 17 + 49 + 20 ***[1 mark]***
= 159, so upper bound = 159 ***[1 mark]***

2 a)

	A	B	C	D	E	F
A	—	90	80	170	210	220
B	90	—	120	220	150	130
C	80	120	—	100	270	170
D	170	220	100	—	240	140
E	210	150	270	240	—	100
F	220	130	170	140	100	—

[2 marks available — 2 marks for all 4 distances correct, 1 mark for 2 or 3 distances correct.]

b) 90+ 120 + 100 + 240 + 100 + 220 ***[1 mark]***
= 870 m ***[1 mark]***

c) ACDFEBA ***[1 mark]***
= 80 + 100 + 140 + 100 + 150 + 90 ***[1 mark]***
= 660 m ***[1 mark]***

d) 660 m ***[1 mark]***

e) Apply either Prim's or Kruskal's algorithm to get MST for reduced network.

E.g.

	(A)	(B)	(C)	D	E	(F)
A		90	80	[illegible]	210	220
B	(90)	—	120	[illegible]	150	130
C	(80)	120		[illegible]	270	170
D	[illegible]	[illegible]	[illegible]	[illegible]	[illegible]	[illegible]
E	210	150	270	[illegible]	—	(100)
F	220	(130)	170	[illegible]	100	

80 + 90 + 130 + 100 = 400 m

[4 marks — 1 mark for attempting to apply appropriate algorithm or 2 marks for correct application, 1 mark for correct edges, 1 mark for MST.]

Shortest edges incident to D = 100 + 140 = 240 m ***[1 mark]***
Lower bound = 400 + 240 = 640 m ***[1 mark]***

Answers

3 a) E.g. ABCDEFGA
[2 marks available — 1 mark for including all vertices exactly once each, 1 mark for returning to start vertex.]

b) There are a few possible MSTs here. Only 1 is shown:

E.g.

	A	B	C	D	E	F	G
A	—	15	10	7	6	7	4
B	15	—	8	5	12	13	14
C	10	8	—	6	15	18	2
D	7	5	6	—	7	5	13
E	6	12	15	7	—	3	10
F	7	13	18	5	3	—	6
G	4	14	2	13	10	6	—

Order added: AG, GC, CD, DB, DF, FE.
[4 marks — 1 mark for attempting to apply appropriate algorithm or 2 marks for correct application, 1 mark for correct edges, 1 mark for order.]

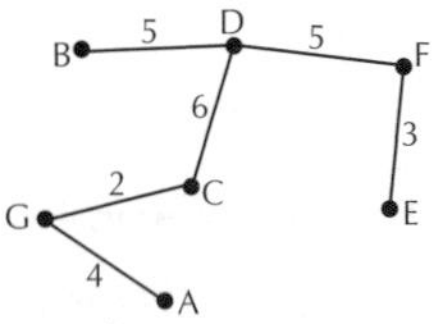

[1 mark]

Weight = 4 + 2 + 6 + 5 + 5 + 3 = 25 mins ***[1 mark]***

You could have included AE (6) or FG (6) instead of CD (6). It doesn't change the weight though. Also, your tree might look completely different, depending on how you arranged your vertices.

c) Weight = 25 – 5 = 20 ***[1 mark]***
Shortest edges incident to B = 5 + 8 = 13 ***[1 mark]***
Lower bound = 20 + 13 = 33 mins ***[1 mark]***

d) FEAGCDBF ***[1 mark]***

	A	B	C	D	E	F	G
A	—	15	10	7	6	7	4
B	15	—	8	5	12	13	14
C	10	8	—	6	15	18	2
D	7	5	6	—	7	5	13
E	6	12	15	7	—	3	10
F	7	13	18	5	3	—	6
G	4	14	2	13	10	6	—

Weight = 3 + 6 + 4 + 2 + 6 + 5 + 13 ***[1 mark]***
= 39 mins, so upper bound = 39 mins ***[1 mark]***

e) 33 mins ≤ optimum weight of tour ***[1 mark]*** ≤ 39 mins ***[1 mark]***

D1 Section 5 — Linear Programming Warm-up Questions

1) a) Decision variables represent the quantities of the things being produced in a linear programming problem.

b) The objective function is the thing you're trying to optimise — a function in terms of the decision variables that you want to maximise or minimise.

c) The optimal solution is a feasible solution that optimises the objective function (i.e. maximises or minimises it).

2) A dotted line represents a strict (< or >) inequality. You don't include this line in the feasible region.

3) An objective line is a line in the form $Z = ax + by$, which represents all solutions of the objective function that have the same value of Z.

4) E.g. A linear programming problem involving liquid doesn't need integer solutions — e.g. mixing vinegar and oil to make vinaigrette. You can have fractions of a litre.
E.g. A linear programming problem about making musical instruments needs integer solutions — you can't make half a trumpet.

5) a) The decision variables are the number of large posters and the number of small posters, so let x = number of large posters and y = number of small posters.
The constraints are the different amounts of time available. Printing a large poster takes 10 minutes, so printing x large posters takes $10x$ minutes. Printing a small poster takes 5 minutes, so printing y small posters takes $5y$ minutes. There are 250 minutes of printing time available, so the inequality is $10x + 5y \leq 250 \Rightarrow 2x + y \leq 50$. Using the same method for laminating time produces the inequality $6x + 4y \leq 200 \Rightarrow 3x + 2y \leq 100$. The company wants to sell at least as many large posters as small posters, so $x \geq y$, and at least 10 small posters, so $y \geq 10$. Also, $x, y \geq 0$ (the non-negativity constraints).
The objective function is $P = 6x + 3.5y$ (profit) which needs to be maximised.

You could use x for the number of small posters and y for the number of large posters instead — so x and y in each inequality would just swap round.

b)

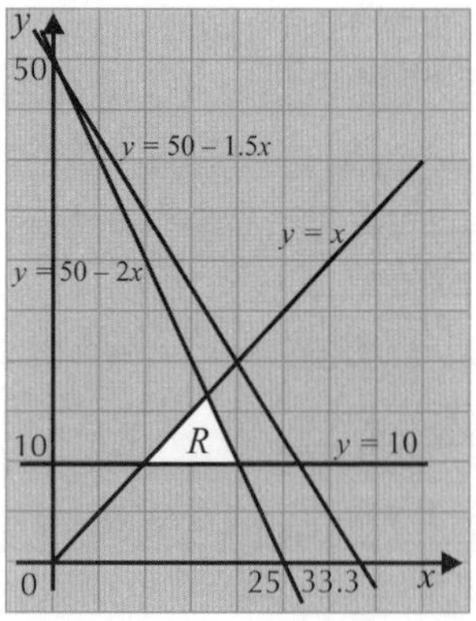

The line y = 50 – 1.5x isn't actually used to form the feasible region.

c) E.g. starting with the objective line for $P = 42$ (which goes through (0, 12) and (7, 0)) and moving it towards R gives:

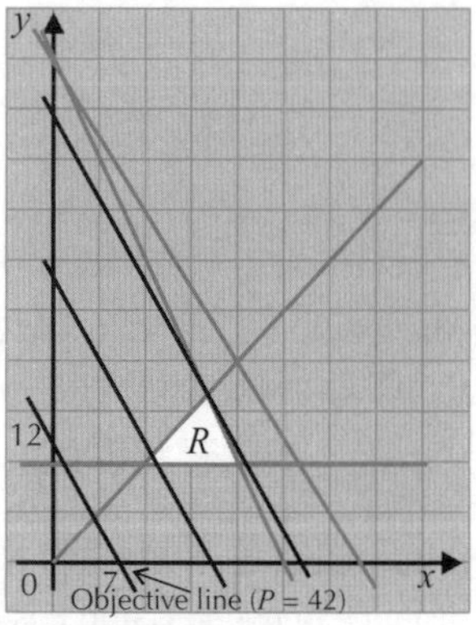

So the final point in the feasible region touched by the

Answers

objective line is the point of intersection of the lines $y = x$ and $y = 50 - 2x$. Solving these simultaneous equations gives the point of intersection as $\left(\frac{50}{3}, \frac{50}{3}\right)$. Putting these values into the objective function gives a maximum profit of £158.33.

You can choose any value of P as a starting value — 42 makes it easy to draw the line. This answer uses the objective line method, but you could have used the vertex method instead.

d) From part c) above, the maximum profit is found at $\left(\frac{50}{3}, \frac{50}{3}\right)$. However, you can't have fractions of a poster, so an integer solution is required. The points with integer coordinates nearby are (16,16), (16,17), (17, 17) and (17, 16). (16, 17) doesn't satisfy the constraint $x \geq y$ and (17, 17) doesn't satisfy the constraint $y \leq 50 - 2x$. The value of the objective function at (16, 16) is £152 and at (17, 16) it's £158, so (17, 16) gives the maximum solution — you need to make 17 large posters and 16 small posters.

Exam Questions

1 The objective function is to minimise the cost ***[1 mark]***, $C = 0.75x + 0.6y$ in £ (or $C = 75x + 60y$ in pence) ***[1 mark]*** (where x is the number of red roses and y is the number of white roses), subject to the constraints:
$x, y > 0$ ***[1 mark]*** (from the statement that she will sell both red and white roses — x and y can't be 0).
$x > y$ ***[1 mark]*** (from the statement that she will sell more red roses than white roses).
$x + y \geq 100$ ***[1 mark]*** (from the statement that she will sell a total of at least 100 flowers).
$x \leq 300$ ***[1 mark]*** and $y \leq 200$ ***[1 mark]*** (from the statement that the wholesaler has 300 red roses and 200 white roses).

Don't waste time trying to solve these inequalities — the question doesn't ask you to find a solution. Just write them down and run.

2 a) There are 6 sheets of foil in a gold pack, so in x gold packs there will be $6x$ sheets of foil. There are 2 sheets of foil in a silver pack, so in y silver packs there will be $2y$ sheets of foil. There is 1 sheet of foil in a bronze pack, so in z bronze packs there will be z sheets of foil. There are 30 sheets of foil available, so the inequality is $6x + 2y + z \leq 30$ ***[2 marks — 1 mark for LHS, 1 mark for correct inequality sign and RHS]***. Using the same method for sugar paper produces the inequality $15x + 9y + 6z \leq 120$ ***[1 mark]***, which simplifies to give $5x + 3y + 2z \leq 40$ ***[1 mark]***. For tissue paper, the inequality is $15x + 4y + z \leq 60$ ***[2 marks — 1 mark for LHS, 1 mark for correct inequality sign and RHS]***. Finally, the amount of foil used is $6x + 2y + z$, and the amount of sugar paper used is $15x + 9y + 6z$. The amount of sugar paper used needs to be at least three times the amount of foil, so the inequality for this constraint is
$15x + 9y + 6z \geq 3(6x + 2y + z)$ ***[1 mark]***
$15x + 9y + 6z \geq 18x + 6y + 3z$
$3y + 3z \geq 3x$
$y + z \geq x$ ***[1 mark]***

b) (i) If the number of silver packs sold is equal to the number of bronze packs, then $y = z$. Substituting this into the inequalities from part (a) gives:
$6x + 2y + y \leq 30 \Rightarrow 6x + 3y \leq 30 \Rightarrow 2x + y \leq 10$
$5x + 3y + 2y \leq 40 \Rightarrow 5x + 5y \leq 40 \Rightarrow x + y \leq 8$
$15x + 4y + y \leq 60 \Rightarrow 15x + 5y \leq 60 \Rightarrow 3x + y \leq 12$
$y + y \geq x \Rightarrow 2y \geq x$. ***[3 marks available — 1 mark for making the correct substitution, 1 mark for correctly forming the inequalities and 1 mark for simplifying the inequalities.]***

(ii) 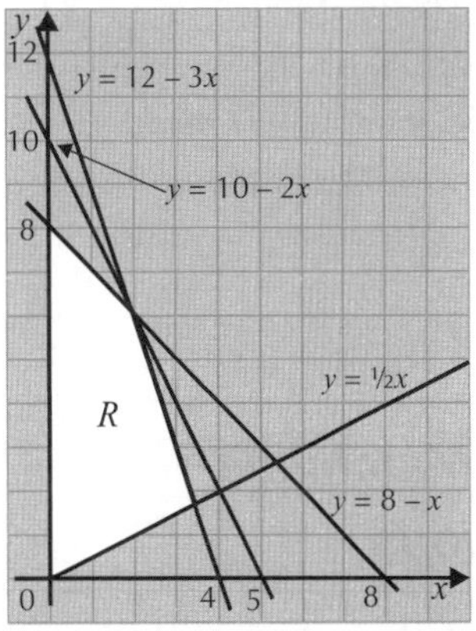

[5 marks available — 1 mark for each of the four inequality lines (with equations as shown on the graph) and 1 mark for correct feasible region]

(iii) Using the vertex method, the feasible region has vertices (0, 0) (the origin), (0, 8) (intersection of the y-axis and $y = 8 - x$), (2, 6) (intersection of $y = 8 - x$ and $y = 10 - 2x$) and $\left(\frac{24}{7}, \frac{12}{7}\right)$ (intersection of $y = \frac{1}{2}x$ and $y = 12 - 3x$) ***[1 mark]***. The number of packs made on Monday is $x + y + z = x + 2y$, so the numbers made at each vertex are 0, 16, 14 and $\frac{48}{7} = 6\frac{6}{7}$, so the maximum number of packs made that day is 16 = 8 silver and 8 bronze ***[1 mark]***.

You could have used the objective line method instead, using the line $Z = x + y + z = x + 2y$ to find the maximum.

(iv) The objective function is $P = 3.5x + 2y + z = 3.5x + 2y + y = 3.5x + 3y$, which needs to be maximised. The value of P for each of the vertices found in part (iii) is £0, £24, £25 and £17.14 ***[1 mark]***. The maximum value is £25 ***[1 mark]***, which occurs at (2, 6), so the company needs to sell 2 gold packs, 6 silver packs and 6 bronze packs (as the number of bronze packs is equal to the number of silver packs) ***[1 mark]***.

3 a) First, the non-negativity constraints are $x, y \geq 0$. The solid line that passes through (0, 6) and (3, 0) has equation $y = 6 - 2x$, and as the area below the line is shaded, the inequality is $2x + y \geq 6$. The dotted line that passes through (2, 0) has equation $y = x - 2$, and as the area below the line is shaded, the inequality is $x - y < 2$. The horizontal solid line that passes through (0, 4) has the equation $y = 4$, and as the area above the line is shaded, the inequality is $y \leq 4$.

Answers

[5 marks available — 1 mark for each line equation and 1 mark for all inequality signs correct]

b) The coordinates of the vertices of R are (1, 4) (the intersection of the lines $y = 4$ and $y = 6 - 2x$) ***[1 mark]***, (6, 4) (the intersection of the lines $y = 4$ and $y = x - 2$) ***[1 mark]*** and $\left(\frac{8}{3}, \frac{2}{3}\right)$ ***[1 mark]*** (the intersection of the lines $y = 6 - 2x$ and $y = x - 2$) ***[1 mark for solving the simultaneous equations].***

c) The value of C at (1, 4) is 8, the value of C at (6, 4) is 28 ***[1 mark if both are correct]*** and the value of C at $\left(\frac{8}{3}, \frac{2}{3}\right)$ is $\frac{34}{3} = 11\frac{1}{3}$ ***[1 mark]***. Hence the minimum value of C is 8, which occurs at the point (1, 4) ***[1 mark]***.

This answer uses the vertex method, but you could also have used the objective line method to answer this question — pick whichever method you prefer.

D1 Section 6 — Matchings

Warm-up Questions

1) a)

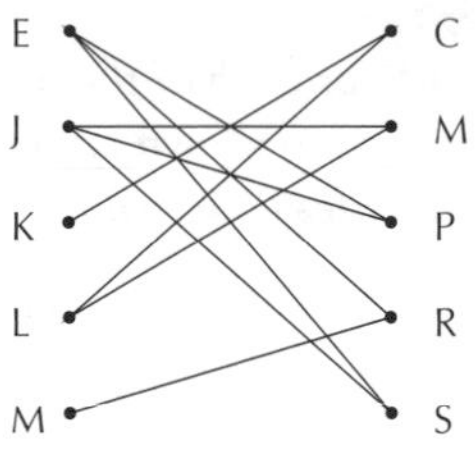

b)

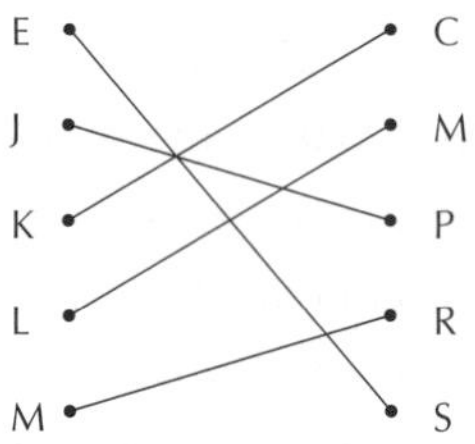

So E = S, J = P, K = C, L = M, M = R.
An alternative complete matching would start with E = P and J = S. Kitty, Lydia and Mary can't change.

These are both complete matchings — it shows that you can have more than one.

2) a)

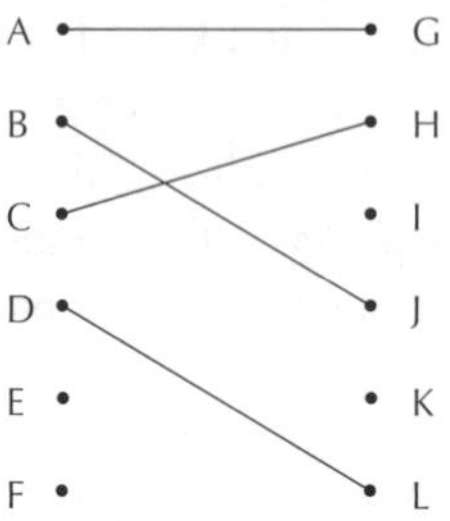

b) Find an alternating path starting from E:
E - - G — A - - I (where - - = not in, — = in).
(You could have done E - - H — C - - K instead — it reaches a breakthrough just as quickly.)
Changing the status of the edges gives:
E — G - - A — I

Construct the improved matching:
A = I, B = J, C = H, D = L, E = G

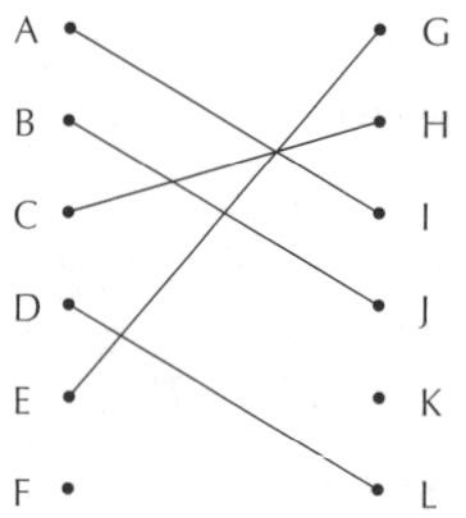

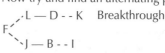

Now try and find an alternating path from F to K:

F ⋰ L — D - - K Breakthrough
F ⋱ J — B - - I

The path F - - L — D - - K reaches a breakthrough first, so use this one (if you found a path from E to K in your first alternating path, the second path would be F - - J — B - - I).
Changing the status of the edges: F — L - - D — K
Construct the improved matching:
A = I, B = J, C = H, D = K, E = G, F = L

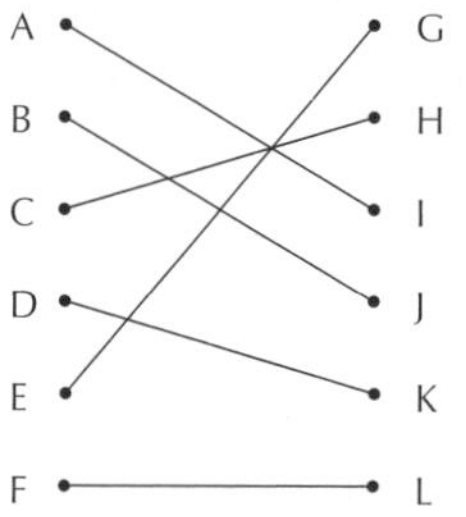

There are no more unmatched vertices, so this is a complete matching.

An alternative complete matching would be A = G, B = I, C = K, D = L, E = H, F = J if you'd used the alternating paths in brackets.

Exam Question

1 a)

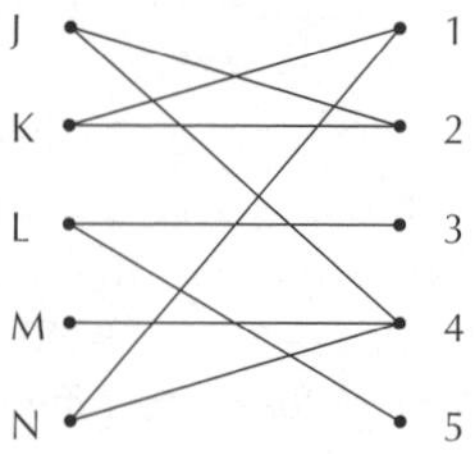

[2 marks for all vertices and edges correct, 1 mark for up to 4 edges correct]

b) The initial matching looks like this:

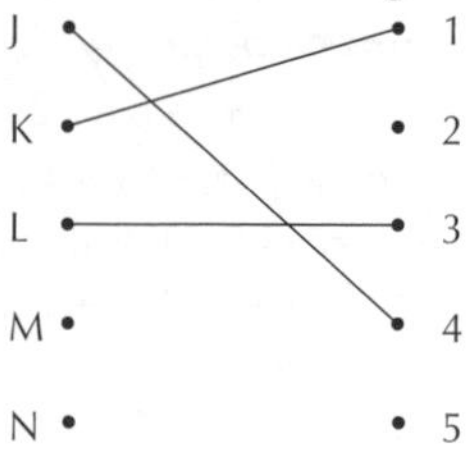

Alternating path is M - - 4 — J - - 2
(where - - = not in, — = in) ***[1 mark]***.

Answers

Change the status of the edges: M — 4 - - J — 2 ***[1 mark]***
So in the new matching, Mia will teach class 4 and James will teach class 2. Lee and Kelly are unchanged.
Construct the improved matching:
J = 2, K = 1, L = 3, M = 4 ***[1 mark]***

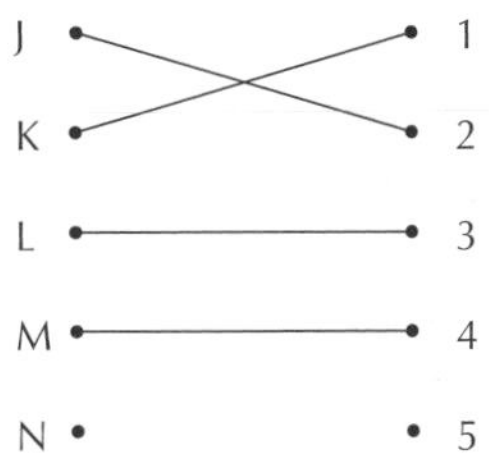

Make sure you list your matchings — you might not get marks for just drawing the graph.

c) Lee is the only person who can teach class 3 and the only person who can teach class 5, so both classes cannot be taught at the same time ***[1 mark]***.

d) Finding a new alternating path from N:

N - - 4 — M - - 3 — L - - 5 Breakthrough
N - - 1 — K - - 2 — J - - 4

The path N - - 4 — M - - 3 — L - - 5 reaches a breakthrough quicker, so use this one ***[1 mark]***.
Change the status of the edges: N — 4 - - M — 3 - - L — 5
Construct the improved matching:
James = 2, Kelly = 1, Lee = 5, Mia = 3 and Nick = 4
[1 mark]

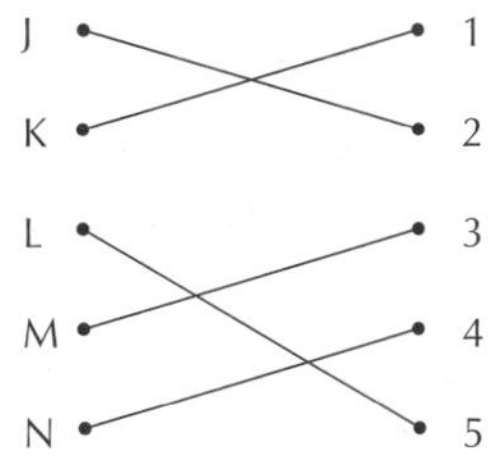

There are no more unmatched vertices, so this is a complete matching ***[1 mark]***.

D1 — Practice Exam 1

1 a) Use the first item (Frederick) as the pivot. The list becomes:
Dexter, Anthony, Frederick, Vicky, Rhiannon, Janet, Nahla, Sophia, Luke, Paul, Samira ***[1 mark]***.
Use Dexter as the pivot in the first list and Vicky as the pivot in the second list. The list becomes:
Anthony, Dexter, Frederick, Rhiannon, Janet, Nahla, Sophia, Luke, Paul, Samira, Vicky ***[1 mark]***.
Anthony is now in a list on his own, so is in the right place.
Use Rhiannon as the next pivot. The list becomes:
Anthony, Dexter, Frederick, Janet, Nahla, Luke, Paul, Rhiannon, Sophia, Samira, Vicky ***[1 mark]***.
Use Janet as the pivot for the first list, and Sophia as the pivot for the second list. The list becomes:
Anthony, Dexter, Frederick, Janet, Nahla, Luke, Paul, Rhiannon, Samira, Sophia, Vicky ***[1 mark]***.
Samira is in a list on her own so she is in the right place.
Use Nahla as the next pivot. The list becomes:
Anthony, Dexter, Frederick, Janet, Luke, Nahla, Paul, Rhiannon, Samira, Sophia, Vicky ***[1 mark]***.
Luke and Paul are in lists on their own, so they are in the right place. The list is now in order.

If you want, you can abbreviate the names (e.g. Fr, De, Ka etc.) to speed things up a bit. It's dead easy to see if you've made a mistake here — if your final list isn't in alphabetical order, something's gone horribly wrong...

b) There are 11 names in the list, so a shuttle sort would need 11 – 1 = 10 passes ***[1 mark]***.

2 a) First iteration: A = 2, N = 1.
P = A ÷ N = 2 ÷ 1 = 2
Q = P + N = 2 + 1 = 3
R = Q ÷ 2 = 3 ÷ 2 = 1.5 ***[1 mark for P, Q and R correct]***
|R – N| = |1.5 – 1| = 0.5 > 0.01, so go to Line 80
N = R = 1.5 ***[1 mark]***.
Second iteration: A = 2, N = 1.5.
P = 2 ÷ 1.5 = 1.33333
Q = 1.33333 + 1.5 = 2.83333
R = 2.83333 ÷ 2 = 1.41667 ***[1 mark for P, Q and R correct]***
|R – N| = |1.41667 – 1.5| = 0.08333 > 0.01, so go to Line 80.
N = R = 1.41667 ***[1 mark for correct iteration]***.
Third iteration: A = 2, N = 1.41667
P = 2 ÷ 1.41667 = 1.41176
Q = 1.41176 + 1.41667 = 2.82843
R = 2.82843 ÷ 2 = 1.41422 ***[1 mark for P, Q and R correct]***
|R – N| = |1.41422 – 1.41667| = 0.00245 < 0.01
So R = 1.41 ***[1 mark]***.
The square root of 2 to 2 d.p. is 1.41.

b) The algorithm calculates square roots to 2 decimal places, so for A = 9, the output would be 3.00 ***[1 mark]***.

3 a)

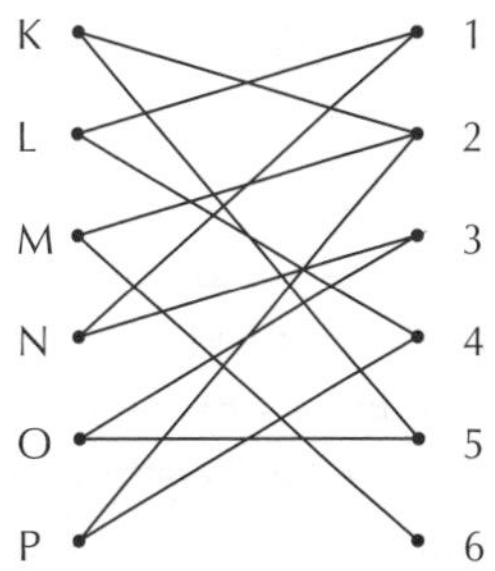

[2 marks for all edges correct, or 1 mark for 6 - 11 correct edges and no incorrect edges]

b) Initial matching:

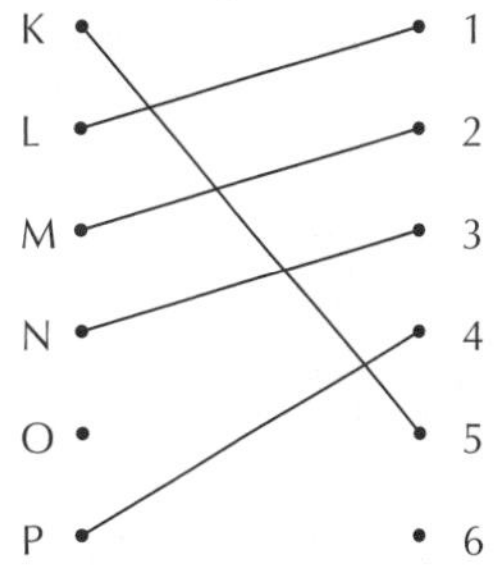

[1 mark]

Answers

Find an alternating path from Olivia to job 6:
O - - 5 — K - - 2 — M - - 6 ***[1 mark]***
(the other route, which starts O - - 3 is longer, so shouldn't be used). L, N and P remain unchanged.
The matching is now:

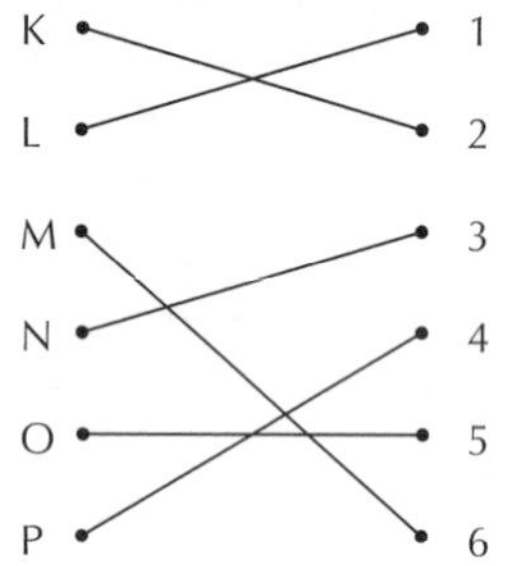

So K = 2, L = 1, M = 6, N = 3, O = 5, P = 4 ***[1 mark]***.
There are no more unmatched vertices so the matching is complete ***[1 mark]***.

4 a) $0.4x + 0.5y \leq 8$ or $4x + 5y \leq 80$ ***[1 mark]***.

b) Natalie can make no more than 8 blank cards ***[1 mark]***. Also, the number of blank cards made must be no more than number of birthday cards made ***[1 mark]***.

c)

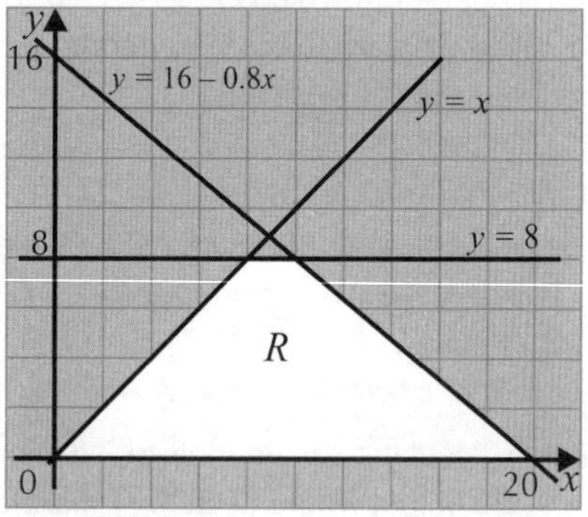

[4 marks available — 1 for each correct line (with equations as shown on graph), 1 for correct feasible region]

d) $P = 0.75x + 1.5y$ ***[1 mark]***.

e) Using the vertex method, the vertices of the feasible region are (0,0), (8, 8) (the intersection of $y = x$ and $y = 8$), (10, 8) (the intersection of $y = 16 - 0.8x$ and $y = 8$) and (20, 0) ***[2 marks for all 4 correct, or 1 mark for 2 correct vertices]***. Putting these values into the objective functions gives P values of £0, £18, £19.50 and £15 ***[1 mark]***, so the maximum profit is £19.50 ***[1 mark]***, which occurs at (10, 8), so Natalie must make 10 birthday cards and 8 blank cards ***[1 mark]*** to maximise her profit.

You could have used the objective line method instead, using the line P = 0.75x + 1.5y to find the maximum.

5 a)

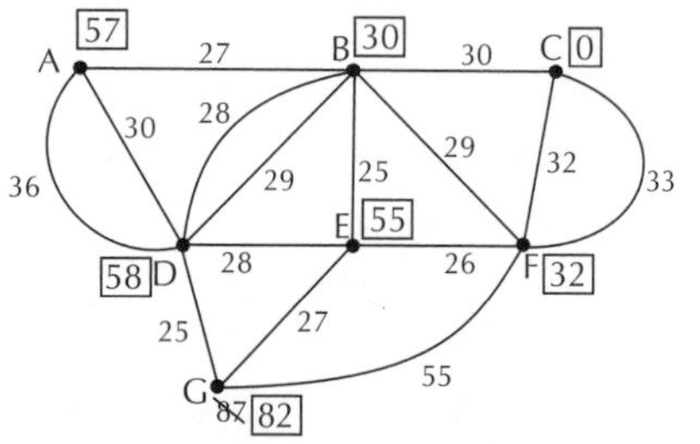

Route = CBEG, length = 82 miles

[7 marks available — 1 mark for route, 1 mark for length, 5 marks for all vertices correctly completed, alternatively 4 marks for 6 vertices correct, 3 marks for 5 vertices, 2 marks for 4 vertices, 1 mark for 2 or 3 vertices.]

b) (i) AB and BF ***[1 mark]***

This is the shortest route between the other odd vertices, A and F.

(ii) 460 + 56 = 516 miles ***[1 mark]***

c) Odd vertices are A, C, F, G ***[1 mark]***

Pairings: AC + FG = 57 + 53 = 110
AG + CF = 55 + 32 = 87
AF + CG = 56 + 82 = 138

[1 mark for pairings, 1 mark for lengths]
minimum pairing = AG + CF ***[1 mark]***
route length = 460 + 87 = 547 miles ***[1 mark]***

6 a) Order paths added:
GF, FB, BC, CA, GH/CH, HI, ID, IK, KL, KJ, IE

[5 marks for answer fully correct. Lose 1 mark for each mistake.]

b)

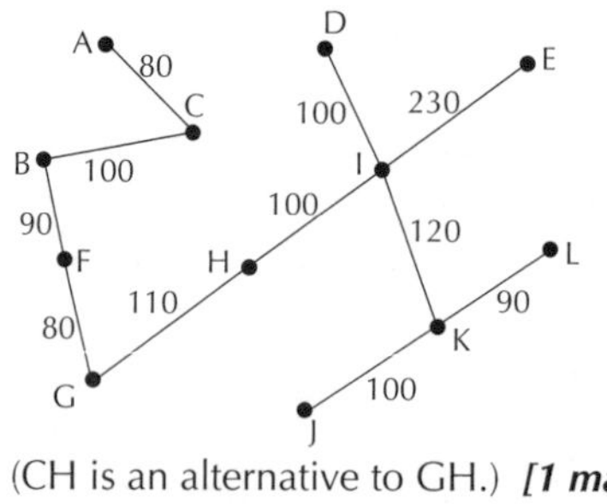

(CH is an alternative to GH.) ***[1 mark for 7 or 8 correct edges, 2 marks for 9 or 10 correct, 3 marks for 11 correct]***
Length = 1200 metres ***[1 mark]***

c) [AC – add; FG – add]; [BF – add; KL – add]; [BC – add; DI – add; HI – add; JK – add]; [AB – don't add; GH/CH – add; GH/CH – don't add]; IK – add; [HJ – don't add; HK – don't add; IL – don't add]; CD – don't add; EI – 230 – add; GJ – 250 – don't add; DE – 270 – don't add; EL – 280 – don't add; FJ – 300 – don't add. (The edges in square brackets can be considered in any order.)

The tenth and eleventh edges are IK and EI.

[3 marks available — 1 mark for each correct edge, 1 mark for evidence that Kruskal's algorithm has been applied.]

7 a) (i) 213 ***[1 mark]***

(ii) 217 ***[1 mark]***

Remember — you want the biggest lower bound and the smallest upper bound.

(iii) 213 ≤ weight of optimum tour ≤ 217 ***[1 mark]***

b) (i) e.g. ABCDEFA ***[1 mark]***, which has weight
6 + 4 + 9 + 3 + 1 + 2 = 25 ***[1 mark]***.

Answers

(ii) BCAFEDB *[1 mark]*

	A	B	C	D	E	F
A	–	6	7	5	11	2
B	6	–	4	12	8	6
C	7	4	–	9	10	13
D	5	12	9	–	3	7
E	11	8	10	3	–	1
F	2	6	13	7	1	–

Weight = 4 + 7 + 2 + 1 + 3 + 12 = 29 ***[1 mark]***, so an upper bound is 29 minutes ***[1 mark]***. This is greater than the upper bound found in part (i), so 25 is the better upper bound ***[1 mark]***.

Your comment will depend on the weight of the network you found in part (i) — if the Hamiltonian cycle you found had a weight greater than 29, you'd use 29 as the upper bound.

(iii) Here is a possible MST for the network:

	A	B	C	D	E	F
A	–	6	7	5	11	2
B	6	–	4	12	8	6
C	7	4	–	9	10	13
D	5	12	9	–	3	7
E	11	8	10	3	–	1
F	2	6	13	7	1	–

[2 marks — 1 mark for applying algorithm correctly, 1 mark for correct edges]

Order edges added: CB, BA, AF, FE, ED ***[1 mark]***.

The MST looks like this:

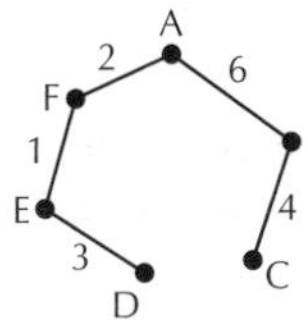

[1 mark]

Weight = 4 + 6 + 2 + 1 + 3 = 16 mins ***[1 mark]***

You could have added edges BF (6) and FA (2) instead of BA (6) and AF (2). This wouldn't change the weight of the MST.

(iv) Weight = 16 – 3 = 13 ***[1 mark]***
Shortest edges incident to D = 3 + 5 = 8 ***[1 mark]***
Lower bound = 13 + 8 = 21 mins ***[1 mark]***

D1 — Practice Exam 2

1 a) There are 9 numbers so you want 9/2 = 4.5 = 4 subsets:
First subset: 281 263 271
Second subset: 276 287
Third subset: 255 296
Fourth subset: 290 259 ***[1 mark]***
Sorting these subsets into order gives:
First subset: 263 271 281
Second subset: 276 287
Third subset: 255 296
Fourth subset: 259 290 ***[1 mark]***
Putting the list back together: 263, 276, 255, 259, 271, 287, 296, 290, 281 ***[1 mark]***. This is the order after the first pass.

For the second pass, you want 4/2 = 2 subsets:
First subset: 263 255 271 296 281
Second subset: 276 259 287 290
Sorting these subsets into order gives:
First subset: 255 263 271 281 296
Second subset: 259 276 287 290
Put the list back together: 255, 259, 263, 276, 271, 287, 281, 290, 296 ***[1 mark]***. This is the order after the 2nd pass.
For the third pass, you want 2/1 = 1 subset. Sort this using a shuttle sort:

255, 259, 263, 276, 271, 287, 281, 290, 296 no swap
255, 259, 263, 276, 271, 287, 281, 290, 296 no swap
255, 259, 263, 276, 271, 287, 281, 290, 296 no swap
255, 259, 263, 276, 271, 287, 281, 290, 296 swap
255, 259, 263, 271, 276, 287, 281, 290, 296 no swap
255, 259, 263, 271, 276, 287, 281, 290, 296 no swap
255, 259, 263, 271, 276, 287, 281, 290, 296 swap
255, 259, 263, 271, 276, 281, 287, 290, 296 no swap
255, 259, 263, 271, 276, 281, 287, 290, 296 no swap
255, 259, 263, 271, 276, 281, 287, 290, 296 no swap

The final order is 255, 259, 263, 271, 276, 281, 287, 290, 296 ***[1 mark]***.

b) On the first pass, there were 5 comparisons ***[1 mark]*** and 3 swaps ***[1 mark]***.

c) Using a bubble sort on 9 numbers, the maximum number of comparisons needed would be ½ × 8 × 9 = 36 ***[1 mark]*** and the maximum number of passes would be 9 ***[1 mark]***.

2 a) Alternating path is E - - 3 — F - - 2 — D - - 1 ***[1 mark]***.
Switching status gives E — 3 - - F — 2 - - D — 1 ***[1 mark]***.
So in the new improved matching, Edgar will do task 3, Fatima will do task 2 and Diego will do task 1. Carlos and Gino remain unchanged. This is the improved matching:

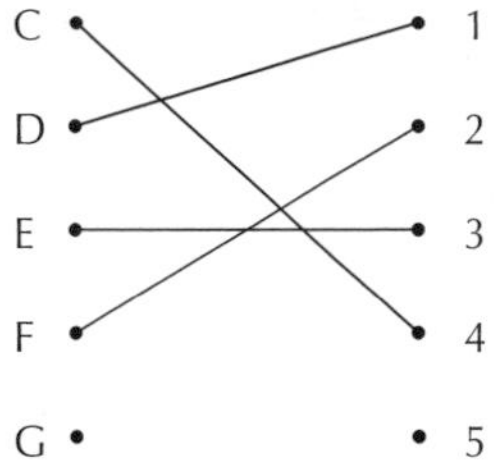

[1 mark]

b) Edgar and Gino can both only do task 3, so a complete matching is not possible (other answers are possible) ***[1 mark]***.

c) Alternating path is G — 3 - - E — 5 ***[1 mark]***.
Switching status gives G - - 3 — E - - 5.
So in the new matching, Gino does task 3 and Edgar does task 5. The improved matching is:
C = 4, D = 1, E = 5, F = 2, G = 3 ***[1 mark]***

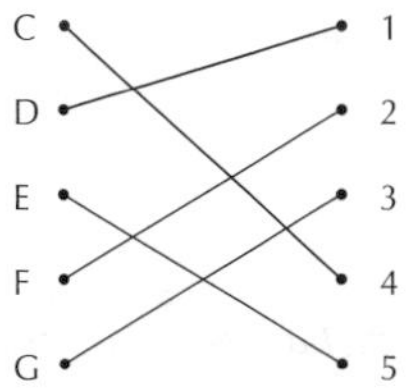

There are no more unmatched vertices, so this is a complete matching.

Answers

3 a)

A	*B*	*B* = 1 digit?	*B* = 9?	Output
977	23	No		
	5	Yes	No	
978	24	No		
	6	Yes	No	
979	25	No		
	7	Yes	No	
980	17	No		
	8	Yes	No	
981	18	No		
	9	Yes	Yes	981

The result is 981.

[4 marks available in total. 4 marks if all correct, lose 1 mark for each mistake.]

b) 981 ***[1 mark]***. This starting value would produce the same trace table as in part (a), but starting one row down.

c) The next multiple of 9 is at most 8 numbers away from the starting value, so no more than 9 passes will be needed ***[1 mark]***.

4 a) The graph is Eulerian ***[1 mark]*** because all its vertices are of even degree ***[1 mark]***.

b) To make the graph semi-Eulerian, Scott would have to make two of the vertices odd. He could do this by removing one of the edges ***[1 mark]***.

c) 8 edges ***[1 mark]***.

d) 8 – 1 = 7 edges ***[1 mark]***.

5 a) Odd vertices are B, D, F, G ***[1 mark]***

Pairings: BF + DG = 190 + 100 = 290

BD + FG = 180 + 110 = 290

BG + DF = 80 + 210 = 290

[1 mark for pairings, 1 mark for lengths]

minimum pairing = any of above ***[1 mark]***

route length = 2200 + 290

= 2490 ***[1 mark]*** metres ***[1 mark]***

b) (i) D and F ***[1 mark]***. B and G are the pair of odd vertices with the shortest path between them, so that's the path that should be repeated, which means the route must start and finish at the other odd vertices ***[1 mark]***.

(ii) 2200 + 80 = 2280 m ***[1 mark]***

c)

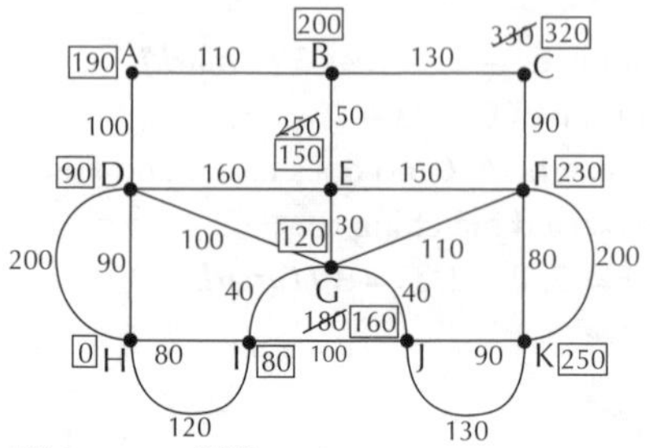

Distance = 320 metres

Route = HIGFC

[6 marks available — 1 mark for distance, 1 mark for route, 4 marks for all vertices correctly completed, alternatively 3 marks for 9 or 10 vertices correct, 2 marks for 7 or 8 vertices, 1 mark for 5 or 6 vertices.]

6 a)

	Ⓐ	Ⓑ	Ⓒ	Ⓓ	Ⓔ	F
~~A~~	~~–~~	~~140~~	~~167~~	~~205~~	~~150~~	~~173~~
~~B~~	~~(140)~~	~~–~~	~~145~~	~~148~~	~~159~~	~~170~~
~~C~~	~~167~~	~~(145)~~	~~–~~	~~210~~	~~195~~	~~180~~
~~D~~	~~205~~	~~(148)~~	~~210~~	~~–~~	~~155~~	~~178~~
~~E~~	~~(150)~~	~~159~~	~~195~~	~~155~~	~~–~~	~~185~~
F	173	(170)	180	178	185	–

Order of edges = AB, BC, BD, AE, BF

[3 marks available — 1 mark for first 3 vertices in correct order or 2 marks for all 5 vertices in correct order. 1 mark for correct crossing out and circling on matrix.]

Once you get into the swing of this — cross out, circle, cross out circle — it's easy. Check back to D1 Section 2 if you've forgotten what you're crossing out and circling.

b)

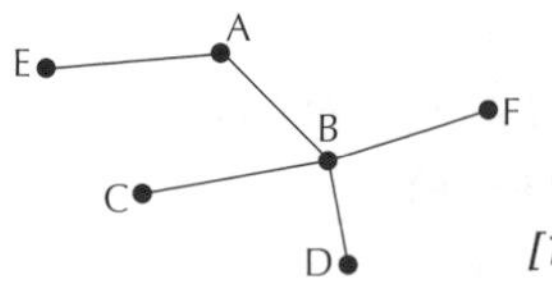

[1 mark]

753 metres ***[1 mark]***

c) Cycles are only formed when joining vertices already in the tree ***[1 mark]***. Prim's algorithm only considers connections to vertices not already in the tree, so a cycle can't form ***[1 mark]***.

7 a)

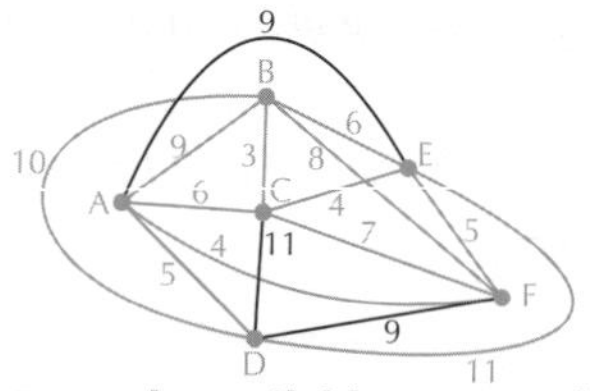

[2 marks available — 1 mark for adding 3 edges (AE, CD and DF) and 1 mark for correct weights (AE = 9, CD = 11, DF = 9)]

b) AFECBDA ***[1 mark]***

Weight = 4 + 5 + 4 + 3 + 10 + 5 ***[1 mark]*** = 31,

so the upper bound is 31 ***[1 mark]***.

c) BC (3) — add; CE (4) — add; AF (4) — add; AD (5) — add; EF (5) — add ***[2 marks — 1 mark for correct edges, 1 mark for adding edges in correct order]***.

All the vertices are joined so stop. Edges with equal weight can be added in any order (i.e. you could have added CE and AF the other way round, and AD and EF the other way round.

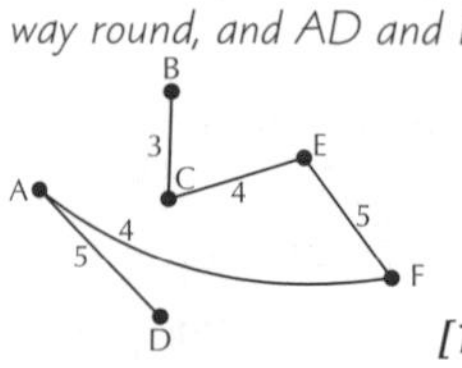

[1 mark]

Weight = 3 + 4 + 4 + 5 + 5 = 21 ***[1 mark]***.

d) Weight of reduced network = 21 – 3 = 18 ***[1 mark]***

Shortest edges incident to B = 3 + 6 = 9 ***[1 mark]***

Lower bound = 18 + 9 = 27 ***[1 mark]***

e) 27 ≤ duration of optimum tour ≤ 31 ***[1 mark]***

8 a) (i) Number of tickets: $5x + 10y \leq 300 \Rightarrow x + 2y \leq 60$ ***[1 mark]***. Number of bottles of wine:

$2x + 8y \leq 160 \Rightarrow x + 4y \leq 80$ ***[1 mark]***.

Answers

(ii) $x \geq 5$, $y \geq 5$ ***[1 mark]***.

There have to be at least 5 business class packages and at least 5 premier packages sold.

$x + y \geq 20$ ***[1 mark]***.

The total number of packages sold has to be more than 20.

b) (i)

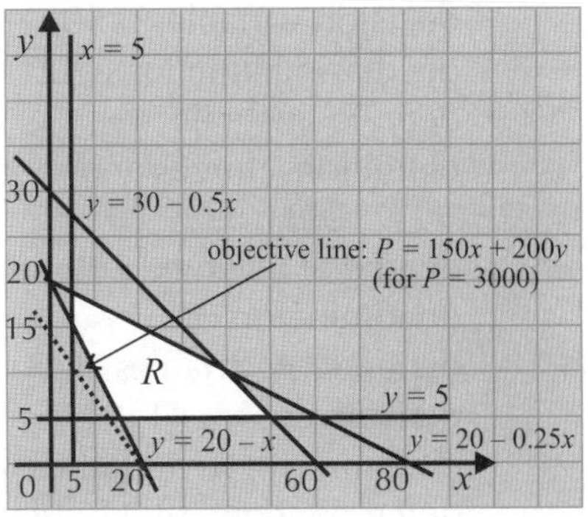

[6 marks available — 1 for each correct line (with equations as shown on graph — 4 in total), 1 for correct feasible region and 1 for objective line with correct gradient.]

(ii) Using the objective line, the last point in the feasible region it touches is the point (50, 5), the intersection of the lines $y = 5$ and $y = 30 - 0.5x$. At this point, $P =$ £8500 ***[1 mark]***, so this is the maximum profit possible, achieved when 50 business class packages and 5 premier packages are sold ***[1 mark]***.

(iii) Using the objective line, the first point in the feasible region it touches is the point (15, 5), the intersection of the lines $y = 5$ and $y = 20 - x$. At this point, $P =$ £3250 ***[1 mark]***, so this is the minimum profit possible, achieved when 15 business class packages and 5 premier packages are sold ***[1 mark]***.

Index

A

acceleration 39, 158-160, 163, 165, 170, 180-185, 189-191
acceleration-time graphs 160
adjacency matrices 210
air resistance 168, 181, 183, 189
algebraic division 27
algebraic fractions 5
algorithms 199-204, 211-215, 219, 220, 225-227, 239
alternating paths 238
approximation of area under curve 96
arc length 76, 77
arcs 207
area 43, 160
  between curves 50
  normal distribution graph 128
  of sectors 76, 77
  under a curve 49, 95, 96
  under a graph 160
arithmetic progressions 62, 63, 67
arrows to represent vectors 162
asymptotes 78
averages 104

B

bases 87-89
beam 168
bearings 164, 166
bell-shaped graph 128
bias 134
binomial coefficients 121
binomial distributions 122-125
  mean 124
  parameters of distribution 122
  standard deviation 124, 125
  the 5 conditions 122
  using tables 123, 125
  variance 124, 125
binomial distribution tables 154-157
binomial expansions 68-70
binomial formula 70
binomial probability function 121
bipartite graphs 207, 208, 237-239
Bond villain 162
book on a table 169
box of biscuits 114
brackets 3, 6, 9
breakthrough 239
bubble sort 201

C

cake 199
cancelling 6
carrots 49
CAST diagram 80, 82
central limit theorem 136
change of base 87
chessboard 64
Chinese postman problem 219
chords 35
circles 34, 35
coalescing particles 178
coefficient of friction 174, 181, 184, 185
collisions 178, 179
column vectors 179
common denominators 5
common difference 62
common factors 4
common ratio 64, 66, 67
complements of events 113, 114
complete matching 239
completing the square 11, 12, 35
components
  of a vector 163, 165, 171, 172
  of motion 189
  of velocity 189
conditional probability 115
confidence intervals 137, 138
connected particles 183-185
conservation of momentum 178, 179
constant acceleration equations 158, 159, 165, 181, 185, 189-191
constant of integration 47, 48
constants 1
constraints (linear programming) 231
convergent sequences 61, 65
convergent series 65
converting angles 76
coordinate geometry 30-35
correlation 141, 142
correlation coefficient 141, 142
cosine rule 74, 75
cows 16
cubics 26-28
  cubic equations 48
  cubic graphs 33, 41
cumulative distribution functions 120, 123
cumulative distribution function tables 154-157
cunning plan 68
curve sketching 8, 33
cycle 207

D

data values 104
daydreaming 43
decay 89
deceleration 161
decision variables 230, 234
decreasing functions 42
definite integrals 49
definitions (forces and modelling) 168
degree of a vertex 209
denominators 2
dependent variables 143
derivatives 38, 41, 92
difference of two squares 83
differentiation 38-43, 48, 92, 93
digraphs 208
Dijkstra's algorithm 214, 215
direction of a vector 162, 163
discrete random variables 120, 122
discriminant 14, 15
dispersion 106-108
displacement 158-160, 162, 164, 191
displacement-time graphs 160
distance 158, 162
distance matrices 210, 213, 226
distance-time graphs 160
divergent sequences 61, 65
divergent series 65, 66
division 1
dog 180
drag 168

E

elimination method of solving simultaneous equations 19
equation of a circle 34
equation of a line 30, 32
equations 1
equilibrium 168, 172, 174, 182
Ermintrude 16
estimated mean 105
estimator of population means 135
Eulerian graphs 218, 219
events 112-114, 116, 117
exact answers 2
exclusive events 113
expected value 120, 124, 125, 128
experiment 112
experimental error 144
explanatory variable 143

Index

exponentials 88, 89
  exponential graphs 88
  exponential growth and decay 89
extrapolation 144
extreme values 106

F

factorials 68, 69, 121
factorising 69
  cubics 26, 28
  quadratics 9, 10
factor theorem 28
feasible region 231-234
feasible solutions 231
feet 141
finite sequence 61
flow charts 200
forces 163, 168-173, 180-185
fractions 5
frequency 104, 108
friction 168, 169, 173, 174, 181, 182, 184, 185
functions 1
funny thing 66

G

g (acceleration due to gravity) 158
geometric
  interpretation 21
  progressions 64- 67
  series 65-67
German Shepherd 220
global warming 233
gradients 30, 32, 35, 38-42, 160, 161
  gradient rule 35
graphs
  algorithms 211-215, 219-221, 225-227
  motion 160
  sketching 41
  theory 207-210, 218
  transformations 58
  trig functions 78-81
gravity 158, 168, 170, 189, 191
greater than 22
greater than or equal to 22
greedy algorithm 211
grouped data 105
growth 89

H

Hamiltonian cycle 207, 224, 225
Hercules 179

I

identities 1, 74, 82, 83
i + j vectors 162, 163, 165, 173, 180
important angles 76, 77
inclined planes 173, 181, 182
increasing functions 42
indefinite integrals 47
independent events 116
independent trials 122, 125
independent variables 143
indices 57, 87
inequalities 22, 23, 230, 231
  linear 22
  quadratic 23
inextensible 168, 169, 183, 185
infectious diseases 116
infinite sequence 61
initial matching 238
input (algorithms) 199
insomnia 41
instructions 199
integer solutions 234
integration 47-50, 95
  formula 47, 95
interest rate 66
interpolation 144
interquartile range 106, 108
intersections (Venn diagrams) 112
intervals 80-82

J

James J Sylvester 10

K

Kruskal's algorithm 211, 227

L

last term 63
laws of indices 57
laws of logarithms 87, 89
laws of motion 180
least squares regression 144
less than 22
less than or equal to 22
light 168, 169, 183
limiting equilibrium 182
limiting friction 174, 182
limits (sequences) 61, 65
limits of integration 47-49
linear programming 230-234
linear regression 143, 144
linear scaling 109, 142, 144
lines of best fit 143
location 104, 105, 108
logs 87-89
lower bound algorithm 227
lower class boundary 105
lowest common multiple 19, 30

M

magic formula 13
magnitude 162-164, 171, 178
marvellous thing 130
masses on strings 173
matchings 237-239
matrices (graphs) 210, 213, 226
maximum
  height 159, 190
  points of graph 8
  stationary points 41, 42,
  volume 43
maximum matching algorithm 239
mean 104, 105, 107-109, 120
  binomial distribution 124
  normal distribution 128, 130
median 104, 106
mid-class value 105, 108
mid-points 31
minimum
  points of graph 8, 12
  stationary points 41
minimum spanning trees 211, 212, 227
mode 104, 105
modelling 168-170, 189-191
  assumptions 159, 168-170, 183
modulus 65
momentum 178, 179
motion 180, 181, 189, 191
  graphs 160, 161
  under gravity 158
Mr Men 212
multiplying out brackets 3
mutilating soft toys 234
mutually exclusive events 113

Index

N

natural numbers 63
nature of stationary points 41
nearest neighbour algorithm 225, 226
negative correlation 141
networks 208
Newton's Laws of Motion 180
no correlation 141
non-uniform 168
normal contact force, N 173
normal distribution 128-131, 136
normal distribution tables 153
normal reaction 168, 170, 173, 174, 180, 181
normals 32, 35, 40
nose to tail vectors 162, 171
n-shaped graphs 8, 23, 33
nth term 60, 62

O

objective function 232
objective line method 232, 234
observed value 144
old chestnuts 169
optimal solutions 232-234
 integer solutions 234
order of events 117
ordinates 96, 97
outcomes 112
outliers 108, 144

P

parallel lines 32
parameters (binomial distribution) 122, 124, 125
paranormal distribution 129
particles 159, 165, 168-171, 174, 178-180, 183, 185, 191
Pascal's triangle 68, 70
paths 207
pegs 183, 184
percentage points 129
periodic sequences 61
perpendicular directions 163, 180
perpendicular lines 32, 35
pivots (quick sort) 204
planes 168, 170, 173, 179-182, 184, 185, 191
Plato 10
PMCC 141, 142
points of contact 168
points of inflection 41
points of intersection 21
polygons of forces 172
polynomials 1, 26-28
populations 134
Porsche 66
position vectors 165
positive correlation 141, 142
potato printing 239
powers 57, 87, 92
Prim's algorithm 212, 213, 227
probability 112-118, 120-123, 125
 binomial distribution 122, 124, 125
 distribution 120, 122, 134
 function 120, 121
 normal distribution 128, 129
product-moment correlation coefficient 141, 142
projectiles 189-191
pseudo-code 200
pulleys 183-185
Pythagoras 31, 34, 77, 163, 171, 179

Q

quadratic
 equations 67, 83
 formula 13, 20
 graphs 8
 inequalities 23
quantitative data 105
quartiles 106, 108
quick sort 204

R

radians 76, 77, 97
radioactivity 89
radius 34, 35
random events 112, 113
random scattering of points 141
range 106
rates of change 38, 39
rationalising the denominator 2
reaction force 168-170
real-life problems (differentiation) 43
recurrence relations 60, 62
reflections 58
regression 143, 144
regression line 143, 144
remainder theorem 28
replacement (sampling with) 114
residuals 144
resolving
 forces 172-174, 180-184
 vectors 163, 164
 velocity 189-191
response variable 143
resultant force 162-164, 171, 172, 180, 181
rigid 168, 169
rod 168
rollercoaster 88
roots
 of cubics 33
 of quadratics 12, 14
 powers 57
roughness and friction 168, 174, 181, 184, 185
route inspection algorithm 219-221, 225
rules of degrees 209
running total 104
Russian Peasant algorithm 199

S

sample 134, 136
sample mean 136
sampling bias 134
sampling distribution 134
sampling with replacement 114
satnavs 214
scalars 162
scatter diagrams 141-144
Scitardauq Gnisirotcaf 9
second order derivatives 39
sectors 76, 77
semi-Eulerian graphs 218
sequences 60-67
 arithmetic 60-63, 67
 geometric 64-67
series 62-67
 arithmetic 62, 63
 geometric 64-67
Shakespeare 201
shell sort 203
shortest route 214
shuttle sort 202
sigma notation 63
simple random sampling 134
simplifying
 expressions 3-6
 surds 2
simultaneous equations 19-21, 67, 233
 elimination method 19

Index

geometric interpretation 21
normal distribution 131
substitution method 20
with quadratics 20
sine rule 74, 75, 173
sledge 169
smooth 168, 169, 179-181, 183
SOH CAH TOA 77
sorting 201-204
spanning trees 209
speed 158, 160, 162, 178, 180, 183, 190, 191
speed-time graphs 160
squashes (trig graphs) 79
squeaking 218
standard deviation 107, 109, 120, 130, 135
standard error 136
standardising equation 131
standard normal distribution 129
static 168
stationary points 41
statistical tables 153-157
statistics (definition) 134
stretches of graphs 58, 79
string 169, 173, 183-185
strips 97
subgraphs 209
substitution method of solving simultaneous equations 20
summarised data 109
sum of squares 107
sum to infinity 65-67
superpowers 237
surds 2
swaps
bubble sort 201
shuttle sort 202
sweeping statements 142, 144
symmetrical distributions 128, 131
symmetry (trig graphs) 78

T

tables 104
tangents 21, 35, 39, 40, 93
tea 234
tension 168, 173, 182-185
theoretical mean 124
thin 168
thrust 168
Time Warp 215
trace table 200
transformations 58, 79, 81
translations 34, 58, 79
Trapezium rule 96, 97
travelling salesperson problem 224-227
traversable graphs 218
tree diagrams 114, 115, 239
trees 209
trials 112, 122
trigonometry 163, 171, 189
trig equations 80-83
trig formulas 74, 75
trig functions 78, 79, 97
trig identities 74, 82, 83
trig values 77
true love 213
turning points 41
two-way tables 112

U

uniform 168
union (Venn diagrams) 112
unit vectors 162, 163, 165, 173, 180
upper class boundary 105
u-shaped graphs 8, 23, 33
"uvast" variables 158, 159

V

valency 209
variables 1
variance 107, 108, 120, 128, 130, 135
binomial distribution 124
vectors 162-165, 171-173, 178-180
velocity 158, 160-164, 178, 179, 189-191
velocity-time graphs 160
Venn diagram 112, 113, 116
vertex method 233, 234
vertices 8, 207-215, 218-221, 224-227, 233, 237-239
vertigo 233
volume 43

W

weight 168, 170, 221

Z

Z-distribution 130, 131

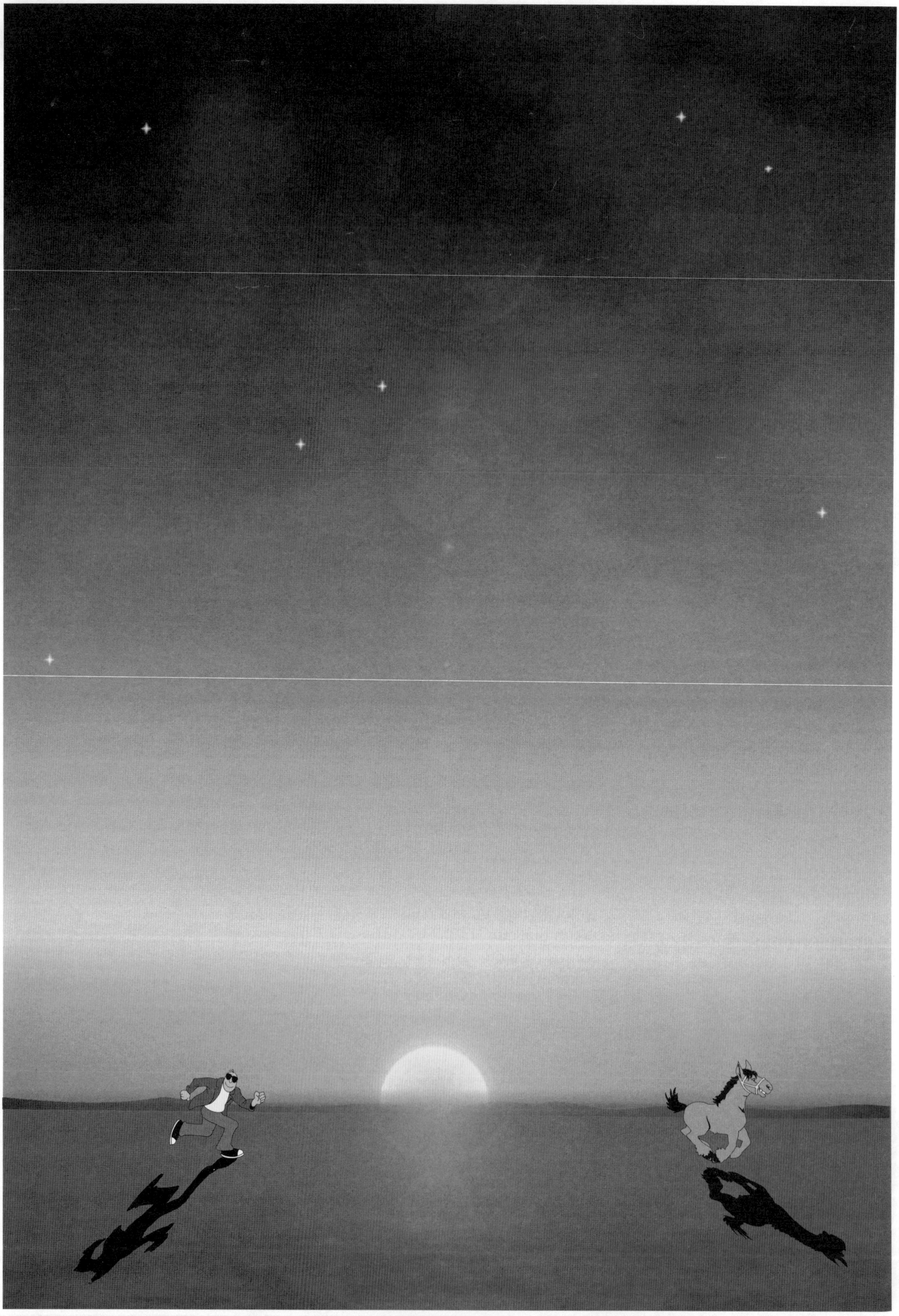